From the Scottish Borders to the Pacific Rim in search of plants...

从苏格兰边境到太平洋沿岸地区，搜寻植物的踪迹……

Robert Fortune—Plant Hunter

罗伯特·福琼——植物猎人

David Kay Ferguson

北京师范大学出版集团
BEIJING NORMAL UNIVERSITY PUBLISHING GROUP
安徽大学出版社

图书在版编目(CIP)数据

罗伯特·福琼:植物猎人=Robert Fortune——Plant Hunter:英文/(奥)戴维·弗格森(David Ferguson)著.一合肥:安徽大学出版社,2017.10

ISBN 978-7-5664-1451-9

Ⅰ.①罗… Ⅱ.①戴… Ⅲ.①罗伯特·福琼(1812—1880))—传记—英文 Ⅳ.①K835.616.15

中国版本图书馆CIP数据核字(2017)第229089号

罗伯特·福琼——植物猎人　　(奥)戴维·弗格森 著

出版发行:北京师范大学出版集团
安徽大学出版社
(安徽省合肥市肥西路3号 邮编230039)
www.bnupg.com.cn
www.ahupress.com.cn

印　　刷:安徽昶颉包装印务有限责任公司
经　　销:全国新华书店
开　　本:184mm×260mm
印　　张:28
字　　数:681千字
版　　次:2017年10月第1版
印　　次:2017年10月第1次印刷
定　　价:139.00元(含光盘)
ISBN 978-7-5664-1451-9

策划编辑:李　梅　李　雪
责任编辑:李　雪
责任印制:赵明炎
装帧设计:李　军
美术编辑:李　军

CONTENTS

Preface ········ 001

Acknowledgements ········ 001

Chapter 1 From Childhood to Manhood in Scotland ········ 001

1.1 The World at War ········ 001

1.2 Life in the Scottish Borders ········ 005

1.3 Employment in the Scottish Capital ········ 011

Chapter 2 Early Professional Career ········ 015

2.1 The Move to London ········ 015

2.2 Preparing for the Trip to China ········ 024

Chapter 3 Fortune's First Visit to China ········ 030

3.1 Fortune's First Year in the Flowery Land ········ 030

3.2 Fortune's Second Year in China: New People, Places and Experiences ········ 042

3.3 A Short Trip to the Philippines ········ 050

3.4 Back in China: Encounters with Mandarins, Locals and Pirates ········ 054

3.5 The Results of Fortune's Forays in the Flowery Land ········ 061

Chapter 4 Chelsea Interlude ········ 069

4.1 Curator of the Chelsea Physic Garden ········ 069

4.2 Preparing for Service in the East India Company ········ 075

Chapter 5 In the Service of the East India Company ········ 081

5.1 The Outward Journey ········ 081

5.2 An Illegal Journey in Search of Green Tea ········ 085

5.3 More Teas and Temples ········ 091

5.4 A Secret Journey to the Black Tea District of Fújiàn ········ 097

5.5 Preparing for India ········ 109

5.6 Tour of Inspection in Northern India ········ 112

5.7 Home for a Year ········ 123

5.8 Searching for Tea in Troubled Times ······ 135
5.9 A Jaunt with Friends ······ 144
5.10 A Visit to the Silk District ······ 158
5.11 Taking Leave of China ······ 166
5.12 Second Tour of Inspection in Northorn India ······ 168

Chapter 6 Intermezzo: Moving Up in the World ······ 171
6.1 Back to London ······ 171
6.2 Working on the Book ······ 176

Chapter 7 Working for the Americans ······ 181
7.1 The History of Tea in the USA ······ 181
7.2 Fortune Employed and Dismissed ······ 186
7.3 Other Plants Introduced to the States by Fortune ······ 187

Chapter 8 Two Trips to Japan ······ 191
8.1 The Race for Japan's Botanical Riches ······ 191
8.2 Fortune Travels North ······ 201
8.3 Fortune's Second Season in Japan ······ 216

Chapter 9 A Final Farewell to China ······ 232
9.1 En Route to the Chinese Capital ······ 232
9.2 Excursions In and Around Běijīng ······ 243

Chapter 10 Out of the Limelight ······ 262
10.1 Life in London ······ 262
10.2 The Emergence of Modern Japan ······ 276
10.3 China's Continuing Contentions ······ 280
10.4 Prizing the Chinese Interior Open ······ 290
10.5 Decline and Death ······ 300
10.6 Fortune's Personality ······ 303

Chapter 11 Epilogue ······ 308

Index to Place Names ······ 314
Index to Persons, Firms and Vessels ······ 361
Index to Plants and Animals ······ 400
Illustration Credits ······ 433

Preface

China is known as a botanical bonanza with more than 33,000 species of flowering plants, many of which are of ornamental value. Without them our western gardens would be all the poorer. Robert Fortune (1812 – 1880) was one of the first professional plant hunters to recognize the richness of the Chinese flora. Although he also travelled to India, Japan and the Philippines, he spent most of his time between 1843 and 1861 on mainland China. Probably because Fortune gave a full account of his exploits in the four books he wrote (1847, 1852, 1857, 1863), most biographers have restricted their accounts of the plant hunter to a resumé of his books.

Although he commented on all he saw in the Far East, Fortune was in many ways a very private person. His books give no indication that he was married and sired six children. Some authors have assumed that he was an inveterate bachelor! Unfortunately for the prospective biographer, his personal papers seem to have been destroyed by his family after his death in 1880. By travelling in Robert Fortune's footsteps and documenting what is left of the fast-disappearing local culture, I have attempted to get closer to the man and the emotions he must have experienced.

The present book aims to:

(1) Summarize Fortune's four books, which come to over 1,640 pages.

(2) Identify those places which Fortune visited. As Fortune was writing before the Wade-Giles system of transliteration of Chinese names had been invented, he used his own phonetic system. In the present book the official Pinyin System, which is currently used in Chinese publications and maps, has been employed, so the modern traveller should experience no difficulty in following in Fortune's footsteps.

(3) Give the presently accepted names of the plants and animals which Fortune mentioned. In this way it should be possible to reassess the number of plants Fortune was responsible for introducing.

(4) Provide background information on the people and events, which Fortune mentioned. Fortune only referred to people by their surname, partly on account of Victorian etiquette, but also because many of these people were well known to his mid-19th Century readership. Throughout the book some 70 potted biographies of Fortune's acquaintances/friends and other personalities are included, along with boxes dealing with important events, and information about six of the ships Fortune travelled on.

(5) Over the past ten years I have attempted to discover as much as possible about Fortune and his times. While most readers will implicitly assume that this biography has been fully researched, for those critical minds I have documented the sources of information used in the form of numbered notes. These notes and the references are included on the CD-ROM inside the cover of this book. In this way, the reader can pursue certain leads, expose inherent weaknesses, and pinpoint what remains to be done. The author cannot claim to be a specialist in all the fields covered.

(6) Fortune's books contain only a few line drawings. By providing a wide range of illustrations, the present book aims at giving those readers, who have not had an opportunity to visit the Far East, a taste of the countryside through which the plant hunter travelled. Unfortunately, few authors writing about Fortune have ever been to China, Japan, India and the Philippines!

Robert Fortune (1812–1880)

Acknowledgements

The idea for this book started to crystallize in 2006. The next year the author travelled to Scotland to work at the Royal Botanic Garden Edinburgh (RBGE). With the support of Dr. Ian Hedge and the aid of librarian Graham Hardy, he consulted various sources of information. James McCarthy took him to the National Library of Scotland, where with the help of Rachel Thomas he viewed the ledgers and Fortune correspondence in The John Murray Archive. He also visited Fortune's birthplace, where Rev. Duncan Murray (Chirnside) showed him round the area, and helped him search for relevant tombstones in a number of graveyards. A visit was also paid to Tranent Parish Churchyard to look for the grave of Fortune's youngest daughter, Alice Durie. There Rev. Tom Hogg supplied him with copies of old ordnance survey maps, and suggested he visit Haddington Library. This library proved to be an excellent source of information on the Durie family, thanks to Craig Statham, Bill Wilson and other members of staff. Two other books on local history were kindly lent by Sally and Guy Turner (Elphinstone). I am also grateful to Eric Imery (Edinburgh) for scouring the records for information on Fortune's family.

In London the author was given access to the Minutes of the Garden Committee of the Chelsea Physic Garden at Apothecaries' Hall thanks to archivist Dee Cook, and with the help of Marijke Booth was able to consult the day—books in the archives of the auctioneers Christie's. A visit was also paid to Brompton Cemetery to search for Robert Fortune's grave. With the aid of Jay Roos the weathered stone was finally located. In 2010 this limestone slab was covered by a newly inscribed granite slab.

In the autumn of 2007 the first of a number of excursions was undertaken in China with members of Professor LǏ Chéngsēn's team from the Institute of Botany in Běijīng. I should particularly like to thank Dr. WÁNG Yǔfēi for financing and leading the trip to Níngbō and the Zhōushān Islands, and LǏ Jīnfēng and LǏ Yàméng, who in addition showed me some of the places Fortune visited in and around Běijīng. In 2008 Yàméng accompanied me to various places in Tiānjīn Province, while Jīnfēng acted as driver and guide during another trip to the Níngbō area in 2010. This trip was financed by Dr. LĚNG Qín [Nánjīng Institute of Geology and Palaeontology, now Bryant University, with the help of the CAS/SAFEA International Partnership Program for Creative Research Teams, and The Pilot Project of Knowledge Innovation of CAS (Grant No. KZCX2-YW-105)]. In May 2008 Dr. LĚNG and the University of Vienna jointly paid for an excursion to the tea districts in Zhèjiāng and Fújiàn, which I undertook with my department's photographer, Rudolf Gold. Our guide was WÁNG Lì (Nánjīng Institute of Geology and Palaeontology, now Xīshuāngbǎnnà Tropical Botanical Garden). We should like to acknowledge the help of JIĀNG Yúyīng (Xīn'ānjiāng) during this trip. Thanks are also due to Professor WU Chia-li (Tamkang University) for inviting the author to Táiwān in 2010. There with the help of LIAN Su-chiu and WANG Chao-ping he first set eyes on *Lilium formosanum* and *Tetrapanax papyrifer* respectively.

In April 2009 it was time to visit some of the places in the Lesser Himalayas where Fortune's tea plants had been cultivated. The excursion was organized by Dr. Sudha Gupta (Kolkata) and Mr. Sandip Aditya (Chandannager). We were accompanied part of the way by Dr. Gupta, her sister Deepa and Dr. Ruby Ghosh (Lucknow). Later that year the author was invited to Japan by Professor OHSAWA Masahiko and his wife Mutsuko, who allowed him to stay in their apartment, showed him round the Tokyo area, and took him to Yokohama and Kamakura. In Yokohama we were guided by KITAMURA Keiichi, in Kamakura by TAMABAYASHI Yoshio. That left Nagasaki, which Fortune briefly visited in 1860 and 1861. In 2012 Professor WANG San-lang (Tamkang University) put the author in touch with Professor MATSUDA Masako (Nagasaki University), who arranged a wonderful team consisting of her colleague Professor MORINAGA Haruno, a Nagasaki guide YAMAGUCHI Fumiko, and a local historian HARADA Hiroji, which enabled him to see more of the port than he had expected in just over two days. MISUMI Koyo kindly showed us round Myo-gyo-ji where Fortune had stayed. The author would also like to thank KUWABARA Setsuko (Berlin) for suggesting a number of improvements to an earlier version of Chapter 8. He also acknowledges the help of JIǍ Huì, who searched the Chinese Internet for additional information, checked the Pinyin spelling of various persons and places, and created maps of India and Japan. LǏ Jīnfēng was responsible for making the maps of China and London.

In July 2014 the author made another trip to the UK to see those people and places for which there had not been enough time in 2007. In London he visited Battersea Park, what remains of the Cremorne Pleasure Gardens, and thanks to the then curator Christopher Bailes saw round the Chelsea Physic Garden, which Fortune had once run. In Brompton Cemetery he examined Fortune's revamped gravestone, and with the help of Jay Roos looked for the graves of other personalities. John Murray VII kindly showed him the room in Albemarle Street where so many famous writers met. Teresa Magner then drove him to Cranleigh and Woking to see where some of Robert Fortune's descendants had lived, and consult various documents at the Surrey History Centre. He examined the letters the plantsman had written to the 13th Earl of Derby in Liverpool's Central Library, and with the help of Dr. Tony Parker (Liverpool World Museum) photographed the birds Fortune had collected in the Philippines in 1845. In Scotland he had the good fortune to meet Thomas A. Dykes, Ian & Sarah Russell and Martin Steven, the owners of a number of farms once run by Fortune's descendants. They were able to provide him with useful background information. A meeting with Kenneth McLean, Ronald Morrison, and Elizabeth Snow of the Dunse History Society was also highly productive. Alina Varjoghe gave him permission to look round Carberry Tower and Gardens, the seat of the Elphinstone Estate, for which Fortune's eldest son once worked. The trip was rounded off by a week spent in the RBGE Library. I am especially indebted to Graham Hardy for his help.

This by no means exhausts the list of contributors. To save space, other people who supplied specific information have been acknowledged at the relevant place in the text.

It would have been impossible to write this book without the help of many, often anonymous, contributors to the Internet.

(1) The Internet provides a wide range of information in such diverse fields such as genealogy, Asian history, and plant nomenclature. If handled with care it is an invaluable source of

knowledge.

(2) Good web pages such as Wikipedia often cite their sources which can be consulted in libraries, or sometimes found on the Internet itself (e.g. googlebooks, JSTOR).

(3) The Internet has speeded up the search for the necessary literature. One no longer has to passively wait for an interesting book to be offered in a catalogue, but can actively locate a bookseller with the desired title using one of the search machines (e.g. AbeBooks). I am indebted to Mike Park for drawing my attention to this online bookmarket.

The illustrations for this book were selected with the help of my good friend Rudolf Gold. Finally, I should like to thank Dr. YǏN Jiànlóng (Héféi) for contacting Anhui University Press on my behalf. From the very start the branch proprieter, LǏ Méi and her team proved very supportive.

Chapter 1 From Childhood to Manhood in Scotland

1.1 The World at War

At the dawn of the 19th Century, Europe was in a state of turmoil. By a combination of brute force and some heavy-handed diplomacy, Napoleon Bonaparte (1769 – 1821; **Fig. 1.1**) had subjugated large parts of continental Europe between the Russian border and the Atlantic.[1] If it had not been for the Royal Navy, his plans to invade Great Britain in 1803 – 1805 might well have proved successful.[2] After the Battle of Trafalgar on 21 October 1805, Britain ruled the waves, and brought pressure to bear on Napoleon by blockading European ports from Brest to the Elbe ("Fox's Blockade", on 16 May 1806).[3] Napoleon countered with his "Continental System" in 1806, whereby British goods were banned from those parts of continental Europe he controlled.[4] Although Britain's exports suffered, it did not bring the UK to its knees because the system also acted against the interests of many of the countries in Napoleon's alliance.[5] As a result, it was often blatantly flaunted.[6]

Portugal, an old ally of Great Britain was unwilling to apply Napoleon's blockade, so the French felt obliged to invade this country.[7] With the help of the Spanish, who were interested in regaining at least partial control over Portugal, an attack was launched in 1807.[8] General Jean-Androche Junot ("The Tempest", 1771 – 1813) entered Lisbon on 30 November 1807, just too late to capture the regent Prince João (1767 – 1826) and the Portuguese fleet.[9] French control of Portugal could have continued indefinitely as long as Spain was France's ally.[10] Underestimating their national pride, Napoleon assumed the Spanish would prefer an "enlightened" regime to their own autocratic monarchy.[11] However, when the French turned on their one-time ally in 1808, placing the Spanish royal family under house-arrest, and installing Napoleon's elder brother Joseph (1768 – 1844) on the throne, the Spanish started a long, drawn-out campaign to rid their land of the French.[12] What the Spanish soldiers lacked in training they compensated by their patriotism.[13] Although they were beaten countless times, they always managed to bounce back. The French policy of requisition, to save bringing the troops' rations from France, did not help to win the Spanish hearts and minds, and foraging parties often suffered a horrible fate at the hands of partisans (**Fig. 1.2**).[14] The Spanish and

Fig. 1.1 Napoleon Bonaparte (1769 – 1821). In the year that Fortune was born, Napoleon made the fateful decision to invade Russia. This led to his downfall.

Portuguese forces were backed by a number of British expeditionary forces under a succession of generals. However, because they were small compared with Napoleon's Grande Armée, they had to act cautiously to keep casualties low.[15]

In the meantime the Russians, having lost an important source of revenue since they joined Napoleon's Continental System (Treaty of Tilsit, on 7 July 1807), started to reexport shipbuilding materials (wood for masts and planking, hemp for caulking, linseed and pitch as protectants, iron for cannons, and potassium for gunpowder) to Britain on 31 December 1810.[16] Aware that others might follow the czar's example, Napoleon felt obliged to invade Russia to teach the czar a lesson.[17] In order to assemble a sufficiently large army, he was forced to withdraw 30,000 men from Spain and fight a war on yet one more front.[18]

Fig. 1.2 The brutality of the Peninsular War (1808 – 1814) was portrayed in Francisco Goya's "Los desastres de la guerra".

Napoleon's invasion of Russia was doomed to failure from the outset. Due to the immense size of the army, the logistics of organizing the campaign proved immense.[19] By the time the campaign got underway it was already midsummer.[20] Owing to their inferior numbers, the Russians dared not face Napoleon's army, but retreated eastwards, drawing the Grand Armée further and further inland with the result that its line of supply became severely attenuated, with increasing numbers of soldiers being transferred to protecting the depots and garrisoning the cities captured.[21] The very day Napoleon entered Moscow, that is, 15 September 1812, the Russians torched the city, destroying most of its centre.[22] By the time Napoleon realized that Czar Alexander (1777 – 1825) was not going to capitulate it was mid-October, and winter was not far off.[23] On 17 October Napoleon ordered his army to start retreating in two days' time.[24] A break-out in a southerly direction was contained by the Russians, so the Grand Armée was forced to follow the impoverished route they had taken on the way to Moscow. By mid-November with night temperatures falling as low as -30℃ frostbite was common, and many soldiers perished of exposure.[25] Others suffered a worse fate at the hands of the peasants, who had been maltreated by the soldiers on their outward journey.[26] By the time the Grand Armée finally reached Poland, it had lost at least 83% of its troops.[27]

1812 also represented the turning point in the war on the Iberian Peninsula. By the late summer the Spanish, with British backing, were able to push the French out of half the area they had occupied since 1808.[28] However, it took until 1814 to completely free the country of the French.[29] After fighting a number of rearguard actions, the sheer weight of enemy numbers finally forced Napoleon to abdicate on 6 April 1814.[30] He was appointed sovereign of Elba, a remarkably mild sentence for someone who was responsible for millions of deaths.[31] Napoleon's escape from Elba, his defeat at Waterloo, incarceration on St. Helena, and slow death by arsenic poisoning are too well known to require repetition.[32]

While the Spanish were fighting their War of Independence, their colonies in America were largely left to their own devices. In 1809 when it looked as though the French would gain complete control of the Iberian Peninsula, there was a fear that the Spanish dominions would also fall into their hands.[33] To prevent this from happening, the American colonies established local juntas (committees) to govern the viceroyalties in the name of the deposed King Ferdinand VII (1784 – 1833; **Fig. 1.3**).[34] Although some dominions became autonomous, at this stage there was little popular desire for complete independence from Spain.[35] However, once Ferdinand was freed from his exile in Valençay (C France) in 1814 at the end of the Napoleonic Wars, he reestablished an autocratic rule, reinstated the Inquisition, and organized the reconquest of Latin America by General Pablo Morillo (1775 – 1837) using 10,000+ hardened troops that had fought Napoleon.[36] By the end of 1816 only N Argentina remained in the hands of the American "rebels".[37] However, not only had the colonies grown accustomed to home rule, but the reconquest was accompanied by such brutality that it turned the Latin Americans against the monarchy.[38] With the antimonarchists outnumbering the monarchists in Latin America, it was only a matter of time before the various viceroyalties gained their independence.[39] Without a navy since Trafalgar, without funds following the Napoleonic War, and political wrangling at home, Spain was incapable of restoring its control over Latin America.[40] When some of the soldiers being assembled for the reconquest of Argentina mutinied in Cádiz on 1 January 1820, the revolt spread to other disgruntled units that considered they had been poorly treated for their role in freeing Spain and restoring Ferdinand VII to power.[41] The uprising effectively ended all hopes of a reconquest of the overseas dominions.[42] By late 1821 all territories except Quito (Ecuador), Peru, Upper Peru (Bolivia), Cuba and Puerto Rico were independent.[43] By February 1826 only Cuba and Puerto Rico with their strong Spanish garrisons were still under Spanish rule.[44]

Fig. 1. 3 King Ferdinand VII (1784 – 1833). Most of Spain's South American colonies were lost during Ferdinand's reign.

Great Britain also came to blows with its former colony, the USA. The Americans were unhappy that the British were arming the native Americans, with the restrictions placed on their trade with continental Europe caused by the Royal Navy blockades, and the fact that some of its citizens were being impressed into His Majesty's Service.[45] Impressment took place not only in British ports but on the high seas, leaving some American ships dangerously undermanned.[46] After the British boarded the USS "Chesapeake" on 22 June 1807 to retake three seamen who had deserted from HMS "Melampus", British ships were banned from entering US ports.[47] The USA imposed an embargo on Britain on 22 December 1807, hoping to bring the UK to its knees. While Lancashire cotton mills were forced to cut back or close down, and the Royal Navy had once more to seek new sources of timber for its shipbuilding programme, the loss of trade harmed the US more than Britain.[48] At the beginning of March 1809, at the end of President Thomas Jefferson's term, the Embargo Act was finally revoked and replaced by a Non-Intercourse Act closing trade (both exports and imports) with both the British Empire and those areas controlled by France, and prohibiting their armed vessels from entering US ports.[49] Thousands of British firms failed, Parliament was petitioned by the manufacturers to rescind the Orders-in-Council, and rioting swept the Midlands in the Spring of 1812.[50] Although the British had finally agreed in 1811 to release the two surviving American sailors and pay reparations for the Chesapeake incident, the "War Hawks" in the US Government continued to advocate hostilities once the time was ripe.[51] For the bellicose "War Hawks" and "Scarecrows", who advocated intimidating Britain with sabre-rattling, there was no way back without loss of face.[52] Moreover, US President James Madison (1751 – 1836; **Fig. 1.4**) was unwilling to admit that he had been tricked into believing that Napoleon had rescinded his Continental System.[53] With mixed messages coming out of Washington, nobody in Britain took the US bluster seriously.[54] Opposition spear-headed by Henry Brougham (1778 – 1868) eventually led the British Cabinet to revoke its Orders-in-Council on 23 June 1812, blissfully unaware that the USA had already declared war on 18 June 1812.[55]

With the Napoleonic Wars raging in Europe, war with the US only worsened an already bad situation.[56] Many British thought that the USA was taking advantage of their preoccupation in Europe to conquer British North America (Canada).[57] Britain did not declare war on the USA until 9 January 1813, hoping no doubt that President Madison would rescind his declaration of war.[58] In a tit for tat both sides captured one another's ships.[59] While the Royal Navy's losses wounded its sense of invincibility, the USA simply had too few ships to gain the upper hand.[60] On 1 June 1813 the British won a moral victory when they captured the USS "Chesapeake", which henceforth became HMS "Chesapeake", before being broken up in 1820.[61]

Fig. 1. 4 US President James Madison (1751 – 1836), who declared war on Great Britain in the year that Robert Fortune was born. In 1814 he had to flee the White House.

On land, with few troops available and little chance of reinforcements arriving from war-ridden Europe, the Canadian Governor-General Sir George Prévost (1767 – 1816) was forced to take a largely defensive stand.[62] On the other hand, with US Presidential elections looming on 3 December 1812, the Republicans did not want to raise taxes to pay for a well-equipped force to invade Canada.[63] The dream of driving the British out of North America soon evaporated, as the French Canadians and the US settlers failed to welcome the US invaders and rise against the British.[64] Burning and pillaging by the US troops did not help to endear them to the local populace.[65] Drafty tents, poor food, disease, irregular pay, and incompetent leadership lowered the soldiers' morale.[66] To complicate the issue, some of the US generals like Stephen Van Rensselaer (1764 – 1839) were Federalists, who were set on proving the folly of the "Republican" war.[67] Moreover, many US militiamen, believing that their service was restricted to defending US territory, refused to cross into British North America.[68] As a result the war became a bit of a farce, fighting being largely restricted to skirmishes along the US-Canadian border, during which cross-border trade was temporarily suspended.[69]

Once peace was established in Europe in 1814, more hardened British troops were sent to Canada.[70] It now became possible for Britain to go on the offensive.[71] However, by then the Americans had become more experienced in combat.[72] In an attempt to divert US troops away from the Canadian front, Admiral Sir John Borlase Warren (1753 – 1822) and his successor Vice-Admiral Sir Alexander Cochrane (1758 – 1832) were ordered to launch a series of raids along the US East Coast in addition to stepping up blockades.[73] Undoubtedly the most daring of these exploits was an attack against the US capital Washington in August 1814.[74] In the event resistance proved minimal.[75] President Madison had to flee the White House, leaving the dining-room table set for 40 guests, and some excellent Madeira, which the British officers drank.[76] By the time President Madison rode back into Washington on 28 August, the White House was a roofless shell.[77]

With Napoleon's defeat, it was no longer necessary for the British to blockade the French and other ports or impress American sailors into the Royal Navy, so two of the reasons for the war were no longer relevant.[78] Both sides were weary of the war which affected their respective economies and had brought no clear gains.[79] In November 1813 the British Foreign Secretary Robert Stewart, Lord Castlereagh (1769 – 1822) sent a letter to the US President proposing peace negotiations.[80] In January 1814 the US Congress agreed to negotiate a settlement, and sent a delegation to Europe.[81] After some stalling by the British, negotiations were started in Flanders on 8 August 1814.[82] With the costs of the 1815 campaign threatening to bankrupt the US Government, and New England likely to secede from the USA, the American commissioners gave up their claim to British North America.[83] Likewise, the Duke of Wellington advised the British side to make a quick peace, as he saw no prospects of a decisive victory in 1815.[84] After much wrangling the Treaty of Ghent on 24 December 1814 finally terminated hostilities between USA and Britain, but simultaneously dispelled any aspirations the native Americans may have had of possessing their own state.[85]

1.2 Life in the Scottish Borders

The various wars had a certain impact on rural life in the Scottish Borders, which had become a peaceful region ever since the last Jacobite Rebellion had been crushed in 1746. Now the country was recruiting soldiers to fight Napoleon, and lists of men aged between 18 – 23 were posted on the doors of the local churches.[86] As a result, manpower on the farms was in short supply.[87] The presence of French prisoners of war between 1811 and 1814 was a further reminder that the world was not as peaceful as it seemed.[88] A few families knew this at first-hand, having lost a father or son, but on the whole, life in the villages went on much as before.[89] With foodstuffs in short supply, there was a drive to drain and fertilize unutilized land to grow crops and raise cattle.[90]

While the landowners begrudged the taxes, that had been levied during the latest wars, many prospered during the period of the Napoleonic Wars because they were able to sell their farm produce at inflated prices.[91] For instance, by 1809 – 1815 oats cost twice as much as in 1770 – 1790, while the price of sheep had tripled.[92] The inflation did not affect their farm servants, because they were paid almost entirely in kind, receiving such allowances as oats (for porridge), barley, peas, the keep of a cow (for milk) and ground for planting potatoes.[93] Because of import restrictions, there was a certain amount of smuggling of brandy, gin and teas at the turn of the century.[94]

Once the soldiers returned in the aftermath of the various wars (1815), manpower again became more plentiful. However, even before the final showdown at the Battle of Waterloo on 18 June 1815, another disaster struck. On 10 April 1815 Mount Tambora in the Dutch East Indies (now Indonesia) blew its top, ejecting large quantities of sulphur dioxide into the stratosphere and cutting out the sunlight. In 1816 cool temperatures and heavy rains resulted in failed harvests in the British Isles.[95] Although Scotland was not as badly affected as some parts of Europe, Rev. John Hastie (1762 – 1822) wrote in his diary on 4 October 1816 that "incessant rains with frost occasionally have checked vegetation and retarded all farming operations. This year too much resembles 1799—stock of every kind and almost no market; turnips have failed completely in consequence of the rains and cold—almost NO Sun shine through the whole of what should have been summer and none now—the demon of drizzle ever hangs over our heads in shape of dense black clouds. What farmers will do it is scarcely possible to say."[96] There was little need to hire extra labourers to bring in the poor harvest, so many mouths went hungry. Moreover, those landowners who had invested in new technologies, such as threshing machines, required fewer labourers.[97] As prices for agricultural products fell after the wars, some farmers went bankrupt, and employment became even more difficult to come by.[98] The economic slump, which started in 1818, led to a commercial crash in 1825.[99] Alexander Somerville (1811 – 1885) was unable to find a job in November 1827, although he attended the hiring markets in Duns, Dunbar and Haddington.[100] In November 1828 he fared no better.[101] This unemployment led some people to search for work in the bigger centres, or even emigrate to North America.[102] Others turned to poaching.[103]

Thomas Lawful Fortune (1786 – 1847), who was born in Ayton, a village 10 km NW of Berwick-upon-Tweed, probably came to Chirnside 7 km away looking for work.[104] He was hired by the Boswalls of Blackadder House.[105] As he was paid in kind, Thomas would have had little money to spend in a public house or at a fair. The servants' simple amusements, such as singing and dancing, tended to take place within the confines of the farmhouse.[106] It is possible that he met Agnes Redpath (Ridpath; 1784 – 1858), a young woman from nearby Edrom, while she was working for the Boswalls. As Agnes was already in her mid-twenties, she was probably keen to get married and start her own family. One thing led to another. What started with a stolen kiss ended in heavy petting. During the celebrations at the end of 1811, the inevitable happened. By the time Thomas and Agnes married on Wednesday 24 June 1812, it would have been obvious to the minister, Rev. John Hastie, that Agnes was pregnant.[107] It is possible that they were called to stand before the parish for "antenuptial fornication".[108] In the Church of Scotland this involved the couple having to confess in front of the assembled congregation, and being subjected to formal rebukes and sometimes a fine.[109] As Thomas' employers not only attended Edrom Parish Church, but were members of the Kirk Session (Church Court), the Boswalls would have been aware of his misbehaviour.[110]

The grand old lady, Elizabeth Boswall (ca. 1741 – 1830; **Fig. 1.5**), was probably scandalized, and demanded that the culprit, who had brought so much shame on their family, be dismissed once his 6-month contract expired.[111] The couple were allowed to stay in one of the cottages of Blackadder Toun, a "town" consisting of no more than a few cottages near Blackadder House, until their child was born.[112] However, after their son was born on 16 September 1812, Thomas Lawful Fortune lost his job on the Blackadder Estate. As cottages were reserved for farm servants, being unemployed meant that the Fortunes no longer had a roof over their heads.[113] Thomas was therefore all the more grateful to George Buchan Jr. of the neighbouring Kelloe Estate (**Box 1.1**), who had just returned from India, for giving him a job as a hedger in 1813. As a member of the Kirk Session, George Buchan would have been aware of the Fortunes' plight.[114] Thomas was allotted a cottage near Kelloe House (**Fig. 1.11**) on the condition that his wife helped during harvest-time.[115] As a married man, he would have been hired on a yearly basis.[116] He must have carried out his work to the entire satisfaction of George Buchan Jr., who in turn must have treated him well. Thankful Thomas worked for George Buchan for the rest of his life, making sure the enclosures were well kept.[117]

Fig. 1.5 Elizabeth Boswall (ca. 1741 – 1830) of Blackadder House. She probably played a role in the dismissal of Thomas Fortune.

Box 1.1

George Buchan Jr. (1775 – 1856) Eldest son of George Buchan Sr. (1751 – 1813) and Anne Dundas (1752 – 1792), fourth daughter of Robert Dundas of Arniston, the younger (1713 – 1787).[1] He would have been educated by private tutors, including John Hastie.[2] At the end of 1790 he went to France to study French in preparation for a career with the East India Company.[3] On 2 May 1792 he sailed to India in the "Winterton" (1781) commanded by Captain George Dundas, Laird of Dundas (1752 – 1792).[4] The captain and many of the crew drowned when the ship broke up off the coast of SW Madagascar on 22 August 1792.[5] After waiting for 7 months for help to arrive from Mozambique, it took the passengers another 10 months to reach Madras.[6] By then only a third of the original 280 were still alive.[7] Buchan worked for almost 16 years in India, finally becoming Chief Secretary to the Madras Government.[8] While he was in India he had a number of close escapes, surviving a second shipwreck, and narrowly avoiding being murdered by Malays.[9] His father's declining health required him to return to Scotland in 1810.[10] He managed the Kelloe Estate for the rest of his life, taking an especial interest in the wellbeing of his subserviants.[11] In the mid 1830s he fell into an ice-pit, and was lame thereafter.[12] His lucky escapes made him believe in Providence, and he served the church in a number of functions.[13] For many years he took part in the General Assembly of the Church of Scotland.[14] When internal strife about the right of a patron to appoint a minister of his choice resulted in the Disruption of May 1843, he joined the Free Church of Scotland.[15] He served as representative for Dunse and Chirnside at the Free Assemblies of 1844 and 1845.[16] When he died he bequeathed some of his money to support aged and infirm ministers, and the missionary schemes of the Free Church.[17]

The harrowing experience of standing before the parish in Edrom Church (or simply their anti-establishment feelings?) could explain why the Fortunes started to attend the Relief Church (1761 – 1847), an off-shoot of the Church of Scotland, whose congregation appointed its own minister.[118] It also meant that Thomas Fortune no longer had to face his former employer. The Fortunes' son was baptized before the Relief congregation in Duns South Church (**Fig 1.6**) by Rev. John Ralston (being Minister between 1801 to 1838) on 10 December 1812.[119] Following Scottish tradition, he was named Robert after his paternal grandfather (1764 – 1846).[120] As a child he would have been exposed to numerous ailments. A streptococcal infection, such as scarlet fever or tonsillitis (*Streptococcus pyogenes*), could have been responsible for an attack of acute rheumatic fever, a condition which was to plague him for the rest of his life.[121]

Fig. 1.6 Duns South Church (1763), where Robert Fortune was baptized. The church, which was rebuilt in 1851, is now a carpet warehouse.

Fig. 1.7 Edrom Churchyard. The obelisk on the right marks the grave of Rev. John Hastie (1762–1822), who married Robert Fortune's parents in June 1812.

Fig. 1.8 Thomas Fortune's gravestone in Edrom Churchyard.

Robert enjoyed the undivided attention of his parents for two years, until his sister Isabella was born on 11 October 1814.[122] The young Robert probably got his first lessons in practical botany from his father, as the latter did his rounds of the Kelloe Estate checking on the state of the hedges. In the days before barbed wire, this was an important job.[123] From him Robert would have learned how to distinguish the different trees and shrubs, and utilize those hedgerow species with flexible twigs to weave an impenetrable fence.[124] Robert respected his father and later erected an imposing tombstone to his parents in Edrom Churchyard (**Figs. 1.7, 1.8**; the Relief Church in Duns did not have a cemetery).

When Robert was between five and eight, he started to attend Edrom Parish School (**Fig. 1.9**).[125] His teacher was the young George Peacock, who succeeded the long-serving William Knox (1759 – 1816) on 16 June 1816.[126] The standard at parish schools like Edrom was high, with a wide range of subjects including English reading and writing, arithmetic and geography, and sometimes even French, Latin, and Greek.[127] In the true spirit of the Enlightenment, the teachers tried to give their pupils a greater awareness of the world.[128] No doubt, as an incentive to improve their lot, Peacock would also have drummed the fear of God into his pupils ("God helps those that help themselves"). After all, there was ample evidence in the district of the biblical deluge in which so many unbelievers had been drowned.

Fig. 1.9 Edrom Parish School with its gothic windows where Fortune went to school.

> There are in this district, the strongest proofs of the effects of the deluge, that vast flood by which our mountains were submerged, and the plains strewed with the records of its violence.[129]

Be this as it may, Fortune referred to himself as a "consistent Protestant".[130]

After school Robert and his friends would have indulged in a variety of activities such as swimming and climbing trees.

> Like all boys who have the opportunity, I had been fond enough of this occupation in my youthful days, but I little thought that the knowledge I acquired in climbing for the nests of wood pigeons and rooks, when my more sober elders consoled me with the prophecy that I would break my neck, would turn out so useful as it afterwards often did. Throwing off my coat and shoes, I clasped the noble tree with my arms, and soon found that, although I had not climbed for years, I had not forgotten my early lessons in the art. I soon reached the branches [at 12 – 15 m] and procured a fine supply of ripe seeds ...[131]

The rook and wood pigeon eggs could have been sold to "egglers", or may have been used to supplement the Fortunes' frugal diet of porridge and skimmed milk for breakfast and supper.[132] Likewise, any fish that Fortune was able to catch in the Blackadder (Blackwater) (**Fig. 1.10**), which was known for its fine trout [133], would have made a welcome addition to the rations as the family grew.[134]

With two adults and nine children life in the 1-2-roomed cottage must have been cramped. If Robert's parents wanted to be alone, they would have had to send their children out. Conversely, if Robert wished to get away from his squabbling siblings, he probably went for a walk in the countryside. These walks made him aware of the beauty of nature. Many years later, when he was in China, he was reminded of his native Scotland.[135]

Fig. 1.10 The Blackadder from Kelloe Bridge. As a boy Fortune swam and fished in the river.

> I now looked on creation itself. The air was cool, soft, and refreshing, as it blows at this time of the year from the south, and consequently comes over the sea. The dew was sparkling on the grass, and the birds were just beginning their morning song of praise (Fortune, 1852: 343).

As he looked around himself, his youthful imagination would have converted the clouds, hills and trees into fanciful objects. In later life, he admitted that he liked to day-dream.

> I sat down on a ledge of rock, and my eyes wandered over these remarkable objects. Was it a reality or a dream, or was I in some fairy land? The longer I looked the more indistinct the objects became, and fancy seemed inclined to convert the rocks and trees into strange living forms. In circumstances of this kind I like to let imagination roam uncontrolled, and if now and then I built a few castles in the air they were not very expensive and easily pulled down again (Fortune, 1852: 235).

Robert's mother probably did not mind if he was out, as it meant one less child to keep an eye on. During the day she would have had her hands full looking after the tiny tots, and cooking porridge for the many mouths. She probably did not get a chance to sew or repair clothes by candlelight or oil lamp until most of the children had gone to bed. Being the eldest child, Robert was at a distinct advantage, as he did not have to wear cast-off clothing. During the main school holiday, he was expected to help with the harvest on the Kelloe Estate.[136] At first he would have been given a light job, such as serving the midday meal of beer, and bean and barley bread to the harvesters.[137] Extra hands were hired to help with the work of reaping, binding and stacking the cereals. These casual labourers came from as far afield as Ireland, where a rapid expansion of the population had led to a vast surplus of labour.[138] Robert probably enjoyed the opportunity to talk to people with different backgrounds, and learn from them.

Fig. 1. 11 Kelloe House, an illustration made before it was demolished.

Once his schooling was over, Robert became apprenticed to his father's employer as a garden boy.[139] This involved digging, planting, sowing and hoeing in the garden adjoining Kelloe House (**Fig. 1.11**).[140] George Buchan, who took an interest in his employees, probably occasionally stopped to have a chat as he passed. No doubt, he told the wide-eyed Robert about some of the adventures he had experienced while he was abroad (**Box 1.1**; Another source of information about the outside world were soldiers returning from the wars). Fortune's questions and comments may have given Buchan a favourable impression of the teenager. It is possible that Buchan thought that Robert's talents were being wasted on his estate, and suggested to his father that he move elsewhere. At the same time, this would help to relieve the cramped conditions in the cottage.

1.3 Employment in the Scottish Capital

In the 1830s Robert became a gardener to the Montcreiffes at their Moredun Estate, just south of Edinburgh.[141] Since the family of George Buchan's mother—the Dundases—and the Montcreiffes were interrelated by marriage[142], it seems likely that the Buchans arranged the move. If so, he may have been given a lift on George Buchan's carriage, when the owner travelled to Edinburgh.[143] As this is the shortest route, they probably travelled over the lonely Lammermuir Hills (**Fig. 1.12**) via Millburn Bridge and Cranshaws to Gifford, rather than go by the easier but roundabout route taken by the Dunse stagecoach via Polwarth, Greenlaw, and Lauder.[144] The carriage would have been reduced to a crawl on some of the steep inclines. Once they reached Moss Law they got their first view of the fertile Forth Valley with the Lomonds in the distance.

Fig. 1.12 The Lammermuir Hills north of Longformacus. The poor soil of these hills makes them only suitable for grazing sheep and their lambs, hence their name.

When Baron David Steuart Moncreiffe (1710 – 1790), one of the founding members of the Royal Society of Edinburgh, purchased the Goodtrees Estate in 1769 he renamed it Moredun after the fort on the summit of Moncreiffe Hill, SE of Perth.[145] It must have been a pleasant place to live, commanding a view of Edinburgh and adjacent country, with the hills of Fife and Perthshire in the distance.[146] Rev. Thomas Whyte, who was Minister of Liberton from 1752 to 1789, has left us a description of the gardens in their heyday.[147] The grounds had been landscaped. Stately trees were dotted throughout the park. In the south the grounds were terraced, and in the north a shady avenue of lime trees and accompanying shrubberies ran parallel to the house. On lower ground paths bordering a meandering rivulet gave ample opportunities for pleasant promenades. Splendid statues had been placed at focal points, and there was even a Chinese temple! On the walled gardens fruit trees had been trained, and there were a number of hothouses in which pineapples, peaches, nectarines and grapes were grown. Fortune would have been in his element!

While Robert was working for the Moncreiffes he met Jane Penny (1816 – 1901) from Swinton ("Swine Town"), about 10 km south of Edrom. He could have met her while he was visiting his parents, e.g. at the Swinton October Fair, in the Relief Church or the Wednesday corn market in Duns, or on the way to or from Edinburgh.[148] When they married on 1 October 1838 her address was given as 43 Albany Street (**Fig. 1.13**), a rather exclusive street in Edinburgh's New Town. As her father Henry was a farm servant[149], it seems likely that she was employed as a domestic servant. If this conjecture is correct, the chances of meeting in Edinburgh itself would have been minimal (Jane was on the north side of the city, while Robert was in the south). By the time of their marriage Robert was staying at 52 Broughton Street (**Fig. 1.14**), not far from the Royal Botanic Garden, so it would seem that he was no longer working for the Moncreiffes.[150]

Fig. 1. 13　43 Albany Street, Edinburgh where Jane Penny was employed.

Fig. 1. 14　52 Broughton Street, Edinburgh with its blue door. Fortune lived here at the time of his marriage to Jane Penny.

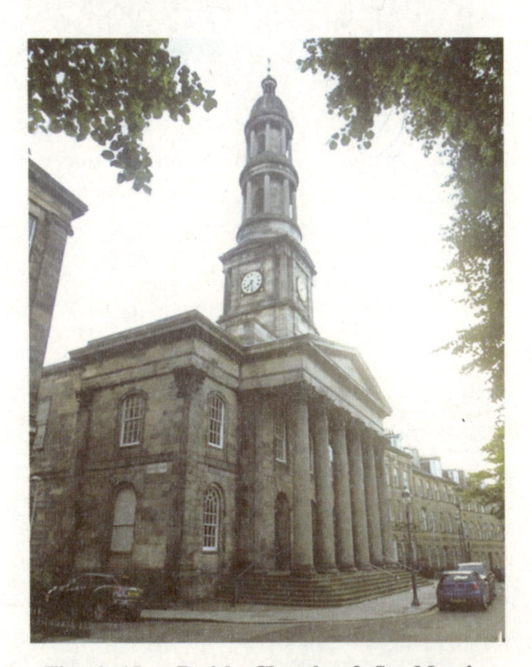

Fig. 1. 15　Parish Church of St. Mary's, Bellevue Crescent, Edinburgh where Robert Fortune and Jane Penny were married on 1 October 1838.

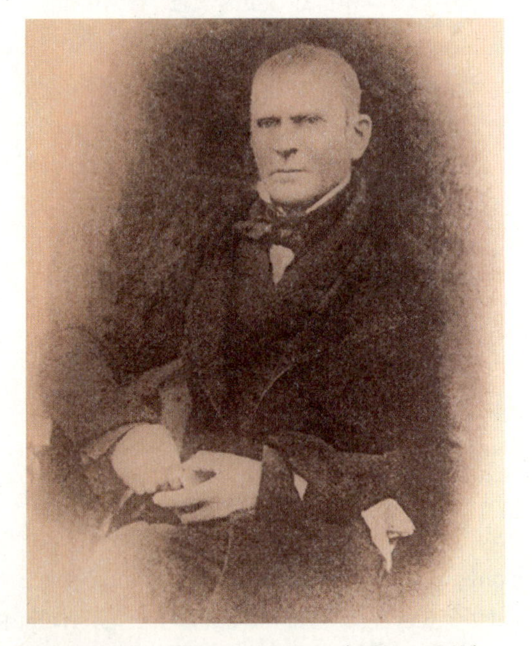

Fig. 1. 16　William McNab (1780 – 1848). Although he was considered hard to please, he wrote a letter to the Horticultural Society of London on Fortune's behalf.

Why the couple decided to marry in Edinburgh is a bit of a mystery. It could have nothing to do with a class difference, as both fathers were farm servants. It seems unlikely that it was a matter of faith, as the Parish Church of St. Mary's (1824) in Bellevue Crescent (**Fig. 1.15**) where they married belonged to the Church of Scotland.[151] Maybe Jane's parents simply felt that she could have made a better match. However, by then Jane would have been in her early twenties (21 or 22), and therefore did not need her parents' consent.[152] The simplest explanation is that Robert may not have been given enough time off work to travel to Swinton. The first of Robert and Jane's six children was a daughter [Helen Jane ("Ellen") Fortune] born on the 6 January 1840 while Robert was working at the Royal Botanic Garden under the famous William McNab (**Box 1.2; Fig. 1.16**).[153]

Box 1.2

Mr. William McNab (1780 – 1848) Son of James McNab, who had a small farm in southern Ayrshire.[1] Although herding cows and sheep was a lonely occupation, he took an interest in the insects, birds and, more particularly, the diversity of wild flowers around him.[2] His father finally agreed to an apprenticeship with Mr. Thomas Kennedy of Dunure (1759 – 1819) in 1796.[3] After three years Kennedy introduced him to Walter Dickson, a nurseryman in Edinburgh, who got him a job with Charles Hamilton (1753 – 1828), the 8th Earl of Haddington, at Tyninghame House on the estuary of Tyne Mouth in Haddingtonshire.[4] In 1801 with a testimonial in his pocket from Rev. John Thomson of Dailly, McNab travelled to London, where he presented himself to William Townsend Aiton (1766 – 1849), the Superintendent of the Royal Botanic Gardens, Kew.[5] He was to work at Kew for the next ten years from 1801 to 1810.[6] After three years in different departments, he was promoted to the position of foreman to replace William Kerr, who had been sent to Guǎngzhōu (Canton) in 1803.[7] A few years later he married a local girl, Elizabeth Whiteman (ca.1778 – 1844), by whom he had nine children.[8] By this time, his diligence and intelligence had already been brought to the attention of Sir Joseph Banks (1743 – 1820), the Director of Kew Gardens.[9] When Banks was contacted by Professor Daniel Rutherford (1749 – 1819) in connection with the vacant curatorship at the Royal Botanic Garden Edinburgh (RBGE), he recommended McNab wholeheartedly.[10] The major stumbling block to McNab's acceptance was the poor pay at the RBGE.[11] Once this had been clarified, McNab willingly returned to Scotland. From May 1810 until his death at the end of 1848 McNab was Curator at the RBGE.[12] As Curator he was expected to have a working knowledge of the Scottish flora and to help students identify the plants they encountered during excursions.[13] These fieldtrips enabled McNab to collect wild plants for the living collection at the RBGE.[14] Through his net of contacts, he very quickly increased the number of exotic plants at the RBGE as well.[15] Between 1820 and 1822 the RBGE was transferred from Leith Walk to its present site on Inverleith Row.[16] This entailed moving many large specimen trees, for which he designed a special cart. That so few of these trees died is a tribute to McNab's practical genius.[17] McNab was very much a hands-on gardener, who passed on decades of experience largely by personal example.[18] He offered the following advice to his gardeners: "Believe nothing implicitly on my authority; exercise your own judgements; take every opportunity which you can possibly command to put to the test of experiment the statements I have made, and abide by the decision of facts."[19] What he did commit to paper now appears somewhat stilted.[20] He himself admitted that "it is no easy matter for a man to communicate in writing what he has acquired by long and continued experience, and more particularly, for one who has not been much accustomed to use the pen ..."[21] One of McNab's duties was to recommend gardeners to prospective employers. He backed Fortune's candidacy for the job at the Horticultural Society of London. When Fortune left for Chiswick, he no doubt gave him a bit of his invariable advice: "Serving your employer well and faithfully, is the best way to serve yourself."[22]

It must have been an exciting time, as the Royal Botanic Garden (RBGE) had comparatively recently been moved from the Leith Walk site to the former estate of the Rocheid family, and so much still required to be landscaped (**Fig. 1.17**).[154] In 1834 a new palm stove, the largest of its kind in Britain, was erected.[155] Based on his subsequent appointment, it would seem that Fortune must have spent at least some of his time working with the exotic plants which McNab was importing. However, the Keeper of the RBGE, Professor Robert Graham (1786 – 1845), also placed great emphasis on the need to study the local flora. In August and September he and McNab would take some students to various parts of the British Isles in search of plants.[156] These field trips also enabled them to expand the living collection at the RBGE. Naturally, they would have needed the help of some of the gardeners to dig up and carry the material. In this way Fortune must have become acquainted with the Scottish Highlands, which he referred to in "Three years' wanderings in the northern provinces of China".[157] During these trips Fortune may have proved to be particularly diligent, or to have been good at spotting the desired plants. Be this as it may, McNab, who was considered hard to please[158], must have been impressed by Fortune's ability, for in 1840 he wrote a letter of recommendation to the Horticultural Society of London backing him for the post of Superintendent of their hothouses.[159]

Fig. 1.17 Inverleith House (1774), the only edifice in the grounds of the Royal Botanic Garden Edinburgh, which Fortune would still recognize.

Chapter 2 Early Professional Career

2.1 The Move to London

Fortune had to wait until 1842 before he was finally appointed to the job at Chiswick.[1] The delay was a consequence of the precarious financial situation in which the Horticultural Society found itself.[2] That Fortune was given the post means that he must have met the Horticultural Society's stringent criteria. As of 1836 student gardeners had to pass a test in literacy, geography, botany and plant physiology.[3] Since schooling did not become compulsory in England until after the Education Act of 1870[4], this entrance qualification gave Scottish gardeners an edge over their English counterparts. As such they were in demand throughout Britain.[5] His sound training under the renowned William McNab (**Box 1.2**) would have done the rest.

At the time Professor John Lindley (**Box 2.1; Fig. 2.1**), was Vice-Secretary of the Horticultural Society, so was no doubt aware of the new employee.

Fig. 2.1 John Lindley (1799 – 1865). The hard-working Secretary of the Horticultural Society of London expected the same dedication from others. Many found him a difficult person to work with. Fortune admired him, and called his first son after him.

Box 2.1

Professor John Lindley FRS (1799 – 1865) Eldest son of the nurseryman and pomologist George Lindley (ca. 1769 – 1835) and his wife Mary Moore.[1] Although he lost the use of one eye in childhood, this did not prevent him from developing his artistic talents.[2] He was educated at Norwich Grammer School, 6 km from home.[3] Since his father could not afford to send him to university or buy him a commission in the army, when he was 16 he got a job with Mr. Wrench, a London seed merchant, who imported plants and seeds from Belgium.[4] His knowledge of French, which he had learned from a French refugee in Norwich[5], would have stood him in good stead. In the winter of 1818 or 1819 William J. Hooker (1785 – 1865) introduced him to Sir Joseph Banks (1747 – 1820), who employed him as assistant to the Scottish botanist Robert Brown (1773 – 1858).[6] This was a very productive period in his life.[7] After Banks' death, he was briefly employed by William Cattley of Barnet (1787 – 1835) before becoming Assistant Secretary of the Chiswick Garden of the Horticultural Society of London in 1822.[8] His income assured, he married Sarah Freestone (1797 – 1869) the next year.[9] They had four children: George (1823 – 1832), Sarah (1826 – 1922), Nathaniel (1828 – 1921) and Barbara (1830 – 1927).[10] In order to support his family he applied for and was appointed (May 1828) to the newly established Professorship of Botany at London University, which had been turned down by Brown and his friend Hooker.[11] That same year he was elected a Fellow of the Royal

Society of London.[12] Not bad for someone who had never been to university! He was to receive more honours, including Honorary Doctorates from the Universities of Munich (1833) and Basel (1839).[13] Although he himself had never attended university, he seems to have had a natural gift for lecturing. His lectures were described as clear, concise, but so replete with well-digested information that not a sentence could be lost with impunity.[14] So when the position of botanical demonstrator (Praefectus Horti) became available at the Chelsea Physic Garden in 1835, he could not resist the challenge.[15] With no more than a piece of chalk and a blackboard "at Chelsea the dry bones were made to live."[16] It was here that he crossed swords with the Curator, William Anderson (1766 – 1846) about the lack of systematic order in the garden.[17] Like others before and after him, Lindley was fascinated by the diversity of orchids. Many of the 286 orchid genera which he described still stand.[18] As he became renowned, his professional advice was sought on matters as diverse as the potato blight, the planting of Ascension Island, and the future of the Royal Botanic Gardens at Kew.[19] He apparently liked to have a finger in every pie, which meant he found it difficult to refuse any opportunity for collaboration. He must have worked flat-out teaching, writing textbooks and scientific papers, editing journals and still finding time to describe the plants which Robert Fortune sent from China.[20] While others showed their respect by calling plants after him, Fortune displayed his appreciation by naming his first son John Lindley Fortune. Fortune probably saw in Lindley an example worth emulating ("rags to riches" through hard work).

No doubt, William McNab's letter of recommendation had kindled his interest in the latest Scot to join their ranks.[5] However, he probably became particularly interested in Fortune as a member of the Horticultural Society's Chinese Committee. This committee was set up after the First Opium War (= First Anglo-Chinese War; **Box 2.2**), when the establishment of five Treaty Ports in China offered the opportunity to send a plant collector to many parts of mainland China.

Box 2.2

Events leading up to the First Opium War. Tea is made from the leaves of the shrub *Camellia sinensis* (Theaceae). It has been used as a medicinal drink by the Chinese for thousands of years.[1] It was first introduced into Europe by the Dutch East Indian Company (VOC) in 1610.[2] From there the drink conquered Britain.[3] As the Chinese were not interested in bartering it for any Western products, it had to be paid for in silver.[4] As more and more tea was drunk, this constituted a serious drain on the British purse.[5] The East India Company (EIC) finally discovered a product for which there was a market in China, namely opium. This is produced from the latex that is exuded by the immature capsules of the opium poppy, *Papaver somniferum*. Although the opium poppy originated in the Mediterranean region[6], it was suitable for cultivation in British India.[7] From Kolkata (Calcutta) it could be exported to China. There was only one snag: the sale and smoking of opium had been banned by the Chinese in 1729.[8] This edict had been followed by others in 1796, 1799 and 1800, which prohibited the import of opium and the growing of the opium poppy in China.[9] As a result, the EIC, which had been growing the opium poppy on a large scale in NE India since 1773 stood to lose an important source of revenue.[10] Since the EIC considered itself an "Honourable Company" acting on behalf of the British government, it could not be seen to be involved in an illegal trade. The dirty work had to be done by agents. These agents were responsible for shipping the opium from British India, where it was legal, to southern China, where it was not. The only Chinese port which had permission to trade with foreigners was Guǎngzhōu (Canton).[11] The foreigners had to act through a merchant guild, the Cohong (Gōngháng, "public profession"), who were responsible to the Chinese government for the foreigners' good behaviour.[12] This included giving guarantees that the foreign ships did not have opium on board.[13] By bribing the Cohong and the local officials, everyone (the addicts, the traders, the local government officials, and the Indian and British governments) was happy. However, when the EIC's monopoly of the China Trade ended in 1834, the situation got out of hand. The private traders, such as the British Jardine, Matheson & Co. and Dent & Co. (**Box 2.4**), as well as the American Russell & Co., greedy for more profit, pushed the opium trade so hard that the balance of payments swung the other way.[14] Instead of having an easy source of specie, the Chinese government found it was now losing money to the foreigners. This situation had to stop. After discussing the pros and cons of legalizing the drug, it was finally decided in the capital Běijīng (Peking) to crack down on the illegal trade.[15] At the beginning of 1839 Commissioner LÍN Zéxú (**Fig. 2.2**) was sent to Guǎngzhōu to root out the evil.[16] He demanded that the foreign firms deliver their stock of opium to him for destruction. When this was not forthcoming, the foreigners' servants were withdrawn, and all trade stopped.[17] In order to break the deadlock, the British Superintendant of Trade, Captain Charles Elliot (1801 – 1875), agreed to surrender the opium and recompense the British traffickers for their losses.[18] More than 20,000 chests of opium were delivered by 21 May 1839 and destroyed by mixing it with salt and lime and washing it out to sea.[19] Sixteen opium barons were ordered to leave China forever.[20] William Jardine (1784 – 1843), who was on his way back to Britain from China, was informed of the developments by his partner James Matheson (1796 – 1878). Once back in London, he put pressure on the British government to act.[21] Although Lord Palmerston (1784 – 1865), the British Foreign Secretary, was sympathetic, he had to consider the feelings of the electorate, which was divided on the idea of going to war to maintain a disreputable trade.[22] Troops and warships were secretly assembled in India.[23] After journalists discovered what was happening, the issue was finally debated in parliament at the beginning of April 1840.[24] By this time open hostilities had already broken out in China.[25] It took until the end of June 1840 before the official naval force of 16 warships, 4 armed steamers and 28 transport ships with 4,000 soldiers was finally assembled off Macao.[26] While a small detachment was left to blockade Guǎngzhōu, the rest sailed north to the Bóhǎi Sea, bombarding Xiàmén (Amoy; 3 July 1840) and capturing Zhōushān Dǎo (Chusan Island; 5 July 1840; **Fig. 3.12**) on the way.[27] A half-hearted attempt was made to blockade the Hǎi Hé (Peiho River) and threaten Běijīng.[28] By September 1840 it was clear that LÍN Zéxú was unable to prevent the British aggression, so he was dismissed and exiled to the Ili District in NW Xīnjiāng.[29] He was replaced by Qíshàn (Kishen, 1790 – 1854), the Governor-General of Héběi Province, who played for time.[30] Finally, on 14 September 1840 he told Elliot that if the British forces withdrew to Guǎngzhōu, all outstanding issues would be settled to their satisfaction.[31] The gullible Elliot pulled his forces out of northern China. Qíshàn did not arrive in Guǎngzhōu until 29 November 1840.[32] Negotiations dragged on through December and into January 1841. When Elliot finally realized that he had been tricked, he attacked and captured the Hǔmén (Bogue) forts guarding the entrance to Guǎngzhōu on 7 January 1841.[33] As Guǎngzhōu itself was now directly threatened, Qíshàn signed what is known as the Convention of Chuānbí (Chuenpi) on 20 January 1841.[34] The terms of the agreement were[35]:

- Britain to receive an indemnity of $6,000,000 to be paid over a period of 6 years
- Hong Kong Island to be ceded to Britain
- The British merchants were free to return to Guǎngzhōu
- China and Britain to establish relations on a basis of equality

- The British to evacuate Zhōushān Dǎo

As Qíshàn had no right to make such concessions without consulting the Emperor, this convention was repudiated and Qíshàn sent into exile in Hēilóngjiāng (Amur).[36] Lord Palmerston was also furious. He pointed out that Elliot had not asked for enough indemnity to cover the cost of the war and the Cohong's debts, and a new Plenipotentiary, Sir Henry Pottinger (1789 – 1856), was sent out to replace him.[37] Pottinger did not waste time. He arrived on 9 August 1841 and spent only a few days in Hong Kong before moving north on 21 August.[38] Xiàmén was captured by Sir Hugh Gough (1779 – 1869) on 25 August and a garrison left on the nearby island of Gǔlàngyǔ (Kulangsu).[39] Bad weather delayed progress, but on 1 October Zhōushān Dǎo was retaken.[40] On 7 October the expeditionary force then headed for the port of Zhènhǎi (Chinhai; **Fig. 5.149**), which guarded the city of Níngbō (Ningpo). After capturing Zhènhǎi, Níngbō fell without a shot on 13 October 1841.[41] With only 700 active servicemen left, an attack on Hángzhōu, the capital of Zhèjiāng, was out of the question.[42] For the next 7 months troop activity was limited to patrolling Níngbō, a sortie to Yúyáo (**Figs. 5.38, 5.39**) and an assault on a Chinese position in the hills north of Cíchéng (Tzeki), following an attempt by the Chinese to retake Níngbō and Zhènhǎi on 9 March 1842.[43] Without waiting for reinforcements, Gough and the marine commander Sir William Parker (1781 – 1866) attacked Zhàpǔ (Chapu), the port of Hángzhōu on 16 May 1842.[44] However, the ships were unable to navigate the treacherous estuary of the Qiántáng Jiāng with its tidal bore[45], so it was decided to skip Hángzhōu and move inland along the Cháng Jiāng (Yangtze River). The forts at Wúsōng (Woosung), which guarded the entrance to the Cháng Jiāng, were silenced on 16 June, followed by the fall of Shànghǎi on 19 June.[46] At the end of June Pottinger, who had returned to Hong Kong in February 1842, arrived with reinforcements.[47] The British armada then edged its way up the Cháng Jiāng to Zhènjiāng (Chen-chiang, Chinkiang), where a decisive battle took place on 21 July 1842.[48] By capturing the city of Zhènjiāng, the British effectively blocked the Grand Canal, thereby cutting off food supplies to northern China.[49] Leaving Major-General James Holmes Schoedde (1786 – 1861, **Box 3.2**) in charge of Zhènjiāng, Pottinger and his army pressed on towards Nánjīng (Nanking), the capital of Jiāngsū Province.[50] With a garrison of no more than 1,600 Manchu troops and fewer than a thousand poorly-armed Chinese soldiers to defend it, the fate of this city was sealed.[51] Negotiations took place in the Jìng Hǎi Temple outside the walls of Nánjīng[52], and the Treaty of Nánjīng signed on board HMS "Cornwallis" on 29 August 1842.[53] The terms of the Treaty of Nánjīng were even more humiliating for the Chinese than the Convention of Chuānbí.[54]

- China to pay $6,000,000 for property [opium] confiscated
- An additional $3,000,000 to be paid to settle Cohong debts with the British
- $12,000,000 to cover the costs of the war
- Zhōushān Dǎo and Gǔlàngyǔ to remain under British control until the $21,000,000 had been paid
- Hong Kong Island to be ceded to Britain
- The "Treaty Ports" of Guǎngzhōu, Xiámèn, Fúzhōu (Foochow), Níngbō and Shànghǎi to be opened for foreign trade and residence
- The Cohong monopoly was revoked, allowing foreigners to trade with anybody
- Consuls to be permitted at the Treaty Ports
- Merchants and their families could reside permanently in these ports

No mention was made of the opium traffic, which had been the cause of the hostilities. The Treaty of Nánjīng was finally ratified at Hong Kong on 26 June 1843[55], just 10 days before Fortune arrived in the Crown Colony.[56]

This was not the first time the Horticultural Society had sent a plant collector to China. In the early 1820s it had sent John Potts and, following his death in 1822, John Damper Parks (ca. 1792 – 1866) was likewise sent to Guǎngdōng. However, at that time the plantsmen were confined to Macao and Guǎngzhōu (Canton), so most of their material consisted of cultivated plants from the Huādì ("Flowery field" or "Flowery land") Nurseries. If they wanted plants from further afield, they had to employ Chinese collectors.[6] As a result their "catch" was modest.[7] However, Lindley (**Box 2. 1**), who was responsible for identifying much of the material, discovered enough new taxa to suggest that an additional trip would be worthwhile.[8] Moreover, it could act as a bait to win more members, as the fellows would receive seed of the exotic plants encountered.

Fig. 2.2 LÍN Zéxú (1785 – 1850). LÍN's destruction of the Indian opium smuggled into China led to the First Opium War.

On Friday, 25 November 1842, the Council of the Horticultural Society:

> Resolved [...] that the Secretary [Dr. Alexander Henderson, 1780 – 1863] be authorized to look out for a competent person to proceed to China as collector to the Society, and to draw up the proper instructions to be submitted to the next Council.[9]

However, as the financial situation of the Horticultural Society was far from rosy[10], the Horticultural Society could not offer to pay the collector more than the £ 100 per annum, which Potts had earned.[11] Only a young gardener would be willing to undertake the difficult task for this salary. For this reason, a newcomer like Fortune was probably considered more suitable for the job. Lindley probably thought that a young man would also have the advantage of being easier to control. Fortune may have seen the special mission to China as an opportunity of proving his worth in the eyes of Jane's parents. Once it got the go-ahead from the Council, the Garden Committee acted quickly. It put forward Fortune's name, and exactly two weeks later, on 9 December 1842:

> It was resolved that the Council do approve the recommendation of the Garden Committee, and confirm the appointment of Mr Robert Fortune as the collector for the Society in China.[12]

On 14 December 1842 Lindley (**Box 2.1**) wrote to the Foreign Secretary, George Hamilton-Gordon (Lord Aberdeen, 1784 – 1860) to inform him of the Horticultural Society's intention. This must have caused quite a stir, for two days later the Under-Secretary of State for Foreign Affairs,

Charles Canning (1812 - 1862) sent this reply[13]:

> I am directed by the Earl of Aberdeen to acknowledge the receipt of your letter of the 14th instant; and I am to acquaint you in reply, that His Lordship would suggest that the Horticultural Society should defer sending any person to China for the purpose alluded to in your letter, until the officers to be appointed by Her Majesty, to reside in the Chinese Ports, shall have entered upon the duties of their offices, and be in a situation to protect the person who may be employed by the Society.

Notwithstanding this warning, the Horticultural Society pushed ahead with its plans. On 28 December 1842 a Chinese Committee Meeting was convened, with the Horticultural Society's Vice-President, George Loddiges (1786 - 1846), in the Chair.[14] The nurseryman Loddiges had been involved in the development of the Wardian Cases (**Box 2.6**), which were to contribute to the success of Fortune's venture. The meeting was also attended by Dr. Lindley (**Box 2.1**), the Secretary Dr. Alexander Henderson (1780 - 1863), Henderson's successor James Robert Gowen, and John Reeves Sr. (**Box 2.3**; **Fig. 2.3**), a former Inspector of Tea for the East India Company, who had himself collected plants while working in southern China.[15]

Fig. 2.3 John Reeves (1774 - 1856). The former tea inspector was the driving force behind the Horticultural Society's Chinese Committee. He believed that green and black teas originated from different species, something Fortune was able to disprove.

Box 2.3

Mr. John Reeves Sr. FRS (1774 - 1856) Youngest son of Rev. Jonathan Reeves, who was initially chaplain to the London Magdalen Hospital for Penitent Prostitutes in 1758 - 1764 before becoming preacher at All Saints Church, West Ham.[1] John Reeves was educated at Christ's Hospital School in London from 1780 to 1789.[2] In September 1790 he found employment in the accounts department of the tea broker, Richard Pinchbeck.[3] Before long he had become a tea taster, and was then employed as a tea broker at the East India Company's tea auctions.[4] His expertise led to his appointment as tea inspector for the East India Company (EIC) in 1808.[5] In 1810 his wife Sarah Russell died, leaving him four young children to look after.[6] He needed a well-paid job. When he was appointed Assistant Inspector of Tea in China in 1812, he left his sister Rachel to take care of the children.[7] Before leaving for China, Sir Joseph Banks (1743 - 1820), the President of the Royal Society and Director of the Royal Botanic Gardens Kew asked him to collect Chinese plants and ship them back to Britain.[8] With the exception of two breaks in England [from 1816 to 1817 to get remarried to Isabella Andrew (1774 - 1840), and from 1824 to 1826], he lived at Macao from 1812 until 1831, working in Guǎngzhōu (Canton) during the tea season.[9] He seems to have been a kind person and was liked by his fellow merchants.[10] He eventually rose to become Chief Inspector of Tea.[11] As a keen amateur naturalist, who was a Fellow of the Horticultural-and Zoological Societies of London[12], he took a great interest in the exotic flora and fauna to be found in southern China. He collected dried specimens for Sir Joseph Banks, the British Museum in London and Kolkata (Calcutta) Botanical

Garden, and on John Lindley's instigation commissioned Chinese artists to make more than 2,000 scientifically accurate watercolours of the plants and animals.[13] In recognition of his work, he was made a Fellow of the Linnean Society and of the Royal Society in 1817.[14] He sent thousands of plants to Britain on board EIC ships.[15] While most of these died or had to be thrown overboard during the 4- to 6-month journey via South Africa, his persistence was ultimately crowned with success.[16] He was responsible, directly or indirectly, for introducing a number of ornamental plants into Europe, e.g. azaleas, camellias, chrysanthemums, paeonies, *Cerasus serrulata*, *Primula sinensis*, *Spiraea cantoniensis*, roses, and last but not least wisteria (*Wisteria sinensis*) in 1816.[17] Moreover, when John Potts and John Damper Parks (ca. 1792 – 1866) were sent to China in the early 1820s by the Horticultural Society, he gave them advice on how to pack and transport their plants.[18] In his honour John Lindley (**Box 2.1**) coined the name *Reevesia* for a small genus of attractive Asiatic plants.[19] When Reeves retired to England in 1831, he wrote several articles on Chinese fishes, frogs, tortoises, birds and mammals.[20] He also continued to be an enthusiastic member of the Horticultural Society, and was still examining plants just a week before his death.[21] The opportunities for plant collection in China offered by the Treaty of Nánjīng in 1842 were not lost on Reeves.[22] Had he not been there to press for a plant collector to be sent to the Treaty Ports, it could have taken years before our gardens were enriched by the ornamental plants brought back by Robert Fortune.

The letter from Canning was "laid before the Committee", as was a message from Admiral Algernon Percy (Lord Prudhoe, 4th Duke of Northumberland, 1792 – 1865) suggesting that Fortune be given letters of introduction.[16] It was resolved to make enquiries regarding a passage on board the sailing ship "Emu".[17] Reeves (**Box 2.3**) stated that Fortune could draw money through Dent & Co. (**Box 2.4**) at Hong Kong.[18]

Box 2.4

Dent & Co. (1824 – 1867) When the East India Company's monopoly of the China Trade ended in 1834, this trading house became second only to Jardine, Matheson & Co.[1] At the time it was run by Lancelot (1799 – 1853), the fifth son of William Dent (1762 – 1801) and his wife Jane Wilkinson (1763 –1840).[2] He was not only a meticulous businessman[3] but a born diplomat, who believed that any pressure brought to bear on the Chinese could only be counterproductive.[4] For this and a number of other reasons[5], he was at loggerheads with William Jardine (1784 – 1843), who considered him a weakling for not standing up to the Chinese.[6] When the Chinese government decided to crack down on the opium trade in 1839, Lancelot Dent, as senior British merchant, was summoned to appear before Commissioner LÍN Zéxú (**Box 2.2**). While he trusted the Chinese and was willing to go, his younger brother Wilkinson (1800 – 1886), James Matheson (1796 – 1878) and other merchants, who were worried that he would be taken hostage, restrained him.[7] Once the opium was delivered to Commissioner LÍN, Dent along with 15 other opium traffickers, was expelled from China.[8] The Dents then used Manila and later Hong Kong as a base for their illicit trade.[9] In 1842 Lancelot returned to England, where he died, unmarried, on 28 November 1853.[10] By the time Fortune arrived in Hong Kong on 6 July 1843, Wilkinson Dent was in charge. As Fortune pointed out[11], the firm went out of its way to make his first expedition a success. They "not only gave me a room in their house, but placed their gardens at Macao and Hong Kong entirely at my service, giving me leave to take from them any plant I might wish to send to England, and to use them for depositing any of my collections in, until an opportunity occurred of sending them home."[12] Fortune repaid the kindness by supervising the planting of some large trees and shrubs in their Hong Kong garden.[13]

After that the Committee considered which tools Fortune should take with him.[19] These included a number of thermometers, two hygrometers, a spade, a couple of trowels, a small pickaxe, a small bill hook (for cutting back shrubs), a geological hammer, some glazing tools [for the Wardian Cases], and a "life preserver" (club to protect himself). They then turned their attention to the plants which Fortune should look out for. These included[20]:

> The True Mandarin Orange called Song-pee-leen [No. 11 in final list]
> The orange called Cum-quat [No. 12 in final list]
> The white Lilies of Fokien eaten as Chesnuts [sic] when Boiled [No. 13 in final list]
> Oxalis sensitiva [No. 14 in final list]
> Lycopodium cernuum Man-neen-chang [No. 15 in final list]
> The Azalea from Lo-fou-Shan a mountain in the province of Canton [No. 16 in final list]
> Cocoons of the Atlas Moth from the Same Place called...[gap; No. 17 in final list]
> The fingered Citron called Haong-Yune or the Fuh-show [No. 8 in final list]
> The Canes of Commerce [No. 18 in final list]
> Varieties of Celosia & Amaranthus [No. 19 in final list]
> Tree Paonies [sic; compare No. 20 in final list]
> Varieties of Illicium [No. 21 in final list]

Fortune was also to look into the possibility of purchasing Japanese plants at the ports of Fúzhōu and Zhàpǔ.[21] Finally, it was decided to ask the Foreign Secretary that a small piece of land on Hong Kong Island be placed at Fortune's disposal for the purpose of depositing his plants prior to their dispatch.[22] Lindley informed Fortune of the Chinese Committee's decisions.

For Fortune the biggest problem was the lack of firearms with which to protect himself in a country still recovering from defeat at the hands of the "foreign devils" (**Box 2.2**). On 1 January 1843 he wrote to Lindley[23]:

> I am much disappointed at the resolution of the Committee with regard to fire-arms, and I still hope that you will endeavour to make them alter their minds upon this subject. I think that Mr. Reeves is perfectly right in the majority of cases—that a stick is the best defence—but we must not forget that China has been the seat of war for some time past, and that many of the inhabitants will bear the English no good-will. Besides, I may have an opportunity, some time, to get a little into the country, and a stick will scarcely frighten an armed Chinaman. You may rest assured that I should be extremely cautious in their use, and if I found that they were not required they should be allowed to remain at home.

At the next meeting of the Chinese Committee on 12 January 1843 it was "Ordered that Mr. Fortune be supplied with a Fowling piece and Pistols."[24] Lindley stated that Mr. Reeves had kindly placed at the disposal of the Committee one of Adie's original sympiezometers and a Chinese Vocabulary.[25] For his part, Reeves stated that Mr. Dent (**Box 2.4**) had liberally offered Mr. Fortune a room in his house at Macao.[26]

In the meantime, Sir Edward Smith-Stanley (**Box 2.5**; **Fig. 2.4**) got wind of Fortune's forthcoming mission to China (possibly through John Reeves Sr., a Fellow of the Zoological Society of London), and asked him to bring back animals.

Fig. 2.4 Sir Edward Smith-Stanley (1775–1851), the 13th Earl of Derby. An avid collector of animals, especially birds.

Fig. 2.5 Liverpool World Museum where some of the Earl of Derby's collections are now housed.

Box 2.5

Sir Edward Smith-Stanley Jr. (1775 – 1851) Only son of the passionate sportsman and socialite Edward Smith-Stanley Sr., 12th Earl of Derby (1752 – 1834)[1] and his attractive but fickle first wife Elizabeth Hamilton (1753 – 1797).[2] He was educated at Eton and Trinity College, Cambridge (MA 1795), before becoming a Whig MP for Preston and later for Lancashire.[3] Unlike his father, and probably as a result of the emotional turmoil in the family as a child, he was known for his quiet and unobtrusive manner, a humble man, who donated much of his wealth to schools and the Church of England.[4] On 30 June 1798 he married his cousin Charlotte Margaret Hornby (1778 – 1817), whose father, the Rev. Geoffrey Hornby narrowly missed being executed in 1806 for homosexual behaviour.[5] Edward and Charlotte had three sons and four daughters.[6] While he inherited his father's love of animals, he was not interested in the sporty aspects but in the animals themselves. He was a Fellow of the Linnean Society (being President from 1828 to 1834) and the Zoological Society of London (being President from 1831 to 1851).[7] In 1834, when his father died, he became 13th Earl of Derby and due to increasing deafness withdrew from politics in order to concentrate on his large menagerie and aviary at Knowsley Hall, 13 km NE of Liverpool.[8] The Earl got to know Edward Lear (1812 – 1888) after the latter applied for permission to draw parrots at London Zoo in 1830.[9] From 1832 to 1837 he hired Lear to make drawings of the animals in his menagerie.[10] Lear's "Book of Nonsense" was written for his grandchildren.[11]

"There was an Old Derry down Derrry, who loved to see little folks merry.

So he made them a book, and with laughter they shook, at the fun of that Derry down Derry [Lear's pseudonym]."

Lear's failing eyesight caused him to quit scientific illustrations.[12] With the help of the Earl of

Derby and his nephew, Robert Hornby, Lear was sent to Rome to recover his health and paint landscapes.[13] By employing collectors in various parts of the world the Earl eventually built up a collection of 20,000 zoological specimens, which were to form the nucleus of the Liverpool Museum natural history collection.[14] With a view to expanding his collection, he took a great interest in Robert Fortune's first expedition to China. From their correspondence[15], it is clear that Fortune supplied him with bird skins and a few living birds which Fortune managed to keep alive during the long journey. Some of these are now in the Liverpool World Museum (**Fig. 2.5**).

The Chinese Committee met for the third and last time on 23 February 1843. The ranks were swollen by one of Lindley's predecessors, Edward Barnard, and the Treasurer, Mr. Thomas Edgar.[27] Messrs. Palmer & Co. were thanked for their letter of credit for £ 500 addressed to Dent & Co. (**Box 2.4**) of Macao.[28] In a letter to Lindley, Lord Stanley stated his intention of providing Fortune with a letter of introduction to the military authorities in China, and pointed out that Gǔlàngyǔ and Zhōushān should be visited before these places were restored to the Chinese Government.[29] Other letters of introduction were laid before the committee, including one from the Foreign Secretary, Lord Aberdeen, who had apparently been won round.[30] The final version of Fortune's instructions was then agreed to, and ordered to be reported to the Council of the Horticultural Society.[31]

2.2 Preparing for the Trip to China

With Reeves' advice, Lindley drafted the final version of the instructions (**Fig. 2.6**).[32] It is interesting to note the mix of Lindley's commands and Reeves' suggestions.

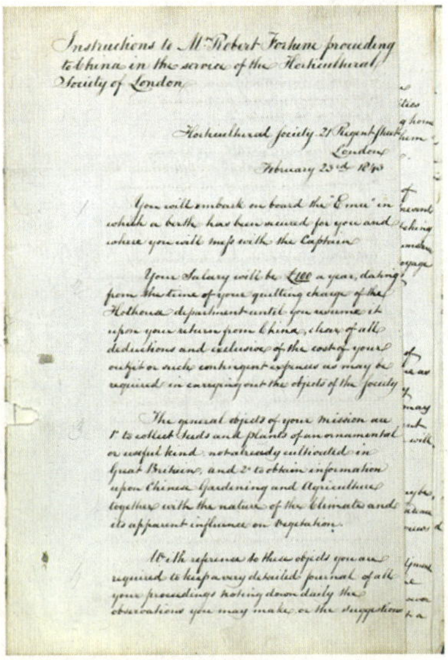
Instructions to Mr Robert Fortune proceeding to China in the service of the Horticultural Society of London

Horticultural Society 21 Regent Street
London
February 23rd 1843

You will embark on board the "Emu" in which a berth has been secured for you and where you will mess with the Captain

Your Salary will be £100 a year, dating from the time of your quitting charge of the Hothouse department until you resume it upon your return from China, clear of all deductions and exclusive of the cost of your outfit or such contingent expenses as may be required in carrying out the objects of the Society

The general objects of your mission are 1st to collect Seeds and plants of an ornamental or useful kind not already cultivated in Great Britain, and 2nd to obtain information upon Chinese Gardening and Agriculture together with the nature of the Climate and its apparent influence on vegetation.

With reference to these objects you are required to keep a very detailed Journal of all your proceedings noting down daily the observations you may make or the suggestions

Fig. 2.6 First page of Fortune's instructions.

You will embark on board the "Emu" in which a berth has been secured for you and where you will mess with the Captain.

Your salary will be £ 100 a year[33], dating from the time of your quitting charge of the Hothouse department until you resume it upon your return from China, clear of all deductions and exclusive of the cost of your outfit or such contingent expences [sic] as may be required in carrying out the objects of the Society.

The general objects of your mission are 1st to collect Seeds and plants of an ornamental or useful kind, not already cultivated in Great Britain, and 2nd to obtain information upon Chinese Gardening and Agriculture together with the nature of the Climate and its apparent influence on vegetation.

With reference to these objects you are required to keep a very detailed Journal of all your proceedings noting down daily the observations you may make, or the suggestions to which the objects you may meet with give rise. This will form the materials out of which a narrative of your expedition will be afterwards prepared for the use of the Fellows of the Society.[34]

You will write home at every opportunity, numbering your letters consecutively, embodying in them as much as you can the materials collected in your Journal. This will enable the Society to judge of the progress you are making in your expedition. All letters must be sent in duplicate, by separate opportunities, so as to guard against the accidents to which a very distant correspondence is liable.

In sending home plants you will always endeavour to ship them on board vessels belonging to Merchants to whom you have introductions. A Bill of lading is to be taken and freight will be paid in England upon the arrival of the packages. You will take care to impress upon the minds of the Captains the indispensible necessity of the glazed boxes[35] being kept in the light, on the poop if possible, or on deck, or failing that, in the Main or Mizen-top. It is also of the first importance that your Seeds should be kept in some well ventilated place. All packages are to be addressed to "the Secretary"[36] 21 Regent Street, London, a letter of advice and bill of lading being in all cases sent by the same conveyance.

Box 2.6

Dr. Nathaniel Bagshaw Ward FRS (1791 – 1868; **Fig. 2.7**) Son of the general practitioner, Stephen Smith Ward (ca. 1751 – 1824), who encouraged his interest in natural history.[1] As a boy Nathaniel expressed a desire to become a sailor, so his father sent him on a voyage to Jamaica when he was only 13 years of age.[2] While this cured him of his passion for seafaring, he was fascinated by the island's tropical vegetation.[3] He then decided to study medicine in London.[4] While he was training to become a doctor, his interest in botany was further stimulated by the charismatic Thomas Wheeler (1754 – 1847), then Demonstrator at the Chelsea Physic Garden, whose lectures and excursions he attended.[5] He became a member of the Royal College of Surgeons in 1814[6] and worked in his father's medical practice in the East End of London, inheriting it when his father retired.[7] He was known as a kind and attentive GP, who won the affection and admiration of his patients.[8] However, he and his wife, Charlotte Elizabeth Witte (1790 – 1857), had difficulty growing their favourite plants in the Wellclose Square garden due to the smog-laden atmosphere.[9] The plants had to be regularly replenished.[10] In 1829 he discovered quite by accident that when he kept moist vegetable mould in a sealed jar, plants of male fern (*Dryopteris filix-mas*) and annual poa (*Poa annua*) germinated and were able to remain alive due to a recycling of the moisture.[11] These plants survived on his window—sill for three to four years until the lid finally rusted through and admitted the acid rain, which caused the plants to rot.[12] Based on this simple principle, and with the help of the skillful nurseryman George Loddiges (1786 – 1846), he developed hermetically-sealed terraria ("Wardian Cases" "Ward's Cases") that were to prove invaluable to generations of plant hunters.[13] In June 1833 Loddiges and he gave his friend Captain Charles Mallard two Wardian Cases with a number of English grasses and ferns to take to Sydney.[14] When these arrived in Australia six months later, almost all the plants were found to be alive and well.[15] Likewise, the Australian plants packed ready for transport to London in February 1834 survived their 8-month confinement with flying colours.[16] As Vice-President of the Horticultural Society, Loddiges made sure

Fig. 2.7 Nathaniel Bagshaw Ward (1791 – 1868), the inventor of the Wardian Cases, which enabled Fortune to keep his plants alive during their month-long transport.

that Fortune had Wardian Cases with him when he went to China for the first time in 1843.[17] The collectors working for Sir William J. Hooker (1785－1865) also made use of the cases to obtain plants for Kew Gardens from far-away places.[18] With Hooker's backing, Ward was elected a Fellow of the Royal Society (FRS) in June 1852.[19] Ward was a founder member of the London (later Royal) Microscopical Society[20] and had a long association with The Worshipful Society of Apothecaries of London.[21] He joined the Garden Committee in 1833.[22] As Master in 1854－1855 and later Treasurer, he was responsible for reviving the Chelsea Physic Garden, which had almost been discarded by the Society of Apothecaries.[23] Fortune would have made his personal acquaintance, while he was Curator of the Chelsea Physic Garden in 1846－1848. At that time Ward was Examiner for Prizes in Botany in 1836－1854.[24] Fortune struck up a friendship with the amiable raconteur and was convinced of the value of the invention of his "old friend"[25], for he used the Wardian Cases on a large scale.[26] There is no doubt that without them, Victorian gardens would have been less colourful, while the tea plantations in India and the rubber plantations in Ceylon and Malaya would have taken much longer to establish.[27]

But it is desired that although you should take advantage of such opportunities as may from time to time arise for sending home collections, yet that duplicates of the best of them be brought back by yourself on your return.

You will take out with you three cases of live plants for the purpose :1st of making presents to those who may be useful to you & 2nd of watching the effect upon the plants of the various circumstances to which they may be exposed during the voyage—The facts relating to this will form part of your report.

You are also provided with a certain quantity of Kitchen Garden Seeds, the object of giving you which is the same as the [second "same as" crossed out] in the case of the plants viz. that you may make presents where desirable & that you may ascertain the effects upon the Seeds of different modes of packing, the result of which you will also embody in your report.[37]

Fig. 2.8 Gravestone of Alexander Robert Campbell-Johnston (1812－1888) in Brompton Cemetery, London. The former Hong Kong administrator lies buried only a short distance from Robert Fortune's grave.

The Society cannot foresee what it may be possible for you to accomplish during your residence in China; which according to their present views they wish to limit to one year [38] and they therefore leave you to act upon your own judgment [sic] as to the details connected with entering the country or forming collections.[39] They are however advised that Chinese may be engaged at a small daily remuneration who will bring you plants &c from the interior in places where you may find it difficult or dangerous to penetrate.[40]

The Council do not feel able to determine what ports you should visit, or in what directions you should conduct your researches, the relations between China and England being at present too uncertain—They are however disposed to believe that Foo-chou-foo [Fúzhōu], if accessible, holds out the greatest promise of valuable results, because it is the Capital of one of the colder provinces of the Chinese Empire[41]—If Chapoo [Zhàpǔ], the place of resort of the Japanese, should be accessible on your arrival, this would be worth an immediate visit.[42]—They also wish that Goolongsoo [Gǔlàngyǔ] should be visited before that place is restored to the Chinese Government[43]—But with regard to this and all such questions the Council are willing to trust to your discretion to act as the information you may receive on your arrival in China may lead you to suppose is most advisable.

If you should find it desirable to make Hong Kong your head quarters it will be necessary that you should have a piece of ground in which to preserve your plants until opportunities for their Shipment arrive. It is hoped that the Lieut. Governor of the Island, Mr Johnstone [**Box 2.7**; **Fig. 2.8**], will enable you to obtain such a spot without expense, in consideration of your stocking it with such European Seeds & plants as you may take out with you[44].

Box 2.7

Lieutenant-Governor Alexander Robert Campbell-Johnston (1812 – 1888) Third son of Sir Alexander Johnston (1775 – 1849) and Louisa Campbell (1766 – 1852).[1] He was in the Mauritian Civil Service (1828 – 1833), before becoming private secretary to his cousin, William John Napier (1786 – 1834) during Napier's ill-fated mission to China in 1834.[2] During the First Opium War he served on the armed paddle steamer "Nemesis".[3] On 22 June 1841 he was appointed Captain Charles Elliot's deputy[4] and in this function made a census of the Hong Kong population.[5] He also started alloting plots of land on Hong Kong Island, even though the Convention of Chuānbí (**Box 2.2**) had not been ratified.[6] When Sir Henry Pottinger (1789 – 1856) arrived to replace Elliot, he ordered that land sales and building be halted.[7] However, no doubt pressurized by the merchants and acting in his own interests, Johnston took advantage of Pottinger's absence during the First Opium War to continue developing Hong Kong.[8] This information appears to have reached the ears of John Reeves at the Horticultural Society. After a vain attempt to become Consul in Shànghǎi, Johnston finally left Hong Kong in September 1852.[9] In 1856 he married Frances Helen ("Ellen") Palliser (1837 – 1893) by whom he had nine sons and two daughters.[10] They went to live in Yoxford, Suffolk[11], but later emigrated to California, where he died on 21 January 1888. His wife had "The Church of the Angels" in Pasadena built in his memory.[12]

The Chinese Committee then went on to summarize some of the plants they were particularly interested in.

It is needless to particularize at much length the plants for which you must enquire—It is however desirable to draw your attention to —

1. The Peaches of Pekin, cultivated in the Emperor's Garden of... [gap] & weighing 2 lbs.[45]
2. The Plants that yield Tea of different qualities[46]
3. The circumstances under which the Enkianthi grow at Hong Kong, where they are found wild in the Mountains[47]
4. The Double Yellow Roses, of which two sorts are said to occur in Chinese Gardens exclusive of the Banksian [rose][48]
5. The Plant which furnishes Rice Paper[49]
6. The Varieties of Nelumbium[50]
7. Pae[o]nies with blue flowers, the existence of which is however doubtful[51]
8. The fingered Citron, called Haong Yune or the Fuh-Show, and other curious varieties of the

Genus Citrus.

9. The Nepenthes, which are different from those in cultivation

10. Camellias with yellow flowers, if such exist.[52]

11. The true Mandarin Orange called Song-pee-leen

12. The Orange called Cum-quat.[53]

13. The Lilies of Fokien, eaten as Chesnuts [sic] when boiled[54]

14. Oxalis sensitiva

15. Lycopodium cernuum called Man-neen-chang[55]

16. The Azalea from Lo-fou-Shan, a Mountain in the province of Canton[56]

17. Cocoons of the Atlas Moth from the same place, called Teen-tsan

18. The Canes of Commerce

19. Varieties of Celosia & Amaranthus[57]

20. Tree & Herbaceous Pae[o]nies

21. Varieties of Illicium[58]

22. The Varieties of Bamboo and the uses to which they are applied[59]

In all cases you will bear in mind that hardy plants are of the first degree of importance to the Society, and that the value of plants diminishes as the heat required to cultivate them is increased. Acquatics [sic], Orchidaceae[60], or plants producing very handsome flowers are the only exceptions to this rule.

You will take care that the packets of Seeds which you send home shall be of sufficient size for general distribution whenever it is possible to procure them. [61]

You will also collect materials for the analysis of Soils by putting up as many varieties of Soils as may seem useful, and noting what plants best thrive in them—A quantity not exceeding 2 lbs of each Soil is sufficient—It is especially desirable to know in what soil the finest specimens of Chinese Camellias, Azaleas, Chrysanthemums, the Enkianthi &c are grown—In general the Soil in which plants are received from China consists of hard lumps of Mud obtained from the bed of the River.[62]

Although we have statements as to the manner pursued by the Chinese in dwarfing trees, more information is wanted upon that curious subject.[63]

It is a general practice for the Chinese to put up their Seeds mixed with burnt bones. Endeavour to learn whether they do this under the idea that it preserves the vitality of the Seeds, or whether they use burnt bones as a Manure, sowing them mixed with the Seed—The last is not improbable[64]—It is desirable to gain information as to the way in which they manage their Manure; especially if it should appear that their processes are at variance with those adopted in Europe.

To all collections of living plants & Seeds the Society lays exclusive claim. You will also prepare for the Society one set of dried Specimens of all plants that you may meet with and have an opportunity of so preserving—But any other collections which you may form will be your private property.[65]

It is however to be understood that the Society is not to incur any expense in forming your private collections and that they are to be altogether subservient to the claims of the Society on your time.

You are supplied with various tools, which you will leave in China on your return; and with fire arms, which, before your Embarkation for England you may have an opportunity of selling to advantage, in which case you will do so and credit the Society with the amount received; otherwise you will restore your fire-arms to the Society on your return.[66]

When in China you will be provided with the Money required for your expenses by application to Messrs. Dent & Co [**Box 2.4**] who will cash your Bills drawn on the Treasurer of the Society to the amount of £ 500.

All letters, seeds, plants, &c are to be addressed to "The Secretary" of the Society, and to no other person, unless they relate to private matters with which the Society has no concern.

With each letter to the Secretary you will send home a cash account of your expenses, and you will preserve vouchers of your expenditure whenever they can be obtained[67], for the purpose of being audited on your return.

The Treasurer will advance you a Sum of £ 50 in Carolus Dollars[68], to be accounted for at the General Settlement of your Account at the end of your Mission.[69]

The letters entrusted to you for various persons should be delivered at the earliest opportunity—A List of them is attached to these Instructions.[70]

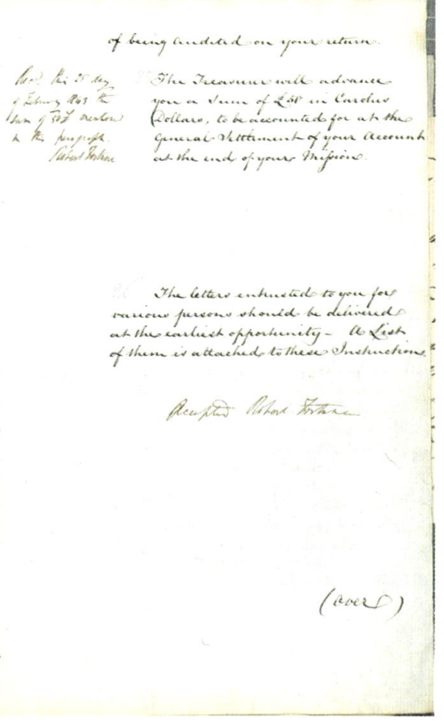
of being audited on your return.

[illegible] Robert Fortune

The Treasurer will advance you a Sum of £50 in Carolus Dollars, to be accounted for at the General Settlement of your Account at the end of your Mission.

The letters entrusted to you for various persons should be delivered at the earliest opportunity— A List of them is attached to these Instructions.

Accepted Robert Fortune

(over)

Fig. 2.9 **Last page of Fortune's instructions with his acceptance, and a note to say that he had received an advanced payment of £ 50.**

Once he had read through the nine pages of instructions, the plantsman scribbled "Accepted Robert Fortune" (**Fig. 2.9**).

When Fortune agreed to go to China on behalf of the Horticultural Society, Jane was eight months pregnant with their second child. It cannot have been easy for him to explain the necessity of going so far away. At least he could comfort her with the knowledge that it was "only to be for a year" (see above). Their first son (John Lindley Fortune) was born at Edrom on 31 December 1842[71] and a few weeks later on 26 February 1843, Robert Fortune sailed away.[72] More than three years were to pass before he finally returned to his native shores on 6 May 1846.[73]

Chapter 3 Fortune's First Visit to China

3.1 Fortune's First Year in the Flowery Land

After a 4-month passage via South Africa, Fortune finally arrived in Hong Kong on 6 July 1843.[1] Fortune did not say anything about the outward journey, possibly because he considered it to be general knowledge at the time, or perhaps because he was confined below deck by sea-sickness. Although he never complained about it, Fortune knew only too well what this was. When he had his first experience of an earthquake in Shànghǎi ("Upper Sea") on 14 April 1853, he compared the feeling to the "sickening sensation not unlike sea-sickness."[2]

As a plantsman, who had seen the lush vegetation of western Java, Fortune was disappointed by the barren granitic hills of Hong Kong ("Fragrant Harbour").[3] So, where was the "flowery land" about which he had heard so much in England?[4] Although he was impressed by Hong Kong Bay with its excellent anchorage, he was of the opinion that Captain Charles Elliot and Sir Henry Pottinger had made the wrong choice for a Crown Colony.[5] However, with a certain foresight, he pointed out that "its importance may yet be acknowledged in the event of another war."[6]

Fig. 3.1 ***Lagerstroemia indica*, one of the three species of crape myrtle found in Hong Kong.**

Fig. 3.2 ***Ixora chinensis*, which Fortune found in Hong Kong "flowering in profusion in the clefts of the rocks..." (Fortune, 1847a: 20)**

He spent the first seven weeks getting orientated in the new environment and preparing for his first journey to the Treaty Ports further north. During this time, he started noticing some of the plants for which Hong Kong is rightly famous. He was impressed by the different crape myrtles (*Lagerstroemia fordii*, *L. indica* and *L. speciosa*; **Fig. 3.1**) on low ground and the flame flower (*Ixora* "*coccinea*", more likely *I. chinensis*; **Fig. 3.2**), which emerged from joints in the granite to flower profusely.[7] He was thrilled to discover *Primulina dryas* (Syn. *Chirita sinensis*), a relative of the African violet, under ever-dripping rocks and sent it back to the Horticultural Society.[8]

He ascertained that most of the ornamental plants, like azaleas, orchids, *Enkianthus quinqueflorus* (Syn. *E. reticulatus*) and *Polyspora axillaris* were to be found at altitudes of 300 – 600 m.[9] He also noted that, with the exception of feral goats, mammals and birds were rare on the island.[10] However, he got an inkling of the diversity of plant and animal life that he was going to encounter on mainland China when he visited the market on Queen's Road, Victoria.[11] There were large numbers of pheasants, partridges, quail, ducks, teal, and sometimes woodcocks and snipes for sale.[12]

Fig. 3.3 Map of some of the places visited by Fortune during his trip to China for the Horticultural Society in 1843–1846.

On 23 August 1843 he sailed for Xiàmén (Amoy; **Fig. 3.3**), but had a high temperature which he attributed to Hong Kong fever (malaria).[13] To make matters worse, the ship had to ride out a storm for three days before finally reaching the rocky island of Nán'ào Dǎo (Namoa Island).[14] Although Nán'ào was not one of the Treaty Ports, the opium traffickers had turned it into a base from which to pursue their contraband trade.[15] A short reconnoitre was sufficient to convince Fortune that the fauna and flora were similar to those of Hong Kong.[16] From Nán'ào the ship followed the indented coast studded with numerous pagodas, which acted as useful landmarks for the crew.[17] As they passed Dōngdìng Dǎo ("East Stone-anchor Island") before entering the Treaty Port of Xiàmén, Fortune noticed how coastal erosion had created a natural arch in the island, which had consequently been dubbed "Chapel Island".[18]

As a country lad, Fortune must have been accustomed to farmyard smells. However in Xiàmén, with its narrow streets thatched over with mats to protect the inhabitants from the sun, his olfactory organs were overwhelmed by the stench.[19] "It is one of the filthiest towns which I have ever seen, either in China or elsewhere; worse even than Shanghae, and that is bad enough."[20] He was glad to escape into the countryside. Near the coast the hills were very barren "with scarcely a vestige of vegetation."[21] However, when he travelled further inland the ground levelled out and was sufficiently fertile to yield good crops of rice, sweet potatoes, peanuts, ginger and sugar cane.[22] The villagers, who had probably never seen a foreigner before, were surprised and somewhat alarmed by his strange appearance.[23] Although he did not have a pigtail, he seemed to eat and drink like them.[24] Fortune took their examination of his clothes and specimens in his stride.[25] He soon established a rapport, and had the boys collecting plants on his behalf.[26] However, unlike their owners, the dogs, with their good sense of smell, remained hostile to Fortune.[27]

While he was in Xiàmén, Fortune made a point of visiting the attractive granitic island of Gǔlàngyǔ ("Drum Wave Islet"), as Reeves had suggested in the instructions (Chapter 2). While there, he had the good fortune to meet Captain Hall, the commander of a detachment of the 41st Madras Native Infantry (MNI), who himself was fond of botany and well acquainted with the plant localities.[28] Together they admired the creeper *Paederia foetida* (Syn. *P. scandens* **Fig. 3.4**), which was colonizing the crevices and hedges, a *Wisteria sinensis* covering an old wall and scrambling up the branches of adjoining trees, and a scentless rose with small double flowers, which Fortune forwarded to the Horticultural Society.[29]

Fig. 3.4 ***Paederia foetida*, which Fortune discovered on Gǔlàngyǔ in hedges and rock crevices. It is "very pretty, but having a most disagreeable odour." (Fortune, 1847a: 43)**

Fig. 3.5 ***Osmanthus fragrans*, an evergreen shrub renowned for its delightful perfume. "When they are in flower in the autumnal months, the air in their vicinity is literally loaded with the most delicious perfume. One tree is enough to scent a whole garden." (Fortune, 1852: 331)**

Fig. 3. 6 Gǎngzīhòu Beach, Gǔlàngyǔ, with Sunlight Rock in the far distance.

Fig. 3.7 David Abeel, the US missionary who accompanied Fortune during his visit to Gǔlàngyǔ.

The gardens were full of chrysanthemums (*Chrysanthemum* spp.), *Hibiscus rosa-sinensis*, *Jasminum sambac* (Arabian jasmine/zambac), *Osmanthus fragrans* (Syn. *Olea fragrans*; **Fig. 3.5**) etc.[30] Hall arranged for Fortune to meet one of the principle mandarins and to see round his garden adjoining the beach (presumably Gǎngzīhòu Beach, "Rear harbour", **Fig. 3.6**). They were accompanied by the American missionary, David Abeel (**Box 3.1**; **Fig. 3.7**), who could speak the Fújiàn dialect.[31] During tea they were bombarded with questions (their names, ages, occupations, and how long they had been away from home) and their clothes inspected.[32] The mandarin was particularly taken by the coloured waistcoats his guests wore. After tea they enjoyed a walk in the garden behind the house with its large banyan trees and sparkling spring.[33] It was dark by the time they finally took their leave.[34]

Box 3.1

Rev. David Abeel Jr. (1804 – 1846) Son of the Dutchman David Abeel Sr., an officer in the US Navy during the American War of Independence, and his pious wife Jane Hassert.[1] As a young man he was a gregarious person and fond of athletics.[2] When his application for the Military Academy at West Point was turned down, he decided to study medicine.[3] During his medical studies he was drawn towards the Christian ministry.[4] He attended the New Brunswick Theological Seminary between 1823 and 1826, and while there got into the habit of retiring to a secluded place in the woods to meditate.[5] He was ordained in October 1826, and served in the unrefined village of Athens (NY) on the Hudson River.[6] He was kept busy preparing sermons, preaching and visiting families.[7] His congregation started to grow.[8] Nonetheless, he often had doubts about the strength of his own faith.[9] He considered testing this by becoming a missionary.[10] However, as an only son, he realized his obligations to his aging parents.[11] By November 1828 this emotional turmoil was affecting his health. He had chronic indigestion, so he decided to quit his parish.[12] His friends advised him to spend some time on the British Virgin Islands (West Indies).[13] Frustrated by a ban imposed on his preaching, he returned to New York in August 1829.[14] He was appointed a chaplain of the Seaman's Friend Society, whose aim was to look after the moral welfare of American sailors abroad, and sailed in the "Roman" from New York to Guǎngzhōu on 14 October 1829 with Elijah C. Bridgman (1801 – 1861), the first missionary sent to China by the American Board of Commissioners for Foreign Missions (ABCFM).[15] Although he was often seasick, he acted as chaplain on board.[16] On 25 February 1830, after 19 weeks at sea, the "Roman" finally reached Guǎngzhōu.[17] The American missionaries were welcomed by the Northumbrian missionary Dr. Robert Morrison Sr. (1782 – 1834).[18] Although Abeel suffered from depression and a certain apathy, he preached every Sunday in either Guǎngzhōu, Huángpǔ or Macao.[19] After 10 months in Guǎngzhōu, Abeel was appointed as a missionary to the ABCFM.[20] Just five days later, on Christmas Day 1830, he left Guǎngzhōu.[21] While en route to Java he again took over the role of chaplain onboard the "Castle Huntley".[22] He stayed with Walter Henry Medhurst Sr. (1796 – 1857) in Jakarta and accompanied him on his rounds.[23] The hard-working Medhurst was to serve as an example.[24] On 4 June 1831 he sailed for Singapore, where he met Rev. Jacob Tomlin before setting sail for the Gulf of Thailand.[25] In marshland south of Bangkok the missionaries set up a clinic and distributed texts to the simple but literate inhabitants.[26] Abeel longed to be able to speak Thai properly.[27] He made a short trip to Singapore and the Chinese College in Malacca (Malay Peninsula) founded by Dr. Morrison at the beginning of 1832, but was back in Thailand on 19 May 1832, this time on his own.[28] After 6 months, ill-health forced him to return to Singapore, where he benefitted from the change of air and constant exercise.[29] When Rev. Robert Burn (1799 – 1833), the British chaplain in Singapore died prematurely, Abeel acted as his replacement for four months.[30] He was kept busy with 3 services a week, missionary meetings, and the study of Chinese and Malay.[31] The overwork affected his health, so he decided to head home on the "Cambridge" on 25 May 1833.[32] He arrived in England on 21 October 1833.[33] On his doctor's advice, he postponed his return to the USA.[34] Between December 1833 and July 1834 he travelled to various parts of continental Europe, where he spoke on behalf of the missions.[35] Before leaving England in August 1834, he and Rev. Baptist Noel (1799 – 1873) helped found the Society for Promoting Female Education in China and the East.[36] He spent his time in the USA from 1834 to 1838 collecting support for the missions, trying to recruit additional missionaries, and increasing his knowledge of medicine.[37] During the cold weather he sought relief in warmer climes.[38] While at St. Thomas (now US Virgin Islands), his disability was diagnosed as an enlargement of the heart, which made breathing difficult.[39] Although by 1838 he was very frail and doctors feared for his life, on 17 October he sailed for China with two missionary couples, Rev. Benjamin and Mrs. Charlotte Keasberry, and Rev. Samuel R. Brown (**Box 8.7**) and his wife Elizabeth.[40] On arrival in Guǎngzhōu on 20 February 1839 Abeel took up Chinese again, but the classes were soon suspended during the crackdown on the opium trade (**Box 2.2**).[41] He managed to escape to Macao before the foreign community became cut off.[42] When the First Opium War (**Box 2.2**) broke out, the restless Abeel decided to visit the Dutch Reformed Churches in Borneo.[43] He was accompanied as far as Singapore by the Browns.[44] Unfortunately, the humid climate on the west coast of Borneo (now Kalimantan) proved bad for his lungs, so after less than a month, on October 1841, he returned via Singapore to Macao.[45] Just six weeks later, on 7 February 1842 he set off with Rev. William Jones Boone Sr. (1811 – 1864) for Gǔlàngyǔ, which was then in British hands, to assess the possibilities of establishing a missionary station there.[46] Their assessment was positive, and after the renovation of a ruined house allocated to them, Boone returned to Macao to pick up his family.[47] The Boones arrived back on 7 June, bringing with them Rev. Thomas Livingston McBryde (1817 – 1863) and Mrs. Mary Williamson McBryde (1820 – 1893) of the American Presbyterian Board, and the independent medical missionary Dr. William Henry Cumming.[48] However, within two months the initial euphoria was dulled when the energetic, efficient and ever cheerful Mrs. Sarah Amelia Boone became feverish and died.[49] This was another blow to Abeel, who had just lost both his parents.[50] After the war, he was appointed chaplain to the British garrison on

Gŭlàngyŭ.[51] In this function he enjoyed good relationships with the Chinese authorities in Xiàmén, one of the New Treaty Ports.[52] This explains how Fortune was able to meet one of the chief mandarins.[53] In 1844, after Fortune's visit, Abeel was finally able to rent premises in Xiàmén itself, thus allowing more people to attend the medical dispensary and religious services.[54] Unfortunately, by this time Abeel's health had deteriorated. In August 1844 he had a short break in Hong Kong, but this did not help to alleviate the symptoms, and in November during a trip to Jīnmén Dăo ("Golden Gate Island") he had to be carried in a sedan chair.[55] On 19 December 1844 he left Xiàmén, and after a few weeks in Hong Kong and Macao, finally sailed for New York on 14 January 1845.[56] By the time he arrived in New York on 3 April 1845, he was worn out.[57] The city's polluted air made his condition worse, so after a few weeks he went further north along the Hudson River.[58] He spent the winter in SE Georgia.[59] In the Spring of 1846 he travelled north to New York and on up the Hudson River to Albany, where he stayed until June.[60] On his return he stopped in Athens. Many of his ex-parishioners wept when they saw how weak Abeel had become.[61] After visiting Rhode Island, he travelled some 500 km to see his relative Rev. Gustavus Abeel (1801 – 1887) in Geneva (NY), before returning to Albany.[62] He had only been in Albany for a few days when he had a relapse.[63] Unwilling to take painkillers, fearing they would benumb his mental facilities, he bore his agony stoically.[64] When he felt his end was near, he asked to be left alone to prepare for his departure to another world.[65] With his passing the world lost a pious, self-effacing gentleman who, notwithstanding his physical disability, devoted his life to the service of mankind.

Fig. 3.8 ***Platycodon grandiflorus*, Chinese bellflower, which Fortune almost lost when his servant was attacked by a group of ruffians at Shēnhù Găng.**

Fig. 3.9 ***Abelia chinensis*, which was nearly destroyed in the scuffle. This ornamental shrub was described by the Scottish botanist Robert Brown in 1818.**

Having surveyed the environs of Xiàmén, Fortune boarded a sailing ship bound for Zhōushān ("Boat Mountain") at the end of September.[35] However, they were caught in a typhoon on 1 October.[36] The decks were awash as they battled against the powerful wind.[37] The ship had to put into Quánzhōu Găng (Chinchew Bay) for repairs and Fortune transferred to another ship commanded by Captain Landers.[38] This vessel fared no better and after being tossed about for a week in the Táiwān Strait, they ended up a few kilometres south of their starting point.[39] Needless to say, the two Wardian cases with plants from Xiàmén had been smashed to pieces.[40] While the ship was being repaired, Fortune decided to reconnoitre the area. Although his Chinese servant and the local boatman warned him that he would probably get robbed or murdered, he headed off in the direction of a dilapidated pagoda.[41] It was not long before he had attracted a crowd, which as time went by became more and more aggressive. Fortune lost some valuables to pick-pockets, while his servant was set upon by some men and the plants they had been so carefully collecting started flying in all directions.[42] Luckily, they were able to retrieve the roots of *Platycodon grandiflorus* (Syn. *Campanula grandiflora* ; **Fig. 3.8**) and *Abelia chinensis* (Syn. *A. rupestris* ; **Fig. 3.9**) before heading back to the ship as quickly as possible.[43]

Fig. 3.10 Sir James Holmes Schoedde (ca. 1788 – 1861) found Fortune somewhere to stay in Zhōushān, once he had read the Earl of Derby's letter of introduction.

After sailing for a further ten days, they reached Zhōushān Dǎo in November.[44] With a letter of introduction from the Earl of Derby, the Commander of the British forces (Sir James Schoedde; **Box 3.2**; **Fig. 3.10**) found him quarters in the town.[45]

Box 3.2

Lieutenant General Sir James Holmes Schoedde (ca. 1788 – 1861), a veteran of the Napoleonic Wars, having served as a teenager in the Egyptian Campaign (on 1801; Gold Medal from the Grand Seignor) and as a young man on the Iberian Peninsula (in 1808 – 1814; Army Gold Medal for Nivelle).[1] From 1833 until 1841 he was the Commanding Officer of the 55th Regiment of Foot.[2] He fought in the First Opium War (**Box 2.2**), commanding a brigade that captured Zhàpǔ, Wúsōng (Shànghǎi's port on the Cháng Jiāng), Shànghǎi and Zhènjiāng, for which he received the China Medal and KCB in 1856 – 1857.[3] After the First Opium War he was in command of the British troops on Zhōushān Dǎo until January 1844, when the 55th Regiment was ordered back to India.[4] He was succeeded by Brigadier Colin Campbell (1792 – 1863).[5]

For Fortune, who had grown accustomed to the barren hills, it was love at first sight. On the slopes above the paddy-fields, azaleas, clematises, honeysuckles, wisteria and wild roses (**Fig. 3.11**) were to be found in abundance. He became convinced that China was indeed the "central flowery land".[46] The succession of hills and glens reminded him somewhat of the Scottish Highlands (**Fig. 3.12**).[47] In Zhōushān Fortune made the acquaintance of Dr. Maxwell, surgeon of the 2nd Madras Native Infantry, who spent his spare time botanizing.[48] Maxwell was able to give Fortune some useful tips about where to go, and the two

Fig. 3.11 Wild roses on Zhōushān Dǎo.

men made a number of excursions together.[49] He also allowed Fortune to sow some of the seeds he had brought from the UK in his small garden.[50] Maxwell made a point of setting aside part of his day to treat Chinese patients.[51] Such actions and the correct behaviour of the occupying force[52], soon helped to establish good relations between the British and the local inhabitants, who had suffered two attacks during the First Opium War (**Box 2.2**). By the time Fortune arrived, there were grocers, silk shops, tailors, porcelain and curiosity shops with fancy English names to cater for the new clients.[53]

Fig. 3.12 Zhōushān Dǎo "is a succession of hills, and glens, presenting an appearance not unlike the scenery in the Highlands of Scotland". (Fortune, 1847a: 62)

Fortune does not tell us whether he first saw fresh kumquat in "Dominie Dobbs, the grocer" or somewhere else. These would have been ripe by the time Fortune arrived on the island. In 1915 Dr. Walter T. Swingle (1871 – 1952) honoured Fortune's memory by establishing the genus *Fortunella* for this minute citrus fruit (**Fig. 3.13**).[54] The white fruits of the tallow tree (*Triadica sebifera*, Syn. *Stillingia sebifera*; **Fig. 3.14**) would also have been ripe in November. These were converted into candles by a complicated process, which was explained to Fortune by Dr. Rawes of the Madras Army.[55] The cake, which remained after the tallow had been extracted, was used to manure the fields.[56] Another source of manure was the clover and crown vetch (*Coronilla*), that was sown on ridges once the rice had been harvested.[57] Tea, which Fortune described as being cultivated "everywhere"[58], has all but disappeared on Zhōushān Dǎo, while the waterproof capes made of hemp-palm (Chusan Palm, *Trachycarpus fortunei*; **Figs. 3.15, 3.16**) have been replaced by polythene ones. What can still be seen is the production of salt by evaporation of sea water (**Figs. 3.17, 3.18**).

Fig. 3.13 Kumquat, a diminutive orange, which is eaten skin and all. "I think if the "Kum-quat" was better known at home it would be highly prized for decorative purposes during the winter months." (Fortune, 1852: 122) In 1870 Fortune gave a talk about it to the Horticultural Society.

Fig. 3.14 *Triadica sebifera*, called the tallow tree, because the white fruits yield a candle-wax. The poplar-like leaves turn bright red in autumn.

Fig. 3.15 The hemp palm, *Trachycarpus fortunei*, one of the hardiest palms in existence. Fortune's wish "that one day we shall see this beautiful palm-tree ornamenting the hill-sides in the south of England, and in other mild European countries." (Fortune, 1852: 58) has been realised.

Fig. 3.16 Detail of Fig. 3.15 showing the fibres which insulate the stem. These were used to make ropes and water-proof clothing.

Fig. 3.17 Salinas on Zhōushān Dǎo. The method of salt extraction has changed little since Fortune's day.

Fig. 3.18 Baskets for scooping up the salt.

At the end of November/beginning of December Fortune crossed to the port of Zhènhǎi (Chinhae, "Calm Sea") and sailed up the polluted Yǒng Jiāng ("Níngbō River") to the walled city of Níngbō (Ningpo, "Peaceful Waves"), a total of 60 km from Zhōushān.[59] On the way they passed a number of ice-houses, which were filled when the flooded paddy-fields froze over in the winter.[60] While he was in Níngbō, Fortune stayed in the same house as the curiously-dressed American missionary Dr. Daniel Jerome Macgowan (**Box 3.3**; **Fig. 3.19**)[61]

Fig. 3.19 Dr. Daniel Jerome Macgowan (1815 - 1893), whose dress amused Fortune.

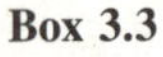

Box 3.3

Dr. Daniel Jerome Macgowan MD (1815 - 1893) After studying medicine at the State University of New York, the little man trained to become a missionary.[1] In 1843 he was sent by the

American Board of Commissioners for Foreign Missions to set up their mission at Níngbō.[2] As a missionary, he clearly felt that more was to be gained by introducing Chinese scholars to Western knowledge of the natural sciences ("natural theology") than bombarding them with Christian tracts or translations of the Bible.[3] His first book, "Bówù Tōngshū" ("Natural Science Almanac"), consisting of explanations of electricity and telegraphy mixed up with an exposition of core Christian doctrines, was published in 1851.[4] Two years later he translated and augmented William Reid's "The progress of the development of the Law of Storms, and of the variable winds" into Chinese as "Hánghǎi jīnzhēn" ("Navigation golden needle") in which the origin and pattern of typhoons in the China Sea was explained.[5] Likewise, he wrote a description of the tidal bore at Hángzhōu for the "Transactions of the China branch of the Royal Asiatic Society", which Fortune quoted.[6] Although he was working in an area frequented by Fortune, there seems to have been little or no contact between the two men. Macgowan may have taken offence at Fortune's remarks about his dress.[7] In 1854 Macgowan offered to gather plants and seed for Dr. James Morrow (1820 – 1865[8]) while the latter was with Commodore Perry's expedition in Japan.[9] In 1859 the Dutch physician Pompe van Meerdervoort (1829 – 1908) invited Macgowan to Nagasaki to teach English and medicine.[10] On 13 February 1860 he attended a meeting of the Royal Geographical Society in London.[11] Unfortunately, ill-health in his family and his own poor state of health forced Macgowan to return to the USA in 1862.[12] During the American Civil War Macgowan served as a surgeon for the North.[13] In 1865 he returned to China and opened a private practice in Shànghǎi.[14] Between visits to his patients he attempted to translate James Dwight Dana's "Manual of Mineralogy" in 1869 and Charles Lyell's "Elements of Geology" in 1871 with the help of HUÁ Héngfāng (1833 – 1902).[15] As HUÁ could not read English and Macgowan's grasp of the Chinese language was far from perfect, HUÁ's translation was helped by watching Macgowan's gestures and facial expressions![16]

The house was not well insulated, so the wind whistled through the cracks and unglazed (papered) windows.[62] As the houses south of the Cháng Jiāng are traditionally unheated, Fortune suffered accordingly. "I never felt so cold in England as I did during this winter in the north of China..." he grumbled.[63] While in Níngbō Fortune visited the delapidated Tiānfèng Tǎ ("Heavenly Order Pagoda"; **Fig. 3.20**) and a number of temples.[64]

Fig. 3.20 Tiānfèng Tǎ, the Heavenly Order Pagoda, Níngbō. From the top of this pagoda, Fortune obtained a good view of Níngbō and the surrounding countryside.

He managed to obtain some plants and cuttings from the gardens of several mandarins, but much to his hosts' surprise declined to take any of their valuable bonsais.[65] An attempt to procure the yellow camellia mentioned in his instructions (Chapter 2) failed, when a Chinaman's business acumen got the better of his moral integrity.[66]

At the end of 1843 Fortune boarded a ship for Shànghǎi. By this time it was miserably cold, and Fortune described how he would often wake up in the morning to find snow on the floor of his bedroom.[67] As Shànghǎi was a centre of trade, there were no mandarin gardens to visit.[68] Moreover, as most of the surrounding area was under cultivation, Fortune did not have much luck finding ornamental plants.[69] One day when he was out shooting birds for the Earl of Derby, he located a nursery garden.[70] Unfortunately, the owner closed and barricaded his premises when he saw the "foreign devil" approaching.[71] Fortune explained the situation to the British Consul, George Balfour (**Box 3.4**), who sent one of his Chinese officers to the nursery to clarify Fortune's intentions.[72]

Box 3.4

Sir George Balfour Jr. (1809 – 1894) Son of Captain George Balfour Sr. and his wife Susan Hume.[1] He was educated at Montrose Academy and the EIC's Military College at Addiscombe, just south of London, entering the Madras Artillery at the age of sixteen.[2] He saw active service in Malacca (in 1832 – 1833), Kurnool (in 1839), Zorapur (Zorapore, in 1839) and during the First Opium War (**Box 2.2**).[3] After the Treaty of Nánjīng in 1842, and his competent handling of the Chinese war indemnity payments, Balfour was selected by Sir Henry Pottinger (1789 – 1856) as Shànghǎi's first British Consul in 1843, much to the disappointment of Alexander Robert Campbell-Johnston (**Box 2.7**).[4] At the time Shànghǎi was not much more than a rural township linked by canals to China's hinterland. However, based on its location near the mouth of the Cháng Jiāng, Balfour foresaw a bright future for the place, with possibilities of trade with the interior of China.[5] He was responsible for planning the foreign settlement to the north of the Chinese township. The acquisition of land proved difficult due to the stubbornness of the inhabitants.[6] Although the area is flat, it was not laid out in the form of a grid, as Balfour took Chinese traditions into consideration.[7] This concern for others would seem to have characterized the man. With his genuine interest in natural history, Balfour offered space in the Consulate grounds for the fruit trees, which Fortune had brought from England for general distribution among the ex-patriots.[8] Fortune related how very supportive he was.[9] He not only sent one of his Chinese officers with Fortune to a nursery to purchase ornamental plants, but got a Chinese employee to procure two pairs of cormorants for the Earl of Derby.[10] After twenty years spent abroad, and the strain of dealing with the Chinese and his own obstreperous countrymen, Balfour felt the need for some rest.[11] He therefore resigned as British Consul in September/October 1846 and returned to Britain.[12] In 1848 he married his cousin Charlotte Isabella Hume (1823 – 1893) from his hometown of Montrose.[13] When Consul F. C. Macgregor (Guǎngzhōu) retired in 1848, he was offered his job, but refused.[14] From 1849 to 1857 he was a member of the Military and Marine Boards at Chennai (Madras), and from 1852 to 1854 was Commissioner of Public Works in Chennai.[15] In 1854 he was awarded the "Companionship of the Bath" (CB) and promoted to lieutenant-colonel.[16] He became colonel in 1856, and was subsequently Inspector-General of Ordnance and Magazines from 1957 to 1859 in Chennai.[17] In the aftermath of the Indian Mutiny, he would have been involved in the rethink to produce weapons and munitions locally, rather than rely on an uncertain supply from Britain.[18] While in India he served in various functions such as the Military Finance Commission in 1859 – 1860, becoming its President and the Chairman of the Military Finance Department from 1860 to 1862.[19] Back in Britain in 1865 he spoke at the Royal Geographical Society about the importance of forest conservation in different British colonies (India, Mauritius and Trinidad).[20] He became major-general in 1865, before joining the Royal Commission on Recruiting for the Army in 1866 – 1867 and becoming Assistant to the Comptroller-in-Chief at the War Office in 1868 – 1871.[21] For his various contributions he was honoured with a Knight Commander of the Bath

(KCB) in 1870.[22] In 1872 he was elected Liberal MP for Kincardineshire in Scotland.[23] Judging from his long term of office, and the fact that a "fine looking" carvel was named after him in 1886[24], he seems to have been well-liked by his electorate. As if being a MP was not enough, he was also Deputy Lieutenant and Justice of the Peace for Kincardineshire.[25] He finally retired from the army as General Balfour in 1879.[26]

In this way Fortune gained entry to this and other nurseries.[73] The "South Garden", about 3 km SW of Shànghǎi, proved to be a wonderful source of ornamental plants.[74] Since it was winter, he could only guess at the identity of the deciduous plants. However, the Chinese names gave an inkling of their affinities.[75] Within a few weeks he had obtained "a collection of plants which, when they flowered, proved not only quite new, but highly ornamental."[76]

At the beginning of 1844 Fortune returned to Zhōushān, which acted as his base in central China.[77] By this time the hills were covered with snow.[78] Having packed his collections, he boarded a vessel bound for Hong Kong, and with a northerly wind blowing, arrived in the Crown Colony in a matter of days.[79] Once the plants were transferred to Wardian Cases and shipped to England, he decided to visit Macao and Guǎngzhōu ("Extensive Prefecture", Canton).[80] As he sailed from Hong Kong to Guǎngzhōu, he was impressed by the grandure of the Zhū Jiāng (Pearl River) Estuary.[81] He must have passed within 5 km of Lìngdìng (Lintin, i.e. solitary nail) Island, which had been used as an opium depot prior to the First Opium War.[82] Where the estuary narrowed, Fortune saw that the Hǔmén ("Tiger gate", Bogue) forts, which had been destroyed during the war (**Box 2.2**), had been rebuilt.[83] North of Hǔmén the mountains receded into the distance and the muddy sediments of the Zhū Jiāng were suitable for growing rice, lotus and sugar cane as well as a range of fruit trees.[84] The ship docked at Huángpǔ (Whampoa), where Fortune would have transferred to a smaller boat, which took him into Guǎngzhōu.[85]

Fortune lost no time in visiting a number of the Huādì (Fa-tee, "Flowery field") nurseries, which had been the source for many of the Chinese plants introduced into Britain in the early 1800s.[86] While Fortune did not find anything new to add to his own collection, he admired the colourful display of azaleas, camellias, paeonies and roses, which reminded him of the exhibitions at the Horticultural Society in Chiswick.[87] The Huādì nurseries supplied local seed merchants such as A-ching with material.[88] Over the years A-ching had got a bad name because his material failed to germinate when sent abroad.[89] It was rumoured that the seed was boiled or poisoned in some way to prevent the Chinese plants from being cultivated in other countries.[90] It had been noticed that the porcelain bottles contained not only seeds but a white ash which was thought by some to be burnt bones.[91] In order to get to the bottom of the matter, Fortune went to see A-ching. When queried about the substance, A-ching replied that it was "burnt lice".[92] This took Fortune aback, until he realized that A-ching meant "burnt rice"! The burnt rice husks were added to prevent maggots from devouring the seeds.[93] Fortune ascribed the loss of viability to the mixing of seed from different years and the long voyage via South Africa.[94]

One day, in the hope of finding something new for the Horticultural Society, Fortune ventured into the hills north of Guǎngzhōu. This nearly proved to be his undoing. When he went into a hilltop cemetery to get a good view of the surrounding countryside, he was set upon by thieves, who stole his money, cap and umbrella.[95] There followed a running battle, until Fortune was finally able to shake off his assailants in the suburbs.[96]

3.2 Fortune's Second Year in China: New People, Places and Experiences

At the end of March 1844, Fortune left Guǎngdōng Province for the north.[97] He stopped in Zhōushān on the way to enjoy the Spring flowers.[98] Beside the azalea and wisteria, he discovered an attractive daphne (*Daphne genkwa*, Syn. *D. fortunei*; **Fig. 3.21**), a new buddlea (*Buddleja lindleyana* Fortune; **Fig. 3.22**) and *Weigela florida* (Syn. *W. rosea*; **Fig. 3.23**).[99]

Fig. 3.21 *Daphne genkwa*, which Fortune discovered on Zhōushān Dǎo.

Fig. 3.22 *Buddleja lindleyana*, one of the most attractive representatives of this genus. Fortune found it growing in hedges on Zhōushān Dǎo in 1844.

From there he set course for Shànghǎi. Unfortunately the schooner in which he was travelling got stuck on one of the sandbanks, and had to be pulled free by a junk which had been following them up the Cháng Jiāng.[100] He finally reached Shànghǎi on 18 April 1844 in time to see the paeonies and azaleas come into flower.[101] The choicest of these were purchased for the Horticultural Society.[102]

From Shànghǎi he travelled to Níngbō. By this time the Glaswegian Robert Thom (**Box 3.5**) had been installed as British Consul there.[103]

Box 3.5

Mr. Robert Thom (1807 – 1846) Youngest of five sons of John Thom, Commissioner of Police in Glasgow, and his wife.[1] He spent a year working in a merchant's office in Glasgow, before becoming an apprentice to Messrs. J. & G. Campbell in Liverpool.[2] During the five years he was there, he came into contact with journalists and men of letters, and contributed articles to the local newspapers.[3] In

June 1828 he went to work for 3 years in La Guaira, the beautiful Venezuelan port 11 km north of Caracas.[4] He quickly learnt Spanish, and was on friendly terms with the Roman Catholic priests there.[5] Afterwards, he spent about 18 months in Mexico, before returning to Britain.[6] He spent the winter and Spring of 1832 and 1833 in England[7], which probably seemed dreary after the exotic places he had been to. In April 1833 he was hired for five years by John Macvicar, the Director of the Manchester Chamber of Commerce and the confidential agent of Jardine, Matheson & Co. in Guǎngzhōu. Macvicar warned his prospective employers that "You will find him awkward at first ... His manners are rather against him, being very Scotch".[8] On the other hand, "He appears to possess good sense and can correspond about Manufactured Goods in a very satisfactory manner, and in this department I look principally to his being useful. He is very industrious and obliging."[9]

Fig. 3.23 *Weigela florida*, one of Fortune's most successful introductions. He discovered it in a mandarin's garden on Zhōushān Dǎo. "This spring, it was loaded with its noble rose-coloured flowers, and was the admiration of all who saw it, both English and Chinese." (Fortune, 1847a: 317)

In July 1833 Thom travelled to Bordeaux (France), where he boarded a ship bound for Guǎngzhōu.[10] He arrived in China in February 1834 and went to work as a clerk for Jardine, Matheson & Co.[11] He learned Chinese in his spare time and within two years had become relatively proficient in the language.[12] He soon became the firm's unofficial interpreter.[13] In 1839 he published under the pseudonym "Sloth" an English translation of a Míng Dynasty story.[14] As the opium crisis reached a head in 1839, he was kept busy translating edicts.[15] In June 1840 he became a British civil servant.[16] During the First Opium War (**Box 2.2**) he saw action and narrowly missed being killed by an arrow at Xiàmén on 3 July 1840.[17] When Zhènhǎi was captured on 10 October 1841, he managed to persuade Sir Hugh Gough (1779 – 1869) to release some 500 Chinese prisoners.[18] In his own words, this gave him more pleasure than if he had been appointed Emperor of China.[19] He was left in charge of the civil administration of the Zhènhǎi district from October 1841 until May 1842.[20] This not only met with the approval of his superiors, but earned him the respect of the Chinese themselves.[21] When he met the Imperial Commissioner, Yīlǐbù Àixīnjuéluó (Ilipu, Elepoo; 1772 – 1843) at the negotiations leading up to the Treaty of Nánjīng in August 1842, the latter is quoted as saying "Luóbódàn [Robert], I thank you for your civil mandarinship at Zhènhǎi—it has gained for you a great name in China."[22] Thom and John Robert Morrison (1814 – 1843) were involved in drafting the Treaty of Nánjīng.[23] The stress and frayed tempers which this involved took its toll on his constitution, which had already been weakened by fever in June 1841.[24]

After the war he was employed as an interpreter by the British occupational force on Zhōushān Dǎo.[25] It was here that Fortune would have first met him. Here he managed to complete his privately published "Vocabulary"[26], which he distributed free to public bodies and interested individuals.[27] When John Robert Morrison died in October 1843, Thom replaced him as Chinese Secretary to Sir Henry Pottinger (1789 – 1856) in Hong Kong.[28] Although it is said that he was not sufficiently versed in idiomatic Chinese to translate official documents properly[29], Pottinger, who generally looked down on his inferiors, wanted to keep him on as his Chinese Secretary.[30] Maybe he was impressed by Thom's war record, or had noticed his failing health.[31] In the event, when the Treaty Ports were opened, Pottinger rewarded him by appointing him British Consul at Níngbō.[32] He arrived there on board HMS "Medusa" on 19 December 1843, just after Fortune's first visit to the city.[33] One of the first things Thom did was to issue a code of good behaviour to the three British residents, a merchant and two spinster missionaries![34] From his consulate at Sānguān táng ("Three officials' hall") on the north (left) bank of the Yúyáo Jiāng (Tzekee River), he was able to keep an eye on the little trade there was[35], since British ships were required to anchor in front of the consulate.[36] It was there that Fortune met him in 1844. Thom, as affable as ever[37], went out of his way to help Fortune.[38] When not involved in consular duties, e.g. confiscating opium[39], complaining about rudeness to foreigners[40], rescuing his right-hand man on Zhōushān Dǎo from an almost certain death in police custody[41], Thom, with the help of his Chinese mistress[42], worked with enthusiasm on a chrestomathy (collection of choice Chinese passages), which was published by the Presbyterian Mission Press shortly after he died of tuberculosis in 1846.[43] Just before he died, Divie Bethune McCartee (1820 – 1900) arranged for his two little children to be taken care of by a friend in Shànghǎi.[44] Although the conduct of many so-called Christians had made Thom sceptical about religion, he was no doubt comforted by the presence of his friend Dr. McCartee at his bedside.[45] He only lived to be 39 years of age.

Thom, his assistant interpreter (Charles Sinclair, **Box 3.6**), and the young Martin Morrison (**Box 3.7**) accompanied Fortune on the first leg of Fortune's visit to the green tea district near Níngbō at the beginning of May.[104]

Box 3.6

Mr. Charles Anthony Sinclair (1818 – 1897) Son of Charles Sinclair and his wife.[1] When orphaned at the age of five, he was brought up in Brussels by a widowed aunt.[2] He worked for five years in the Banque de Belgique in Koblenz before going east in the hope of making a fortune.[3] Using his good connections, he found temporary work at the British Consulate in Guǎngzhōu.[4] In December 1843 he was appointed Assistant Interpreter at the British Consulate there.[5] A month later, he became Assistant Interpreter to Robert Thom (**Box 3.5**) at the consulate in Níngbō.[6] In September 1846 he was promoted to the post of Interpreter at the consulate.[7] Although he complained bitterly about the state of the consulate building, he was still in Níngbō in 1849.[8] In 1850 he moved to the equivalent job in Fúzhōu (Foochow).[9] In one of his reports on 28 November 1850 to Sir Samuel George Bonham (**Box 5.18**), he discussed the trade between Fúzhōu and the Okinawa (Loochoo) Islands.[10] He must have been pleased to note that besides tea, silk and Chinese medicine, British woolen and cotton garments were being exported to Japan. In Fúzhōu he caused an uproar when he suggested that his mistress join him at the consulate, in order to improve knowledge of Chinese language, customs and mentality.[11] The outcome was that he was promoted twice in quick succession![12] In December 1852 he became Interpreter at the Xiàmén Consulate[13], before becoming the Interpreter in the busier Shànghǎi Consulate in March 1855.[14] In August 1855, in his role as Interpreter, he drew up the contract with the tea manufacturers who were to be sent to India.[15] Later that year he acted as interpreter for Commander Edward Westby Vansittart (**Box 5.30**) on one of his forays to eradicate piracy in the Yellow Sea.[16] In April 1856 he became Vice-consul in charge of the British Consulate at Níngbō for a short time, before being appointed British Consul at Fúzhōu in

1861.[17] During his time at Fúzhōu, Sinclair regularly found himself at the centre of the on-going dispute between the stubborn Chinese gentry and the arrogant leaders of the Church Missionary Society (CMS).[18] This lead to the destruction of CMS property by a rabble on a number of occasions in 1864, in 1873 and in 1878.[19] In 1879 he applied on grounds of seniority for the position of consul in Shànghǎi.[20] However, based on his abysmal record of incompetence, the post was given to a younger man.[21] In 1886 he finally decided to retire, to the relief of his superiors and subordinates.[22]

Box 3.7

Martin Crofton Morrison (1827 – 1870) A son of the missionary Rev. Robert Morrison Sr. (1782 – 1834) by his second wife Eliza. As a 16-year-old he arrived in Níngbō on board HMS "Medusa" on 19 December 1843 to study Chinese under Robert Thom (**Box 3.5**).[1] He clearly inherited his father's gift of languages. In 1847 he was appointed Interpreter at the British Consulate in Fúzhōu.[2] Unfortunately, he was ill for much of the time.[3] It was in Fúzhōu that Fortune and he renewed their acquaintance at the beginning of 1849.[4] Later that year he became the officer in change of the interpreters at the British Consulate in Xiàmén.[5] By the beginning of 1854 he had become British Vice-Consul at Guǎngzhōu and experienced at first hand how hated the foreigners were.[6] When out for a walk with two friends, he was attacked by a mob, robbed, beaten and threatened with death.[7] When Fortune was in Guǎngzhōu in 1854 they visited a tea factory on Hénán Island ("South of River", a suburb of Guǎngzhōu south of the Pearl River) to see how caper tea was made.[8] In May 1856 he became Acting Consul in Xiàmén, before being transferred to Fúzhōu in December 1857. Here he quickly rose from Acting Vice-Consul, to Vice-Consul on 17 November 1858 and finally Consul on ca. 21 December 1858.[9] During the Second Opium War he acted as interpreter to Vice-Admiral James Hope ("Fighting Jimmy", 1808 – 1881) and Major-General Sir John Michel (1804 – 1886).[10] When Fortune reached Yāntái (Chefoo) from Shànghǎi on 16 August 1861 after his visit to Japan, his "old and valued friend" was there to greet him.[11] By this time Morrison was British Consul at this port in NE Shāndōng.[12] Although he managed to agree a foreign concession with the Chinese, he spent so much time quarreling with the French over the details, that the British Plenipotentiary, Sir Frederick Bruce (**Box 9.4**), finally decided that a concession in Yāntái was undesirable.[13] Morrison once more became British Consul in Fúzhōu, before being transferred to the port of Pénglái [then known as Dēngzhōu (Tangchow), NE Shāndōng] in 1864.[14] In 1867 he retired on health grounds, and returned to England, where he died at the age of 43.[15]

The first goal of the quartet was the famous Tiāntóng Sì ("Heavenly Child Temple")[105], 23 km east of Níngbō as the crow flies. As it had poured for the greater part of the day, the party was probably in no mood to appreciate the fine avenue leading up to the temple (**Fig. 3.24**).[106] They were glad to dry their clothes, have some dinner and retire to bed early.[107] The next day the group was shown round the temple (**Fig. 3.25**), before viewing the tea plantation, which was one of the sources of income.[108]

Fig. 3.24 Avenue leading to Tiāntóng Sì, the Heavenly Child Temple. "The temple itself is approached by a long avenue of Chinese pine trees." (Fortune, 1847a: 169) In China the pine is a symbol of longevity.

Fig. 3.25 Tiāntóng Sì, an important centre of Chan (Zen) Buddhism, renowned throughout Asia. About 1,225 DŌGEN Zhenji, the founder of the Sōtō School of Buddhism, studied here.

Having seen how the tea was manufactured, Thom, Sinclair and Morrison returned to Níngbō, leaving Fortune to his own devices.[109] After his companions had gone, Fortune felt inexpressively lonely in his strange surroundings, far from friends and home.[110] Fortune spent his days at Tiāntóng enlarging his collections of plants and birds (**Fig. 3.26**)[111]

He was helped in his endeavours by the Buddhist priests, who would carry his plants, drying papers and birds.[112] However, they kept a respectful distance when Fortune was shooting.[113] While Fortune was staying at the temple some wild boars started eating the prized bamboo shoots (**Fig. 5.36**).[114] The priests persuaded Fortune to shoot the offenders.[115] It was a pitch black night, so Fortune was glad that the boars did not put in an appearance, as he was afraid he might have accidentally shot one of the priests.[116] There were two traditional methods employed in protecting the bamboo from unwanted attention. One method was to split a piece of bamboo in half and frighten the animals away by beating the two halves together.[117] The other was to dig a bottle-shaped pit and cover it over with brushwood and grass.[118] Fortune almost fell into these pits on a number of occasions, which reminded him of the fate of David Douglas, one of the Horticultural Society's earlier plant collectors.[119]

Fig. 3.26 *Oriolus chinensis diffusus*, the black-naped oriole, one of the birds Fortune sent to the Earl of Derby. The coin is 22 mm in diameter.

After completing his work at Tiāntóng, Fortune rushed back to Shànghǎi. He had heard about some hills about 30 miles (48 km) to the west of the city and wanted to explore them.[120] On a June morning he set off with a pony belonging to the dàotái (governor) and a pocket compass.[121] Everything went well to start with, but when he got a glimpse of the hills in the distance, he left the road and became lost in a maze of tracks and canals.[122] He almost lost the pony when the bridge over which Fortune was coaxing it collapsed.[123] By the time he reached a small town in the vicinity of the hills it was about 2 p.m.[124] Although he would have liked to stay longer, he had to return the pony to the dàotái. After pony and rider had enjoyed a bowl of boiled rice, Fortune rode back to Shànghǎi.[125] A few days later he set off again, this time in a boat.[126] As they glided along the canals, Fortune was able to take an interest in the crop plants: rice, cotton and a glabrous form of woad (*Isatis tinctoria*, Syn. *I. indigotica* Fortune), which was used for dying cotton garments blue.[127] It took them all day to reach the hills.[128] The vegetation turned out to be a bit of a disappointment, for although he encountered many of the plants he had seen on the Níngbō and Zhōushān hills, he did not meet with a single azalea.[129] Having completed his investigations, Fortune returned to Shànghǎi and shipped some Wardian Cases to the Horticultural Society.[130]

Once this was accomplished, he decided to visit Sūzhōu (Soo-chow-foo; **Fig. 3.27**), renowned for its silk, beautiful gardens and pretty women.[131] Since this city was out-of-bounds to foreigners, he had to go disguised as a Chinaman, pigtail and all.[132] In order not to raise any suspicions, he did not tell the boatmen where he was going until they were well underway.[133] Leaving Shànghǎi by the Sūzhōu Hé (Sūzhōu Creek), he directed the boat in a roundabout way via the walled towns of Jiādìng (Cading) and Tàicāng (Ta-tsong-tseu, "Greatest storehouse").[134] They finally reached Sūzhōu on the fine moonlit evening of 23 June 1844 and moored their boat under the eastern wall of the city.[135]

Fig. 3.27 Shāntáng Canal, one of the many canals for which Sūzhōu is renowned. Although the city was out-of-bounds to foreigners, Fortune decided "to infringe the absurd laws of the Celestial Empire", because he had been led to believe "that it contained a great number of excellent flower gardens and nurseries ..." (Fortune, 1847a: 251) He went disguised as a Chinaman.

Fig. 3.28 *Wisteria sinensis* 'Alba', an albino form of the famous Chinese Wisteria, which Fortune obtained from a nursery in Sūzhōu.

Early the next morning Fortune asked his Chinese servant to find out where the nurseries were.[136] He admitted that he felt a bit apprehensive at the prospect of entering the well-guarded city.[137] However, when nobody took any notice of him, he felt more at ease.[138] He visited a number of nurseries and managed to obtain some new plants, including a white wisteria (**Fig. 3.28**), a double yellow rose (**Fig. 3.29**) and a gardenia with large flowers (**Fig. 3.30**).[139] Having fulfilled his horticultural commitments, he was able to turn his attention to the female population. He approved of the specimens he saw, with the exception of their bound feet and white, powdered faces.[140] Fortune remained in Sūzhōu for a few days before returning to Shànghǎi, where his friend Charles D. Mackenzie (Messrs. Mackenzie, Brothers and Co.), with whom he was staying, did not recognize him at first.[141]

Fig. 3.29 ***Rosa* x *odorata* 'Fortune's Double Yellow' which the plantsman brought back from Sūzhōu.**

Leaving the plants in Mr. Mackenzie's care, Fortune returned to his base at Zhōushān. Having heard about the natural beauties and richness of the vegetation on the neighbouring island of Pǔtuó Shān (Poo-to-san), he determined to visit the place to judge for himself.[142] In July 1844 he and Dr. Maxwell set out for the granitic island with its beautiful sandy beaches (**Fig. 3.31**).[143] During the few days they were there they visited a number of temples including Fǎyǔ Sì ("Buddhist Rain Temple", **Figs. 3.32, 3.33**). Along with the pilgrims they climbed the innumerable steep steps from the beach up to Huìjì Sì ("Intelligent Aid Temple"; **Fig. 3.34**) near the top of Fódǐng Shān ("Buddha's Crown Hill", 291 m).[144]

Fig. 3.30 ***Gardenia jasminoides* var. *fortuneana*, a double form of the sweetly-scented cape jasmine, which Fortune discovered in a nursery in Sūzhōu.**

Fig. 3.31 A beach on Pútuó Shān. Rev. John Lowder, the Anglican chaplain of Shànghǎi, drowned here on 24 September 1849.

Fig. 3.32 Fǎyǔ Sì, the Buddhist Rain Temple, Pútuó Shān. "In order to reach the monastery we crossed a very ornamental bridge built over this pond, which, when viewed in a line with an old tower close by, has a pretty and striking appearance." (Fortune, 1847a: 181).

Fig. 3.33 The main temple of Fǎyǔ Sì.

Fig. 3.34 Huìjì Sì, the Intelligent Aid Temple, near the top of Fódǐng Shān, Pútuó Shān. "There are winding stone steps from the sea beach all the way up to this temple ..." (Fortune, 1847a: 183)

Fig. 3.35 *Carpinus putoensis*, the hornbeam endemic to Pútuó Shān, which Fortune missed.

From the summit the friends admired the hundreds of islands of which the Zhōushān Archipelago is composed lying away to the north in the East China Sea.[145] Being a Buddhist island the vegetation was fairly intact. The two plantsmen found large shrubs (6 – 9 m high) of *Camellia japonica* growing wild in the mixed coniferous/broad-leaved woods.[146] However, they missed a species of hornbeam (*Carpinus putoensis*; **Fig. 3.35**), which is endemic to Pǔtuó Shān, growing next to the temple Huìjì Sì.[147] No doubt, Fortune would have kicked himself, had he known.

Fortune had just returned from the Zhōushān Islands to Níngbō when a typhoon struck on 21 – 22 August 1844.[148] Trees were uprooted, walls blown down and many houses unroofed.[149] When Fortune and his friend Charles D. Mackenzie, at whose house he had been staying, went out to survey the damage they found it next to impossible to stay on their feet.[150]

At the end of the summer of 1844 Fortune returned to Shànghǎi to pack the plants for shipment to Hong Kong.[151] However, he had been weakened by the busy summer and was laid low with fever for a fortnight.[152] Fortune finally arrived in Hong Kong in November 1844. From there he sent a collection of birds to the Earl of Derby (**Box 2.5**), and set about planting the living material in Wardian Cases.[153] These were sent by different ships to minimize the chances of losing valuable plants.[154] When the last shipment was made on 31 December 1844, he boarded a ship bound for Manila in the Philippines.[155]

Liverpool Record Office 920 DER (13) 1/58/1

Hong Kong Dec. 4 1844

My Lord

The enclosed list refers to a case of things which I have collected in the North of China & which I have just shipped in the "Duke of Bedford" [ship; left Hong Kong Dec. 1844] for Your Lordship & addressed to the care of the Secretary of the Horticultural Society.

There are several amongst them which I hope are quite new & certainly extremely rare even in China. Of the "Heavenly Blue Bird" I never saw any more than the specimens I send to your Lordship.

The "Blue Jays" from Chusan are also very rare. What are those Birds with the curious twisted tails & the long slender feathers on the head? I only met with them in one place about 20 miles from Ningpo-foo [Níngbō].

Several in this collection are as you will see duplicates of what I sent last year [i.e. 1843] but only the better kinds. Your Lordship may look at the whole & make what selection you think proper, handing then the remainder to Leadbeater as last year. I am very ignorant about this branch of Natural History but I am inclined to believe that this collection will be more interesting to your Lordship than the last which contained many sp. only sent to give you an idea what kinds of Birds we have in China.

I have to thank your Lordship for the letter you sent me through Dr. Lindley in answer to one I wrote last year. I shall always be glad to have your opinion upon anything I may send home as early as possible.

Have we in England the gold fish with a curious spreading tail which I have observed in various parts of this country & which is very remarkable?

Can I serve your Lordship in any other way during my residence in China. If there is any thing in Chinese Bronzes, Carved Bamboo ornament or any other curiosities which you would like I shall be happy to serve you.

I expect to be able to leave China in 6 or 8 Months & have <u>only</u> [Fortune's emphasis] time to receive an answer to this letter. I hope your Lordship will excuse the freedom I take in frequently writing to you.

I have the honor to be
Your Lordship's Obt. Sert. [obedient servant]
R. Fortune

3.3 A Short Trip to the Philippines

Fortune had probably heard about the Philippines from Hugh Cuming (**Box 3.8; Fig. 3.36**), who acted as his London agent.[156]

Box 3.8

Fig. 3.36 Hugh Cuming (1791–1865), who for some time acted as Fortune's agent in London.

Mr. Hugh Cuming FLS (1791 – 1865) One of three children born to Richard and Mary Cuming.[1] As a boy he developed a passion for natural history from Colonel George Montagu (ca. 1753 – 1815), who was then in 1799 – 1815 living at Knowle Cottage, near Kingsbridge.[2] However, because his parents were poor, Cuming was apprenticed to a sail-maker in 1804.[3] From sailors he must have learned of the opportunities for trade following the collapse of the Spanish Empire on the other side of the Atlantic (**Ch. 1.1**). He travelled to Buenos Aires (Argentina) in 1819[4], where he practised his trade and collected natural history specimens, especially shells.[5] In 1822 he moved to Valparaiso ("Paradise Valley", Chile), which had just been opened to British trade.[6] There he met Maria de los Santos, who bore him two children (Clara Valentina [born in 1825], Hugh Valentine [born in 1830]).[7] His shell-collecting activities were encouraged by the British Consul, Christopher Richard Nugent[8] and several officers of the British Navy including Lieutenant John Frembly, himself a conchologist.[9] He eventually decided to make a living out of the collection and selling of plant and animal specimens. With this in mind, he built a yacht, the "Discoverer", which was specially fitted out for the collection and stowage of samples.[10] During his first expedition with Captain Grimwood in 1827 – 1828 he explored the South Pacific (Juan Fernandez Islands, Easter Island, Tahiti and Pitcairn Island).[11] On Pitcairn he met and befriended John Adams (1767 – 1829), the last surviving mutineer from HMS "Bounty".[12] He then undertook a second voyage in 1828 – 1830 along the west coast of South and Central America, including a visit to the Galapagos Islands.[13] In May 1831 he returned to England, where he exhibited some of his discoveries at the zoo.[14] He must have met Sir Edward Smith-Stanley (**Box 2.5**) there. However, as he wrote to Professor William J. Hooker (1785 – 1865) in 1834, he was "still of a roving mind".[15] Following the intervention of Sir Edward Smith-Stanley with the Spanish Ambassador in London, he was given permission to visit the Philippines, then a Spanish colony.[16] Since he had learned Spanish in South America, he was able to involve the local population in the hunt for plant and animal specimens.[17] Hundreds of schoolchildren were set to work scouring the woods for snails and plants[18], an approach Fortune later applied successfully in China and Japan.[19] In just over three years[20], he visited most of the islands in the archipelago and collected a grand total of 130,000 specimens of dried plants as well as numerous living orchids.[21] His example of shipping live orchids from Manila to England was later followed by Fortune in March 1845.[22] Cuming became a natural history dealer in London shortly after his return on 5 June 1840[23] and for some time acted as Fortune's agent.[24] In the meantime, he continued to add to his collections by purchase and exchange.[25] His house at 80 Gower Street became a mecca for British and foreign scientists.[26] In 1846 he began to suffer chronic bronchitis and asthma, which eventually killed him.[27] He died at home surrounded by the collections which had been the object and solace of his life.[28] He never lived to see his dream fulfilled that his specimens become part of the British Museum's collection and thus accessible to all the scientific world.[29] This occurred in 1866.[30]

Fig. 3.37 A banka, like the one Fortune used to cross the Laguna de Bay (Luzon) in 1845.

Fig. 3.38 Fortune did most of his collecting on Luzon in the shadow of Mount Banahaw.

After a stay in Manila to obtain the necessary permits[157], he headed off in a banka (**Fig. 3.37**) across the large caldera of Laguna de Bay in the direction of San Pablo.[158] He had to be constantly on the alert, as this part of Luzon is not only an active volcanic area[159], but was teeming with bandits.[160] For about three weeks he explored the leech-infested jungle at the foot of the stratovolcano Mt. Banahaw (2,177 m; **Fig. 3.38**) for orchids.[161] By paying for every good specimen, he soon accumulated a large collection of *Phalaenopsis amabilis* (**Fig. 3.39**) and other orchids, which were shipped to the Horticultural Society for distribution among the Fellows.[162] While he was on Luzon he also shot a number of birds for the Earl of Derby (**Box 2.5; Fig. 3.40**).[163] Before leaving the Philippines, he no doubt bought a good supply of his favourite Manila cheroots.[164]

Fig. 3.39 *Phalaenopsis amabilis*, one of the orchids Fortune brought back from the Philippines for distribution among the Fellows of the Horticultural Society of London. "This beautiful species may be well called the 'Queen of Orchids'." (Fortune, 1847a: 337)

Fig. 3. 40A ***Bolpopsittacus lunulatus*, a parrot endemic to the Philippines. The coin is 22 mm in diameter.**

Fig. 3. 40B ***Halcyon smyrnensis fusca*, an Asian tree kingfisher once hunted for its colourful feathers.**

Fig. 3.40C ***Chrysocolaptes lucidus haematribon*** **(male), the Luzon flameback, a woodpecker endemic to the northern Philippines.**

Fig. 3. 40D ***Dryocopus javensis*, the white-bellied woodpecker, widespread in tropical Asia, and represented in the Philippines by a number of subspecies.**

Fig. 3. 40E ***Eurystomus orientalis*, the widespead dollarbird, which occurs from S China to N Australia.**

Fig. 3. 40F ***Merops philippinus*, the blue-tailed bee-eater, widespread in southern Asia. Unlike most of the other birds Fortune collected, it prefers open habitats.**

Fig. 3.40G ***Calidris ruficollis*****, the red-necked stint, a wader, which overwinters in maritime parts of southern Asia and Australasia, but migrates to Beringia in the Spring.**

Fig. 3.40H ***Butastur indicus*****, the gray-faced buzzard, a migratory bird of prey, which overwinters mainly in Indochina, Malaysia and the Philippines, but breeds in NE China, Korea, E Siberia and Japan.**

3.4 Back in China: Encounters with Mandarins, Locals and Pirates

By the 14 March 1845 Fortune was back in Zhèjiāng.[165] As this was to be Fortune's last year in China, he wanted to make a complete collection of the choicest plants/cultivars to take back with him to England.[166] This meant he only had a few weeks in which to check the colour of the spring blooms in Zhōushān, Níngbō and Shànghǎi.[167] Having completed his work in Níngbō, he needed to get to Shànghǎi as quickly as possible. As a foreigner he was expected to travel to Shànghǎi via Zhōushān, but as this might entail waiting 8 – 10 days for a connection, he decided to risk going to Shànghǎi via Zhàpǔ ("Spreading Rivermouth").[168] His Chinese servant YÈ Míngzhū (YE Mingchoo) persuaded an acquaintance to give Fortune a passage on board his junk.[169] Fortune made the cardinal mistake of travelling in his English garb.[170] All went well until the plantsman reached Zhàpǔ on 17 May 1845. As it was 3 years since a Briton had been sighted in Zhàpǔ during the First Opium War (**Box 2.2**), a crowd of civil but inquisitive inhabitants crowded round him.[171] It proved impossible to get on board a Shànghǎi boat, so in his desperation Fortune turned to the mandarins for help.[172] Of course they wanted to know how he came to be in Zhàpǔ. "Where have you come from? Who told you you could travel this way?"[173] With the intention of handing him over to the British authorities, the mandarins offered to take him to Shànghǎi in their own boat.[174] When this ruse did not work, they tried to intimidate him by telling him that the area was infested by robbers and that a Chinese officer would be glad to accompany him.[175] When none of these tactics worked, the mandarins accompanied him to the landing-place and saw him off.[176] When Fortune had left, the mandarins took up the matter with the Dàotái of Shànghǎi, who contacted George Balfour (**Box 3.4**), the British Consul.[177] Luckily the Dàotái's letter was so full of factual errors, that Fortune was able to deny being the person involved, and the case was dropped.[178]

After Fortune had finished his examination of the plants in Shànghǎi, he and Charles Shaw, a Shànghǎi merchant, sailed south to Fúzhōu (Foo-chow-foo, "Good Fortune Prefecture"), the only Treaty Port Fortune had not yet visited. Fúzhōu had been chosen as one of the Treaty Ports, as it lay near the mouth of the Mǐn Jiāng ("Fújiàn River"), which had its source in the black tea country of Wǔyíshān (Bohea Mountains).[179] It had proved to be a bit of a backwater for trade as navigation was problematic due to the strong current during the rainy season and the shifting sandbanks.[180] Because of these problems, Captain Freeman put into the Báiquǎn ("White Dogs") Islands at the entrance to the river to look for a fisherman who could pilot them up the Mǐn Jiāng.[181] When the prospective pilot seemed to hesitate, they simply hijacked him and his crew![182] The Wǔhǔmén (Woo-hoo-mun, "Five Tiger Gate") was successfully negotiated and the ship anchored at a temple just inside the mouth of the river.[183] There Freeman, Shaw and Fortune transferred to a native boat with a shallower draft.[184] It was July and due to the thunderstorms the river was so swollen, that it took them nearly two days to cover the remaining 25 km to Fúzhōu![185] On the way they passed numerous temples and a pretty pagoda on a wooded island.[186] The streets near the Wànshòu Qiáo (Wan-show-jou, "10,000-year-old bridge"; **Fig. 3.41**) in the centre of Fúzhōu were under water, so the companions hired chairs to take them to the British Consulate at Wū Shí Shān Sì, "Black Rock Hill Temple"; **Fig. 3.43**) in the old walled city [more than 3 km to the north of the Wànshòu Qiáo].[187] Fortune was disappointed to learn that Consul George Tradescant Lay (**Box 3.9**; **Fig. 3.44**) had been transferred to Xiàmén and replaced by Rutherford Alcock (**Box 3.10**; **Fig. 3.45**), who was not able to help him find the local nurseries.[188]

Fig. 3.41 Wànshòu Qiáo (1303 AD), Fúzhōu photographed by John Thomson in 1870/1. Although this causeway has been replaced by the Liberation Bridge, a few of the foundations of the old bridge remain.

Fig. 3.42 Liberation Bridge, Fúzhōu, built in 1995-1996.

Fig. 3.43 Wū Shí Shān Sì, the Black Rock Hill Temple, which used to be the British Consulate in Fúzhōu.

Fig. 3.44 George Tradescant Lay (1800 – 1845), the botanist who acted as British Consul in Fúzhōu.

Box 3.9

Mr. George Tradescant Lay (1800 – 1845) Little is known of Lay's origins.[1] Be this as it may, he certainly had developed a "great interest in botanical pursuits" as Robert Fortune put it.[2] In his function as naturalist on board HMS "Blossom" in 1825 – 1828, Lay collected plants in California, Mexico, Okinawa, Bonin Islands, the Philippines and Macao.[3] In fact, Sir William J. Hooker (1785 – 1865) and George Arnott (1799 – 1868) named a genus of Californian composites (*Layia*) after him.[4] When he returned to Britain he did some private tutoring and superintended a Hampstead Sunday school.[5] Being a deeply religious person, he decided to become a missionary. In 1836 he was sent to China by the British and Foreign Bible Society, arriving in Macao in late summer.[6] There he joined the "Himmaleh" on its voyage in 1836 – 1837 to distribute religious tracts to the inhabitants of Singapore, Borneo, Sulawesi, the Spice Islands (Ternate, Tidore) and Mindanao.[7] On his return, he intended to deliver Bibles along the Chinese coast, but his aim was frustrated because all the ships offering a free passage had opium on board.[8] He obviously had a flair for languages, learning Japanese from some shipwrecked sailors stranded at Macao.[9] In 1839 he returned to Britain where he published a book entitled "The Chinese as they are" in 1841.[10] Using this and his knowledge of Chinese as a lever, he returned to China in 1841 as an interpreter with the British forces.[11] While acting as a translator for Sir Henry Pottinger (1789 – 1856) in Wúsōng at the end of the First Opium War, he was slightly wounded in the head, but this did not prevent his officiating at the Treaty of Nánjīng.[12] After the war, Pottinger appointed him British Consul at Guǎngzhōu.[13] At the beginning of 1844 Lay accompanied Fortune into the mountains near this city.[14] Shortly thereafter, in June 1844, he was transferred to Fúzhōu, one of the five Treaty Ports.[15] The local authorities initially allocated him a house standing on piles, which flooded at each high tide, before transferring him to a house which "kept out neither sun nor rain".[16] Although the Chinese tried to prevent foreigners from residing within the walled city, he finally got his way about setting up the consulate at the Buddhist temple of Wū Shí Shān ("Black Rock Hill"; **Fig. 3.43**).[17] However, in order to placate the Chinese he hung the Union Jack at half-mast![18] This caused some consternation among passing Europeans and earned him a veiled reproof from the Governor, Sir John Davis (1795 – 1890).[19] There was little trade at Fúzhōu, so to pass the time he kept a record of the weather and crops.[20] He was transferred to Xiàmén on 19

April 1845 to make way for the ambitious Rutherford Alcock (**Box 3.10**), who had been allocated the position at Fúzhōu by the Foreign Office.[21] Letters written to his eldest son, Horatio Nelson Lay (1832–1898), disclose that he was in low spirits, not having seen his wife Mary Nelson for four years.[22] His wife eventually joined him in Xiàmén, but shortly thereafter he died of a fever[23], leaving his wife pregnant and with not enough money to pay for her passage back home with their four young sons.[24] Donations from the local community raised enough money for her to return to England, where she was paid a lump-sum of £ 300 by the Foreign Office.[25] On her death in 1882, she left less than £ 150.[26]

Box 3.10

Sir John Rutherford Alcock (1809 – 1897) Son of a physician, Dr. Thomas Alcock and his wife.[1] His mother died when he was young, so he was brought up by relatives in Northumberland.[2] He was sent to school in Hexham[3], but returned to London to study medicine.[4] He then became a medical officer in the British Marine Brigade (Miguelite War in Portugal in 1832 – 1834) and British Auxiliary Legion (First Carlist War in Spain in 1835 – 1837). However, he lost the use of his thumbs (rheumatic fever), which put an end to his career as a surgeon.[5] In 1839 he was appointed to a lectureship in surgery at Sydenham College[6] and was Home Office Inspector of Anatomy for a short time in 1842.[7] No doubt bored after the excitement on the Iberian Peninsula, he then explored other career possibilities. With the opening of the Treaty Ports at the end of the First Opium War, he saw a future in diplomacy. Being a tall man (ca. 1.83 m)[8], he must have impressed the Chinese. He was firm but fair in his dealings with them.[9] His diplomatic career started as British Consul at Xiàmén in 1844 – 1845[10]. after which he was appointed Consul at Fúzhōu in 1845 – 1846, quickly followed by the more important consularship at Shànghǎi in 1846 – 1854. He had a low opinion of the British merchants who had settled there, and stood no nonsense from them, as he believed that this would undermine his standing in Chinese eyes.[11] Nonetheless, he was a great believer in commerce as a means of improving mankind's lot, and had no qualms about forcibly opening the Chinese market.[12] Like Sir George Balfour (**Box 3.4**) before him, he saw the possibility of trade with the interior of China.[13] When his wife, Henrietta Mary Bacon, whom he had married in 1841, died in March 1853[14], he was seized by a wanderlust. He asked to be transferred to Europe or the Levant, but was only offered the consularship at Guǎngzhōu which he unenthusiastically accepted because he felt he was no more that a clerk to the Governor of Hong Kong.[15] Luckily, by 1855 the Governor was no longer Sir George Bonham (**Box 5.18**), who had advised the verbose Alcock to save stationary by writing less.[16] In May 1856, after 12 years in China, he went on home leave for two years.[17] When he returned, Guǎngzhōu had become a commercial backwater, so with Sir John Bowring's permission he moved the consulate

Fig. 3.45 Sir John Rutherford Alcock (1809 – 1897), who replaced Lay as British Consul in Fúzhōu. In 1845 he was unable to help Fortune find suitable nurseries, but in 1860 went out of his way to show the plantsman round Edo (now Tokyo).

to Hong Kong.[18] However, this was not to Lord Elgin's liking, so after 3 months in Hong Kong Alcock had to return to Guǎngzhōu.[19] At first he was housed on the "Alligator", but as the floating consulate started to sink, Alcock quickly found premises on terra firma.[20] After a year at Guǎngzhōu, from 1859 to 1862 he was chosen to head the British Legation in Japan.

In the wake of the "Treaty of Yeddo" (26 August 1858), Japan was a dangerous place for foreigners. The British Legation in Edo was attacked on two occasions[21] and Alcock's Japanese interpreter murdered.[22] Nonetheless, making use of his rights, Alcock travelled widely in Japan[23], and he and his Scottish terrier Toby[24] led six of his staff and the horticulturalist John Gould Veitch (**Box 8.1**) to the top of Mt. Fuji on 11 September 1860. The accomplishment was celebrated with a 21-gun salute from revolvers, the hoisting of the Union Jack, and a toast to Queen Victoria with champagne![25] What the Japanese attendants made of this is unfortunately unrecorded. With his interest in biology[26], Alcock invited Fortune to visit Edo in November 1860 and showed him some of the sights from horseback.[27] On Alcock's return to Britain at the end of his first term in Japan, he remarried[28], set up an exhibit of Japanese art at the International Exhibition in 1862, was knighted and received an honorary DCL (Doctor of Civil Law) from Oxford University in 1863.[29] Although he believed Japanese lacquerware had no equal[30], Alcock was unimpressed by Japanese painting and architecture.[31] Alcock returned to Japan for a short time from March to December 1864, before becoming British Plenipotentiary in China between 1865 and 1871 in succession to Sir Frederick Bruce (**Box 9.4**). He continued Bruce's conciliatory policy towards the Chinese and with the "Treaty of Tīanjīn" (**Box 9.1**) due for revision, drew up the "Alcock Convention" on 23 October 1869, in which China was given the right to appoint consuls to British ports and raise customs duties on opium and silk.[32] This was unacceptable to the merchants, who had been lobbying the British government for more concessions.[33] As a result the "Alcock Convention" was never ratified, which soured Anglo-Chinese relations.[34] When he left China at the beginning of 1870, he brought with him more plants for the Royal Botanic Gardens at Kew.[35] Although he was apologetic about his lack of a formal training as a geographer[36], his wide knowledge of the Far East made him an ideal candidate for the presidency of the Geographical Section of the British Association in 1873 and that of the Royal Geographical Society, a function he fulfilled from 1876 to 1878.[37] He continued to serve on the Council of the Royal Geographical Society until May 1893.[38] He was also Vice-president of the Royal Asiatic Society for a few years from 1875 to 1878 and he and his wife actively supported many charitable institutions.[39] With his interest in geopolitics and the promotion of British commerce abroad, he gave his backing for the establishment of the "British North Borneo Company" and acted as its first chairman.[40] This enterprise effectively ruled Sabah for the next sixty years in 1881 – 1941.

Nonetheless, Fortune eventually managed to discover a number of gardens and nurseries in Fúzhōu and environs.[189] He was particularly impressed by the luxuriance of the camellias, hydrangeas, ixoras and the five-fingered citron (*Citrus medica* 'Fingered'; **Fig. 3.46**) and added a few of these to his collection.[190] During these sorties, he saw something of the local agriculture. Banyan and a variety of fruit trees were planted near the villages.[191] In the low-lying fields rice, ginger, sugar cane and tobacco were grown, while sweet potatoes and peanuts were cultivated higher up on the terraced slopes.[192] At an altitude of 600 – 900 m there were tea plantations (**Fig. 3.47**)[193], whose very existence had been denied by the mandarins.[194] By collecting herbarium material and a living plant, he was able to show that the black and green teas originated from the same species.[195] He encountered large conifers (*Cunninghamia lanceolata* [**Fig. 3.48**], *Pinus* *massoniana*, Syn. *P. sinensis*) in the mountains.[196] These were felled and rafted down the Mǐn Jiāng to sawmills and exported to different parts of China.[197] The other major export was tea-oil, which was shipped to Japan in exchange for copper and gold.[198]

Fig. 3.46 Five-fingered citron, *Citrus medica* 'Fingered', once "so common in the shops throughout China" (Fortune, 1847a: 380), was and is used to produce candied peel. Its use as an offering in Buddhist temples earned it its colloquial name of "Buddha's Fingers".

Fig. 3.47 Women and children employed at a tea plantation near Fúzhōu (1871).

With the exception of the tea farmers, Fortune found the local inhabitants impertinent and annoying, so he was quite glad to leave Fúzhōu.[199] He had no difficulty getting a berth on a cargo ship bound for Níngbō.[200] Although these junks were frequently attacked by pirates, they were not allowed to carry guns, so the captain was pleased that Fortune had a double-barrelled shot-gun and two pistols.[201] While the junk was waiting for fair weather to sail, Fortune suffered another spell of fever, going in and out of consciousness.[202] After about a fortnight the weather improved and the junk finally sailed.[203] They had not gone more than 100 km when five pirate ships were sighted.[204] In his feverish state Fortune was called on deck.[205] As the pirates bore down on the junk, Fortune realized that if they were captured, he was a dead man.[206] So with his back to the wall, he had no option but to fight. While the rest of the crew ran below deck, Fortune warned the two helmsmen to stay where they were, if they did not want to be shot.[207] The helmsmen and Fortune ducked as the first pirate ship gave the junk a broadside from less than 20 m.[208] Fortune then raked the buccaneer's decks with his shot-gun, scattering the 40 – 50 pirates and leaving the ship without a helmsman.[209] With her sails flapping idly in the wind, she was soon left far astern.[210] At this stage a second sea-rover took up the chase and started firing.[211] Fortune repeated the procedure, killing the helmsman and disabling the ship.[212] When the other corsairs saw what had happened, they

Fig. 3.48 *Cunninghamia lanceolata*, a conifer renowned for its excellent timber.

gave up the chase.[213]

Two days later, six more pirate ships were sighted.[214] They were similarly surprised by the resistance the junk offered, and finally gave up the pursuit.[215] After two more days Fortune was dropped off in Zhōushān, where he boarded an English vessel bound for Shànghǎi.[216] Once there he was given proper medical treatment by Dr. Thomas Kirk and Dr. William Lockhart (**Box 3.11**; **Fig. 3.49**) and the fever gradually left him.[217]

Fig. 3.49 Dr. William Lockhart (1811–1896), who helped to cure Fortune's fever. People flocked to his hospital in Shànghǎi. After the Anglo-French Expedition (1860), he moved to Běijīng. When Fortune arrived in the Chinese capital, they went sightseeing together.

Box 3.11

Dr. William Lockhart MD (1811 – 1896) Son of Samuel Black Lockhart (1781 – 1848) and his wife Elizabeth Lassel[l].[1] When his mother died in 1816, he was brought up by his pious father and grandmother.[2] After leaving school, he was apprenticed to Mr. Parke, a local apothecary.[3] He then trained to be a doctor at Meath Hospital (Dublin, March-August 1833), passed the examination at The Worshipful Society of Apothecaries of London on his 22nd birthday, and studied at Guy's Hospital in London, between October 1833 and April 1834.[4] He was admitted to the Royal College of Surgeons on 29 April 1834.[5] He returned to Liverpool where he intended to find work in one of the city's hospitals before opening his own practice.[6] However in 1836, inspired by Dr. Walter Henry Medhurst Sr. (1796 – 1857), who was in England on a lecture tour to raise money for the London Missionary Society (LMS), he decided to become a missionary.[7] In 1838 the LMS appointed him medical missionary to Guǎngzhōu.[8] He worked with the American missionary Dr. Peter Parker (1804 – 1888) in Guǎngzhōu for a short time, before reopening Dr. Parker's hospital in Macao on 1 July 1839.[9] When Commissioner LÍN Zéxú (1785 – 1850; **Box 2.2**) threatened the British population in the Portuguese enclave, he escaped to Hong Kong on 30 August 1839 where he embarked on a ship heading for Java.[10] He stayed in Batavia (now Jakarta) with Dr. Medhurst for less than five months, before returning to Macao.[11] On his return, he spent a short time, from September 1840 to February 1841 treating patients on Zhōushān Dǎo ("that pleasant place"), which he much preferred to Macao.[12] There he renewed his acquaintance with Catharine ("Kate") Parkes (1823 – 1918), whom he married on 13 May 1841 after they had been forced to return to Macao following the evacuation of Zhōushān Dǎo (**Box 2.2**).[13] In September 1842, just days after Hong Kong had been ceded to Britain in the Treaty of Nánjīng, Lockhart sailed to the new colony to superintend the establishment of another hospital there.[14] Nine months later the Lockharts were able to return to their beloved Zhōushān Dǎo.[15] From there they moved to Shànghǎi, where Lockhart and Medhurst established a hospital in 1844.[16] Shànghǎi was to remain his base from 1844 until 1857.[17] Lockhart was renowned for his cataract operations, with some patients coming from as far afield as Fúzhōu, Sūzhōu and Nánjīng.[18] In those troubled times, he also had to deal with a stream of casualties caused by the fighting (**Box 5.**

21).[19] In the fourteen years he treated more than 100,000 patients.[20] In his spare time he liked to relax in his excellent garden full of interesting Chinese shrubs and trees.[21] At the end of 1857 domestic circumstances forced him to return to England.[22] The by then portly gentleman took this opportunity to write his memoirs.[23] After the Second Opium War, Dr. Lockhart was posted to Běijīng in 1861 – 1864, where he once more established a hospital and acted as medical officer to the British and French Legations.[24] He enjoyed being in the Chinese capital with all its sights.[25] In 1861 when Fortune arrived in Běijīng, he acted as his guide.[26] In 1864 Lockhart returned to England via Japan and established a practice in Blackheath, Greater London in 1864 – 1896.[27] Lockhart was Chairman of the Board of Directors of the LMS from 1869 to 1870, and donated his library to the society.[28]

Once he had improved, he set about packing his plants.[218] He left Shànghǎi on 10 October 1845 and arrived in Hong Kong a few days later.[219] In Hong Kong he sent off eight of the 26 Wardian Cases by ship, taking the rest with him to Guǎngzhōu.[220] He sailed from Guǎngzhōu in the "John Cooper" on 22 December 1845 and after a long but favourable voyage via South Africa finally docked in London on 6 May 1846.[221]

3.5 The Results of Fortune's Forays in the Flowery Land

Fortune was satisfied by the results of his three years in China.[222] Was this justified? To what extent had he fulfilled the goals of his mission? We can evaluate this by reference to the Horticultural Society's instructions (Chapter 2). Although it was by no means certain whether he would be able to visit all the Treaty Ports, he did so and more. He also visited Sūzhōu and Zhàpǔ, that were out-of-bounds to foreigners. At the various places he stayed, he kept records of the weather, which he expanded with data provided by some of his friends like George Lay (**Box 3.9**) and William Lockhart (**Box 3.11**).[223] In the course of his sojourn, he became aware how climate influenced the vegetation and crops. Thus plants found on the lower slopes in more northerly latitudes were only to be found on higher ground further south.[224] The length of the growing season, also affected the number of crops of rice the Chinese farmers were able to obtain in the course of a year.[225]

With regard to the first of the general objects of his mission, i.e. to collect seeds and plants of an ornamental or useful kind, he seems to have been highly successful. As a result of his first visit to China, he introduced about 40 new taxa, excluding minor colour variants (**Box 3.12**).

Box 3.12

FORTUNE'S INTRODUCTIONS

The name/names in brackets is/are the name(s) under which Fortune's material was once known. Bold type indicates that this was the first time the taxon had been grown outside China. A question mark indicates that Fortune may not have been the first person to introduce the plant.

***Abelia chinensis* R. Brown 1818** (*Abelia rupestris* Lindley 1846)[1]

Aconitum fischeri Reichenbach 1820 (*Aconitum autumnale* Lindley 1847)[2]

Actinidia chinensis Planchon 1847[3]

Adenosma glutinosum (Linnaeus 1753) Druce 1914 (*Digitalis sinensis* Loureiro 1790, *Gerardia glutinosa* Linnaeus 1753, *Pterostigma grandiflora* Bentham 1835)[4]

Aerides spp. (2)[5]

***Akebia quinata* (Houttuyn 1779) Decaisne 1839** (*Rajania quinata* Houttuyn 1779)[6]

Amaranthus tricolor Linnaeus 1753[7]

***Amygdalus persica* Linnaeus 1753 'Alboplena'** [*Prunus persica* (Linnaeus 1753) Batsch 1801 'Alboplena'][8]

***Amygdalus persica* Linnaeus 1753 'Sanguineoplena'** [*Prunus persica* (Linnaeus 1753) Batsch 1801 'Sanguineoplena'][9]

***Amygdalus persica* Linnaeus 1753 'Shanghai Peach'** [*Prunus persica* (Linnaeus 1753) Batsch 1801 'Shanghai Peach'][10]

***Anemone hupehensis* (Lemoine 1908) Lemoine 1910** [*Anemone japonica* (Thunberg 1784) Siebold & Zuccarini 1835, non Houttuyn 1778][11]

Arundina graminifolia (D. Don 1825) Hochreutiner 1910 (*Arundina bambusifolia* Lindley 1831, *Arundina chinensis* Blume 1825)[12]

? *Bambusa vulgaris* Schrader ex J. C. Wendland 1810 'Vittata'[13]

Brassica rapa Linnaeus 1753 var. *chinensis* (Linnaeus 1759) Kitamura 1950 [*Brassica chinensis* Linnaeus 1759, *Brassica rapa* Linnaeus 1753 subsp. *chinensis* (Linnaeus 1759) Hanelt 1986, *Brassica* "*sinensis*"][14]

***Brassica rapa* Linnaeus 1753 var. *glabra* Regel 1860**[15]

***Buddleja lindleyana* Fortune 1844**[16]

***Calystegia pubescens* Lindley 1846 'Flore pleno'**[17]

Camellia japonica Linnaeus 1753 'Myrtifolia' (*Camellia japonica hexangularis* Hort.)[18]

Campanula punctata Lamarck 1785 (*Campanula nobilis* Lindley 1846)[19]

Cannabis sativa Linnaeus 1753 (*Cannabis gigantea* Delile 1849)[20]

***Caryopteris incana* (Thunberg ex Houttuyn 1778) Miquel 1865** [*Barbula sinensis* Loureiro 1790, *Caryopteris mastacanthus* Schauer 1847, *Mastacanthus sinensis* (Loureiro 1790) Endlicher ex Walpers 1845][21]

***Cephalotaxus fortunei* W. J. Hooker 1850** [*Taxus fortunei* (W. J. Hooker 1850) Lawson 1851][22]

***Cerasus glandulosa* (Thunberg 1784) Sokolov 1954 'Alboplena'** (*Prunus glandulosa* Thunberg 1784 'Alboplena', *Prunus sinensis* Persoon 1807 forma *albiplena* Koehne 1893)[23]

Ceratostigma plumbaginoides Bunge 1833 (*Plumbago larpentae* Lindley 1847)[24]

? *Cercis chinensis* Bunge 1833[25]

***Chrysanthemum indicum* Linnaeus 1753 'Chinese Minimum'**[26]

***Chrysanthemum indicum* Linnaeus 1753 'Chusan Daisy'**[27]

***Citrus japonica* Thunberg 1780** [*Citrus margarita* Loureiro 1790, *Fortunella margarita* (Loureiro 1790) Swingle 1915][28]

***Citrus medica* Linnaeus 1753 'Fingered'** [*Citrus medica* Linnaeus 1753 var. *sarcodactylis* (Siebold ex Hoola van Nooten 1863) Swingle 1914][29]

Citrus reticulata Blanco 1837[30]

? *Crepidiastrum denticulatum* (Houttuyn 1779) Pak & Kawano 1992 [*Lactuca denticulata* (Houttuyn 1779) Maximowicz 1874, *Youngia denticulata* (Houttuyn 1779) Kitamura 1942][31]

***Cryptomeria japonica* (Thunberg ex Linnaeus fil. 1782) D. Don 1839**[32]

***Cryptomeria japonica* (Thunberg ex Linnaeus fil. 1782) D. Don 1839 'Nana'**[33]

***Daphne genkwa* Siebold & Zuccarini 1835** (*Daphne fortunei* Lindley 1846, *Daphne genkwa* Siebold

& Zuccarini 1835 var. *fortunei* Franchet 1884)[34]

? *Dendrobium secundum* (Blume 1825) Lindley 1828 [*Callista secunda* (Blume 1825) Kuntze 1891, *Pedilonum secundum* Blume 1825][35]

Dichroa febrifuga Loureiro 1790 [*Adamia versicolor* Fortune 1846, *Dichroa versicolor* (Fortune 1846) D. Hunt 1981][36]

Dioscorea batatas Decaisne 1854[37]

***Edgeworthia chrysantha* Lindley 1846** (*Edgeworthia papyrifera* Zuccarini 1846)[38]

***Forsythia viridissima* Lindley 1846**[39]

? *Fraxinus chinensis* Roxburgh 1820[40]

***Gardenia jasminoides* Ellis 1761 var. *fortuneana* (Lindley 1846) Hara 1952** (*Gardenia florida* Linnaeus 1762 var. *fortuneana* Lindley 1846)[41]

Hydrangea Linnaeus 1753[42]

***Ilex cornuta* Lindley & Paxton 1850**[43]

***Indigofera decora* Lindley 1846**[44]

Isatis tinctoria Linnaeus 1753 [*Isatis indigotica* Fortune in Lindley 1846, *Isatis tinctoria* Linnaeus 1753 var. *indigotica* (Fortune 1846) Cheo & Kuan][45]

***Jasminum nudiflorum* Lindley 1846**[46]

Lamprocapnos spectabilis (Linnaeus 1753) Fukuhara 1997 [*Dicentra spectabilis* (Linnaeus 1753) Lemaire 1847, *Dielytra spectabilis* (Linnaeus 1753) A. De Candolle 1821, *Fumaria spectabilis* Linnaeus 1753][47]

***Lamprocapnos spectabilis* (Linnaeus 1753) Fukuhara 1997 'Alba'** [*Dicentra spectabilis* (Linnaeus 1753) Lemaire 1847 var. *alba* Rollard][48]

***Limonium sinense* (Girard 1844) Kuntze 1891** (*Statice fortunei* Lindley 1845, *Statice sinensis* Girard 1844)[49]

***Lonicera fragrantissima* Lindley & Paxton 1852** [*Caprifolium fragrantissimum* (Lindley & Paxton 1852) Kuntze 1891, *Lonicera fragrantissima* Lindley & Paxton 1852 subsp. *standishii* (Carrière 1858) Hsu & Wang 1984, *Lonicera standishii* Carrière 1858 (non Jacques 1859)][50]

***Lycopodiella cernua* (Linnaeus 1753) Pichi-Sermolli 1968** (*Lycopodium cernuum* Linnaeus 1753)[51]

***Lycoris straminea* Lindley 1848**[52]

***Lysimachia candida* Lindley 1846**[53]

***Mahonia fortunei* (Lindley 1846) Fedde 1910** (*Berberis fortunei* **Lindley 1846**)[54]

***Medicago minima* (Linnaeus 1753) Bartalini 1776**[55]

Morus alba Linnaeus 1753 var. *multicaulis* (Perrottet 1823/24) Loudon 1838 [*Morus alba* Linnaeus 1753 var. *latifolia* (Poiret 1797) Bureau 1873, *Morus latifolia* Poiret 1797][56]

Paeonia suffruticosa Andrews 1804 subsp. *suffruticosa* (*Paeonia moutan* Sims 1808)[57]

***Paeonia suffruticosa* Andrews 1804 'Atropurpurea'**[58]

***Paeonia suffruticosa* Andrews 1804 'Atrosanguinea'**[59]

***Paeonia suffruticosa* Andrews 1804 'Bijou de Chusan'**[60]

***Paeonia suffruticosa* Andrews 1804 'Caroline d'Italie'**[61]

***Paeonia suffruticosa* Andrews 1804 'Colonel Malcolm'**[62]

***Paeonia suffruticosa* Andrews 1804 'Globosa'**[63]

***Paeonia suffruticosa* Andrews 1804 'Glory of Shanghai'**[64]

***Paeonia suffruticosa* Andrews 1804 'Ida'**[65]

***Paeonia suffruticosa* Andrews 1804 'Lilacina'**[66]

***Paeonia suffruticosa* Andrews 1804 'Lord Macartney'**[67]

***Paeonia suffruticosa* Andrews 1804 ‘Osiris’**[68]

***Paeonia suffruticosa* Andrews 1804 ‘Parviflora’**[69]

***Paeonia suffruticosa* Andrews 1804 ‘Picta’**[70]

***Paeonia suffruticosa* Andrews 1804 ‘Pride of Hong Kong’**[71]

***Paeonia suffruticosa* Andrews 1804 ‘Reine des Violettes’**[72]

***Paeonia suffruticosa* Andrews 1804 ‘Robert Fortune’**[73]

***Paeonia suffruticosa* Andrews 1804 ‘Salmonea’**[74]

***Paeonia suffruticosa* Andrews 1804 ‘Versicolor’**[75]

***Paeonia suffruticosa* Andrews 1804 ‘Vivid’**[76]

? *Pecteilis susannae* (Linnaeus 1753) Rafinesque 1837 [*Habenaria susannae* (Linnaeus 1753) R. Brown 1825, *Orchis susannae* Linnaeus 1753, *Platanthera susannae* (Linnaeus 1753) Lindley 1835][77]

Phalaenopsis amabilis Blume 1825[78]

***Pholidota chinensis* Lindley 1847** [*Coelogyne chinensis* (Lindley 1847) Reichenbach 1864][79]

***Pittosporum glabratum* Lindley 1846**[80]

***Platycarya strobilacea* Siebold & Zuccarini 1843** (*Fortunaea chinensis* Lindley 1846)[81]

Platycodon grandiflorus (Jacquin 1776) A. de Candolle 1830 (*Campanula grandiflora* Jacquin 1776)[82]

? *Platycodon grandiflorus* (Jacquin 1776) A. de Candolle 1830 ‘Albus Plenus’ (*Campanula grandiflora* Jacquin 1776, semi-double white)[83]

***Primulina dryas* Möller & Weber 2011** [*Chirita sinensis* Lindley 1844, *Didymocarpus sinensis* (Lindley 1844) Léveillé 1906, *Roettlera sinensis* (Lindley 1844) Kuntze 1891][84]

Rhaphiolepis indica (Linnaeus 1753) Lindley 1820 (*Crataegus indica* Linnaeus 1753)[85]

***Rhododendron farrerae* Sweet 1830** (*Azalea squamata* Lindley 1846)[86]

Rhododendron molle (Blume 1823) G. Don 1834 [*Azalea mollis* Blume 1823, *Azalea sinensis* Loddiges 1824, *Rhododendron sinense* (Loddiges 1824) Sweet 1829][87]

***Rhododendron x obtusum* (Lindley 1846) Planchon 1853/54** (*Azalea obtusa* Lindley 1846)[88]

***Rhododendron x obtusum* (Lindley 1846) Planchon 1853/54 ‘Album’** (*Azalea ramentacea* Lindley 1849)[89]

***Rhododendron ovatum* (Lindley 1846) Planchon ex Maximowicz 1871** (*Azalea ovata* Lindley 1846)[90]

? *Rhododendron simsii* Planchon 1854 ‘Vittatum’ [*Azalea vittata* Dumont de Courset 1811, *Rhododendron vittatum* (Dumont de Courset 1811) Planchon 1854][91]

***Rosa anemoniflora* Fortune ex Lindley 1847** [*Rosa sempervirens* Linnaeus 1753 var. *anemoniflora* (Fortune ex Lindley 1847) Regel 1878; *Rosa triphylla* Roxburgh ex Hemsley 1887][92]

***Rosa x fortuniana* Lindley & Paxton 1851** (*Rosa banksiae* Aiton 1811 x *R. laevigata* Michaux 1803)[93]

***Rosa x odorata* (Andrews 1810) Sweet 1818 ‘Fortune's Double Yellow’** (*Rosa chinensis* Jacquin 1768 x *R. gigantea* Collett ex Crépin 1888; *Rosa chinensis* Jacquin 1768 var. *pseudindica* (Lindley 1820) E. Willmott 1911, *Rosa x fortuniana* Lindley 1852, non Lindley & Paxton 1851, *Rosa x odorata* (Andrews 1810) Sweet 1818 var. *pseudindica* (Lindley 1820) Rehder 1916, *Rosa pseudindica* Lindley 1820][94]

***Rosa x odorata* (Andrews 1810) Sweet 1818 ‘Fortune's Five Colour’**[95]

Scutellaria sp.[96]

? *Selaginella uncinata* (Desvaux ex Poiret 1814) Spring 1843 (*Lycopodium caesium* Hort. ex Anon. 1847, *Lycopodium dilatatum* W. J. Hooker & Greville 1831, *Lycopodium uncinatum* Desvaux ex Poiret 1814)[97]

? *Selaginella willdenowii* (Desvaux ex Poiret 1814) Baker 1867 (*Lycopodium willdenowii* Desvaux ex

Poiret 1814)[98]

? *Spathoglottis pubescens* Lindley 1831 (*Spathoglottis fortunei* Lindley 1845)[99]

***Spiraea chinensis* Maximowicz 1879**[100]

? *Spiraea prunifolia* Siebold & Zuccarini 1840 'Flore pleno' (*Spiraea prunifolia* Siebold & Zuccarini 1840 var. *plena* C. K. Schneider 1905, *Spiraea prunifolia* Siebold & Zuccarini 1840 var. *prunifolia*)[101]

***Torenia concolor* Lindley 1846**[102]

***Trachelospermum jasminoides* (Lindley 1846) Lemaire 1851** (*Rhynchospermum jasminoides* Lindley 1846)[103]

? *Viburnum dilatatum* Thunberg 1784[104]

***Viburnum macrocephalum* Fortune 1847**[105]

***Viburnum plicatum* Thunberg 1794** [*Viburnum tomentosum* Thunberg 1784 non Lamarck 1778 var. *plicatum* (Thunberg 1794) Maximowicz 1880, *Viburnum tomentosum* Thunberg 1784 non Lamarck 1778 β (var.) *sterile* K. Koch 1853 [Hortus dendrologicus p. 301], *Viburnum tomentosum* Thunberg 1784 non Lamarck 1778 forma *sterile* (K. Koch 1853) Zabel 1853][106]

***Weigela florida* (Bunge 1833) A. de Candolle 1839** [*Diervilla florida* (Bunge 1833) Siebold & Zuccarini 1839, *Weigela rosea* Lindley 1846][107]

***Wisteria sinensis* (Sims 1819) Sweet 1827 'Alba'** [*Wisteria alba* Lindley 1849, *Wisteria sinensis* (Sims 1819) Sweet 1826 forma *alba* (Lindley 1849) Rehder & Wilson 1916, *Wisteria sinensis* (Sims 1819) Sweet 1826 var. *albiflora* Lemaire 1858][108]

Fig. 3.50 *Anemone hupehensis*, collected "amongst the graves of the natives, which are round the ramparts of Shanghae; it blooms in November, when other flowers have gone by, and is a most appropriate ornament to the last resting-places of the dead." (Fortune, 1847a: 330)

Fig. 3.51 *Cryptomeria japonica*, the Chinese cedar, sometimes referred to as *Cryptomeria fortunei*.

Fig. 3.52 ***Edgeworthia chrysantha*, an attractive spring-flowering shrub, which Fortune sent to the Horticultural Society in December 1844.**

Fig. 3.53 ***Ilex cornuta* with its characteristic rectangular foliage.**

Fig. 3.54 ***Jasminum nudiflorum*, the widely planted winter jasmine, which Fortune sent to the Horticultural Society in January 1844.**

Fig. 3.55 ***Lamprocapnos spectabilis* 'Alba', a white cultivar of bleeding heart.**

Fig. 3.56 ***Lonicera fragrantissima*, a spring-flowering honeysuckle, sometimes referred to as January Jasmine.**

Fig. 3.57 ***Mahonia fortunei* with its lanceolate leaflets. It is commonly planted in China. Fortune brought it back with him in May 1846. Although it withstands shade, and is deer-proof, it is not widely cultivated in the west.**

Fig. 3.58 ***Trachelospermum jasminoides*, an evergreen climbing plant with sweetly-scented flowers.**

Fig. 3.59 ***Viburnum plicatum*, a species with corrugated leaves, sent to the Horticultural Society ca. 1845.**

Since Fortune obtained most of his plants in nurseries and mandarins' gardens, this increased his chances of finding attractive plants. Thus he eventually found a *Camellia* with yellow petals, whose very existence had been doubted (**Box 5.16**). With his training in horticulture, Fortune seems to have had an eye for a good ornamental plant.[226] Many of the plants which he introduced have become such household objects (e.g. *Anemone hupehensis*, *Forsythia viridissima*, *Jasminum nudiflorum*, *Weigela florida*; **Figs. 3.23**, **3.50 – 3.59**) that we have difficulty realizing that until 170 years ago they were unknown in European and North American gardens. The Chinese gardens and nurseries usually had a good collection of bonsais. Although Fortune does not appear to have particularly liked bonsais, even refusing to take one as a present[227], he took copious notes on how these plants were grown and trained, as stipulated in the Horticultural Society's instructions.

While in China, Fortune devoted a considerable amount of time to economic plants. He described how the various types of teas were manufactured, which led to his discovery that black and green tea originated from the same shrub. He made a point of obtaining various fruit trees (mandarin orange, fingered citron, kumquat, and peaches), which were specifically mentioned in the Horticultural Society's instructions. While he may not have found the gigantic "Peaches of Pekin", he found something very large even by modern standards.[228] He also made a point of introducing a number of crop plants, e.g. amaranth, cabbage, hemp and woad in the hope that they might represent species distinct from those already known in Europe.[229]

There were a certain number of plants mentioned in the Horticultural Society's instructions which seem to have eluded him[230]. Thus he never refers to *Biophytum sensitivum* (Syn. *Oxalis sensitiva*), *Illicium verum* (star anise) or *Nepenthes* (pitcher plant), although the last-named occurs in South China.[231] Likewise, there is no record of the Lilies of Fokien, although there are three species of the genus *Lilium* in Fújiàn Province.[232] Fortune eventually found *Tetrapanax papyrifer* (Syn. *Aralia papyrifera*; **Fig. 5.143**), the plant which furnishes rice paper, when he visited Táiwān in March 1854.[233]

Agriculture has a long tradition in China, so the Horticultural Society had high hopes of learning from the Chinese experience. While Fortune was impressed by some of the methods employed, e.g. the way the Chinese used the square-pallet chain-pump to raise water to irrigate the paddy fields[234], terrace cultivation to increase the area available for crops and lessen erosion of the steep slopes[235], and the spreading of fresh night soil to fertilize the fields[236], his general impression was that few advances had been made in the recent past.[237]

Finally, according to his instructions, Fortune was expected to write up the results for the Fellows of the Horticultural Society. This must have kept him busy on his return to England in 1846. It is possible that Professor Lindley suggested that it might be worthwhile writing up his travelogue in book form for a broader audience.[238] The copyright would earn Fortune some well-needed "pocket money" with which to augment his meagre salary. The manuscript was shown to the renowned publisher John Murray of London[239], who agreed to publish it. Murray seems to have taken a genuine interest in Fortune's introductions, because when he built "Newstead" in Wimbledon in the 1850s, he had a number of Fortune's discoveries planted in the grounds.[240] Fortune received the good sum of £ 105 for the copyright.[241] The 1,000 copies which were originally printed quickly sold out. Since then this classic has been frequently reprinted and still makes good reading.

Chapter 4 Chelsea Interlude

4.1 Curator of the Chelsea Physic Garden

By the time Fortune arrived back in London on 6 May 1846 he had already made his mark as a successful plant hunter. Nevertheless, he went back to his job at the Horticultural Society on his return, as stipulated in Lindley's instructions of 23 February 1843. From there he wrote to the Earl of Derby (**Box 2.5**; **Fig. 4.1**) offering him a pair of Javanese peafowl (*Pavo muticus imperator*) and a white Mandarin duck (*Aix galericulata*) that had survived the long journey via the Cape of Good Hope.

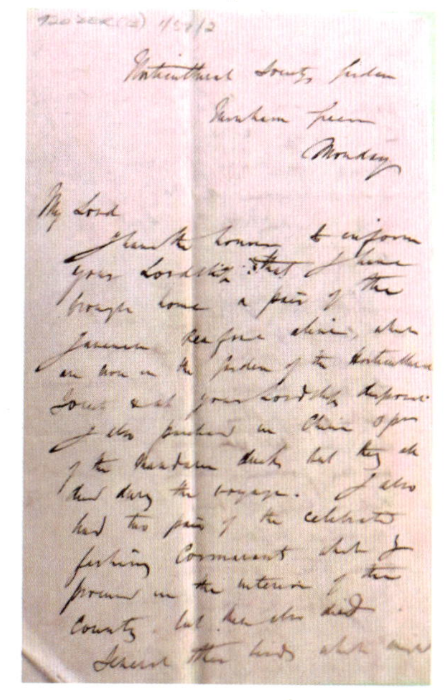

Fig. 4.1 Letter written by Fortune to the 13th Earl of Derby regarding the birds he brought back from the Far East. Note the energetic writing almost free of errors.

Liverpool Record Office 920 DER (13) 1/58/2

Horticultural Society's Garden
Turnham Green
Monday [18 May 1846]
[Rec. 19 May 1846]

My Lord,

I have the honour to inform your Lordship that I have brough[t] home a pair of the Javanese Peafowl alive, which are now in the Garden of the Horticultural Society & at your Lordship's disposal. I also purchased in China 3 pairs of the Mandarin ducks but they all died during the voyage. I also had two pairs of the celebrated fishing Cormorants which I procured in the interior of the country but they also died.

Several other birds which might have been acceptable to your Lordship also met with the same fate. Their passage from China to this country is so long & we have to pass through so many changes of temperature, that it became a most difficult matter to bring certain kinds of birds alive.

I will do myself the honour of calling upon your Lordship probably on Wednesday morning if this will be convenient. I have another case of things containing several which have not been forwarded in my former cases. Then I shall be happy to [offer] them to your Lordship as usual before they are offered to any other person.

I have the honour to be
Your Lordship's obt. sert. [obedient servant]
R. Fortune

P.S. I have also a Chinese white duck alive here. It is very like our common duck but has a tuft on its head & its legs are placed very far back beneath the body, so that when it walks it appears to have the body nearl perpendicular with the legs.

This is probably a common var. but your Lordship will I dare say know it at once from the above description. If it is rare, of course it is also at the Service of your Lordship.

RF

It appears that the Earl of Derby bought all three birds, but there was some disagreement as to whether the Javanese peafowl represented a pair, or two male peacocks (obviously less valuable to someone who was interested in breeding). Fortune tried to remain positive by stating that if they were both males, his lordship was getting two species for the price of one!

Liverpool Record Office 920 DER (13) 1/58/4

21 Regent St. London [May 1846]

My Lord

I have heard by this morning's post from your Lordships Servant Mr. Thompson of the safe arrival of the Birds at Knowsley. Mr. Thompson [John Thompson was the manager of the Earl of Derby's menagerie. After the Earl's death, he was Superintendent of the gardens of the Zoological Society of London from 1852 to 1859, during which time (1856 – 1857) he went to India to make a collection of live Himalayan pheasants for the Zoo] also says that he thinks both the Peafowl males [Fortune's emphasis]. If that be the case, then they most certainly are different species [Fortune's emphasis] for the smaller bird is quite different from the other. For your Lordship's satisfaction I will state the following facts concerning them. I purchased what I considered 2 pair [sic] of peafowl in Canton. One of the males died in a short time after we sailed. I then had one male and two females. One of the females (the one belonging to the male now in your Lordship's possession) was lost overboard off the Scilly Islands—the other two as your Lordship knows arrived in safety. I therefore considered I had a pair [Fortune's emphasis] for your Lordship, but Mr. Thompson who must be a good judge of such matters thinks differently. They must, however, be different species if they are both males.

As I am anxious to wind up my affairs with Mr. Leadbeater I shall feel extremely obliged if Your Lordship will send me an account of the Skins which your Lordship has had the goodluck to select from my different collections, as well as the value which has been put upon them. I will send your Lordship the other Skins when I return from Scotland which will be in the course of two months. Since my return I have been too busy with my living [Fortune's emphasis] plants to get the Skins unpacked.

I hope, although the Peafowl are not a pair that they will be valued by your Lordship for their rarity & may still be considered worth £ 18 the sum offered me by Your Lordship before I left England. I would not make this request but several of the birds bought in China were very expensive & as I was not fortunate enough to bring them home alive, these sums were lost to me. I must beg your Lordship's pardon for this long Note.

Any communication with which your Lordship may honor me may be addressed to me 21 Regent St. [Fortune's emphasis].

I have the honor to be
Your Lordship's obt. sert.
R. Fortune

After spending July and August with his family in Scotland, he was again writing to the Earl of Derby in the hope of recouping the expenses he had incurred in sending the bird skins from China.

Liverpool Record Office 920 DER (13) 1/58/3

Horticultural Society's Garden
Sept. 2nd 1846
[Rec. Sept. 3rd 1846]

My Lord

Having returned from Scotland where I have been for the last two months on account of my health, I am now anxious to have all my accounts settled, & shall feel extremely obliged if your Lordship will order me to be paid for the Skins of Birds &c., which I had the honor of sending to your Lordship from China & Manila. Your Lordship will probably recollect what I formerly said with regard to the value of these things viz "that I was unacquainted with such matters & would leave that to be settled by your Lordship".

As the whole of the contents of each case was sent to your Lordship first, I do not know how many things your Lordship was pleased to select, & how many were returned to Mr. Leadbeather [sic].

I have called upon Mr. Leadbeather [19 Brewer Street, Piccadilly] very frequently since my return but cannot find him at home. Perhaps the Curator of your Lordship's Museum may be able to give me the desired information. I shall put some memoranda on a seperate [sic] sheet which may perhaps assist him.

I mentioned, when I had the honour of seeing your Lordship in London, that I had brought home some more things with me, many of which are duplicates of those already sent home to your Lordship. As I have time to look to these now, perhaps your Lordship might wish to see these before they were offered to any other person.

I find there is a Bill against one of £ 9"10"2 for the freight & charges of 8 Boxes—4 of which Skins sent to your Lordship. I do not know whether your Lordship is in the habit of paying the freight, or whether I have any claim for this sum or part of it, but I mention the circumstances for your Lordship's favourable consideration.

I have the honor to be
Your Lordship's most obt. sert.
R. Fortune

Boxes Containing Birds &c. sent to the R[t]. Honl The Earl of Derby &c&ca. from China.

							Skins
Hong Kong	Jany	1844.	Ship	"Cornwall"	1 Box. No 12	containing	54.
"	April	"	"	"Bombay"	" " " 18[a.]	"	10.
"	Decr	"	"	"Duke of Bedford"	" " H.S. 2	"	29.
"	March	1845	"	"Chusan"	- - H.S. 3	"	60

Living Birds

Delivered May 1846. 2 Java Pea fowl 1 Chinese Duck.

R. Fortune

In the meantime the death of William Anderson (1766 – 1846), the Curator of the Chelsea Physic Garden, left a vacancy for which Fortune was well qualified. When The Worshipful Society of Apothecaries' Garden Committee met on 16 October 1846 to discuss Anderson's successor, Fortune was "very strongly recommended by Dr. Lindley as being particularly well qualified to perform the duties of Curator".[1] There is no indication in the Minutes that alternative candidates were proposed by other members of the committee. Maybe they did not dare. Professor John Lindley (**Box 2.1**), the Praefectus Horti (botanical demonstrator), was not the sort of person one crossed swords with.[2] The duties of the Curator, as stipulated by the Garden Committee, also bear the authoritarian stamp of Professor Lindley. Thus:

(1) He is to be under the orders of the Master, and Wardens, and of the Garden Committee, whose instructions he is bound to obey.

(2) He is also to be under the orders of the Praefectus Horti, whose directions as to the plants to be provided for lectures and the general arrangement of the Garden are to be implicitly obeyed.

(3) He is, upon his own responsibility to hire, and dismiss the labourers required for the proper maintenance of the Garden, the number to be employed being determined by the Praefectus.

(4) He is not to receive any fee or other payment from any persons for any services connected with the Garden, upon any pretence whatever.

(5) He is to prepare for each Meeting of the Garden Committee a return mentioning the following particulars at least, viz.

a The general condition of the Garden.

b The improvements in it which he would suggest.

c The objects required for the next 6 months service.

d The names of all officinal ['officinal' scored out by the Master, John Ridout, and initialed 'JR'] plants that may have been presented to the Garden with the names of the donors.

e The names of all officinal ['officinal' scored out by the Master, John Ridout, and initialed 'JR'] plants that may have been lost to the collection during the previous six months.

(6) He is to form a correct catalogue of the plants actually in the Garden, to see that they are all properly labelled, with names approved of by the Praefectus, and to pay especial regard to the importance of preventing one plant from imperceptibly ['imperceptibly' scored out by the Master, John Ridout, and initialed 'JR'] usurping the place of another.

(7) He is to understand that the great object of the Garden is the preservation of the most complete collection of officinal plants which it may be found possible to collect, and of plants suitable for Botanical instruction—and that their skilful cultivation is to be the object of his assiduous attention. With reference to this point the Praefectus will give directions from time to time.

Although the salary was no more than his predecessor William Anderson had received[3] or Fortune had earned in China (£ 100 per annum), he was "provided with house room and coals and until the Society appropriates it otherwise, allowed to cultivate with vegetables etc. for his own use, a small strip of land next the River."[4]

It should be pointed out that at that time the house was still without piped water or a WC.[5]

When the Court of Assistants convened on 30 October 1846 they rubber-stamped the Garden Committee's recommendation ("Mr. Robert Fortune was then proposed and seconded to fill the office of Curator of the Garden at Chelsea until the annual election of Officers of the Society in the year 1847 upon the terms recommended by the Garden Committee"[6]).

Jane Fortune came to join her husband at the Chelsea Physic Garden[7] and it was there that their third child Agnes was born on 16 April 1847.[8] With a wife and three children under the same roof, it cannot have been easy for Fortune to prepare his report in time for the meeting of the

Garden Committee on 27 May 1847. Professor Lindley probably helped him, as there are a number of scathing comments on the state in which the garden had been left by his predecessor (Lindley and Anderson did not see eye to eye[9]). Thus:

> From various causes, with which the Committee are doubtless acquainted, the Garden has been allowed to get into a most ruinous condition. When I took charge of it last Autumn, I found it overrun with weeds, the Botanical arrangements in confusion, the exotic plants in the Houses in very bad health, and generally in a most unfit state for the purpose for which it was designed. [10]

Fortune goes on to describe what he did to rectify the garden, and suggested a number of changes to modernize it.

> My first business, after getting clear of the weeds, was washing and thoroughly cleansing the Houses and exotic plants. This was done during the winter months when out door operations were in a great measure suspended. At the same time, with the assistance of Dr. Lindley, I entered into correspondence with various public Gardens and Nurserys in different parts of the Country for the purpose of recruiting our collections. I am happy to inform the Committee that everywhere we met with the most willing and liberal aid, and I have the satisfaction of laying upon the table the various list [sic] of plants and seeds which have been presented to the Society. It will be seen upon referring to these lists that we are indebted to the Royal Botanic Gardens at Kew, Edinburgh & Dublin, the Horticultural Society, and various other Botanical Gardens and Nursery's [sic] for many valuable species not formerly in this collection. Besides we have had several Collections of seeds presented by Dr. Lindley, Dr. Royle, Mr. Wailes of Newcastle & others.
>
> I have gone this far without incurring any other expences [sic] than those connected with Carriage and correspondence which are of a trifling nature. Those articles which were indispensable to the service of the Garden, such as Garden tools, Soils etc. have been ordered by the authority of the Master & Wardens. The Garden Tools were completely worn out, and there was no soil to shift the plants with, so that it was absolutely necessary that these articles should be procured without delay. But if the Committee are desirous of renovating the Garden and making it what a Garden of this kind ought to be and worthy of the name it has always until late years born, they must be prepared to spend a very considerable sum of money. The Stoves and Greenhouses are old, badly constructed, and generally in a very bad state of repair. Last winter the smoke was constantly coming out through the sides of the decaying flues and destroying the leaves of the plants. Several of the Pits had to be entirely cleared, otherwise everything would have been killed. The same thing will doubtless happen next winter if something is not done in the meantime to prevent it. Those fine plants which we have had presented to us from the Gardens already named, and which are now growing luxuriantly when little fire heat is required, will be injured or destroyed next winter.
>
> Should the Committee determine to renovate the Garden, I would suggest the propriety of building new houses, instead of going to any great expence [sic] in repairing the old ones. [11]

It seems unlikely that Fortune, who had only been Curator for a few months, would have proposed such radical changes, if he had not had the backing of Professor Lindley (**Box 2.1**) and the Examiner in Botany, Dr. Nathaniel Ward (**Box 2.6**).

Fortune also considered the systematic arrangement of the plants:

With regard to out-door plants and the general arrangement of the Garden I have the following observations to lay before the Committee. There are three Botanical arrangements at present in the Garden with numbers affixed to the plants. Two of these are very incomplete owing to the plants having been lost, and the third is in confusion.

I propose

1st—That the old Medical arrangement at the entrance should remain upon the present plan and be made as complete as possible, but that the plants should be named.

2nd—That the two others be re-arranged & one good one made out of them.

3rd—That the different orders be arranged upon the grouping system, and the species between the groups laid down in grass.

4th—That all the plants be named and the name of the natural order affixed to each group. [12]

Fortune's suggestion to rearrange the collection according to a more natural system also smacks of Lindley, who had an anathema for the Linnaean "sexual system" based on floral parts.[13]

It is symptomic of Fortune's energy that many of the improvements had been made by the time Fortune prepared his next and final report on 31 May 1848.[14]

Fig. 4.2 Swan Walk Gate, Chelsea Physic Garden, which was moved to its present position during Fortune's curatorship.

Since the Meeting of the Committee in August last the following improvements which were then suggested, have been carried into effect.

1st—A new arrangement of medical plants has been made in front of the Class room.

2nd—The under part of the Garden has been re-arranged and the natural orders laid out in groups. This will afford greater facility to the students in their studies-the plants will grow better, as the herbaceous kinds are now removed from under those which form trees; and the garden will have a fresher & better appearance owing to the groups being surrounded with a green sward.

3rd—The naming of the plants is going on as fast as possible-those species in the Medical arrangement are nearly all named.

4th—Two glazed houses are now completed upon the most improved plans, in a neat & substantial manner. These will afford the means of producing excellent specimens of rare medical and other plants for the Lectures, and of an interesting kind to the Members of the Society when they visit the Garden.

Fig. 4.3 Lower Tank Pond, Chelsea Physic Garden, created during Fortune's curatorship.

5th—The entrance gate [Swan Walk gate, **Fig. 4.2**] has been removed a little further down the lane where it will be much cleaner than in its old position, and also improve the appearance of the Garden. [15]

6th—The pond [Lower Tank Pond, **Fig. 4.3**] for water plants is in the course of formation.

7th—Various presents of Plants and Seeds have been received since the last meeting of the Committee. Lists of these, and also of some which have been given out, are laid upon the table for inspection.

In conclusion I am happy to be able to say that everything in the Garden is going on in a most satisfactory manner—we have no difficulty with the students; the men employed are steady and attentive to their business and the plants are growing as well as they can be expected to do. I trust that the change which is about to take place with regard to myself, will not tend in any way to the injury of the men who have all conducted themselves to my satisfaction.

(signed) Robert Fortune

There follows a statement of the estimates tendered and the amounts paid in connection with the improvements:

Woodwork + Slate	£ 534-4-0
Ditto for extra work per account	£ 0-12-8
Brickwork tanks	£ 117-2-0
Building Polmaise heating	£ 80-3-7
New entrance	£ 16-15-10
1 Polmaise Stove	£ 11-10-0
New pond	£ 64-10-0
For extra labour, turf, grass seeds, holly plants &c in improving the Garden	£ 80
Naming the plants	£ 70-0-0
Repairs to old houses	£ 100
Taking down end of old greenhouse & rebuilding it	£ 8-5-0
Laying on water & making closet	£ 43

It will be noted that Fortune managed to convince the Garden Committee and Court of Assistants of the need to spend money on renovation. He got his new glasshouses, which were heated by a Polmaise stove[16], and had running water and a toilet installed in the house. On the other hand, a certain amount of rebuilding was carried out. According to Fortune's obituary "two new houses were built, and heated on the Polmaise system, which was soon abandoned".[17] This form of heating was replaced by a hot water system.[18]

Fig. 4.4 Professor John Forbes Royle (1798 – 1858), a specialist in Himalayan plants, who acted as intermediary between Fortune and the East India Company.

4.2 Preparing for Service in the East India Company

On 12 May 1848 the Garden Committee was convened by the new Master, Edward Bean. He informed them of a letter which he had received from

Professor John Forbes Royle (**Box 4.1**; **Fig. 4.4**).[19]

Sir

The Court of Directors of the East India Company having for some years promoted the cultivation of the Tea plant in the Himalayan Mountains, are desirous of taking such further measures as are calculated to insure every attainable degree of success. Among the most important of these, is the sending of a properly qualified person to the more northern coasts of China for the purpose of collecting the best kinds of Tea-plants and seeds, as well as for making enquiries respecting any peculiarities of Manufacture which may prevail in the districts where the best Teas are made.

Mr. Fortune having casually mentioned to me his willingness to be employed in carrying out an operation for which, from his previous visit to and pursuits in China, he is particularly well qualified I have been authorized by the Court of Directors of the East India Company to enter into communication with him on this subject. But adverting to Mr. Fortune being in your service, I beg leave to inform you that I have done so, and also that I mentioned the subject to Dr. Lindley, your Professor of Botany when the business was first mooted. I trust that I may be excused for requesting that you will be good enough to grant him any facilities in your power for carrying out an object which I am sanguine in thinking will ultimately prove of National importance.

I have the Honor to be, Sir,
Your most obt. sert.
J. Forbes Royle M.D.

East India House
8 May 1848

Box 4.1

Professor John Forbes Royle FRS (1798 – 1858) Only son of Captain William Henry Royle, who died while he was still a child.[1] He was educated at the Royal High School in Edinburgh.[2] Dr. Antony Todd Thomson (1778 – 1849), who was to become Professor of Materia Medica and Therapeutics and of Forensic Medicine at University College, London, kindled his interest in natural history.[3] In 1820 Royle joined the British East India Company (EIC) as an Assistant Surgeon[4], no doubt in the hope that this would give him time for botanical pursuits. He sailed for Kolkata (Calcutta), was first stationed in Dum-Dum, notorious for the production of the first dum-dum bullets, and subsequently in various parts of Bengal and the Northwestern Provinces.[5] In 1823 he succeeded George Govan as Superintendent of the Saharanpur Botanical Garden (**Box 5.11**), a former pleasure ground, which the EIC had purchased in 1817.[6] With the help of native plant collectors, he was able to amass some ten thousand dried specimens from northern India.[7] He took a great interest in the medicinal plants and their efficaciousness, and was one of the first to realize the possibility of growing Chinese tea in the Himalayas.[8] After 12 years in India he returned to England[9] with his herbarium, in order to complete the work on his "Illustrations" during 1833 – 1840[10] with the help of a number of outstanding botanists.[11] In 1837 he succeeded Professor John Ayrton Paris as Professor of Materia Medica at King's College, London.[12] He was elected a Fellow of the Linnean Society in 1833, Royal Society of London in 1837, Geological Society of London, and Horticultural Society, and served in various functions on their councils.[13] He was also the recipient of various honours (MD Munich in 1833; Légion d'Honneur, for his work at the "Exposition Universelle" in Paris in 1855).[14] In the course of this busy life, he found time to publish various tracts on economic plants, and get married on 26 August 1839 to Annette Solly (1816 – 1894), youngest daughter of Edward Solly Sr. (1776 – 1848), the British wood and grain merchant and enthusiatic art collector.[15] They had three sons and a daughter.[16]

Royle would appear to have known Fortune quite well, as indicated by the "casually mentioned" at the beginning of the second paragraph of his letter to the Society of Apothecaries. He apparently gave Fortune six young *Cinchona* plants as a source of quinine to present to his good friend Hugh Falconer (**Box 5.7**) in Kolkata.[17] He may well have presented Fortune with a copy of his "Illustrations", which Fortune referred to.[18] Considering his deep concern for the Indian flora and the need for the conservation of its forests[19], it is fitting that Nathaniel Wallich (1786 – 1854) named a Himalayan genus of Lamiaceae after him (*Roylea* Wallich ex Bentham).

Fortune was then called in. He explained that he wanted leave of absence for two years as he wished to take up the offer of the East India Company.[20] After all, he had been offered a salary of £ 500 per annum.[21] The Garden Committee then considered what to do. It was finally resolved.

> That after a careful consideration of all the circumstances the Garden Committee do not think it expedient to recommend to the Society that Mr. Fortune should have leave of absence for two years, inasmuch as the interests of the Garden as a School of practical Botany would be seriously affected thereby.[22]

They were clearly in a bit of a quandary. While they wanted Fortune to stay, they could not match the East India Company's offer.

> That altho' the Committee do not feel themselves at liberty to grant Mr. Fortune the leave of absence suggested, yet they will make no difficulty respecting Mr. Fortune's acceptance of any offer made to him by the Directors of the East India Company, in case he should find it conducive to his interests, but at the same time the Society would very much regret the loss of the services of a Curator who has fulfilled the duties entrusted to him to their entire satisfaction.
>
> These resolutions were then communicated verbally to Mr. Fortune.

Maybe they thought he would change his mind. In the next fortnight Fortune no doubt reconsidered the "pros" and "cons". Finally, on 29 May 1848 he sat down to compose the letter that was to change his life.

> Sir,
>
> With reference to the communication which I made to you a few days ago and which was laid before the Committee at its last meeting I beg to say that as it was not deemed expedient to grant me leave of absence, it will be necessary for me to resign the appointment which I now hold. I assure you I do so with much regret and only because I think it my duty to accept an appointment which is highly advantageous to myself & my family. I trust you will mention this to the Committee & also say that I feel grateful for the attention which they have given to the subject, and for their kind expression of satisfaction with my services.
>
> As it will not be necessary for me to leave England until the 20th of June I shall have great pleasure in managing the Garden until that time and in giving every information in my power to the person who may be appointed to succeed me.
>
> I am Sir,
> Your obt. sert.
> R. Fortune

To the Worshipfull
The Master
of the Society of Apothecaries

This letter was read at the meeting of the Garden Committee on 31 May 1848. It is clear that the committee was sorry to lose one of their best curators.

That in accepting this resignation the Committee of the Botanic Garden desire to express their regret at losing the services of a Curator whose conduct has on all occasions been deserving of their warmest approbation, at the same time the Committee congratulate Mr. Fortune upon obtaining an appointment not only greatly tending to the advancement of his own prospects in life, but also so creditable to his character as a Scientific Botanist.

It was further Resolved that in consideration of the extra services which Mr. Fortune has rendered in the arrangement of the Garden he be presented with the sum of Thirty guineas.

Ordered, that these resolutions be communicated to Mr. Fortune in writing, and be signed by the Master & Wardens.

Dr. Lindley having been requested to name a successor to Mr. Fortune recommended Mr. Thomas Moore.[23]

It seems likely that the expedition on behalf of the East India Company was planned a long time in advance. Why then did Royle wait until 8 May 1848 to write his letter? Was Fortune too embarrassed to broach the subject to the Garden Committee or was he trying to up the stakes with the East India Company by playing hard to get? Had he hoped by giving short notice that The Society of Apothecaries would be bulldozed into granting him leave of absence? In this way he would have had a job to go back to after the contract with the East India Company expired. Without Fortune's personal papers, we may never know.

Once the Garden Committee's decision was known, Fortune travelled to Scotland to say good-bye to his mother and some of his siblings (his father had died the previous year). He arrived back in London on 13 June 1848, and with a week left before he sailed from Southampton, was again pestering the Earl of Derby (**Box 2.5**) to settle the outstanding bills.

Liverpool Record Office 920 DER (13) 1/58/5

Botanic Garden Chelsea
June 13th 1848

My Lord

On my return from Scotland this morning I had the honor to receive your Lordship's letter & I am greatly obliged by the orders of your Lordship to settle my account. I trust I shall hear from your Lordship's Steward on the subject in a day or two as I must have all my affairs wound up before ["by" is crossed out] Saturday.

With regard to my new Mission I have already informed your Lordship of the countries which I am likely to visit in order that an idea may be formed of the kinds of animals with which I am most likely to meet.

As your Lordship knows almost all the Birds & other animals which exist in China & India perhaps the best way would be to send me a list of those kinds which would be most desirable to get over in a living state.

The great difficulty, as your Lordship well knows, is not the procuring of the birds &c. but the getting them home alive. When I left China for England I had 4 pairs of Mandarin Ducks—1 pair Japan jays—4 beautiful specimens of the Javanese Peafowl & some others, which I was bringing home to your Lordship, & yet I only succeeded in bringing 3 birds alive to this Country after the greatest care & attention. This however shall not discourage me in making another attempt particularly with the overland route which I agree with your Lordship in thinking by far the best for animals as well as for plants.

If your Lordship will make out the list which I have suggested & affix to each Animal or pair of Animals the sum which you would be willing to give, I think this would much simplify matters.

I should also like if your Lordship would agree to pay those sums which are expended in the purchase of the animals & the freight to this Country. It is very discouraging to a Collector when he finds that owing to those difficulties which attend the introduction of living Animals from a very distant Country, & owing sometimes to the Carelessness & indifference of Captains of Ships, he not only does not receive any thing for all his trouble but has some heavy expenses to pay.

I trust your Lordship will do me the honour of communicating your views on these subjects as soon as possible as I have now little time to spare having to sail from Southampton on the 20th.

I have been speaking to two respectable persons either of whom will act as my agent—namely Mr. Cuming & Mr. Hemard—but I have not yet made my final arrangements.

I have the honor to be

Your Lordship's

Very humble Sert.

R. Fortune

Although Fortune proposed collecting more birds for his lordship, it seems not unlikely that the Earl of Derby had had enough of Fortune's straightforward manner. Since he did not hear from him, Fortune wrote again on the day of his departure for China (**Fig. 4.5**).

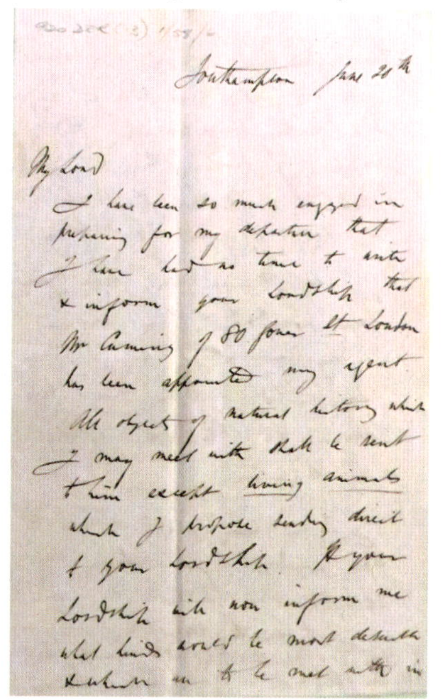

Southampton June 20th

My Lord

I have been so much engaged in preparing for my departure that I have had no time to write & inform your Lordship that Mr Cuming of 80 [illegible] London has been appointed my agent. The objects of natural history which I may meet with shall be sent to him except living animals which I propose sending direct to your Lordship. If your Lordship will now inform me what kinds are to be most desirable & what are to be met with in

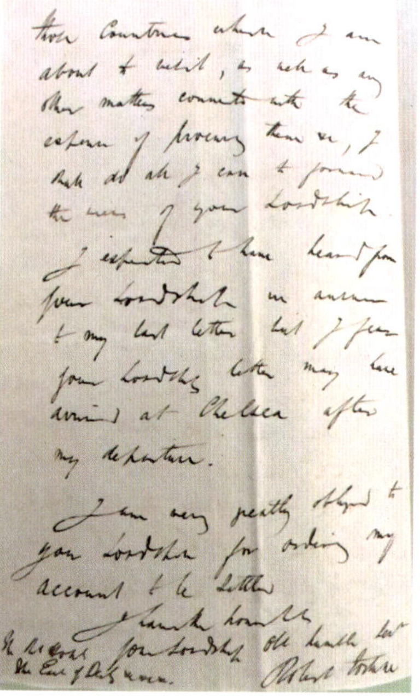

those Countries which I am about to visit, as well as any other matters connected with the expense of procuring them &c, I shall do all I can to forward the views of your Lordship.

I expected to have heard from your Lordship in answer to my last letter but I fear your Lordship's letter may have arrived at Chelsea after my departure.

I am very greatly obliged to your Lordship for ordering my account to be settled.

[illegible] your Lordship [illegible]

Robert Fortune

The Right Honble
The Earl of Derby [illegible]

Fig. 4.5 Letter from Fortune to the Earl of Derby prior to leaving for China on behalf of the East India Company.

Liverpool Record Office 920 DER (13) 1/58/6

Southampton June 20th 1848

My Lord

I have been so much engaged in preparing for my departure that I have had no time to write & inform your Lordship that Mr. Cuming of 80 Gower St London has been appointed my agent. All objects of natural history which I may meet with shall be sent to him except living animals [Fortune's emphasis] which I propose sending direct to your Lordship. If your Lordship will now inform me what kinds would be most desirable & which are to be met with in those Countries which I am about to visit, as well as any other matters connected with the expense of procuring them &c, I shall do all I can to forward the views of your Lordship.

I expected to have heard from Your Lordship in answer to my last latter but I fear Your Lordship's letter may have arrived at Chelsea after my departure.

I am very greatly obliged to Your Lordship for ordering my account to be settled.

I have the honor to be
Your Lordship's obt. humble Sert.
Robert Fortune

Address in China
R. Fortune
care of Messrs. Dent & Co
Hong Kong

Fortune's departure for China was overshadowed by the death of his one-year-old baby daughter Agnes. She was buried in West London Cemetery (now Brompton Cemetery) on 19 April 1848.[24] The service was conducted by Rev. Albert Badger of St. Luke's Church, Sydney Street, Chelsea (**Figs. 4.6, 4.7**) where twelve years earlier Charles Dickens and Catherine Hogarth had been married.[25] For Jane Fortune this must have been a trying time, losing her baby girl and having to say good-bye to her husband in quick succession.

Fig. 4.6 St. Luke's Church, Chelsea, close to Fortune's house in Thistle Grove. A minister from this church conducted the service for Agnes, the first of Fortune's children to die.

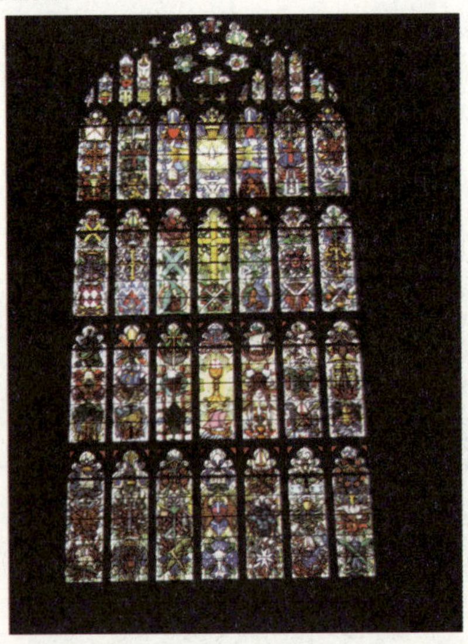

Fig. 4.7 Stained glass window in St. Luke's Church.

Chapter 5 In the Service of the East India Company

5.1 The Outward Journey

Since we do not have his diary to go on[1], we can only imagine what went through Fortune's mind as he travelled to Southampton to catch his ship (**Fig. 5.1**). He must have had mixed feelings about the enterprise he was about to embark upon. He knew he was exchanging the comparative comfort of an existence in London with his family and friends for a life full of danger, dirt and disease. Moreover, being far from the Treaty Ports to steal one of China's major export articles (tea) meant that, if he were caught, he could not expect any leniency. Would he ever see his wife and children again?

Fig. 5.1 Southampton Docks, where Fortune boarded the SS Ripon.

His immediate concern as a landlubber was that he suffered somewhat from seasickness.[2] At least he knew that he would not have to endure it for as long as on the previous occasions, as he was now travelling with the P&O (**Box 5.1**) by the "overland route" through Egypt, which cut the journey from Southampton to Hong Kong to a matter of 55 days.[3]

Box 5.1

The early history of the Peninsular and Oriental Steam Navigation Company (1840 – 2006)

The origins of this shipping company can be traced back to a meeting of a London shipbroker, Brodie McGhie Willcox (1786 – 1862) with a midshipman, Arthur Anderson (1792 – 1868), who had been demobilized in 1815 at the end of the Napoleonic Wars.[1] Willcox must have been very satisfied with his new clerk, as the two men became partners in 1822.[2] In 1826 the partners became the London agents of the City of Dublin Steam Packet Company when its owner, Charles Wye Williams (1779 – 1866), introduced a service from London to Belfast.[3] When they started a steamship service to Portugal, Anderson and Willcox soon became embroiled in Iberian politics.[4] During the succession crisis following the death of King João VI of Portugal in March 1826, Anderson helped Maria II of Portugal (1819 – 1853) to raise money for an expeditionary force to crush her opponents.[5] As a reward, the partners received many royal favours.[6] During the Carlist War from 1833 to 1839, they sided with the Spanish Queen Isabella II (1830 – 1904).[7] It was therefore natural when the Spanish wanted to start a steamship link between Britain and Spain in 1835, that they should turn to Anderson and Willcox.[8] The owner of the Dublin and London Steam Packet Company, Captain Richard Bourne supplied them with the "William Fawcett" in 1828 and joined their company.[9] The three partners inaugurated a regular steamer service between London and four ports on the Iberian Peninsula under the name Peninsular Steam Navigation Company.[10] In 1836 they even started a weekly service to Madeira.[11] However, the great breakthrough came in 1837 when they won a contract from the British Admiralty to deliver mail to the Iberian Peninsula, followed in 1840 by another contract to deliver mail to Alexandria (Egypt).[12] The Peninsular and Oriental Steam Navigation Company (P&O) was born. In 1842 the P&O established a link ("overland route") between Alexandria and Suez, where the brand-new, purpose-built paddle-steamer "Hindostan" commanded by Captain Robert Moresby (**Box 5.3**) was waiting to take the mail and passengers to Kolkata (Calcutta).[13] By 1845 the P&O had extended its operations to Singapore and Hong Kong.[14] In this way, it proved possible to reach the Far East without going via the Cape of Good Hope (South Africa). As a result, the journey from Southampton to Hong Kong was cut to a matter of 55 days.[15] When the Suez Canal was opened in 1869, the overland route became obsolete.[16] Unfortunately, the bureaucrats forgot to change the wording of the mail contract. As a result, until 1888 the mail continued to be unloaded in Alexandria, and carried overland by dromedaries to Suez, where it was reloaded onto the ship which had in the meantime proceeded through the canal![17] Mail contracts continued to be the basis for P&O's prosperity until World War II.[18]

Box 5.2

SS Ripon (1846 – 1870) This paddle-steamer, which was built by Money Wigram (1790 – 1873) at his Blackwall Wharf in London, was named after Frederick John Robinson, the First Earl of Ripon (1782 – 1859), who as President of the India Board was interested in improving the overland route.[1] The original ship, with its novel iron hull, was designed to carry 153 passengers (131 first class, 22 second class) from Southampton to Alexandria.[2] Just four months before Fortune embarked on her, her tonnage had been increased by enhancing her depth and length.[3] During the Crimean War from 1853 to 1856 the "Ripon" was requisitioned as a transport ship.[4] When the hostilities were over, she again plied the Southampton to Alexandria route.[5] In 1861/1862 she was once more enlarged, and new engines fitted.[6] In 1870, when the "Ripon" was sold, the engines were removed, and the ship reduced to a brig.[7] By the time Fortune died (1880), she had been reduced to a hulk, which was eventually scuttled off the island of Trinidad.[8]

Box 5.3

Captain Robert Moresby (1794 – 1854) Son of Fairfax Moresby (1753 – 1820) and his wife Mary Rotton (1767 – 1830).[1] As a young man Moresby joined the Bombay Marine (renamed Indian Navy in 1832).[2] With the coming of steam the overland route to India became a real possibility, as steamships were less affected by the northerly winds that funneled down the Gulf of Aqaba into the Red Sea.[3] Unfortunately, there were only preliminary maps published by the EIC's Scottish hydrographer, James Horsburgh (1762 – 1836) in ca. 1827.[4] Robert Moresby and Thomas Elwon (ca. 1794 – 1835) were assigned the task of drafting detailed maps of the Red Sea.[5] While Elwon (HMS "Benares") worked in the south, Moresby (HMS "Palinurus") initially concentrated on the Gulfs of Aqaba and Suez.[6] Working at temperatures often well in excess of 40°C took its toll on the crews.[7] Elwon was frequently ill, and was eventually transferred to the Persian Gulf in 1833.[8] This left Moresby with the job of completing the whole survey.[9] It took him almost 5 years, from 1829 to 1833, to accomplish the task.[10] The Red Sea Charts were masterpieces of their kind, with information not only on hazardous reefs, but on the availability of water, provisions, and the disposition of the local population to Europeans.[11] After a short break in Bombay, Moresby then undertook surveys of the coral atolls of the Indian Ocean as recommended by Horsburgh.[12] A survey of the Maldives was followed by a map of the Chagos

Archipelago, as these lay in the path of the India to South Africa trade route.[13] Probably his most arduous assignment was the cartography of the Saya de Malha Bank, a vast submerged reef southeast of the Seychelles, as he and his crew were forced to stay on board ship all the time.[14] The years of privation began to tell on his health.[15] He retired from the Indian Navy and joined the P&O.[16] He had the honour of commanding the SS "Hindostan" on her maiden voyage from Southampton to Calcutta at the end of 1842.[17] A few years later he transferred to the Southampton to Alexandria leg of the overland route to India, and it was on this section of the journey that Robert Fortune and Frederick Roberts met him.[18] He was apparently a kind man, who liked to yarn about the difficulties he had experienced while mapping the Red Sea.[19]

The P&O paddle steamer "Ripon" (**Box 5.2**, **Fig. 5.2**) left Southampton on schedule on 20 June 1848.[4] She soon passed the Isle of Wight and headed down the English Channel into the Atlantic Ocean. Once she rounded the rocky coast of Brittany, Captain Robert Moresby (**Box 5.3**) steered her in the direction of Galicia (NW Spain) and along the Portuguese coast. When they docked at Gibraltar, Fortune was able to stretch his legs for a few hours (**Fig. 5.3**). The next leg

Fig. 5.2 The SS Ripon (1846), on which Robert Fortune travelled to Alexandria.

Fig. 5.3 The harbour at Gibraltar, where the SS Ripon docked.

Fig. 5.4 Grand Harbour, Malta, where Fortune's ship took on supplies for the passage to Alexandria. The breakwater at the entrance to the harbour was not there in Fortune's day, being added in the 20th Century.

Fig. 5.5 The Great Pyramid of Giza. Although the pyramids were considered as one of the seven wonders of the ancient world, Fortune felt they paled in comparison with the grandeur of the Himalayas.

of the journey took them to Malta (**Fig. 5.4**), where coal and provisions were taken on board for the 1500 km run to Egypt. It would have been about the 6 July when the passengers disembarked in Alexandria.[5] Here a race against time started. They had just four days in which to make the connection at Suez, or run the risk of missing their ship.[6] In the sweltering heat Fortune and the other on-going passengers were packed into a tug-drawn barge, which took them the 77 km along the Mahmoudieh Canal to Atfeh on the western arm of the Nile.[7] There they transferred to a small river steamer bound for Cairo. Here Fortune first set eyes on the pyramids at Giza (Gizeh; **Fig. 5. 5**), one of the seven wonders of the ancient world.[8]

In order to get to Suez, Fortune and the other passengers had to endure a 135 km journey across the desert in horse-drawn carriages in the middle of summer.[9] Although stops were made to change the horses and allow the passengers to find something to eat and drink, the fly-covered food was far from appetizing and the beer was warm (**Fig. 5.6**).[10] Luckily, their steamship was still waiting for them at Suez (**Fig. 5.7**).[11] They steamed down the Gulf of Suez into the Red Sea, arriving at Aden on 16 July. From Aden they headed eastwards across the Indian Ocean to the port of Galle in southwestern Ceylon (now Sri Lanka). It took them about 11 days to cover the more than 4,000 km, so no doubt Fortune was glad to go onshore for a day.[12] In Galle Fortune transferred to the "Braganza" (**Box 5.4; Fig. 5.8**).

Fig. 5.6 Desert resthouse en route for Suez.

Fig. 5.7 Suez, where a second ship was waiting to take the passengers to Ceylon.

Fig. 5.8 SS Braganza (1837), which Fortune took from Ceylon to Hong Kong.

Box 5.4

SS Braganza (1837 - 1852) This wooden-hulled paddle-steamer was built by Fletcher & Fearnall at Poplar (London) for the Peninsular Steam Navigation Company (**Box 5.1**).[1] She was named after the Portuguese royal house and initially plied between Southampton and Gibraltar.[2] In 1844 she was enlarged, but after two round voyages to the Levant, it was decided to use her on the new route between Galle (Ceylon) and Hong Kong.[3] In February 1851 she was discovered to have dry rot.[4] This sealed the fate of the ship on which Fortune had travelled. She was broken up in Bombay in 1852.[5]

This steamer headed east across the Bay of Bengal, rounded the northern tip of Sumatra and entered the Strait of Malacca. From Singapore it was another 2,700 km across the South China Sea to their final destination, Hong Kong. The "Braganza" made good time, docking in Victoria Harbour on 14 August 1848.[13] As he was in no hurry to disembark, Fortune enjoyed the clear moonlit evening on deck in the company of Captain Potts.[14] When he went ashore the following day he was surprised at the amount of building that had been going on since he was last in Hong Kong in 1845.[15]

5.2 An Illegal Journey in Search of Green Tea

Since he had no business in Hong Kong, he set out for Shànghǎi ("Upper Sea") at the first opportunity.[16] As Fortune sailed up the Huángpǔ Jiāng ("Yellow Riverside River") in September 1848, he was impressed by the forest of masts belonging to Chinese junks and British and US ships.[17] Shànghǎi had also grown since Fortune was last there. In the new town to the north of the Chinese settlement there were now a number of fine houses and gardens including those of his merchant friends Charles D. Mackenzie (Messrs. Mackenzie, Brothers and Co.) and Thomas C. Beale (**Box 5.5**).[18]

Box 5.5

Mr. Thomas Chaye Beale (1805 – 1857) Son of Thomas Beale Senior (ca. 1775 – 1841), who came to Guǎngzhōu in ca. 1791 while still a teenager.[1] Beale Jr. joined Magniac & Co. in Guǎngzhōu as early as 1826.[2] He severed his connections with this firm in the early 1830s when this was taken over by Jardine, Matheson & Co., and operated on his own until 1845 when he established a "house of agency" (acting as an intermediary between merchants in different countries) in Shànghǎi with Lancelot Dent (**Box 2.4**) under the name of Dent, Beale & Co.[3] In 1851 he was Portuguese Consul and Vice-Consul for the Netherlands at Shànghǎi.[4] Beale had a large two-storied mansion between the Chinese walled city and Sūzhōu Creek.[5] In front of the house was a fine lawn extending down towards the Huángpǔ Jiāng.[6] Behind the house was another lawn separated from a shrubbery by a low ornamental wall.[7] It was presumably here that Fortune was allowed to stockpile his plants prior to shipping them abroad. As the groundlevel had been raised by adding soil, the garden was well enough drained to enable a large variety of Chinese shrubs and trees to grow.[8] These included some of the plants which Fortune had collected on his travels (*Cupressus funebris*, *Gardenia jasminoides* var. *fortuneana*, *Mahonia bealei* etc.).[9] In spring and early summer the air was filled with the fragrance of the *Gardenia*, while the powerful perfume of *Osmanthus fragrans* would have been spread throughout the garden in the autumn.[10] The idyllic setting was completed by an aviary with colourful pheasants, something which brought back Beale's childhood memories of Macao.[11]

Having been duped by Chinese on a number of occasions, Fortune decided against sending Chinese collectors to obtain tea plants and seeds from Huángshān ("Yellow Mountain", Ānhuī

Province) on his behalf.[19] Instead he employed an interpreter, one of the many Mr. WÁNGs ("Mr. King") to be found in China, and a coolie from the area, who were prepared to accompany him on his secret mission.[20] At the start of the trip Xìng Huā (i.e. "Fortune flower", Fortune's Chinese name) had his head shaved and adopted Chinese dress, so as to be less obvious.[21] The trio hired a boat and set off for Hángzhōu ("Hang Prefecture"), the capital of Zhèjiāng Province (**Fig. 5.9**). Near Sōngjiāng ("Pine River") some hills broke the monotony of the plains (**Fig. 5.10**).[22]

Fig. 5.9 Map of places Fortune visited in eastern China during his trips on behalf of the East India Company. The peaks represent holy mountains, while the little house symbolizes a temple.

Fig. 5.10 Hill near Sōngjiāng. "During this three days' journey we had been passing through a perfectly level country, having seen only three or four small hills near the city of Sung-kiang-foo." (Fortune, 1852: 30)

Once they got to Jiāxīng ("Fine Prosperity") they sailed down the Dà Yùnhé ("Grand Canal"), passed the ancient city of Shímén ("Stone Entrance") with its ruinous ramparts, and the bustling town of Tángqī ("Dyke Dwelling"), arriving on the northern outskirts of Hángzhōu on the evening of 22 October 1848.[23] In order to reach another boat on the Qiántáng Jiāng ("Money Embankment River"), Fortune either had to go through Hángzhōu or make a detour round the city. As Hángzhōu was out-of-bounds to foreigners, Fortune's men advised him against taking a sedan-chair through the city.[24] However, Fortune soon found to his horror that the bearers had decided to take a short-cut through the centre of the capital.[25] As if this was not bad enough, they only took him part of the way.[26] There ensued a quarrel over pay with the bearers who had been hired to take him the rest of the way.[27] Fortune was obliged to make up the difference in order to get out of the city before his presence became known.[28] He was taken to an inn and having had no breakfast, looked forward to getting a meal.[29] However he had not used chopsticks for three years, so rather than give himself away, he did without lunch.[30] Even though his servants brought him some cooked kumara (sweet potato, *Ipomoea batatas*), he still felt famished.[31]

Fig. 5.11 Léifēng Pagoda, Hángzhōu. This pagoda was in a ruinous state when Fortune saw it. "Wild briers and other weeds were growing out of its walls, even up to its very summit, and it was evidently fast going to decay." (Fortune, 1852: 48) It finally collapsed in 1924. The present pagoda was built in 2000–2002.

Fig. 5.12 Liùhé Pagoda, Hángzhōu, which acted as a lighthouse to aid boatmen to moor their boats on the nearby river.

Fig. 5.13 The tallow tree, *Triadica sebifera*, displaying its autumn colouring. "Tallow-trees were still in extensive cultivation, and at this season of the year, being clothed in their autumnal hues, they produced a striking effect upon the varied landscape." (Fortune, 1852: 61)

At dawn on 24 October 1848, Fortune and his companions left Hángzhōu on their mission to the green tea country.[32] Soon the Léifēng ("Thunder Peak") and Liùhé ("Six Harmony") pagodas **(Figs. 5.**

11, 5.12) were left astern, as the boat glided rapidly upriver.[33] Fortune was entranced by the wooded hills lit up by the red autumnal hues of the tallow tree (*Triadica sebifera*; **Fig. 5. 13**).[34] Upstream from Fùyáng ("Wealthy place north of the river") the hills confine the river (here referred to as the Fùchūn Jiāng = "Nourishing life-giving river") to a narrow thalweg. At the walled town of Méichéng ("Plum City"), the easterly flowing Xīn'ān Jiāng ("New Peaceful River") enters the Fùchūn River.[35] Due to its strategic position at the confluence of the two rivers, this city became an important trading centre in the Táng Dynasty (618 – 906 AD). Each of the banks of the Xīn'ān River was crowned by a pretty pagoda, as Fortune noted.[36] As the Xīn'ān River was known for its rapids, it was necessary to hire more crew to manoeuvre the boat upstream.[37] The two days spent at Méichéng gave Fortune an opportunity to explore the city and the surrounding countryside (**Fig. 5.14**).[38]

Fig. 5.14 Countryside near Méichéng. "The hills about Yen-chow-foo are barren, but the valleys and low lands are rich and fertile." (Fortune, 1852: 53) Note the aquaculture of freshwater pearls in the foreground.

Fig. 5. 15 *Cupressus funebris*, which Fortune optimistically assumed would "one day produce a striking and beautiful effect in our English landscape and in our cemeteries." (Fortune, 1852: vi) However, it is too tender for most parts of the UK.

Fig. 5.16 *Vernicia fordii*, the source of tung oil, "a valuable oil which is used in mixing with the celebrated varnish of the country, and hence this tree is often called the varnish-tree." (Fortune, 1852: 119) In China it is still used for varnishing wood.

Slow progress up the Xīn'ān River gave the plant hunter more time for exploration.[39] He would normally rise at dawn and spend the morning inspecting the nearby hills and valleys before returning to the boat for breakfast.[40] Once this was over he and his two servants would climb the nearest hill to judge how much progress the boat was likely to make in the course of the day.[41] If there were many meanders and rapids to be traversed, they were able to spend most of the day on shore. During these rambles they saw a number of birds, and met many of the plants, introduced during his first visit in 1843 – 1845, growing in their natural habitats.[42] It was here that Fortune first set eyes on the weeping cypress (*Cupressus funebris*; **Fig. 5.15**) growing in the grounds of a country inn.[43] Fortune also encountered a number of economic plants cultivated for their oil (tóng = tung tree, *Vernicia fordii* (Hemsley) Airy-Shaw, Syn. *Aleurites fordii* Hemsley, *Dryandra cordata* Thunberg; **Fig. 5.16**), the multi-purpose hemp-palm (*Trachycarpus fortunei*; **Figs. 3.15, 3.16**) and two conifers (*Cunninghamia lanceolata*, **Fig.3.48**; *Pinus massoniana*) valued for their timber.[44]

After a day spent in the woods the trio would often sit down on a hill-top to enjoy the picturesque scenery around them.[45] At the end of October (29-30 October 1848) the boat reached Cháyuán ("Tea garden"), Zǒuyábù ("Visit ivory pier"), Gǎngkǒu Zhèn (Kang'koo, "Harbour town") and Chúnān ("Pure calm") on the edge of the green tea district.[46] In 1959 all of these towns disappeared below water when the Xīn'ān River was dammed.[47] The main function of the Xīn'ān Jiāng Reservoir or Qiāndǎo Hú ("Thousand-Island Lake"), which thus came into being, is hydroelectricity, although it is also used for irrigation, fish breeding and the prevention of floods.[48] Because of the unpolluted air and the clear water, this lake with its 1,078 islands has become a tourist attraction.[49] However, the barren shores of the islands, caused by a strongly fluctuating water-level, betray its man-made character.

On the evening of the 31 October 1848 the boat reached the walled city of Wēipíng Zhèn ("Powerful Level Ground Garrison"), just before the border between Zhèjiāng and Ānhuī Provinces.[50] Being border country, it was full of soldiers and robbers, neither of whom Fortune was particularly keen to meet.[51] They had to be on their guard. While they did have "visitors" in the night, these left as soon as they discovered that Fortune and his companions were awake.[52] However, once they were well within Ānhuī, Fortune was able to start roaming the countryside again (**Fig. 5.17**).[53] On 2 November 1848, as the river became shallower, all the passengers had to disembark temporarily.[54] The captain finally decided that it was necessary to transfer part of the cargo to a second boat.[55] At this stage Fortune became aware that he had been sleeping above two coffins with the remains of Chinamen who had died in Hángzhōu![56]

Fig. 5.17 View along the Xīn'ān River in Ānhuī. "Nearly all the way from Yen-chow-foo the river was bounded by high hills on each side." (Fortune, 1852: 84)

Fig. 5.18 Old Street in Túnxī.

Fig. 5.19 ***Mahonia bealei*, named after Fortune's friend, Thomas Chaye Beale. Fortune first encountered it in a deserted garden near Túnxī. He took some cuttings, as the shrub was too large to move. The upright inflorescences are characteristic of this species.**

Fig. 5.20 Tea plantations at Sōngluó. "It is famous in China as being the place where the green-tea shrub was first discovered, and where green tea was first manufactured." (Fortune, 1852: 86)

A few days later they reached their destination, the thriving town of Huángshān Shì ("Yellow Mountain City" alias Túnxī, Tung-che, "Village Rivulet"; **Fig. 5.18**).[57] While there Fortune caught sight of a very fine specimen of *Cupressus funebris* in an overgrown walled garden.[58] Having collected more seed of this beautiful conifer, Fortune was about to leave when he spotted a large specimen of *Mahonia bealei* (**Fig. 5.19**).[59] The attractive shrub had only one drawback: it was too big to dig up and take away![60] At one of Huángshān's inns they hired chairs for Sōngluó (Sung-lo, "Soft Trailing Plants"; **Fig. 5.20**).[61] Although Fortune had been complimenting himself about his disguise[62], one of the customers clearly detected something unusual about his appearance and kept asking questions. The awkward situation was finally defused by the innkeeper, who explained that Fortune could only speak Mandarin and did not understand the local dialect.[63]

By the time Fortune arrived at Huángshān, the tea fruits were ripe. He collected some seeds and noted how they were kept viable until the following spring by mixing them with sand and damp earth.[64] He also watched as the tea growers added a mixture of Prussian blue and anhydrite to the tea leaves to make them more attractive for the foreign market.[65] Fortune calculated that for every hundred pounds of coloured green tea, the consumer drank more than half a pound of Prussian blue and anhydrite![66] "No wonder that the Chinese consider the natives of the west to be a race of 'barbarians'", he commented.[67]

Fig. 5.21 Bore in Hángzhōu Bay. "I went out also to see what was going on, and observed a large wave coming rolling up towards us." (Fortune, 1852: 109)

Fortune spent a week in Sōngluó with WÁNG's parents, where he was not only able to obtain a good collection of tea-seeds and young plants, but also three specimens of *Mahonia bealei*, even though it rained for much of the time.[68] On 20 November 1848 he returned to Huángshān Shì, where he boarded a small boat bound for Hángzhōu.[69] On the way Fortune stopped to collect more seeds of tea and *Cupressus funebris*, and obtain cuttings of the *Mahonia bealei* he had discovered in the walled garden.[70] Thirty-four days after he left Hángzhōu, he was back in Yìqiáo (Nechow, "Significant Bridge"), a few miles upriver from the capital of Zhèjiāng. Here he experienced at first hand the bore for which the funnel-shaped Hángzhōu Bay is famous (**Fig. 5.21**).[71]

5.3 More Teas and Temples

As Fortune was keen to visit the island of Jīntáng Dǎo ("Metal Dyke Island"; **Fig. 5.22**) to obtain more tea-seeds, he hired a boat to take WÁNG, the coolie, and himself to the town of Cáo'é ("Cao Pretty Woman").[72] On the way they spent a morning in the walled city of Shàoxīng ("Continuous Prosperity"; **Fig. 5.23**), famous for its virtuous women and literary men, before pressing on to Cáo'é, which they reached at 3 p.m.[73] Here they had to walk to the straggling town of Shàngyú (Bǎiguān) to catch a boat to Níngbō ("Peaceful Waves").[74] Although Fortune was immediately recognized as a foreigner, he experienced no difficulty in hiring a boat.[75] Between Shàngyú and Yúyáo the water-level changed twice. Since there were no locks, the boat had to be lowered down an embankment by huge capstans (**Fig. 5.24**).[76] Fortune arrived in Níngbō the next morning, where he engaged a boat to take him to Jīntáng Dǎo.[77]

Fig. 5.22 Map of the places between Shànghǎi and Pǔtuó Shān, which Fortune visited during his trips on behalf of the East India Company.

As Fortune had been led to believe that the island was full of fierce Chinese soldiers and embittered farmers, he was pleasantly surprised by the kindness he received (**Fig. 5.25**).[78] At the time the tea shrub was extensively cultivated in the interior of the island[79], so he was able to obtain all the tea-seed he required from the farmers.[80] Once the tea-seeds and those of the tung [*Vernicia fordii* (Hemsley) Airy-Shaw, Syn. *Dryandra cordata*] and tallow (*Triadica sebifera*, Syn. *Stillingia sebifera*) had been packed, Fortune returned to Shànghǎi via Zhàpǔ ("Spreading Rivermouth"), a route he had pioneered in 1845.[81]

Fig. 5.23 One of Shàoxīng's many canals. "On the sides of the canal the houses have a somewhat mean and poor appearance..." (Fortune, 1852: 111)

Fig. 5.24 "We were drawn over the embankment by means of a windlass and an inclined plane. This mode of getting from a higher to a lower level, or *vice versâ*, is common in China, where locks, such as those seen in Europe, do not seem to be used." (Fortune, 1852: 113)

When he got to Shànghǎi it was the middle of January 1849.[82] By this time preparations for the Chinese New Year on 24 January 1849 were in full swing.[83] Although Fortune was busy filling Wardian Cases with tea plants[84], he could not avoid noticing the forced flowers of *Amygdalus persica*, *Camellia*, *Cerasus glandulosa* "Alboplena" (Syn. *Prunus sinensis alba*), *Paeonia suffruticosa*, and *Yulania liliiflora* (Syn. *Magnolia purpurea*) in the florists' shops.[85] As there was no ship going direct to Kolkata (Calcutta) from Shànghǎi, Fortune decided to take the cases to Hong Kong and ship them to India from there.[86] On the vessel the Wardian Cases had the company of game birds, which Shànghǎi merchants were sending to their friends in Hong Kong and Guǎngzhōu.[87] As a result the poop of the ship resembled a poulterer's shop at Christmas![88] With a northerly wind blowing the ship covered the 1,500 km from Shànghǎi to Hong Kong in just four days.[89] Once in

Fig. 5.25 Jīntáng Dǎo: "When I reached the top of the first ridge of hills, and looked down on the other side, a most charming view presented itself." (Fortune, 1852: 117)

Hong Kong Fortune lost no time in getting the Wardian Cases transferred to vessels bound for India, where they arrived in excellent condition.[90] While in Hong Kong Fortune made a number of plant-hunting forays in the company of Captain John George Champion (**Box 5.6**), whom he considered to be one of the best botanists he had met in China.[91]

Box 5.6

Major John George Champion (1815 – 1854) Eldest son of Major John Carey Champion (ca. 1790 – 1825), 21st Royal North Britain Fusiliers, and Elizabeth Harries Urquhart (1792 – 1858).[1] As a boy he was fascinated by insects and made a complete collection of Scottish bees.[2] Although his personal preference was for the ministry[3], his widowed mother thought a military career was better for him. He was educated at the military academy Sandhurst (from 1828 to 1831) and then served with the 95th Regiment in various parts of the British Empire.[4] While stationed on the British protectorate Corfu, he made a large collection of insects from the Ionian Islands.[5] When his regiment was transferred to Ceylon in 1838, he continued to collect beetles and other insects.[6] In 1839 he returned to Britain to recover from a severe fever.[7] While there he met and married Frances Mary Carnegie in 1841[8], who stimulated him to collect and dissect plants. On their return to Ceylon, he also received help from fellow Scot Dr. George Gardner (1809 – 1849), who became superintendent of the Peradeniya Botanic Garden in Kandy in 1844.[9] In 1846 Gardner named *Championia*, an African violet from Ceylon, in his honour.[10] With his artistic talent, Champion made a large number of drawings of both plants and animals for a projected "Natural History of Ceylon".[11] In 1847 the 95th Regiment was posted to Hong Kong. In the company of his wife, daughter and enthusiasts like John Charles Bowring (1821 – 1893) he thoroughly investigated the botany and entomology of the island in just three years.[12] Many of the species were new to science.[13] When Fortune was in Hong Kong to ship his material at the end of each year, the two men often undertook rambles together.[14] It seems likely that it was Champion who encouraged Fortune to collect insects as well as plants. Champion returned with his regiment to England in 1851, and brought with him a large collection of dried plants, which formed the basis for George Bentham's "Flora Hongkongensis" in 1861.[15] Champion's promising career was cut short when he was fatally wounded at the Battle of Inkerman on 5 November 1854 during the Crimean War.[16] He died in the notorious military hospital at the British base near Istanbul made famous by Florence Nightingale (1820 – 1910).[17]

Having some time to spare before commencing his second campaign in eastern China, Fortune spent a few days in Guǎngzhōu ("Extensive Prefecture"), before setting off for Fúzhōu (Foo-chow-foo, "Good Fortune Prefecture").[92]

Although Fúzhōu was one of the five Treaty Ports, it was a bit of a backwater as far as trade was concerned.[93] In fact, Charles Spencer Compton (1799 – 1869) was the only foreign merchant permanently living there.[94] One evening he invited Fortune to have dinner in the walled city 3 km north of the Mǐn Jiāng ("Fújiàn River"). After a pleasant evening, Fortune set off back to his own lodgings, only to find the city gate closed for the night, so along with other late-comers he had to clamber over the ramparts![95] While he was in Fúzhōu, Fortune stayed at the house of F. G. Hely, captain of the "Denia" in 1840, an opium schooner operated by Dent & Co. (**Box 2.4**) from 1846 onwards.[96] The Chinese authorities no doubt kept a close eye on this house, so Fortune should not have been surprised when a mandarin called and quizzed the servants about his plans.[97] The same mandarin shadowed him, when he went to visit Martin C. Morrison (**Box 3.7**), the interpreter at the British Consulate.[98] While looking around for souvenirs, Fortune spied some vases, cups and figurines, which had been carved from quartz.[99] As quartz is a comparatively hard mineral, he was duly impressed. One day he headed for the Yǒngquán ("Gushing Spring") Temple on Gǔshān (Koo-shan, "Drum Mountain") to the east of Fúzhōu.[100]

The mountain gets its name from a piece of granite (**Fig. 5.26**) which is said to emit the sound of a drum when struck by lightning.[101] Today it is possible to take a car or cable car up to the temple. In Fortune's time the temple could only be reached on foot (**Fig. 5.27**). The effort this took was however amply rewarded by the magnificent view across to Fúzhōu and up the Mǐn Valley (**Fig. 5.28**).[102] Fortune was taken round the temple, which was founded

Fig. 5.26 Granite boulder on Gǔshān, which is said to emit a drum-like sound when struck by lightning.

Fig. 5. 27 Steps leading to Yǒngquán (Gushing Spring) Temple on Gǔshān. "A well-paved path, about six feet in width, has been made the whole way up to the temple." (Fortune, 1852: 136)

Fig. 5. 28 View of Fúzhōu from Gǔshān. "As the traveller ascends by this winding causeway ... a glorious view is spread before him. It is the wide and fertile valley of the Min, intersected everywhere by rivers and canals, and teeming with a numerous and industrious population." (Fortune, 1852: 136)

Fig. 5.29 The entrance to the Gushing Spring Temple on Gŭshān.

Fig. 5.30 Gushing Spring Temple dining room. "The dining-room is a large square building, having a number of tables placed across it at which the priests eat their frugal meals." (Fortune, 1852: 137)

in 783 AD. In the dining room (**Fig. 5.30**) he was invited to join the monks at their meal.[103] He declined, but visited the kitchen to see the enormous cauldron needed to cook up to 250 kg rice and three smaller cauldrons for vegetables for over 100 monks (**Fig. 5.31**).[104] In the 17th Century Library, which contains more than 20, 000 books from the Míng Dynasty (1368 – 1644 AD) and Japan, there are 657 written in blood to emphasize the importance of the message they contain.[105] Based on the amount of dust they had collected, Fortune got the impression that these had not been read for a long time.[106] While in the library Fortune examined a crystal vase and one of "Buddha's" teeth.[107] The vase can still be viewed in a shrine (**Figs. 5.32, 5.33**), but the tooth has unfortunately gone missing in the past 160 years. Fortune's drawing (**Fig. 5.34**) would suggest that the tooth was the molar of a Cenozoic elephant (*Stegodon*).[108] After visiting the temple, Fortune and his men hired a boat, and travelled up the Mǐn Jiāng as far as Nánpíng ("Southern Peace", Suiy-kow).[109] The river is tidal, so during ebb the boat was made fast, and Fortune was free to explore the countryside. He noticed that the natives were miserably poor, although they had a lot of fruit trees and grew crops such as sugarcane and tobacco.[110] Fortune sent his two servants (Sing-Hoo and WÁNG "Sōngluó") on to Wǔyíshān (Woo-e-shan), while he returned to Fúzhōu and took a Portuguese lorcha to Níngbō.[111]

Fig. 5.31 Gushing Spring Temple kitchen. "In the kitchen the wonders shown to the visitors are some enormously large coppers in which the rice is boiled." (Fortune, 1852: 138) The huge cauldrons were used for preparing food for over 100 monks.

Fig. 5.32 Shrine with crystal vase in the Library of Gushing Spring Temple.

Fig. 5.33 Crystal vase as viewed through the hole in the shrine. "It appeared to be a small piece of crystal cut in the form of a little vase, with a curious-looking substance inside." (Fortune, 1852: 140)

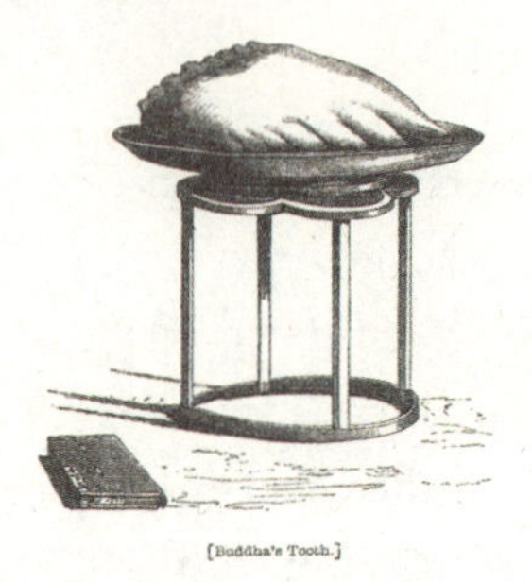

Fig. 5.34 "Buddha's Tooth". "I had heard that in this part of the building there was a precious relic, nothing less than one of Buddha's teeth ... " (Fortune, 1852: 138) This tooth would seem to have belonged to a fossil elephant.

While Fortune was waiting for his servants to return from their travels, he went deer shooting near Tiāntóng Sì ("Heavenly Child Temple") with his friend Charles Wills and some Chinese sportsmen armed with matchlocks.[112] Once the hunters had bagged enough deer, Fortune walked across the hills to Tiāntóng Temple.[113] He noted that tea was grown on the lower slopes, while the valley bottoms had been converted into paddy fields.[114] Being the beginning of May, the hillsides were lit up by the yellow flowers of *Rhododendron molle* (Syn. *Azalea sinensis*, *A*. "*chinensis*") and the pure white blossoms of the ornamental shrub, *Exochorda racemosa* (Syn. *Amelanchier racemosa*; **Fig. 5.35**).[115]

Fig. 5.35 *Exochorda racemosa* with its dazzling white flowers. "There was another shrub which is new to botanists, and scarcely yet known in Europe, called *Amelanchier racemosa*, not less beautiful than the azalea, and rivalling it in its masses of flowers of the purest snowy white." (Fortune, 1852: 154)

In the immediate surroundings of the temple, fine trees of *Cryptomeria japonica* and giant bamboo were growing.[116] Bamboo was and is widely used in China, and

Fig. 5.36 Grove of *Phyllostachys edulis* at Tiāntóng Temple. "There are also some fine bamboo woods here, which deserve more than a passing glance." (Fortune, 1852: 155) These had to be protected against marauding wild boars.

visitors to the Tiāntóng Temple can still see a grove of giant bamboo (*Phyllostachys edulis*, **Fig. 5.36**). On the other hand, in the past 160 years the light-demanding *Cryptomeria* has suffered as a result of competition from the evergreen chestnut, *Castanopsis fargesii*.

When Fortune arrived at Tiāntóng, he bumped into his old acquaintances, Mr. Bowman and Dr. Kirk from Shànghǎi.[117] Dr. Kirk was trying to explain the principle of a siphon to the sceptical monks.[118] After a day or two spent at Tiāntóng, the trio set off for Dōngqián Hú (Tung-hoo, "East Money Lake"; **Fig. 5.37**), the largest natural lake in Zhèjiāng.[119] Fortune busied himself collecting plants and insects.[120] While he was there, he received a message to say that Sing-Hoo had returned from Wǔyíshān.[121] He therefore went back to Níngbō at the double.[122]

5.4 A Secret Journey to the Black Tea District of Fújiàn

Fig. 5.37 Dōngqián Hú, the largest natural lake in Zhèjiāng. "The shores of the lakes were rich in plants, and richer still in insects." (Fortune, 1852: 158)

In the meantime, Fortune had decided to visit Wǔyíshān himself.[123] So he got dressed up as a Chinaman with a pigtail down almost to his heels, and on the 15 May 1849 he and Sing-Hoo set off from Níngbō in a boat with destination Shàngyú.[124] At the walled city of Yúyáo, Fortune noted the many Buddhist temples on Lóngquán Shān ("Dragon Spring Hill"; **Fig. 5.38**).[125]

Once they got to Shàngyú, Fortune and Sing-Hoo walked to another canal and hired a boat to take them to Shàoxīng and Yìqiáo.[126] From there they sailed up the Fùchūn Jiāng as far as Méichéng (**Fig. 5.14**).[127] On the way Fortune enjoyed the lively green colours of the Spring vegetation.[128] They reached Méichéng on the 20 May 1849, and set off the very next morning up the Qú Jiāng ("Thoroughfare River").[129] That day they only went as far as the small town of Dàyáng, as the crew were tired from hauling the boat over a series of rapids.[130] Fortune used the opportunity to go hunting for plants.[131] It was the beginning of the summer monsoon, so after a downpour the river became swollen.[132] On the morning of the 22 May the river was still too wild to proceed, so Fortune went on shore after breakfast.[133] He noted that a variety of cereals were grown near the town: rice on the low-lying land,

and wheat, barley, maize and millet at higher elevations.[134] Although he encountered some camphor, mulberry, tea and tung plants, the tallow tree was clearly the staple product of the district.[135] No wonder, for the waxy fruits were then widely used to produce candles.[136] The evergreen weeping cypress and pine trees marked the sites of the ancestral graves.[137] By the afternoon the water-level in the Qú Jiāng had fallen sufficiently for the boat to proceed.[138] However, the travellers did not get very far, before the boat was entered by a gang of four men.[139] It transpired that the captain still owed the boarding party money for rice that they had supplied him with on a previous occasion.[140] When he showed no inclination to pay them, they removed the sail, which immediately immobilized the boat.[141] Sing-Hoo suggested going 15 km upriver to Lánxī (Lanchee, "Orchid Stream"; Nan-che) to hire another boat.[142] Sing-Hoo returned at midday on 24 May 1849, and after settling their account with the wife of the owner of the immobilized boat, they proceeded in their new craft to the attractive town of Lánxī (**Fig. 5.40**).[143]

Fig. 5.38 Yúyáo: View from Lóngquán Shān (Dragon Spring Hill) towards Tōngjì Bridge in the middle distance. "The walls and ramparts enclose a hill of considerable extent, on whose summit many Buddhist temples have been erected." (Fortune, 1852: 161)

Fig. 5.39 Tōngjì Bridge, Yúyáo, which stopped the advance of the paddle steamers "Nemesis" and "Phlegethon" during the First Opium War.

The next morning Fortune visited three nurseries in the suburbs of Lánxī in the hope of finding some novelties, but was disappointed that these had the same collection of Arabian jasmine, azaleas, camellias, clerodendrons, lotus and roses, which he had encountered elsewhere.[144] So having hired another boat and stocked up with provisions for the trip to Jiāngshān (Chang-shan, "River Mountain"), Fortune and Sing-Hoo set off up the Lánxī Valley with its hills, meandering river and clumps of trees dotted here and there.[145] Fortune compared the landscape to a beautiful garden full of tallow trees.[146] After travelling for 45 km, they reached Lóngyóu (Long-yeou, "Dragon River") with its pagodas and numerous camphor trees.[147] Like so many cities in China, where transport was traditionally by river, Lóngyóu was established at the junction of two rivers, the Qú Jiāng and the Língshān Gǎng ("Fairy Mountain Stream"). Here Fortune noticed that in addition to the usual crops, such as wheat, buckwheat, barley, maize, millet and soybeans, barberries were cultivated.[148] Fortune guessed that the berries were either used as a dye or medicine.[149] In fact, the berries of this shrub were used to treat gastroenteritis and diarrhoea.[150] Between Lóngyóu and Qúzhōu (Chu-chu-foo), the next city on the Qú Jiāng, there were many rapids and disused water-wheels.[151] Fortune and Sing-Hoo reached the walled city of Qúzhōu ("Thoroughfare Prefecture"; **Fig. 5.41**) on 1 June 1849.[152]

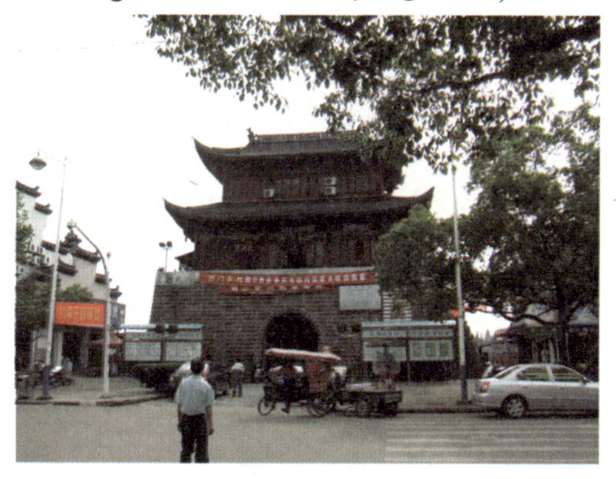

Fig. 5.40 A gate on Lánxī's town wall, which was rebuilt in 1995. Lánxī "is built along the banks of the river, and has a picturesque hill behind it: an old tower or pagoda in ruins heightens the general effect of the scene." (Fortune, 1852: 174)

Fig. 5.41 Qúzhōu, where Fortune stayed only long enough to procure provisions. "It is not large, its walls are scarcely more than two miles in circumference, and there are many large spaces inside on which there are no buildings." (Fortune, 1852: 178)

Fortune remarked on the oranges and tea which thrive in the mild climate, as well as the peanuts and soybeans growing on the sandy soil, washed off the barren hills.[153] Today these hills are not quite as bare as they once were, as cooking no longer relies exclusively on a supply of firewood. Fortune was always pressing on, so his party only stayed in Qúzhōu long enough to procure provisions for the next stage of their journey.[154] To save weight, Fortune had not taken any gauze curtains with him.[155] As a result, he was bothered by the mosquitoes, and could not sleep.[156] The crew advised him to burn mosquito tobacco, after which he slept well.[157] He was full of praise for the long-forgotten Chinese inventor.[158]

On the next day, 2 June 1849, the travellers arrived in Jiāngshān, the last city before the Zhèjiāng/Jiāngxī border.[159] It had rained heavily during the night, and Fortune regretted not hiring mountain chairs with an oil-paper cover for the 48 km cross-country trek to Yùshān (Yuk-shan, "Jade Mountain").[160] He comforted himself in the knowledge that he obtained a better view of the countryside by holding an umbrella.[161] As the chair-bearers pressed on, Fortune enjoyed the scenery, which reminded him of the countryside of the Zhōushān (Chusan, "Boat Mountain") Archipelago.[162]

At the border between Zhèjiāng and Jiāngxī they had to pass under a huge stone gateway, but there were no border guards to check their identity.[163] In fact, the military stations on either side of the border were in a ruinous state.[164] At midday the party stopped at an inn in Xiàzhèn ("Lower Garrison Town").[165] As Fortune smoked his Chinese pipe and sipped his tea, he was eyed by two Cantonese merchants, one of whom he had frequently seen in Shànghǎi.[166] As he was by now well outside the limits to which foreigners were allowed to travel, he pretended not to notice their interest.[167] The merchants prompted the innkeeper to ask Fortune where he had started from, where he was going, and the object of his journey.[168] Fortune managed to evade the questions, but when Sing-Hoo went to pay the bill, he was quizzed about his master.[169] As always, he replied that Fortune was "from some country beyond the Great Wall".[170] After this incident, Fortune was glad to get on the move again. The road was full of an unbroken line of coolies carrying tea from Yùshān to Jiāngshān and returning with other merchandise.[171] At about 4 p.m. on 3 June 1849 Fortune and his men crossed the fine stone bridge over the Xìn Jiāng (Kin-keang, "Aimless River"), and entered the gate into the bustling town of Yùshān (**Fig. 5.42**).[172]

Fig. 5.42 Old town wall at Yùshān. "Passing over a fine stone bridge, we were soon at the walls of the city. Having entered the gates, we proceeded along one of the principal streets." (Fortune, 1852: 193)

Fig. 5.43 Lóng Tán Pagoda in the town of Shàngráo, where to Fortune's dismay, his assistant bought 18 kg of grass cloth.

Within 30 minutes Sing-Hoo had hired a boat to take them down the Xìn Jiāng as far as Shàngráo ("Superior Rich", Quan-sin-foo).[173] They arrived at this opulent walled city (**Fig. 5.43**) early on 4 June 1849, but the restless Fortune did not intend to stay there for long.[174] However, Sing-Hoo took longer than expected to hire a boat for Qiānshān ("Lead Mountain", Hékǒu, Hokow) because he went shopping for grass-cloth![175] Fortune was annoyed, because he had tried all along to keep the baggage to a minimum.[176] However, no amount of argument could convince Sing-Hoo to leave his 18 kg of grass-cloth behind.[177] He even suggested that Fortune's status would be enhanced if he employed more porters![178] Fortune momentarily forgot the quarrel, as the boat sped down the Xìn Jiāng between magical sandstone hills, which had been moulded by time into curious forms (**Fig. 5.44**).[179]

Fig. 5.44 Rounded sandstone hills on the Xìn Jiāng at Qiānshān. "I observed many curious rocks, shaped like little hills, but without a vestige of vegetation of any kind upon them. They stood in the midst of the plain like rude monuments, and had a curious and strange appearance." (Fortune, 1852: 195)

The party arrived at Qiānshān, famed as a black tea trading post, in the afternoon of 4 June 1849.[180] Here Sing-Hoo engaged a comfortable mountain chair with padded seat and back for his master, and hired coolies for the four-day journey across the Bohea Mountains.[181] The chair-bearers set off at a cracking pace, which left Sing-Hoo well behind.[182] Unfortunately, as soon as they got into hilly country, Fortune's chair was besieged by a host of beggars.[183] As Sing-Hoo had the small change with him, Fortune feigned sleep.[184] When the beggars in desperation grabbed his clothes, he roused himself and almost capsized the chair.[185] His bearers shooed the beggars away.[186]

Fig. 5.45 Bridge at Yǒngpíng (Perpetual Peace). "It is about 60 le distant from Hokow, and stands on the banks of the mountain stream. Though not large, it seems a flourishing place." (Fortune, 1852: 201)

At midday on 5 June he finally arrived at the beautiful town of Yǒngpíng ("Perpetual Peace", Yuen-shan; **Fig. 5.45**), where tea was shipped to Qiānshān.[187] By now he was in veritable tea country.[188] Here he noticed that the best tea was transported in such a fashion that it never touched the ground (**Fig. 5.46**).[189] Fortune and his men spent the night in Zǐxī ("Violet Rivulet"), where he passed for a Chinaman.[190] He liked the bamboo he had for dinner and continued to order it during the rest of his trip.[191] Just before Fortune crossed into Fújiàn Province, he looked up at the peaks of the Bohea Mountains.[192] The guards at the border pass took no interest in the traveller, so he was able to proceed towards Wǔyíshān.[193] On the way, he encountered a number of interesting plants

Fig. 5.46 In Fortune's day the best tea was carried in such a way that it never touched the ground.

which he dug up and took with him.[194] The coolies must have wondered whether his native country had any plants at all.[195] Close to the border he encountered a magnificent *Cryptomeria japonica* (36+ m), which was revered, having been planted by a former emperor.[196] On crossing one of the many passes, which the modern traveller avoids by driving through a series of tunnels, he was forced by strong gale-force winds and a downpour to seek shelter in a couple of tea-houses.[197] The "mandarin from a far country beyond the Great Wall" was given a fine supper.[198] At Sāngǎng ("Three harbours"), where the black teas were sorted and packed, he avoided the Cantonese merchants, who were more likely to realize he was a foreigner in disguise.[199] Once they had left Sāngǎng Fortune and his men were again on their own.[200]

The party finally got their first glimpse of Wǔyíshān (**Fig. 5.47**), which was renowned for its hundreds of temples.[201] After toiling up a hill, they at long last arrived at the Tiānxìn Yǒnglè Chán Sì ("Heavenly Heart, Eternally Happy Buddhist Temple", usually abbreviated as "Temple of Eternal Happiness" or "Ever Happy Temple"; **Figs. 5.48, 5.49**) founded in the Míng Dynasty (1368 – 1644 AD).[202]

Fig. 5.47 Paved track through tea country, Wǔyíshān. "I had expected to see a wonderful sight when I reached this place, but I must confess the scene far surpassed any ideas I had formed respecting it." (Fortune, 1852: 224)

Fig. 5.48 Monks' quarters at Tiānxìn Yǒnglè Chán Sì (Ever Happy Temple), Wǔyíshān. This is where Fortune would have stayed.

Although HÁN Dòngxū meant it as a place for Daoist guests to stay, by the time Fortune lodged there, it had become a Buddhist temple.[203] As a token of welcome the chief priest gave Fortune some tobacco for his pipe.[204] He was invited to dinner, and was given the place of honour on the left-hand side of the priest.[205] The "mandarin" and his Buddhist friends toasted each other.[206] The priests pressed him to try all sorts of dishes, and picked out tasty morsels with their chopsticks and presented them to Fortune.[207] Although he had an aversion to eating from someone else's chopsticks for hygenic reasons, he felt obliged to do so.[208] Fortune investigated the countryside near the temple. He took a particular interest in the geology of Wǔyíshān, and collected samples to show to Hugh Falconer (**Box 5.7; Fig. 5.50**) and William Jameson (**Box 5.8**) in India.[209]

Fig. 5.49 One of the temples at Tiānxìn Yǒnglè Chán Sì, Wǔyíshān. "The temple, or collection of temples, which we now approached, was situated on the sloping side of a small valley, or basin, on the top of Woo-e-shan, which seemed as if it had been scooped out for the purpose." (Fortune, 1852: 226)

Fig. 5.50 Professor Hugh Falconer (1808 – 1865) was not only Superintendent of the Botanical Garden in Kolkata, an expert in vertebrate palaeontology, but also an outstanding geologist. "Specimens of these rocks were brought away by me and submitted both to Dr. Falconer of Calcutta and Dr. Jameson of Saharunpore, who are well known as excellent geologists." (Fortune, 1852: 233)

Box 5.7

Professor Hugh Falconer MD, FRS (1808 – 1865) One of five sons and two daughters of David Falconer and Isabel Mcrae.[1] He was educated at Forres Academy, Aberdeen University and the University of Edinburgh.[2] He was then appointed as Assistant Surgeon to the Bengal branch of the EIC, but because he was not yet 22, he first assisted Dr. Nathaniel Wallich (1786 – 1854) with the distribution of his Indian herbarium, William Lonsdale (1794 – 1871), the new curator of the Geological Society of London's Museum, and worked on the fossil mammals collected in 1827 by John Crawfurd (1783 – 1868) from the banks of the Irrawaddy (Burma, now Myanmar).[3] Falconer finally arrived in Kolkata (Calcutta) in September 1830[4] and was posted to Meerut, 60 km NE of Delhi in 1831.[5] On his way to Landour in the Himalayan foothills with a group of invalids, he met Dr. John Forbes Royle (**Box 4.1**), then Superintendent of the Saharanpur Botanical Garden (**Box 5.11**). The two men struck up a lifelong friendship.[6] When Royle returned to England later that year, Falconer succeeded him as superintendent in 1832.[7] In this function, he was consulted by the government about the fitness of India for the cultivation of tea.[8] He considered the Lesser Himalayas had a suitable climate, and suggested that Mussoorie would be a good place to undertake the trials.[9] Following Falconer's positive assessment, plants were imported and placed under his charge.[10] He was responsible for establishing tea nurseries at Bhurtpur (Bhurtpore) and Lachmeshwar (Luchmaiser) in 1835.[11] As Saharanpur is only 50 km from the Himalayan foothills, it was not long before Falconer and his friend Sir Proby T. Cautley (1802 – 1871) were investigating the Siwaliks, piedmont sediments formed by the erosion of the Himalayas. In recognition of the diverse fauna they uncovered, and the revised age of the sediments this implied (not Triassic but Neogene), they were jointly awarded the Wollaston Medal by the Geological Society of London in 1837.[12] As they were both in India, Royle accepted the medals on their behalf.[13] In the same year Falconer accompanied the ill-fated Scot Sir Alexander Burns (1805 – 1841: murdered in Afghanistan) on a mission to Afghanistan to meet Dost Mohammad (1793 – 1863).[14] They arrived in Kabul to a rapturous welcome on 20 September 1837.[15] On his way back from Kabul, Falconer travelled NNE from Rawalpindi until he reached the Indus, and followed the river some of the way towards its source before returning through the mountains to Saharanpur.[16] During this epic journey, he became completely worn out and was forced to return to Britain in 1842 to recover.[17] He brought with him some 70 – 80 large chests of dried plants, which were deposited in the East India Company's Museum and forgotten until Sir Joseph Dalton Hooker (1817 – 1911) started to prepare his "Flora of British India".[18] He also had with him 48 cases of Siwalik fossils plus 4 cases of comparative material[19], which were to form the basis of "Fauna Antiqua Sivalensis" from 1846 to 1849.[20]

When Nathaniel Wallich retired as superintendent of the Kolkata Botanical Garden in 1847, Falconer was appointed to replace him.[21] He also became Professor of Botany at the Medical College in the Bengal metropole.[22] In the seven years that he was in Kolkata, he promoted the conservation of teak forests, the introduction of cinchonas as a source of quinine, and tea plantations.[23] When Fortune arrived in India on 15 March 1851 to supervise the introduction of his tea plants, he stayed with Falconer.[24] Fortune's tea manufacturers showed the superintendent how to prepare the tea leaves[25], and knowing Falconer's reputation as an excellent geologist, Fortune got him to examine the rock specimens from Wǔyíshān in Fújiàn.[26]

In 1855 illness forced Falconer to retire.[27] He returned to England via Afghanistan, Palestine, and Syria, while taking advantage of the Crimean War to visit this part of Russia.[28] He now turned his interest to the Plio-Pleistocene of Europe, examining fossils in European musea, and investigating cave deposits at home and abroad (France, Gibraltar, Malta, Sicily).[29] In the course of this work, he discovered human artefacts, which were much older than those known at the time. He started writing a book on "Primeval Man, and his contemporaries", but Charles Lyell (1797 – 1875) published his "Geological evidences of the antiquity of man" in 1863, before Falconer's book was anywhere near complete.[30] Falconer, no doubt furious at being pre-empted, wrongly accused Lyell of "lifting" some of "his" data, which sparked an acrimonious exchange in the "Athenaeum".[31] Nonetheless, Falconer was a respected scientist with a prodigious memory, an eye for detail, original ideas, who was cautious to a fault[32], which explains why he was both Foreign Secretary of the Geological Society of London and Vice-president of the Royal Society at the time of his death.[33] His premature death caused Charles Darwin and Joseph Hooker to reflect on the futility of human life.[34]

Box 5.8

Dr. William Jameson MD (1815 – 1882) Son of Laurence Jameson (1783 – 1827) and his wife Jane Watson.[1] He attended Edinburgh High School and studied medicine at the University of Edinburgh.[2] After his medical studies Jameson was appointed as an Assistant Surgeon to the Bengal Medical Service on 30 August 1838.[3] In those days knowledge of medicinal plants was considered an integral part of medicine, so it was hardly surprising that after a temporary curatorship of the Museum of the Asiatic Society of Bengal, he became Superintendent of the Botanical Gardens in Saharanpur in 1842 on Hugh Falconer's recommendation.[4] He built up a large collection of ornamental and medicinal plants (esp. *Hyoscyamus*), which were supplied to those requiring them.[5] Jameson commissioned a number of local artists to make watercolours of over 300 of these plants (*Saharunpore Gardens Art Collection* in Natural History Museum, London).[6] He also experimented with growing flax, and introduced maize to the Himalayan foothills.[7] In April 1843 he paid his first visit to the tea nurseries established by Falconer in 1835, and established an additional nursery at Gadoli near Paurie.[8] Jameson was responsible for looking after Fortune's tea plants on their way to their final destination in the Himalayas. The first batch of seedlings failed because a curious official at Allahabad opened the Wardian Cases prematurely.[9] By the time they got to Saharanpur only 1,000 of the 13,000 seedlings were still alive.[10] Falconer, who had already had doubts about Jameson's ability[11], gave him the blame, as he clearly did not understand how Wardian Cases worked.[12] This dispute meant that when Falconer retired, he did not recommend Jameson to succeed him.[13] Nonetheless, Fortune had a high opinion of Jameson's abilities.[14] Fortune also considered him to be an excellent geologist, and asked his opinion about the rocks he had collected on Wǔyíshān.[15] During Fortune's tour of the tea plantations, he was accompanied by Jameson, who was in charge of the government's tea plantations, and his wife, Emily Field.[16] The Jamesons later had a daughter Janet Jane Helen (1852 – 1876), who was to marry Russell Richard Pulford (1845 – 1920) of the Royal Engineers on 27 November 1875, the year in which Jameson retired.[17] He was then able to devote himself wholeheartedly to running a tea plantation in Dehradun.[18]

He also examined the soil in which the tea shrubs had been grown since the Míng Dynasty (**Fig. 5.51**).[210] According to a legend a scholar who drank the tea at the temple during the Qīng Dynasty was immediately cured of his stomach-ache and passed the official examination in Běijīng with flying colours.[211] While Fortune was there, the monks were busy drying tea leaves for sale to passing merchants.[212] After getting a panoramic view of the area as far away as the Bohea Mountains, he returned to the temple.[213] As the sun set and the moon started to shine, a lake glistened as if covered by gems. He imagined he was in some sort of fairyland in which the trees came alive and the rugged rocks reared their heads high above the temples.[214] His reverie was broken by the arrival of Sing-Hoo to say that dinner was served.[215]

Fig. 5.51 Red robe tea plants grown since the Míng Dynasty. The tea from these shrubs fetches record prices.

Fig. 5.52 Remains of the Daoist Temple of Zhǐ Zhǐ on the inclined slope. Fortune stayed here for a few days. "The top of the rock overhung the little building, and the water from it continually dripping on the roof of the house gave the impression that it was raining." (Fortune, 1852: 241)

Fig. 5.53 The Jiǔqū Xī (Nine Bend Brook), a famous tourist attraction at Wǔyíshān, then and now. "The stream of 'nine windings' flowed past the front of the temple. Numerous boats were plying up and down, many of which, I was told, contained parties of pleasure, who had come to see the strange scenery amongst these hills." (Fortune, 1852: 241)

On the morning of the third day he bade farewell to his hosts and proceeded in the direction of Xīngcūn (Tsin-tsun, "Star Village").[216] He was given lodgings by an old monk in charge of the Daoist temple of Zhǐ Zhǐ ("Stop Temple") beside the Jiǔqū Xī ("Nine Bend Brook" or "River of the Nine Windings"; **Fig. 5.52**).[217] Then as now this river was a popular tourist attraction, with numerous pleasure boats plying up and down (**Fig. 5.53**).[218] The temple, which was founded during the Jìn Dynasty, was home to many famous Daoists in the Jìn (265 - 420 AD), Táng (618 - 906 AD) and Sòng (960 - 1279 AD) Dynasties.[219] It was repaired and expanded in the Sòng-, Míng- (1368 - 1644 AD) and Qīng (1644 - 1911 AD) Dynasties, but by the time Fortune visited it, it was in a poor state of repair.[220]

Fig. 5.54 Vertical channels on the hills at Wǔyíshān, gouged out by running water. "At a distance they seemed as if they were the impress of some gigantic hands. I did not get very near these marks, but I believe that many of them have been formed by the water oozing out and trickling down the surface." (Fortune, 1852: 242)

Being built under an overhanging rock, over which a stream flowed during the monsoon period, it was damp.[221] One night when it rained Sing-Hoo and Fortune were alarmed to see a figure entering their room on the second floor.[222] It was none other than the old monk, who was unable to sleep, because the rain was dripping through the leaky roof onto his bed below.[223] While he was at the Daoist temple, Fortune was visited by country people, who were curious to see the mandarin "from Tartary".[224] In the two days that Fortune stayed at the temple, he was able to reconnoitre the area and obtain 400 young tea plants.[225] He observed the vertical channels in the rocks caused by running water (**Fig. 5.54**), but was unable to locate any of the boat burials in the cliffs mentioned by Jean-Baptiste Du Halde (**Box 5.9**).[226] These have since been radiocarbon dated as 3750-3295 years old.[227]

Box 5.9

Father Jean-Baptiste Du Halde (1674 – 1743) Nothing is known of his childhood. He entered the Jesuit Order (Compagnie de Jésus) on 8 September 1692.[1] He became an instructor at their seminary in La Flèche, some 220 km SW of Paris in 1698, before teaching literature at the Collège Louis-le-Grand in Paris where Voltaire was a pupil.[2] He had a natural talent for writing. He composed poems in Latin to celebrate the births of Louis d'Orléans and the short-lived Louis, Duke of Brittany (1704 – 1705), and in 1702 wrote a play about King Midas, which was performed by the pupils in 1702.[3] This was followed in 1707 by a tragedy entitled "Narcisse".[4] In 1708 he was ordained.[5] The following year he succeeded Charles Le Gobien (1653 – 1708) as editor of the "Lettres édifiantes et curieuses écrites des missions étrangères par quelques missionnaires de la Compagnie de Jésus" (ca. 1702 – 1776), which was a major source of information about the Catholic missions.[6] The subjects covered ranged from missionary work and local culture to geography, natural history and the production of porcelain.[7] In 1724 he published "Le Sage chrétien" (The Christian Sage) on how to lead a Christian life.[8] This probably led to his appointment as confessor to the pious Louis d'Orléans (1703 – 1752), the Duke of Orléans.[9] He had a superb ability to synthesize information, publishing a four-volume treatise on the Chinese Empire without ever leaving Paris![10] Although there were those who were disappointed by the text, most of Du Halde's contemporaries were favourably impressed by both the contents and the production of these tomes.[11] They were translated into a number of European languages (English, German, Russian ...)[12] As such, this magnum opus led to an increasing interest in all things Chinese, and was to have an impact on the philosophers of the Enlightenment.[13] The accuracy of the maps, based on surveys by a number of Jesuit fathers from 1708 to 1717 and earlier reports, was so good that these were still being used well into the Twentieth Century![14]

After only two days spent at the Daoist temple, Fortune and Sing-Hoo continued on to Xīngcūn.[228] Here Fortune considered the option of going down the Mǐn River to Fúzhōu.[229] He decided against this plan, as it would have meant that he would not cover much new ground and would have difficulty finding a ship to take him on his onward journey.[230] Instead he opted to travel in the direction of Pǔchéng ("Riverside City"), and cross the Bohea Mountains into Zhèjiāng Province.[231] Even today the road network in this part of the country is poorly developed. He travelled from Chóng'ān (Tsong-gan, "Sublime Calm") to Shípí ("Rock Surface"), and toiled up a "gigantic mountain".[232] They encountered dense woods of tall *Cunninghamia lanceolata* plus abundant *Pinus massoniana* (Syn. *P. sinensis*), *Cinnamomum camphora*, and

Triadica sebifera, as well as eugenias, naturalized guavas, a deciduous oak, evergreen "oaks" (*Lithocarpus glaber* (Syn. *Quercus inversa*), *Castanopsis sclerophylla* (Syn. *Quercus sclerophylla*), azaleas and at least one *Rhododendron*.[233] This was an untamed and lonely place where wild boars and tigers roamed.[234] He regretted not having his shot-gun with him to add some specimens to his collection of birds.[235] His party stayed the night at Shípí surrounded by fields of lotus and tobacco.[236] They encountered tea plantations as they approached the rundown market town of Pǔchéng (**Fig. 5.55**) set in a wide and beautiful valley.[237] Fortune noted that "A considerable trade in tea is carried on here."[238] He and his men were glad to find an inn cum opium den to get out of the rain.[239]

Fig. 5. 55 Pǔchéng, where Fortune's assistant got into an argument with the chair-bearers. "It might have been about midnight when I was awakened by the sounds of angry voices, and amongst them I could distinguish those of my chair-bearers and Sing-Hoo." (Fortune, 1852: 289)

In Pǔchéng Fortune told Sing-Hoo to pay off the chair-bearers and coolie he had hired at Wǔyíshān.[240] Sing-Hoo said he had done so. However, in the middle of the night Fortune was awoken by a heated argument between Sing-Hoo and the porters.[241] Sing-Hoo was up against a wall trying to ward off his attackers with a large joss stick, which reminded Fortune of the way Bailie Nicol Jarvie had defended himself with the red-hot coulter of a ploughshare in "Rob Roy", one of the Waverley novels published in 1817 by Sir Walter Scott (1771 – 1832).[242] Fortune was able to separate the combatants and discovered that Sing-Hoo had not paid the chair-bearers in full.[243] Although Fortune ordered Sing-Hoo to pay the difference, the chair-bearers were so incensed that they made sure that Sing-Hoo was unable to hire any new porters in Pǔchéng.[244] As a result Sing-Hoo and Fortune had to carry their own belongings until they were well out of the township.[245] On the way the bamboo cane which Sing-Hoo used as a yoke to carry the luggage snapped.[246] Since they were in the middle of nowhere, Sing-Hoo had to go back to another part of Pǔchéng to find coolies and another chair.[247] This took about an hour.[248] In the meantime Fortune found shelter from the rain in a shed crowded with beggars, a situation hardly to his liking.[249]

Once the new men had arrived, the party set off in the direction of Xiānyáng ("Immortal Town South of Hills").[250] From his chair Fortune noted that rice and tobacco were the two main crops, while *Triadica sebifera* was cultivated for its tallow.[251] That evening they arrived at Tsong-so (presumably Zhōngxìn).[252] The next day they crossed 3 passes with Buddhist temples and adjoining tea-houses.[253] At the border they were queried by a mandarin, who wanted to know where they had come from and where they were going.[254] The answers he received from Sing-Hoo seem to have satisfied him, as the party was allowed to proceed into Zhèjiāng Province.[255] Just after crossing the border one of the bamboo levers supporting Fortune's chair snapped.[256] While the chair-bearers went to some cottages to get a replacement, Fortune walked ahead, admiring a red-flowered *Spiraea* (*S. japonica*) on the way.[257] Fortune waited on the pass for the party to catch up with him.[258] They reached Èrshíbādū ("Twenty-eighth Metropolis", **Fig. 5. 56**) just as a thunderstorm broke.[259]

Fig. 5.56 Street scene in Èrshíbādū, where Fortune sought shelter from a thunderstorm.

The next day they arrived at the Buddhist temple of Guāndì Miào ("Guān Yǔ Temple") perched on a hillside among tall *Cryptomerias*, large *Cinnamomum camphora*, evergreen "oaks" (*Castanopsis sclerophylla* and *Lithocarpus glaber*), fine thickets of bamboo and many shrubs (azaleas, hydrangeas, roses and *Spiraea japonica* (Syn. *S. callosa*) and *S. cantoniensis* (Syn. *S. reevesiana*).[260] Here Fortune observed a Buddhist service, and here he pondered the Christianization of China.[261] Among the rice paddies at the bottom of the hill, Fortune observed the manufacture of paper from bamboo stems.[262] They spent the night in the main inn at Xiákǒu ("Gorge Entrance"), where they were entertained by songs sung by the owner's daughter.[263] Between Xiákǒu and Qīnghú ("Clear Lake") Fortune's party was held up for some time by a mandarin and his entourage on their way from Běijīng to Fúzhōu.[264]

Qīnghú, which lies on a major artery, the Jiāngshān Gǎng, was then a bustling town concerned with the transshipment of basket teas.[265] Here Fortune took care to pay the chair-bearers and coolie himself, before engaging a boat to take Sing-Hoo and himself downstream to Yìqiáo.[266] On the way they stopped at Lánxī (**Fig. 5.40**) and spent a day at Méichéng (**Fig. 5.14**) to procure some plants of the weeping cypress (*Cupressus funebris*) for Beale's garden in Shànghǎi.[267]

5.5 Preparing for India

After his journey of almost 3 months, Fortune relaxed at Beale's house until the end of September.[268] The next two months were spent preparing the tea seeds and young tea plants from Sōngluó (Ānhuī), Wǔyíshān (Fújiàn), and various places in Zhèjiāng (Jīntáng, Zhōushān, Níngbō area) for their long journey to India.[269] They were packed in Wardian Cases and sent by four different ships from Hong Kong to Kolkata (Calcutta).[270] The first few months of 1850 must have been a trying time, as he waited anxiously for news of his tea plants. It was to be the summer of 1850 before he learned that his plants had arrived safely in the Himalayas.[271]

Fortune returned from Hong Kong to Shànghǎi in April 1850 to engage some first-rate tea manufacturers, procure a supply of implements used in tea manufacture, and make another large collection of tea plants.[272] It would have been easy to hire people from the coastal regions of China (Guǎngzhōu, Xiàmén), but it was an entirely different thing with men from inland parts.[273] He left the matter in the hands of Beale's Chinese compradore.[274] In June 1850 Fortune went to the Níngbō area to make arrangements for another supply of seeds and young plants from this part of Zhèjiāng.[275] He stayed at a monastery (presumably Tiāntóng).[276] At the end of June, as the weather became hotter, he moved to Zhōushān Dǎo.[277] The capital, which had reverted to Chinese soverainty, showed no evidence of the quaint English signboards in front of the shops.[278] In the meantime, it had expanded greatly, so Fortune had difficulty locating some of the old houses.[279] Eventually he discovered the large British hospital, which had been converted into a kind of customhouse.[280] He found an old mandarin, who offered him a room next to his.[281] Unfortunately, the mandarin was an inveterate opium-smoker, so Fortune had to suffer the sickly fumes.[282] During his time on Zhōushān, he paid a visit to a yángméi plantation in central Zhōushān.[283]

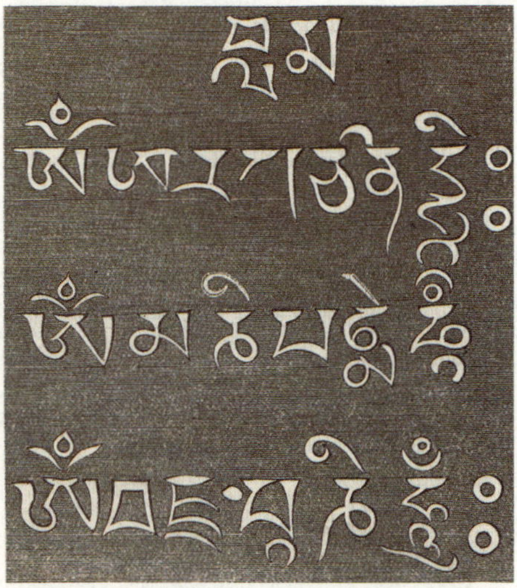

After staying a few days on Zhōushān he decided to visit the Buddhist island of Pǔtuó to escape the heat and investigate a couple of carved stones he had seen by the roadside (**Fig. 5.57**).[284]

Fig. 5.57 Stone inscribed with Tibetan sutras on Pútuó. "I applied to some of the *most learned* priests in Poo-too, but without success. They could neither read them, nor could they give me the slightest information as to how they came to be placed there." (Fortune, 1852: 347). In 2007 the priests at Pǔjì Sì did not know of its whereabouts.

None of the learned priests could decipher the inscriptions.[285] Fortune rightly surmised that these were written in "some northern Indian language".[286] They are, in fact, Tibetan sutras, presumably brought to Pǔtuó some time after Buddhism was established on the island in the Táng Dynasty.[287] It is said that an Indian monk visited Pǔtuó in the late 9th Century.[288] Repeated inquiries on Pǔtuó in 2007 failed to discover the present location of these stones.[289] Having made copies of the inscriptions, Fortune proceeded to the Pǔjì Sì ("Universal Salvation Temple"; **Fig. 5.58**), one of the three major temples on Pǔtuó Shān.[290]

This temple, which was originally built in 916 AD, received numerous donations of land and money during the Sòng and Yuán (1260 - 1368 AD) Dynasties, but fared less well during the turbulent Míng Dynasty.[291] Attacks on the temple during the Míng Dynasty may be the real reason why the main gate has only been opened on rare occasions. However, there is a popular legend that the main gate has been kept closed ever since Emperor Qiánlóng (1711 - 1799), disguised as a pilgrim, was denied entry.[292] Fortune saw a Chinaman surreptitiously fishing in the lotus-filled lake in front of the temple.[293] The fisherman was soon discovered by some of the monks who chased him away with blows inflicted by a bamboo

Fig. 5. 58 Pǔjì Sì (Universal Salvation Temple), Pǔtuó. According to a legend, the main gate has been kept closed since Emperor Qiánlóng (1711 - 1799) was refused entry.

cane.[294] Since Fortune's time the quality of the water has deteriorated to such an extent that the living lotus plants have had to be replaced by plastic counterparts (**Fig. 5.59**).[295]

Fig. 5.59 The lotus lake in front of Pǔjì Sì. "As I stood on the little romantic bridge I looked to the right and left; my eye rested on thousands of these flowers, some of which were white, others red, and all were rising out of the water and standing above the beautiful clear green foliage." (Fortune, 1852: 350) No doubt the plantsman would have been horrified by their plastic replacements.

Fig. 5. 60 Paddle-steamer "Lady Mary Wood", which transported Fortune and his tea manufacturers from Hong Kong to Kolkata.

At the beginning of September 1850 Fortune returned to Níngbō.[296] "Having procured a large quantity of tea-seeds and young plants", he left Níngbō for Shànghǎi at the end of December.[297] Beale's compradore had been successful in hiring six first-rate tea manufacturers, two lead-men, and assembling a large assortment of implements for the manufacture of tea, so nothing remained for him to do but pack the Wardian Cases, which had been prepared.[298] Fortune sowed the tea seeds between rows of young plants, a method which he had successfully employed in 1849.[299] On the 16 February 1851 Fortune left Shànghǎi for Hong Kong with 16 Wardian Cases, his tea manufacturers and their implements.[300] With a northeasterly wind blowing, the sailing ship "Island Queen" (Captain McFarlane) reached Hong Kong in four days, where Fortune and his men immediately transferred to a paddle-steamer, the "Lady Mary Wood" (**Box 5.10**; **Fig. 5.60**) with destination Kolkata.[301]

Box 5.10

SS Lady Mary Wood (1842 – 1858) This paddle-steamer, specially built for the Peninsular and Oriental Steam Navigation Company (P&O) by Thomas Wilson & Co. (Liverpool), was named after Lady Mary Wood (1807 – 1884), the wife of Sir Charles Wood (1800 – 1885), Secretary to the Admiralty from 1835 to 1839.[1] The ship was designed to carry 200 tons of cargo in addition to 110 passengers (60 first class, 50 second class) and 89 members of crew.[2] As a protection against pirates, she was "efficiently armed".[3] During her 17 years of service for P&O, she witnessed some high drama. During her maiden voyage in 1842, she was damaged on rocks off Cape Tourinan (Torinana, NW Spain), and in 1844 narrowly avoided being run down at night by a sailing ship.[4] The novelist William Makepeace Thackeray (1811 – 1863) travelled in her as far as Gibraltar during his all-expenses-paid tour of the Mediterranean in 1844.[5] That same year the "Lady Mary Wood" left Southampton to open a mail service between Galle (Ceylon) and Hong Kong.[6] The ship was at Galle at the end of July 1848 when the Sinhalese rose against the oppressive taxes, which the British Government had imposed on dogs, carts, guns and shops on 1 July 1848.[7] With her engines running at top speed (12 knots) the "Lady Mary Wood" was rushed to Madras to pick up a few hundred soldiers. She returned to Trincomalee (NE Ceylon) in a matter of days and thus helped to contain the revolt.[8] Shortly before Fortune and the tea manufacturers boarded her, the "Lady Mary Wood" had seen service between Shànghǎi and Hong Kong.[9] Once the ship reached Kolkata on 15 March 1851, she was laid up for almost eight months.[10] In October 1854 she unsuccessfully searched for a Frenchwoman known as Fanny Loviot, who had been abducted by pirates after the Chilean barque "Caldera" was wrecked by a typhoon south of Macao.[11] Finally in December 1858 the "Lady Mary Wood" was sold to E. C. Wermuth, C. S. van Heeckeren & Co. (Semarang, Java) and renamed the "Oenarang".[12] She ran briefly between the Dutch East Indies (now Indonesia) and China, before being sold to W. Cores de Vries (Batavia) in 1862.[13] She was finally broken up in 1867.[14]

5.6 Tour of Inspection in Northern India

They arrived in Kolkata on 15 March 1851 and stayed with Hugh Falconer (**Box 5.7**), the Scottish superintendent of the renowned botanical garden (**Fig. 5.62**).[302] When the Wardian Cases were opened, "the young tea plants were found to be in a good condition. The seeds which had been sown between the rows were also just beginning to germinate." [303] It was here that the tea manufacturers used *Pongamia pinnata* (Syn. *P. glabra*, **Fig. 5.63**) to show Falconer, Mr. Bethune and others how tea was prepared from fresh leaves (**Figs. 5.64 – 5.66**).[304] Fortune was enthralled by the beauty of *Amherstia nobilis* and *Saraca asoca* (Syn. *Jonesia asoca*), which were in full bloom at the time[305], but lamented that the landscaping of the garden by William Griffith (1810 – 1845; **Fig. 5.67**) prior to Falconer's appointment had resulted in the destruction of a number of valuable trees planted between 1793 and 1842 when William Roxburgh (1751 – 1815) and Nathaniel Wallich (1786 – 1854) were in charge.[306] Curiously enough, he makes no mention of the world-renowned banyan tree (*Ficus benghalensis*, **Fig. 5.68**), which was already about one hundred years old at the time of his visit.[307]

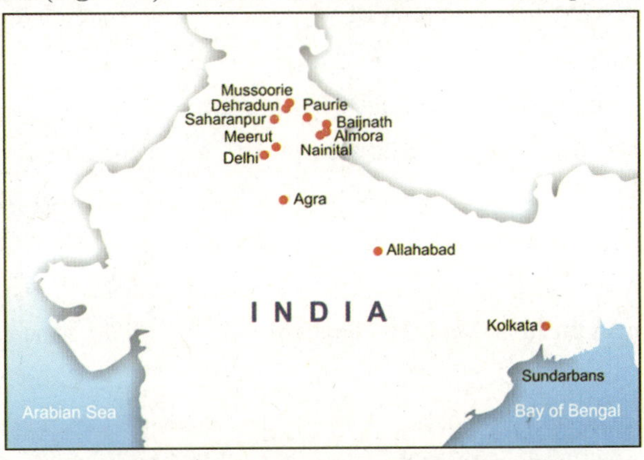

Fig. 5.61 Map of the places in India visited by Fortune during his first tour of inspection. The dot for Nainital also covers Bhimtal.

Fig. 5.62 Kolkata Botanical Garden.

Fig. 5.63 *Pongamia pinnata*, which was used in lieu of *Camellia sinensis* to show how tea is prepared.

Fig. 5.64~5.66 The traditional method of drying tea by hand, which the Chinese tea manufacturers would have employed. To remove the moisture, the leaves were first dried in the sun, then roasted and kneaded in a pan.

Fig. 5.67 William Griffith (1810 - 1845), who, during his short term as Superintendent of Kolkata Botanical Garden, attempted to create more systematic order. This meant felling a number of valuable trees, something which appalled Fortune.

Once an additional 14 Wardian Cases had been prepared for the tea seeds which had germinated in the two cases with camellias, Fortune boarded a small river steamer and set off for Allahabad on 25 March 1851.[308] Because the Hooghly River (**Fig. 5.69**) is shallow during the dry season, they had to travel downstream to the Bay of Bengal, before entering the Ganga (Ganges).[309] In this way Fortune was able to see the tiger-infested Sundarbans (Sunderbunds, **Fig. 5.70**), with their extensive mangrove vegetation.[310] The steamer finally docked in Allahabad (**Fig. 5.71**) on 14 April 1851.[311] Once Fortune had made arrangements for the Wardian Cases to be sent on, and the Chinese tea manufacturers and their belongings had been loaded onto nine bullock carts, he set off for the Saharanpur (Saharunpore) Botanical Garden (**Box 5.11**; **Figs. 5.72**, **5.73**) in an Indian government carriage.[312]

Fig. 5.68 The world-renowned banyan tree (*Ficus benghalensis*) in Kolkata Botanical Garden.

Fig. 5.69 The Hooghly River, a branch of the mighty Ganges.

Fig. 5.70 Tiger-infested Sundarbans. "I was much struck with the dense vegetation of the Sunderbunds. The trees are low and shrubby in appearance; they grow close to the water's edge, and many dip their branches into the stream." (Fortune, 1852: 361)

Fig. 5.71 Allahabad, which Fortune reached on 14 April 1851.

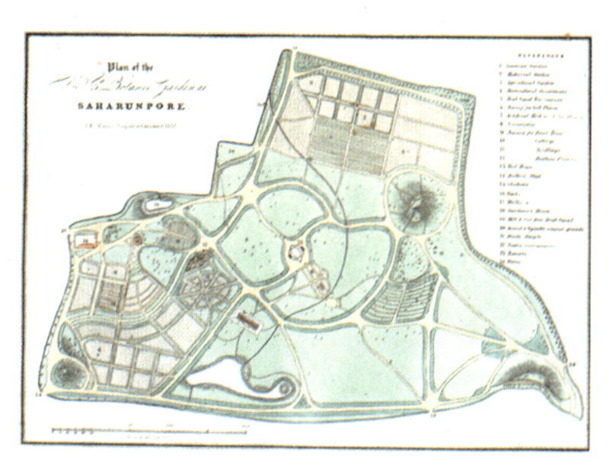

Fig. 5.72 Royle's plan of Saharanpur Botanical Garden (1831). "It contains a large collection of ornamental and useful plants suited to the climate of this part of India, and they are propagated and distributed in the most liberal manner to all applicants." (Fortune, 1852: 363)

Fig. 5.73 Potted plants at Saharanpur Botanical Garden. Note the *Bougainvillea* in the background.

Box 5.11

Saharanpur Botanical Garden The agricultural centre of Saharanpur lies on the North Indian Plain, some 50 km from the Himalayan foothills. The town has a long and turbulent history. As the Mughal Empire declined at the beginning of the 18th Century, it fell into the hands of the Sikhs.[1] Many of those connected with the garden in its early years met an unnatural death. Minister Intizam ud Daula (Intizam ud Dowlah), who founded the garden some time before 1750, was strangled to death on 30 November 1759.[2] This pleasure garden was further developed by the avaricious Rohilla chief Ghulam Qadir Khan, who took the Mughal emperor Shah Alam II (1728 – 1806) hostage in July 1788 in the hope of obtaining his treasure.[3] Qadir himself was captured by the Maratha warlord Mahadji Sindhia (Mahadaji Shinde, Mahaji Sindhia; 1730 – 1794) in December 1788, tortured, and executed shortly thereafter.[4] Internal fighting between the various factions hastened the takeover by the British forces. Saharanpur became incorporated into the British Raj in 1803.[5] In 1817 the East India Company (EIC) purchased the 64 ha garden for 900 rupees, and appointed Dr. George Govan (1787 – 1865) as its first Superintendant in 1819.[6] Under the guidance of a succession of able superintendants (e.g. John Forbes Royle in 1823 – 1831; Hugh Falconer in 1832 – 1841; William Jameson in 1842 – 1875; John Firminger Duthie, 1875 – 1903; Amos C. Hartless, ca. 1906 – 1920) seeds of northern Indian plants were sent all over the world in exchange for those of exotic plants which were tested for their suitability for the Indian Subcontinent.[7] As a result, the Saharanpur Botanical Garden, or Company Bagh as it is sometimes called, acquired a rich collection of exotic trees and became the most important garden after the Botanical Garden in Kolkata. Unfortunately, many of its trees were destroyed by Indian troops during World War II.[8] After Independence from Britain in 1947 the local government decided to establish the Plain Fruit Research Station here.[9] In 1957 this station became a fully fledged Horticultural Research Institute.[10] Since 1976 horticultural courses are given, and the institute is now officially referred to as the Horticultural Experiment and Training Centre.[11] It presently specializes in vegetables, fruit plants, and certain ornamentals (e.g. *Bougainvillea* cultivars).

Soon after his arrival in Saharanpur Fortune received orders to draw up a report on all the existing tea plantations in Garhwal (Gurhwal) and Kumaon.[313] With this object in mind, he, Dr. William Jameson (**Box 5.8**) and Jameson's wife Emily first travelled to Dehradun (Deyra Doon) at the foot of the Himalayas.[314] They visited the tea estate of Kaolagir (**Fig. 5.74**), which is situated on a piece of flat, low-lying land, that does not receive much moisture during the dry season.[315]

As Fortune pointed out, the clayey soil tends to get baked during dry weather and therefore had to be irrigated.[316] To facilitate this, the tea shrubs were planted in trenches.[317] Because of the far from ideal microclimate and edaphic conditions, the tea plants did not display the fresh and vigorous appearance that Fortune was accustomed to seeing in China.[318]

Fig. 5.74 Kaolagir Plantation, Dehradun, where it is necessary to irrigate the tea plants. "The plants generally did not appear to me to be in that fresh and vigorous condition which I had been accustomed to see in good Chinese plantations." (Fortune, 1852: 370)

From Dehradun Fortune and the Jamesons followed the mule trail ("Kipling Road") from Dehradun to the hill station of Mussoorie (Mussooree; **Fig. 5.75**) at about 2,000 m.[319] As they ascended the steep slopes, the dense sal (*Shorea robusta*) forests gave way to a mixed deciduous monsoon woodland vegetation.[320] Mussoorie is not only cooler, but also wetter than Dehradun, as indicated by the lichen-draped trees (**Fig. 5.76**). Just below Mussoorie the slopes had been terrassed for the cultivation of crops.[321] Today wheat is grown on these terraces.

Fig. 5.75 View of the hills at Mussoorie.

Fig. 5.76 Lichen-covered tree at Mussoorie (2,000 m), reflecting a humid microclimate suitable for growing tea.

While Fortune was in Mussoorie, he visited the small botanical garden (**Fig. 5.77**) which had been established by the Indian Government in 1826 at the suggestion of John Forbes Royle (**Box 4.1**).[322] The idea was to have a garden at high altitude in which to test the suitability of Himalayan trees and shrubs for the European market, and at the same time to provide a centre from which to distribute useful European plants to the Indian population.[323] By the time Fortune visited the garden in May 1851 it could boast a fair collection of native conifers (*Cedrus deodara*, *Cupressus torulosa*, *Picea smithiana* [Syn. *Abies smithiana*], *Pinus* spp.), fruit trees introduced by Royle (**Box 4.1**) and Falconer (**Box 5.7**) from Kashmir, as well as apples, pears, plums and almonds collected by Jameson (**Box 5.8**) in Kabul.[324] In addition, there were many English fruit trees and bushes (apples, pears, plums, raspberries etc.), and some apple and pear trees, which had been sent from the USA in an ice freighter.[325] Although it was not intended as an ornamental garden, some garden flowers (e.g. dahlias, fuchsias, mimuluses, pelargoniums, pinks and violets) were also being propagated.[326] Both Fortune and Royle would be shocked to learn that today these gardens with their playgrounds and boating lake belong to the municipality, and are of little botanical interest.

Fig. 5.77 Botanical Garden Mussoorie, "the celebrated garden from which so many interesting Himalayan trees and shrubs had found their way to Europe." (Fortune, 1852: 364)

Fig. 5.78 Terraces near Dhanaulti, on the route from Mussoorie to Paurie traversed by Fortune in 1851.

From Mussoorie, the party headed through some spectacular country (**Fig. 5.78**) towards the township of Paurie (Paorie). Progress was slow. Even today with better roads and cars, it is frequently impossible to drive faster than 25 km/hour. Just below Paurie they visited the Gadoli (Guddowli) Plantation (**Fig. 5.79**), which had been established by Jameson in 1844, with the help of Chinese tea manufacturers originally from Guǎngzhōu (Canton) in southern China.[327] Fortune described the plantation as "most promising", because the soil was suitable, drainage was good, and only a little irrigation of the terraces was necessary during the dry season.[328] The tea manufacturers, who had accompanied Fortune all the way from China, were settled here.[329] When

Fig. 5.79 Abandoned terraces at the Gadoli Tea Plantation established by William Jameson in 1844. "This plantation is a most promising one, and I have no doubt will be very valuable in a few years. The plants are growing admirably, and evidently like their situation." (Fortune, 1852: 373)

Fortune left, they got up early to see him off.[330] Although tea is no longer grown near Paurie, the mongoloid features of some of the local population, the overgrown terraces, and the ruins of a tea factory (**Fig. 5.80**) bear silent witness to a once-thriving industry. Production stopped in the 1920s as a result of a drainage of manpower during World War I.[331]

From Paurie Fortune and the Jamesons headed on to the tea plantations in the vicinity of Almora (Almorah).[332] They arrived at the Hawalbagh (Hawulbaugh) Plantation (1844; **Fig. 5.81**), 10 km northwest of Almora, on 29 June 1851.[333] The tea shrubs, which had been planted in a rich sandy loam on undulating ground, only required a limited amount of irrigation.[334] In general the plants appeared to be healthy, although there had been a certain amount of overplucking (**Fig. 5.82**).[335] In 1864 Henry Ramsay (**Box 5.13**; **Fig. 5.83**) transferred this governmental plantation to the leper asylum in Almora as a source of income.[336] This once-thriving plantation became less productive in the course of time. In 1908 it only produced 3,326 kg of black tea and 543 kg green tea.[337] In 1952 it was taken over by the Bengali plant physiologist and mystic Boshi Sen (1887 – 1971) for agricultural research.[338] The Boshi Sen Field Research Laboratory is part of the Vivekananda Institute of Hill Agriculture since 1974.[339] Fortune also visited two small plantations below Almora, namely the Lachmeshwar (Lutchmisser) and Kuppeena Plantations. Although he commented favourably on the good drainage and light sandy soil, these plantations no longer exist. Lachmeshwar Plantation, which was established by Hugh Falconer (**Box 5.7**) in 1835, ceased production between 1920 and 1925.[340]

Fig. 5.80 Ruins of the tea factory, Gadoli Tea Plantation.

Fig. 5.81 Hawalbagh Plantation (established in 1844), now the Vivekananda Institute of Hill Agriculture. "The land is of an undulating character, consisting of gentle slopes and terraces, and reminded me of some of the best tea-districts in China." (Fortune, 1852: 375)

Fig. 5.82 Only the immature leaves are plucked to produce the best tea.

Having examined the Indian government's plantations, Fortune went on to examine some estates belonging to the local landlords (zamindars), who were patronized by the Commissioner, John H. Batten (**Box 5.12**), and the Assistant Commissioner (Henry Ramsay, **Box 5.13**) of Garhwal and Kumaon.[341]

Box 5.12

Sir John Hallet Batten FRGS (1811 – 1886) Eldest son of Rev. Joseph Hallet Batten FRS (1778 – 1837), Principal of the East India Company College (Haileybury, Hertfordshire), and his wife Catherine Maxwell (1786 – 1862).[1] He was educated at Haileybury, Charterhouse in Surrey and Trinity College, Cambridge.[2] He entered the Bengal Civil Service in 1828 as a writer.[3] He was Assistant to the Collector and Magistrate at Azamgarh in 1833, spent a short time in the same function at Saharanpur in 1835, before becoming Assistant Commissioner to the callous Colonel G. E. Gowan, who tacitly approved of the ordeal by hot iron, in 1836.[4] In 1839 George Thomas Lushington (1806 – 1849) replaced Gowan as Commissioner of Kumaon.[5] In 1842 Batten was appointed Settlement Officer for Garhwal.[6] This included the compilation of a detailed revenue and rent roll based on a personal inspection of the villages.[7] His talent for organization was rewarded, when he was put in charge of the settlement of Kumaon.[8] Shortly before Lushington's death in 1849, Batten became Commissioner of Kumaon.[9] As commissioner from 1848 to 1855, he had a strong impact on the administration, making sure that the laws were fully implemented.[10] During the Indian Mutiny (**Box 6.4**) he was persuaded to stay at Nainital by his successor, Henry Ramsay (**Box 5.13**).[11] After the Indian Mutiny he was Judge of Kanpur in Cawnpore, 1858, then Judge at Mainpuri in Mynpoory, 1860, before becoming Commissioner of Agra from 1863 to 1866).[12] After he retired from the Bengal Civil Service in 1866, he returned to Penzance, where he became Honorary Secretary of the Royal Geological Society of Cornwall.[13] He finally moved to Exeter, where he got involved in various projects designed to improve the city and county.[14]

Box 5.13

Major General Sir Henry Ramsay (1816 – 1893). Fifth son of Lt.-Gen. Hon. John Ramsay (1775 – 1842) and Mary Delise (1780 – 1843).[1] He was educated at the Edinburgh Academy (1829 – 1832).[2] In 1834 he was nominated for the Bengal Infantry by John Lock on behalf of his mother.[3] He served in the army until 8 August 1840, when he embarked on an administrative career as Junior Assistant-Commissioner, Kumaon Division.[4] One of the first things he did in this function was to have some stone huts built in Almora for the lepers.[5] Realizing as a Scot the importance of a basic education, he and Rev. John Henry Budden (1813 – 1890) established a high school in Almora in 1850.[6] On 11 November 1850 he married Laura Lushington (1829 – 1914), the daughter of the former Commissioner of Kumaon, George Thomas Lushington (**Box 5.12**) and his wife Marianne Gordon.[7] They had two sons and three daughters.[8] In 1850 he was put in charge of the development of the mosquito- and robber-infested

Fig. 5. 83 Sir Henry Ramsay (1816 – 1893), who encouraged the local tea industry.

Terai at the base of the Himalayan foothills.[9] He set about making this area habitable by supervising the clearing of land for fields, the draining of marshes, the digging of irrigation canals, and the building of roads between the villages.[10] In this way the Terai became the granary of Uttar Pradesh.[11] He also advocated the development of tea plantations in Kumaon.[12] When Fortune came to Kumaon in 1851 Ramsay, who was then Senior Assistant-Commissioner of Kumaon, showed him two tea estates which he was in charge of. Here the villagers were not required to pay revenue tax, but help with ploughing, preparation of the soil and manuring of the plants.[13] As a man concerned with the welfare of the population, he organized smallpox vaccinations all over Kumaon in 1854.[14] When Batten (**Box 5.12**) resigned as Commissioner of Kumaon, Ramsay reigned from 22 December 1855, to 10 February 1856.[15] As freshly appointed Commissioner, he felt obliged to issue an order prohibiting the indiscriminate felling of trees in his district.[16]

He had been Commissioner for little over a year when the Indian Mutiny (**Box 6.4**) broke out. He snuffed out any insurrection on the home front by proclaiming martial law.[17] There were some attacks from the south, but these were repulsed by Ramsay's Gurkhas.[18] As a whole Kumaon remained quiet throughout the hostilities, which led to a large influx of people into the area.[19] This put even more pressure on the natural resources. With the expansion of the railway network from 1855 onwards, sal trees (*Shorea robusta*) were in demand for sleepers, while in the hill country the forest was making way for tea estates.[20] From 1859 to 1880 the land under tea cultivation increased 24-fold from 166 acres (= 67 ha) to 3,977 acres (= 1,610 ha).[21] Unfortunately, the success of the tea plantations and the consequent expansion in the population led to the further destruction of the native forest and erosion of the slopes.[22] In order to protect the forests, Ramsay placed these under his direct jurisdiction from 1858 to 1868. [23] Ramsay took his work seriously, and visited nearly all the forests in person in order to acquaint himself with the status quo.[24] No doubt because he could foresee the consequences, Ramsay opposed the settlement of Kumaon by British subjects. Unfortunately, this was not to be. Inevitably the presence of the hill stations and the consequent demand for building materials, fuel and food led to the importation of fast-growing eucalypts and acacias and the introduction of exotic animals to replace those hunted to extinction.[25] In this way the appearance and ecology of the region were transformed within decades.

During the more than 28 years he was Commissioner of Kumaon, Ramsay became something of a figurehead. The inhabitants referred to him as the "King of Kumaon".[26] While he was certainly autocratic, he was by no means dictatorial, but rather more like a patriarch.[27] Having learnt the language of the hill people (Paharis), he was able to communicate freely with them.[28] His hands-on approach to problems, meant that he had little repect for the rules and regulations drawn up by bureaucrats.[29] Consequently he tended to judge legal cases brought before him somewhat subjectively.[30]

When Ramsay retired in 1884, it was his wish to stay in Kumaon for the rest of his life.[31] He bought the Khali Estate between Almora and Bageshwar and had a cottage erected in the forest.[32] Yet another of his dreams was dashed, when his two sons (Henry Lushington Ramsay, 1854 – 1928 and John Ramsay, 1862 – 1942) came to take him back to Britain in 1892.[33] From his enforced exile he yearned for his beloved Kumaon.[34] He died heartbroken in 1893.

One of these was at a place called Lohba, 80 km west of Almora.[342] Although the undulating slopes were ideally suited for growing tea, the plants had been liberally watered, with the result that nearly all the plants had died.[343] The second set of Zemindaree plantations were in the hills near Baijnath (Byznath). Encouraged by Henry Ramsay (**Box 5.13**) the villagers ploughed the land, prepared it for the tea shrubs and manured the bushes in lieu of paying revenue tax.[344] As a result, the two plantations were flourishing. Fortune said that he "never saw, even in the most favoured districts of China, any plantations looking better than these."[345] Tea plantations are still to be found in this area (**Fig. 5.84**).

Fig. 5. 84 Picking tea on the Ritha Tea Estate between Gwaldam and Baijnath. The tea plants are protected from direct sunlight by a canopy of *Pinus roxburghii*.

In the middle of July Fortune proceeded on to Bhimtal (Bheem Tal) to examine four plantations.[346] The Anoo- and adjoining Kooasur Plantations occupied low flat land, which was unsuitable for tea.[347] Today the Anoo Plantation (**Figs. 5. 85, 5.86**) is used for growing sweet chestnut, which is fed to moth larvae to produce tussah silk.[348] The Bhurtpur (Bhurtpoor) Plantation (**Fig. 5.87**), the second plantation established by Hugh Falconer (**Box 5.7**) in 1835, on the other hand, was situated on a hillock with a light loamy soil containing pieces of Precambrian Nagthat Schist. Both the habitat and soil were ideal, and tea thrived here up until Independence from Britain in 1947.[349] The final plantation Fortune visited in Bhimtal was the Russia Plantation. The land had been terraced, but the plants were unfortunately suffering from a combination of overwatering and overplucking.[350]

Fig. 5.85 The Anoo Plantation. "The plants do not seem healthy or vigorous; many of them have died out, and few are in that state which tea-plants ought to be in. Such situations never ought to be chosen for tea-cultivation." (Fortune, 1852: 382) The tea plants have since been replaced by sweet chestnut with which the moth larvae used to produce tussah silk are fed.

Fig. 5. 86 Detail of a pollarded sweet chestnut at the Anoo Plantation.

In Bhimtal Fortune parted company from the Jamesons, who had to get back to the Hawalbagh Plantation near Almora.[351] Fortune was joined by John H. Batten (**Box 5. 12**),

Commissioner of Kumaon.[352] Together they crossed the hills to Nainital (Nainee Tal, **Fig. 5.88**), "one of the prettiest stations I have seen in the Himalayas." [353] On the way Fortune pointed out to the Commissioner many tracts of land suitable for tea cultivation.[354]

On 28 July 1851 Fortune finally set off on his homeward journey.[355] He visited Meerut (**Fig. 5.89**), the scene of the Sepoy Uprising in 1857 (**Box 6.4**), and passed through Delhi on his way to Agra.[356] He presumably saw the Taj Mahal, but to Fortune such edifaces paled in comparison with the grandeur of mountain ranges like the snow-capped Himalayas.[357] After a trip down the Ganga, he arrived back in Kolkata on the 29 August 1851.[358] He again stayed with Hugh Falconer (**Box 5.7**), until his ship was ready to depart for England.[359] While he was waiting at the Botanical Gardens, he admired the giant water-lily *Victoria* (**Fig. 5.90**) flowering outdoors in one of the ponds.[360] The plant would have had a special meaning for Fortune, as the genus was first described by his mentor, John Lindley (**Box 2.1**).[361] The enormous leaves with their upturned margins and stout veins can support the weight of a child, as Joseph Paxton (**Box 5.14**; **Fig. 5.91**) was quick to show.[362] This caught the public's imagination, so most botanic gardens felt obliged to grow it.[363] Although Fortune does not mention this, it seems likely that he would have taken with him some seeds and plants from India for propagation in the UK. Hugh Falconer may have presented him with some of his favourites from the Botanical Garden.

Fig. 5.87 Bhurtpur Plantation, established by Hugh Falconer in 1835. "Both the situation and soil of this plantation are well adapted to the requirements of the tea-shrub, and consequently we find it succeeding here as well as at Guddowli, Hawulbaugh, Almorah, and other places where it is planted on the slopes of the hills." (Fortune, 1852: 383) Tea was grown here until 1947.

Fig. 5.88 Nainital, "one of the prettiest stations I have seen in the Himalayas. Its romantic-looking lake is almost surrounded by richly wooded mountains." (Fortune, 1852: 397)

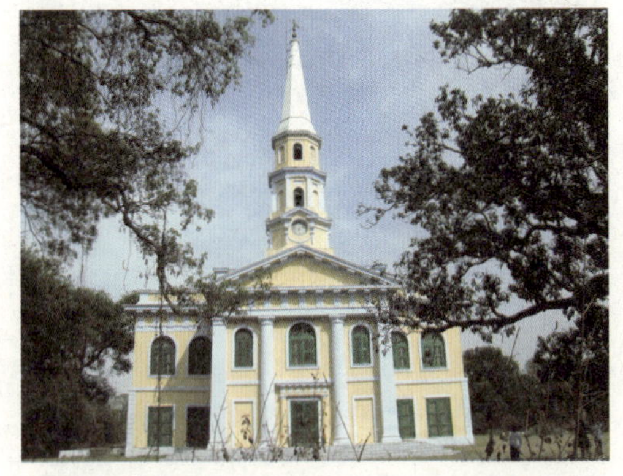

Fig. 5.89 St. John's Church (1819–1822), which Fortune must have seen when he passed through Meerut.

Fig. 5.90 ***Victoria amazonica*** **in Kolkata Botanical Garden. "On the 5th of September I had the pleasure of seeing the *Victoria regia* flower for the first time in India. It was growing luxuriantly in one of the ponds in the botanic garden, and no doubt will soon be a great ornament to Indian gardens." (Fortune, 1852: 398)**

Fig. 5.91 Sir Joseph Paxton (1803 – 1865), the gardener who, inspired by the *Victoria* waterlily, became an architect.

5.7 Home for a Year

Fortune arrived back in London in the autumn of 1851 just too late to visit what was described as "the major event of the century" [364], the Great Exhibition from 1 May to 11 October 1851. However, he would have heard about the success of the exhibition, and seen Paxton's "Crystal Palace" (**Box 5.14**) in Hyde Park, when he went into London to consult his publisher or visit Christie & Manson's auction house. He was probably chuffed that a "simple gardener" like himself had achieved such renown.

Box 5.14

Sir Joseph Paxton FLS (1803 – 1865) Seventh son and last of nine children of the farmer William Paxton (1759 – 1810) and his wife Ann (1761 – 1823).[1] At the age of 15 he worked for the eccentric Sir Gregory Osborne Page-Turner (1785 – 1843) of Battlesden House, near Woburn.[2] Two years later he became apprenticed to William Griffin, a skilful fruit-grower, gardener to the banker and MP Samuel Smith (1754 – 1834) of Woodhall Park (Wotton, Herts.).[3] In 1821 he returned to Battlesden until 1823, when he started work with the Duke of Somerset at Wimbledon.[4] However, when William George Spencer Cavendish, 6th Duke of Devonshire (1790 – 1858) leased the Chiswick Gardens to the Horticultural Society of London in 1823, Paxton moved there.[5] The energetic Paxton became foreman in the arboretum in 1824, but when the Duke of Devonshire spontaneously offered Paxton the post of Head Gardener (Superintendent) at his Derbyshire retreat, Chatsworth Estate in April 1826, Paxton jumped at the opportunity.[6] He fell in love with and married the housekeeper's niece, Sarah Bown (1800 – 1871), who managed their home with seven children so efficiently, that Paxton was able to devote all his time to work.[7] When the Duke returned from the coronation of Czar Nicholas I (1796 – 1855) in December 1826, he was impressed by the changes Paxton had already made to his run-down policies.[8] Some of the enthusiasm for horticulture and arboriculture rubbed off on the Duke, who then invested vast sums of money on a pinetum during 1829 – 1830s, arboretum in 1835, a rockery during 1843 – 1846 and various conservatories during 1834 – 1848, which he filled with exotic plants.[9] In 1849 a specially heated tank was built to house the South American water-lily, *Victoria regia*, which Sir William J. Hooker, the Director of Kew Gardens, presented to Chatsworth.[10] Paxton and the Duke became close friends, and undertook a number of tours together.[11] In 1838 – 1839 after the Duke and his mistress separated, they visited Switzerland, Italy, Malta, Greece and Turkey.[12] From the Duke Paxton gained an appreciation of art.[13] The Duke gave him a free hand to get involved in other projects.[14] The hyperactive Paxton published a number of journals, such as "The Horticultural Register and General Magazine" during 1831 – 1836, "Paxton's Magazine of Botany and Register of Flowering Plants" during 1833 – 1849, "Paxton's Flower Garden" during 1850 – 1853, and founded the long-running "Gardeners' Chronicle" in 1841 with John Lindley (**Box 2.1**), Charles Wentworth Dilke (1810 – 1869) and the publisher William Bradbury (1800 – 1869) to which Fortune contributed a large number of articles.[15] Paxton helped to establish the "Daily News" in 1846, and was author or co-author of three books.[16] With his innovative mind, Paxton became interested in railways early on.[17] In 1835 he started investing in railway shares, played a role in the formation of the Manchester, Buxton, Matlock and Midlands Junction Railway in 1845, and became a director of the Furness and Midland Railways in 1848.[18] Later, he even became involved in railway schemes in Spain, India, Mauritius and Argentina.[19] In addition, he designed various parks in N England, e.g. Prince's Park in Liverpool, Birkenhead Park, Tatton Park, Knutsford, Cheshire, People's Park in Halifax, Yorkshire, and S Scotland, such as Kelvingrove Park in Glasgow (originally West End Park), Queen's Park in Glasgow, Baxter Park in Dundee, and a number of stately houses, including Mentmore Towers and Château de Ferrières for the Rothschilds.[20] However, he is best remembered for designing the building to house the Great Exhibition of 1851 after more than 230 other designs had been rejected.[21] This huge prefabricated iron, wood and glass structure, which was based on the "Water-lily house" at Chatsworth, was completed in Hyde Park in about 7 months, and dubbed the "Palace of very crystal" by the writer Douglas William Jerrold (1803 – 1857).[22] The novel design earned him a knighthood.[23] In 1852 – 1854 Paxton supervised the dismantling and re-erection of the Crystal Palace on Sydenham Hill, and acquired most of the stock from Loddiges' Nursery for the Crystal Palace Park of which he was director.[24] He was elected Liberal Member of Parliament for Coventry, and as such helped silk workers, who had lost their jobs as a result of cheaper French imports, find alternative employment.[25] As MP he was responsible for sending 3700 navvies as an Army Work Corps to build roads, railways and depots in the Crimea, and designing better accomodation for the soldiers during the war with Russia.[26] During the "Great Stink" in 1858 he seconded a proposal to deal with London's sewage problem, and became Chairman of the Select Committee to decide on how to finance the scheme.[27] He did not always toe the party line, and helped to bring down Palmerston's government in 1857 over the bombardment of Guǎngzhōu during the "Arrow Incident" (**Box 6.1**).[28] When the Duke of Devonshire died on 18 January 1858, Paxton relinquished his position at Chatsworth.[29] As a sign of his respect, the next duke allowed him to stay in his house on the estate and granted him an annual salary of £ 500.[30] Plagued by gout and rheumatism, Paxton continued to work on various projects until heart and liver failure put an end to his busy life.[31] He bequeathed his books to the Royal Horticultural Society.[32]

After the warm weather in India, it must have felt chilly in London. The change cannot have done his rheumatism much good.

One of Fortune's first tasks would have been to check how the plants and seeds he had sent from China between 1848 and 1851 were doing. This meant going to see John Standish (**Box 5.15**; **Fig. 5.92**) at the Sunningdale Nursery just east of Ascot.

Fig. 5.92 John Standish (1814–1875), one of the nurserymen to whom Fortune entrusted his plant material.

Box 5.15

Mr. John Standish (1814 – 1875) Born on the Falconer's Hall Estate, near Foxholes, while his father was working for the flamboyant sportsman, Colonel Thomas Thornton (1757 – 1823).[1] At the age of twelve his father became forester to the moderate Whig politician, Henry Petty-Fitzmaurice (1780 – 1863), the Third Marquis of Lansdowne at Bowood House in Wiltshire.[2] The Bowood policies had been landscaped in 1761 – 1769 by Lancelot "Capability" Brown (1716 – 1783).[3] It was in this wonderful environment that young Standish trained to be a gardener.[4] Once he finished his apprenticeship, he became foreman at Bagshot Park (Surrey) under Andrew Toward, the gardener to Mary, the Duchess of Gloucester (1776 – 1857), George III's fourth daughter.[5] In the late 1830s Standish started his own nursery at Bagshot Bridge across the road from Bagshot Park, where he began hybridizing various plants.[6] In order to concentrate on his horticultural work, he joined forces with the businessman Charles Noble (1817 – 1898) in 1846, at which time the nursery became known as Standish & Noble. In 1847 with their joint capital, Standish & Noble bought a second nursery at Sunningdale (Berkshire), a few km from Bagshot, where the great variety of soils and terrain allowed many different trees and shrubs to be grown.[7] With seeds and plants pouring in from many parts of the world in the late 1840s and early 1850s, the firm flourished. Their descriptive "Catalogue of select hardy ornamental plants" included a number of the taxa Fortune had sent between 1848 and 1850.[8] The partners also wrote a bestseller on growing ornamental plants in 1852, which included information on the cultivation of rhododendrons.[9] They specialized in raising and marketing the plants from Fortune's second expedition to China, and the rhododendrons which Joseph Dalton Hooker (1817 – 1911) had discovered in Sikkim (1848 – 1851).[10] Standish found that he could induce some of the novel rhododendrons to flower prematurely by grafting scions onto old standard rhododendrons.[11] This reduced the time needed to produce hybrids. Unfortunately, Noble's bossy nature antagonized his partner, their staff and customers, so after little more than ten years the partnership was dissolved.[12] Standish established an entirely new nursery ("The Royal Nursery") on a more extensive scale (80 acres) at Ascot.[13] On 2 March 1861 he proudly advertised that he had "received all the plants collected by Mr. Fortune during his last voyage to China".[14] He certainly raised most of the plants that Fortune had collected in China and Japan.[15] Standish suffered from diabetes and this may have forced him to go into partnership once more. In 1875, the year in which he died, the firm was listed as "Standish & Ashby".[16] When he died, he left the little he had to his wife Lucy.[17]

As Standish & Noble's "Catalogue of select hardy ornamental plants" written in 1850 included a number of the taxa Fortune had sent from China, it would seem that at least some of his plants had taken root.[365] Among the plants were a number of novelties (**Box 5.16**; **Figs. 5.93 – 5.101**).

Fig. 5.93 ***Citrus trifoliata*, a thorny shrub. Unlike other members of the genus, it has compound, deciduous leaves.**

Fig. 5.94 ***Citrus trifoliata* in fruit.**

Fig. 5.95 ***Clematis lanuginosa*, a species introduced by Fortune, and used in breeding large-flowered cultivars.**

Fig. 5.96 ***Gentiana scabra*, a perennial gentian belonging to the Section Pneumonanthe with rough leaf margins and veins.**

Fig. 5.97 ***Rhododendron indicum*** **'Crispiflorum', an azalea with corrugated petals.**

Fig. 5.98 ***Rhododendron*** **x** ***obtusum*** **'Amoenum', a charming cultivar of this azalea hybrid.**

Fig. 5.99 ***Rhododendron simsii*** **'Bealei', an azalea named after Fortune's friend, Thomas Chaye Beale.**

Fig. 5.100 ***Rhododendron simsii*** **'Vittatum Punctatum', an azalea with striped and spotted petals.**

Fig. 5.101 ***Spiraea japonica*** **'Fortunei'. Fortune probably found this attractive shrub on his return from Wǔyíshān.**

Box 5.16

Fortune's Plant Introductions from China 1848 – 1851

The name/names in brackets is/are the name(s) under which Fortune's material was once known. Bold type indicates that this was the first time the taxon was grown outside China. A question mark indicates that Fortune may not have been the first person to introduce the plant.

Abelia uniflora R. Brown 1830[1]

***Camellia japonica* Linnaeus 1753 'Anemoniflora'**[2]

***Camellia japonica* Linnaeus 1753 'Fortune's Yellow'**[3]

***Castanopsis sclerophylla* (Lindley & Paxton 1850) Schottky 1912** (*Quercus sclerophylla* Lindley & Paxton 1850)[4]

Cephalotaxus fortunei* W. J. Hooker 1850 var. *fortunei (*Taxus fortunei* (W. J. Hooker 1850) C. Lawson 1851)[5]

? *Chionanthus retusus* Lindley & Paxton 1852[6]

***Citrus trifoliata* Linnaeus 1763** ("*Limonia trifoliata*", *Poncirus trifoliata* (Linnaeus 1863) Rafinesque 1838)[7]

***Clematis lanuginosa* Lindley 1853**[8]

Clerodendrum bungei Steudel 1840 (*Clerodendrum foetidum* Bunge 1833 non D. Don 1825)[9]

Cupressus funebris Endlicher 1847 (*Chamaecyparis funebris* (Endlicher 1847) Franco 1941)[10]

? *Eupatorium fortunei* Turczaninov 1851[11]

***Exochorda racemosa* (Fortune ex Lindley 1847) Rehder 1913** (*Amelanchier racemosa* Fortune ex Lindley 1847, *Spiraea grandiflora* W. J. Hooker 1854)[12]

***Gentiana scabra* Bunge 1835** (*Gentiana scabra* Bunge 1835 var. *fortunei* (J. D. Hooker 1854) Maximowicz 1888, *Gentiana fortunei* J. D. Hooker 1854)[13]

Ilex cornuta Lindley & Paxton 1850 (*Ilex fortunei* Lindley 1857)[14]

Ilex latifolia Thunberg 1784[15]

Juniperus chinensis Linnaeus 1767 var. *chinensis* (*Juniperus fortunei* Carrière 1855, *Juniperus sphaerica* Lindley 1850)[16]

? *Keteleeria fortunei* (A. Murray 1862) Carrière 1866 (*Abies fortunei* (A. Murray 1862) A. Murray 1863, *Picea fortunei* A. Murray 1862, *Pinus fortunei* (A. Murray 1862) Parlatore 1868, *Pseudotsuga fortunei* (A Murray 1862) McNab 1877)[17]

Lilium concolor Salisbury 1806 var. *concolor* (*Lilium sinicum* Lindley & Paxton 1851, *Lilium concolor* Salisbury 1806 var. *sinicum* (Lindley & Paxton 1851) J. D. Hooker 1872)[18]

Mahonia bealei (Fortune 1850) Carrière 1854 (*Berberis bealei* Fortune 1850, *Berberis consanguinea* Fortune 1852, *Berberis japonica* (Thunberg ex Murray 1784) R. Brown 1818 var. *bealei* (Fortune 1850) Skeels 1912)[19]

Mahonia trifurca (Lindley 1852) Loudon 1855 (*Berberis trifurca* Lindley 1852)[20]

***Rhododendron indicum* (Linnaeus 1753) Sweet 1830 'Crispiflorum'** (*Azalea crispiflora* W. J. Hooker 1853)[21]

Rhododendron mucronatum (Blume 1823) G. Don 1834 'Narcissiflorum' [*Azalea narcissiflora* Fortune 1850 ined., *Rhododendron narcissiflorum* (Fortune 1850) Planchon 1854][22]

***Rhododendron* x *obtusum* (Lindley 1846) Planchon 1853 'Amoenum'** (*Azalea amoena* Fortune ex Lindley 1852)[23]

***Rhododendron simsii* Planchon 1854 'Bealei'** [*Azalea bealei* Fortune 1852 [p.330: nomen nudum], *Azalea vittata bealei* Morren 1866, *Rhododendron simsii* Planchon 1854 var. *vittatum* (Planchon 1854) Wilson 1921 forma *bealei* Wilson 1921, *Rhododendron vittatum* Planchon 1854 var. *bealei* Hort. in Planchon 1854][24]

***Rhododendron simsii* Planchon 1854 'Vittatum Punctatum'** (*Azalea indica vittata punctata* Van Houtte 1854, *Azalea vittato-punctata* Lemaire 1854, *Rhododendron vittatum* Planchon 1854 var. *punctatum* Planchon 1854)[25]

***Skimmia reevesiana* (Fortune 1851) Fortune 1852** [*Ilex reevesiana* Fortune 1851, *Skimmia fortunei* Masters 1889, *Skimmia japonica* Thunberg 1783 subsp. *reevesiana* (Fortune 1851) Taylor & Airy-Shaw

1987][26]

Spiraea cantoniensis Loureiro 1790 'Lanceata' (*Spiraea cantoniensis* Loureiro 1790 'Flore Pleno')[27]

***Spiraea japonica* Linnaeus fil. 1782 'Fortunei'** (*Spiraea fortunei* Planchon 1853, *Spiraea japonica* Linnaeus fil. 1782 var. *fortunei* (Planchon 1853) Rehder 1902)[28]

Trachycarpus fortunei (W. J. Hooker 1860) Wendland 1862 (*Chamaerops fortunei* W. J. Hooker 1860)[29]

? *Vernonia solanifolia* Bentham 1842 (*Vernonia fortunei* Schultz-Bipontinus 1852)[30]

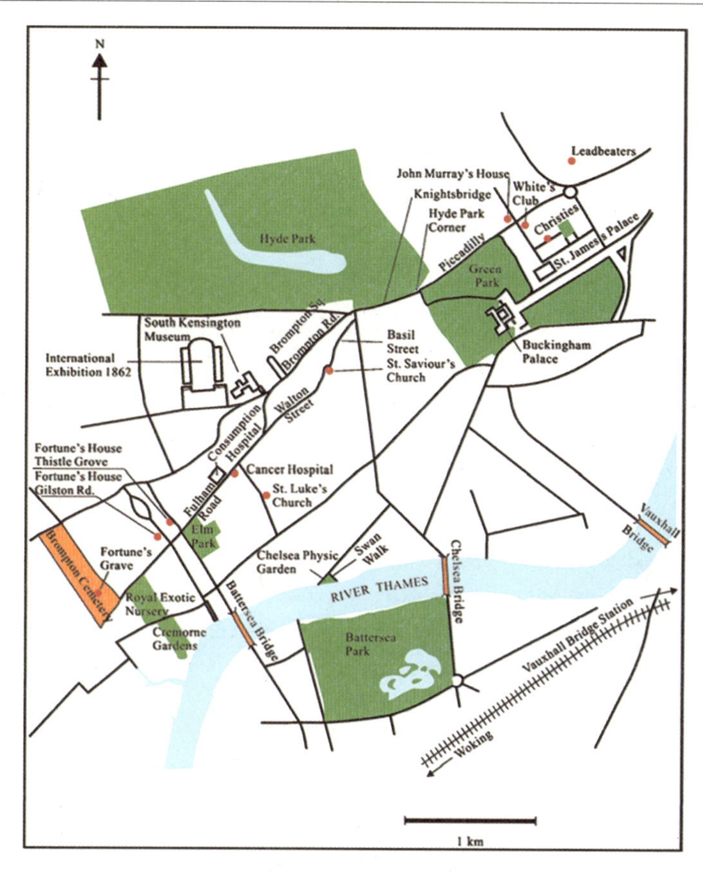

Fig. 5.102 Fortune's London.

After the success of "Three years' wanderings in the northern provinces of China", Fortune had probably already decided in China to write up his latest experiences for publication. He was able to make some use of the "Notes of a traveller" published in the "The Gardeners' Chronicle" during 1849 – 1851 as a backbone for his book. However, he must have spent the best part of three months, thus, November 1851 until January 1852, at home in Thistle Grove Lane (**Fig. 5. 103**) working on his "Journey to the Tea Countries".

Fig. 5. 103 Fortune's house at 8 Thistle Grove, London SW10.

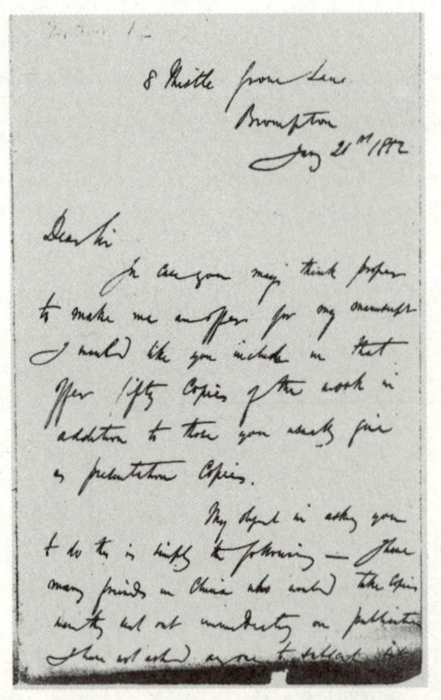

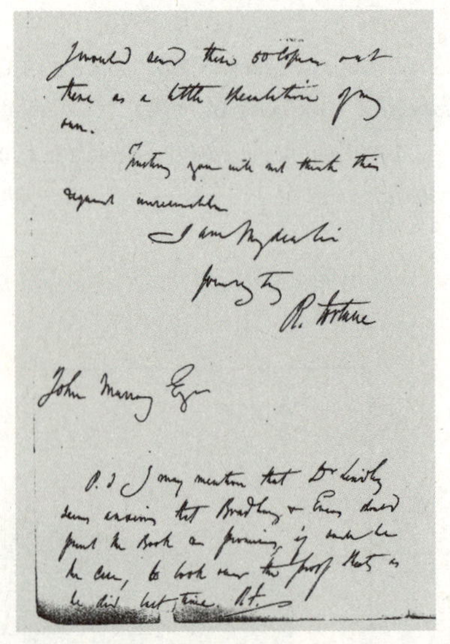

Fig. 5.104 Fortune's letter of 21 January 1852 to the publisher John Murray III.

Already on 21 January 1852 he was probing the publisher John Murray in a business-like tone (**Fig. 5.104**).[366]

Dear Sir

In case you may think proper to make me an offer for my manuscript I would like you [to] include in that offer fifty copies of the work in addition to those you usually give as presentation copies.

My object in asking you to do this is simply the following—I have many friends in China who would take copies were they sent out immediately on publication. I have not asked anyone to subscribe but I would send these 50 copies out there as a little speculation of my own.

Trusting you will not think this request unreasonable.

I am[,] My dear Sir
Yours truly
R. Fortune

In order to exert a little pressure on Murray, he mentioned in a postscript that Professor Lindley (**Box 2.1**) wanted to see the book published by Bradbury and Evans, the publishers of "Punch", "The Illustrated London News" and "The Gardeners' Chronicle".[367]

P.S. I may mention that Dr. Lindley seems anxious that Bradbury & Evans should print the Book & promises, if such be the case, to look over the proof sheets as he did last time. R.F.

On 2 February John Murray offered Fortune £ 200 for the copyright of his second work on China.[368] This was almost double the amount he had paid for Fortune's first travelogue. However, for some reason Fortune did not take up the offer immediately. Was he seriously considering Bradbury and Evans, and had to wait until they made a comparable offer, or was he busy unpacking crates for the auction at Christie and Manson's at the beginning of June? Another possibility is that he was laid low by an attack of rheumatic fever brought on by the cold and damp weather.

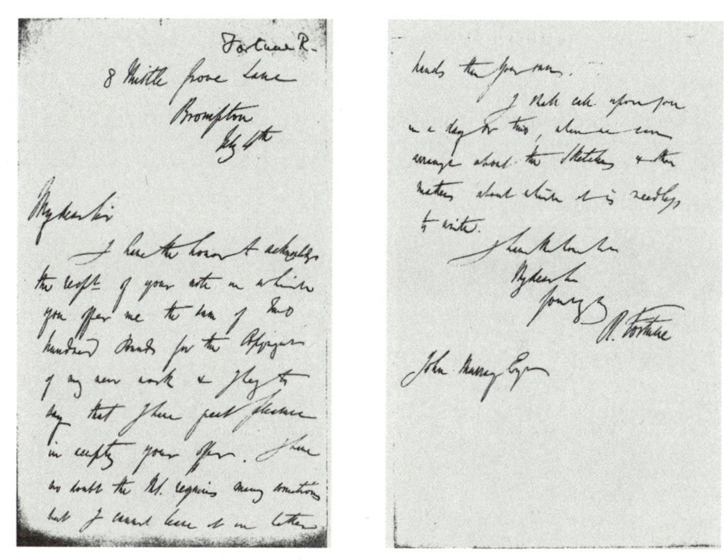

Fig. 5.105 Fortune's letter of 4 July 1852 to John Murray III.

1852 had turned out to be a particularly wet year in England and Wales, with one of the highest rainfalls on record.[369] In April alone the precipitation was three times the long term average.[370] May was not much better. June was not only wet but cold into the bargain.[371] By contrast, July proved to be exceptionally warm.[372] For someone who was sensitive to changes in the barometric pressure, it must have been a trying time. Although experts disagree about the extent to which the different meteorological factors affect rheumatoid arthritis [373], all agree that patients believe that the weather plays a significant role, hence sayings like "feeling under the weather", "aches and pains, coming rains".[374] Cold, wet weather seems to cause an increase in pain in the joints, which eases off under dry conditions.

It may be significant that Fortune did not reply until the 4 July by which time the weather had improved (**Fig. 5.105**).[375]

My dear Sir,

I have the honor to acknowledge the recpt. of your note in which you offer me the sum of Two hundred Pounds for the Copyright of my new work & I beg to say that I have great pleasure in accepting your offer. I have no doubt the MS. requires many corrections but I cannot leave it in better hands than your own.

I shall call upon you in a day or two, when we can arrange about the Sketches & other matters about which it is needless to write.

I have the honor to be
My dear Sir
Yours truly
R. Fortune

Everything was arranged satisfactorarily, and Fortune's latest travelogue duly appeared at the end of the year.

Another source of income was the auction of some of the porcelain, jade, steatite carvings, carvings in wood, lacquerwork and bronzes he had brought back from China. The auction was held at Christie & Manson's premises at 8 King Street on the 8th and 9th of June 1852 under the title "Chinese Works of Art".[376] As it was the first auction of *objets d'art* that he had brought back from China, it seems likely that Fortune would have attended however he felt. Unfortunately, the weather was far from enticing (still no sign of summer at the end of June), so he would have had to wear waterproof clothing and take an umbrella.[377]

Fig. 5.106 A horse-drawn omnibus in London. Contrast the top hats worn by the affluent passengers with the caps of the boys on the street.

Fig. 5.107 Consumption Hospital (1844–1854) on Fulham Road, now "The Bromptons", an exclusive residence.

Fig. 5.108 Cancer Hospital (1851), now the Royal Marsden Hospital, founded by William Marsden in memory of his first wife, who died of cancer in 1846.

Fig. 5.109 Green Park, a Royal Park, which was opened to the general public in 1826.

To prevent getting spattered with dirt he probably took a horse-drawn omnibus (**Fig. 5.106**) from Fulham Road to King Street. On the way Fortune would have passed the Consumption Hospital (**Fig. 5.107**) between Cranley Place and Sumner Place, and the new Cancer Hospital (1851; **Fig. 5.108**) founded by William Marsden in memory of his first wife.[378] The omnibus would have continued on to Hyde Park, where people liked to parade themselves and take their dogs for a walk.[379] By Fortune's day there were no longer any deer in the park, and duelling was a thing of the past.[380] From Hyde Park Corner the omnibus would have headed for Piccadilly, which derived its name from "Pickadilly Hall", built in 1611 – 1612 by Robert Baker, a successful tailor, who had specialized in piccadillies (pickadelles) or turnover collars.[381] Fortune would have got the omnibus to stop at the northeast corner of Green Park, a one-time pleasure garden, which had opened to the general public in 1826 (**Fig. 5.109**).[382]

Fig. 5.110 St. James's Palace, which Fortune would have seen when he turned into St. James's Street.

Fig. 5.111 Facade of Christie's Auctioneers, 8 King Street. This survived the "Blitz" in 1940–1941.

He then turned right into St. James's Street (**Fig. 5.110**), eulogized by Frederick Locker (1821 – 1895) in "London Lyrics" (1857):

> St. James's Street, of classic fame!
> The finest people throng it!
> St. James's Street? I know the name!
> I think I've passed along it!
> Why, that's where Sacharissa sighed
> When Waller read his ditty;
> Where Byron lived, and Gibbon died,
> And Alvanley was witty.

The third street on the left was King Street, where the auctions were held (**Fig. 5.111**).

Fortune would have had his favourite items, that he may have been loathe to part with, and for which he was hoping to get a good price. On the other hand, to his surprise, some of the items may have gone for more than he had expected, as the dealers vied against each other. There were quite a number of antique dealers present in the famous Octagonal Room. In just two days the 237 items fetched almost half of Fortune's annual salary with the EIC, i.e. £ 238 – 9 – 0, or the equivalent of £ 23,370 in 2015.[383] A fine piece of agate went for £ 10 – 10 – 0, while a jar of grey crackle, with lions' heads and ornaments sold for £ 8 – 12 – 6. Even allowing for the freight costs and Christie's 8 – 10% commission[384], this was still a reasonable amount. No doubt, Fortune was well satisfied, because during subsequent trips to China he collected *objets d'art* for sale in London.

The sale of these works of art helped to support Fortune's growing family. By December 1851 Jane was expecting their fourth child, Mary, who was born in August 1852.[385] Mary may have had a weak constitution, but certainly the cold, damp weather did not help. Jane would have been up every few hours to feed her, so one wonders how much sleep Fortune got after her birth. At times he probably longed for the peace and quiet of a Chinese temple! He did not have to wait for long. Pleased with the results of his first expedition on their behalf, the EIC decided to send Fortune back to China.[386]

Shortly before Fortune left on his latest trip, he travelled to Scotland to see his mother Agnes. He presumably travelled as far as York by the Great Northern Railway, which had just opened its large terminus at King's Cross Station (**Fig. 5.112**) on 14 October.[387] He seems to have stayed in Duns with his brother James, who celebrated his 35th birthday on December 1. Writing to John Murray from Duns on Tuesday, 7 December, he had to explain that he was unable to accept the publisher's invitation to dinner in London the very next day (**Fig. 5. 113**).[388] In his letter he suggests calling on the publisher at the beginning of the following week, i.e. on or after Monday 13 December 1852. This was just a week before he was due to leave Southampton!

Fig. 5.112 Kings Cross Station, where Fortune would have caught the train to York.

My dear Sir

I am now sojourning in a little town in the south of Scotland where your letter of the 5th has just found me. This will be a sufficient reason why I cannot accept your kind invitation to dinner tomorrow.

I shall call on you as soon as I come to London—say some morning in the beginning of next week. I sent to Mr. Cooke by yesterday's post a copy of what I think a good title page for the new Edition of my Chinese Book.

Yours sincerely
R. Fortune

The title page (**Fig. 5.114**) shows an over-steepened montane landscape with a cottage surrounded by tea plantations. In the foreground a Chinaman is praying in front of an ancestral grave.

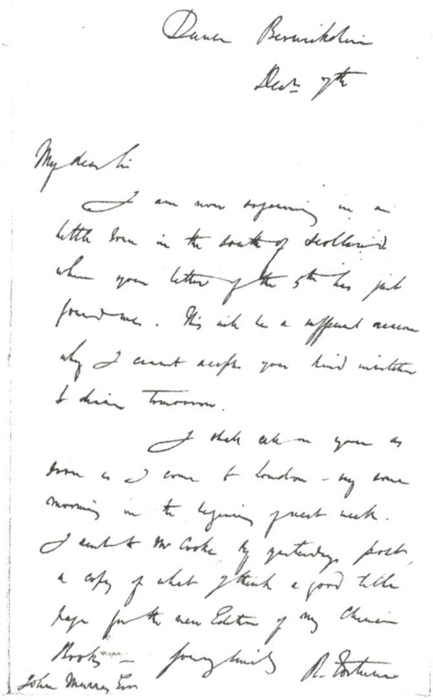

Fig. 5.113 Fortune's letter to John Murray from Duns, 7 December 1852.

Fig. 5.114 Title page of "Journey to the Tea Countries", Fortune's second book.

By the time Fortune went into London to see John Murray, there was still no let-up in the bad weather. It had been raining for the past few months, and many low-lying parts of England were still under water.[389] The Thames had been unable to cope with the excess water, and had overflowed its banks, causing widespread flooding all the way from Oxford (**Fig. 5.115**) to Dartford.[390] As a deeply religious man, Fortune was probably reminded of the biblical flood. With the area around Woking also under water[391], Fortune may have wondered whether the train would get through to Southampton in time to catch his ship, which was due to depart on 20 December 1852. Although it could not have been easy to leave his wife and children just before Christmas, the bad weather, which would have aggrevated his rheumatics, may have lessened the pain of parting.

Fig. 5.115 Flooding of the River Thames near Oxford in 1852.

5.8 Searching for Tea in Troubled Times

Fortune got away in the nick of time. Over the Christmas period during 25 – 27 December 1852, a violent storm hit the UK and many ships were wrecked.[392]

Fortune was scheduled to arrive in Hong Kong on 15 February 1853.[393] In fact, he was almost a month late, so he probably missed his connection in Suez.[394] We can imagine how bored and frustrated the restless plant hunter must have been in Egypt. After all, he had left family and friends before Christmas in order to arrive in China on time for the new tea season. When he finally reached Hong Kong, he managed to obtain a berth on the P&O paddle-steamer "Ganges" (**Box 5.17**) bound for Shànghǎi ("Upper Sea").[395]

Box 5.17

SS Ganges (1850 – 1871) This iron-hulled paddle-steamer, built for the Peninsular and Oriental Steam Navigation Company (P&O) by Tod & McGregor (Glasgow) was named after the holy Indian river.[1] On her maiden voyage from Southampton to Istanbul, she broke the records for both the Southampton to Malta, and the Malta to Istanbul legs.[2] She took a detachment of troops to Cape Town on her way for service in the Far East in 1851.[3] In 1853 she plied between Hong Kong and Shànghǎi.[4] Sold to Rennie & Co. (Shànghǎi) on 1 June 1871, and resold to the Shanghai Steam Navigation Co. the very same year.[5]

Although the ship encountered strong headwinds and heavy seas, it covered the 1500 km in just four days.[396] When he disembarked, Fortune went to live with his friend Thomas C. Beale (**Box 5.5**), who already had a number of visitors including the Governor of Hong Kong, Sir George Bonham (**Box 5.18; Fig. 5.116**).[397]

Fig. 5. 116 Sir Samuel George Bonham (1803 – 1863), the Governor of Hong Kong, met Fortune in 1853 when both were staying with Thomas Beale in Shànghǎi. They would seem to have got on well.

Box 5.18

Sir Samuel George Bonham (1803 – 1863) Son of Captain George Bonham (ca. 1775 – 1810) of the East India Company (EIC) and his second wife Isabella Baines Woodgate (1777 – 1852).[1] When he was about 7 years old he lost his father, when the East Indiaman "True Briton" (1790) sank in a typhoon in the China seas.[2] This did not prevent him from joining the EIC. After serving with the EIC, Samuel was appointed Governor of the Straits Settlements (Singapore, Malacca and Penang; during 1837 – 1847).[3] In 1846 he married Ellen Emelia Barnard (1828 – 1859), a pleasant personality.[4] They had one son, George Francis (1847 – 1927).[5] Although Bonham was Governor of Hong Kong from 21 March 1848 to April 1854 at a time of serious cut-backs imposed by London, he saved money by reducing the number and salaries of civil servants.[6] In fact, he managed to balance the 1848 budget by delaying the payment of his own salary until 1849.[7] His cordial, outgoing nature and his rescinding of most of the petty taxes levied by his predecessor Sir John Davis (1795 – 1890) made him popular with most of the inhabitants.[8] Although it was impossible to be on cordial terms with all his subordinates, he tried to be even-handed.[9] Unfortunately, squabbling among his opinionated subordinates did not make his job any easier.[10] As a result of the instability caused by the Tàipíng

Rebellion (**Box 5.19**), there was an influx of Chinese into the Crown Colony.[11] While some found work in the gold mines of California and Australia, others stayed.[12] Since many of the Chinese had no knowledge of English, Bonham permitted the headmen to settle any internal differences.[13] While he obviously believed in treating the Chinese population with respect, he was of the opinion that a knowledge of the Chinese language would make civil servants open to compromise.[14] As a result, he tended to appoint and promote men who did *not* know the language.[15] Fortune met Bonham on 21 March 1853 at Thomas Beale's house in Shànghǎi, shortly before Bonham departed on board HMS "Hermes" on 22 April 1853 to meet the Tàipíng rebels, who had just conquered Nánjīng.[16] The two men seem to have got on well, for Bonham showed Fortune some of the dispatches.[17] Although pressurized by the Chinese Governor of Shànghǎi, WÚ Jiànzhāng (WU Chien-chang), to attack and extirpate the rebels, Bonham maintained his neutral stance.[18] However, on his return from Nánjīng, he was in no doubt that the Tàipíngs were as haughty and xenophobic as the Qīng Government they were attempting to oust.[19] If Bonham had expected this first-hand information to be welcomed by the British Government, he was sorely disappointed. Instead he was reprimanded for undertaking the mission.[20] The following year, after six years in office, he was forced to retire as Governor by ill-health.[21] He returned to England to recuperate, and does not appear to have worked again.[22]

Fortune arrived back in China at a turbulent time. A strong earthquake was felt in Shànghǎi on 14 April 1853, less than a month after Fortune had arrived in the city.[398] Rumour had it that a village had disappeared into an abyss.[399] However, this was a minor incident compared with the political turmoil caused by the capture of all the major cities along the Cháng Jiāng by a comparatively unknown group referred to as the Tàipíngs (**Box 5.19**; **Fig. 5.117**). The Tàipíngs overran Nánjīng (Nanking, "Southern Capital"), the major city in southern China on 21 March 1853. Many believed that their next goal would be Shànghǎi.[400] Everybody agreed that this had to be hindered at all costs, as Shànghǎi had become the commercial hub for the foreign traders. In haste, some fortifications were erected.[401]

Fig. 5.117 Battle between the Tàipíng and imperial troops on the left.

Box 5.19

Tàipíng Rebellion (1850 – 1864) The mid 19th Century was a period of social unrest in many parts of China (see **Ch. 10**). The Qīng Dynasty had been weakened by the First Opium War (**Box 2.2**), so the government was in no position to combat the corruption of landlords and the underpaid local officials, who extorted money from the rural population.[1] In order to pay their debts many of the farmers were obliged to give up their land, which led to increasing numbers of dispossessed people.[2] The poor therefore became increasingly disenchanted with their foreign (Manchu) rulers and the state religion of Confucianism, which taught them to respect their superiors.[3] Many turned to banditry or piracy.[4] When HÓNG Xiùquán (**Box 5.20**) offered to return the land to the people, and eliminate a number of social evils, such as slavery, footbinding, opium smoking, and judicial torture, he soon had a large following among the poorer classes, who had nothing to lose.[5] By July 1850 he could count on between 10,000 and 30,000 adherents, who were organized along military lines in preparation for an uprising in the southwestern province of Guǎngxī.[6] After some initial successes against the Qīng army in 1850 and 1851, the Tàipíng stronghold in Zǐjīn Shān ("Purple Metal Mountain", commonly referred to as "Thistle Mountain") became untenable.[7] A breakout was successfully carried out, and the pursuing Qīng troops were unable to prevent them from capturing the walled city of Yǒng'ān ("Forever peace"; since 1915 called Méngshān) on 25 September 1851.[8] The Tàipíngs were able to hold onto the city for just over six months, but as Qīng forces finally severed the supply routes and sources of food, another breakout became necessary.[9] This did not go quite as smoothly as the last one with 2,000 of the 40,000-strong force being killed by the Qīng Army.[10] A surprise attack on Guìlín ("Laurel forest"), the capital of Guǎngxī, was thwarted, and a month-long siege ensued.[11] Although the Tàipíngs were unable to capture the city, they seized 40+ large river vessels.[12] With these craft, they continued on their journey in a northerly direction. At Quánzhōu ("Perfect Prefecture") another 200+ vessels of different sizes were added to the fleet.[13] Flushed with their success, the Tàipíngs rushed headlong into an ambush at Suōyī Ford near the Guǎngxī/Húnán border.[14] Having blocked the Xiāng River by trees and logs held together by huge iron spikes, the "Chu Braves" under JIĀNG Zhōngyuán (CHIANG Chungyüan, 1812 – 1854) only had to wait for the inevitable pile-up of hundreds of boats.[15] These boats were set alight and some 10,000 trapped troops and sailors massacred.[16] After the debacle the Tàipíng forces took a well-deserved pauze at Dàozhōu (Dàoxiàn) to regroup and recruit new soldiers before moving further north.[17] The Húnán capital Chángshā ("Long sand") was unsuccessfully besieged for two months during September-November 1852, before the Tàipíngs pressed on into Húběi (Hupeh).[18] In early 1853 the cities along the Cháng Jiāng from Wǔhàn to Wúhú (Ānhuī) were captured in quick succession by the Tàipíng forces.[19] Finally, on 21 March 1853 after 2 days' siege Nánjīng fell and the Manchu garrison was put to the sword.[20] Nánjīng became the Tàipíng capital and was renamed Tiānjīng ("Heavenly Capital").[21] While it is understandable that the Tàipíngs wanted to settle down, they thereby lost the opportunity of capturing Běijīng and usurping the Qīng Dynasty.[22] Since it was necessary to keep part of the army in Nánjīng to protect the sprawling city, it was only possible to send a much smaller force of 75,000 men to capture Běijīng.[23] This was more easily contained by the Imperial forces with their cavalry. After a long detour, one of the columns reached the outskirts of Tiānjīn, some 110 km from the capital, on 30 October 1853, but without supplies and winter clothing, the campaign ground to a halt.[24] Qīng forces pushed them back along the Grand Canal as far as Liánzhèn (Héběi), where after an 8-month siege the Tàipíng Northern Army was annihilated.[25] Another part of the Tàipíng forces was sent back up the Cháng Jiāng to secure those cities either bypassed or captured and subsequently abandoned during the Spring Campaign, but this had no lasting effect.[26] JIĀNG Zhōngyuán's friend ZĒNG Guófān (TSENG Kuo-fan, 1811 – 1872) and his "Húnán Braves" eventually pushed the Tàipíngs eastwards, where they faced British and French forces intent on defending the commercial city of Shànghǎi.[27] After all, had not the Tàipíngs stated their intention to suppress private trade and prevent the import of opium?[28] Had the Tàipíng taken over the city during the Small Sword Uprising (**Box 5.21**), their stabilizing effect would have been welcomed by the foreign community.[29]

In hindsight, the failure of the Tàipíng Rebellion can be attributed to a combination of poor public relations[30], infighting among the leadership[31], and a number of missed opportunities.[32] Once the Treaty of Tiānjīn (**Box 9.1**) was finally ratified in October 1860, the foreign powers had *de facto* committed themselves to supporting the Manchu regime.[33] So even though the Tàipíng went to great lengths not to aggravate the foreign community, they soon found themselves fighting French and British forces.[34] Moreover, with the war with the foreigners now over, the Qīng government was able to put more effort into eradicating the rebels.[35] In the early 1860s the rebel-held cities fell one by one to the various forces.[36] On 19 July 1864 the Tàipíng capital fell, and with it the dream of a Heavenly Kingdom evaporated.[37] However, the conflict dragged on for a number of years. It took until 1871 before the last pocket of resistance was wiped out.[38] It has been estimated that in the course of the conflict between 10 and 30 million people lost their lives in battle or from starvation.[39] Many more were displaced. It took years before the economy of the provinces occupied by the Tàipíngs recovered.

Box 5.20

HÓNG Xiùquán (1814 – 1864; **Fig. 5.118**) The youngest son of a Hakka farmer, HÓNG Jìngyáng (ca. 1774 – 1848) and his second wife WÁNG Shì.[1] Although his schooling was limited, Xiùquán showed great promise.[2] He worked as a village schoolteacher while studying for the Imperial Examinations, which would have enabled him to join the Qīng bureaucracy and thus improve his family's lot.[3] He blamed his failure on four occasions to discrimination against the Southern Chinese by the Manchu government in Běijīng.[4] In 1836, while in Guǎngzhōu (Canton) to take the examinations, he heard a foreign missionary (Rev. Edwin Stevens, 1802 – 1836) preaching in the streets.[5] He was given a copy of extracts from the Bible published in 1832 by LIÁNG Fā (1789 – 1855) under the title Quànshì Liángyán ("Good words for exhorting the Age"), but only gave it a cursory glance.[6] After failing to pass the Imperial Examinations again in 1837, he became critically ill and was confined to bed.[7] While his family watched over him for forty days, he had a vision that he had ascended to Heaven.[8] There he was cleansed in a river by his "Mother", before going to see his "Father", a tall man with a golden beard reaching down to his belly.[9] When his "Father" told him how devils not only filled the world, but had even infiltrated Heaven, HÓNG offered to destroy them.[10] While his "Elder brother" held a golden seal to dazzle the devils, HÓNG fought them with a great sword ("Snow-in-the-Clouds").[11] The battle over, HÓNG rested in Heaven for a while, before returning to Earth as Xiùquán ("Excellently complete").[12] HÓNG's vision remained unexplained until 1843 when his cousin LǏ Jìngfāng (LI Ching-fang) encouraged HÓNG to read LIÁNG Fā's book, which he had had since 1836.[13] He realized for the first time that the "Father" in his vision was Jehovah and his "Elder brother" none other than Jesus Christ.[14] He underwent an immediate conversion to Christianity. From then on he believed he had a mission to found a theocratic state (Tàipíng Tiān Guó, literally "Great Peace Heavenly Kingdom") and drive the Manchu rulers out of China.[15] Having lost his job for removing the "heathen" (Confucian) tablets in his school, HÓNG and three converts set off on a journey to spread the Christian message in Guǎngdōng and Guǎngxī.[16] While FÉNG Yúnshān (1815 –

Fig. 5.118 HÓNG Xiùquán (1814 – 1864), the Tàipíng leader.

1852) stayed on in Guǎngxī, HÓNG returned to his family and a new teaching post in Guānlùbù at the end of 1844.[17] In 1847 HÓNG spent a few months in Guǎngzhōu studying the Bible with the idiosyncratic Rev. Issachar Jacox Roberts (1802 – 1871), who described him as a rather handsome, round-faced, well-built gentleman about 164 cm in height.[18] After leaving Roberts prematurely on 12 July 1847, he joined FÉNG and his Bài Shàngdì Huì ("God-worshipping Society") in Zǐjīn Shān ("Purple Metal Mountain") in Guǎngxī.[19] It was here that HÓNG and FÉNG destroyed the image of King Gàn at the temple in Xiàngzhōu ("Elephant Prefecture").[20] HÓNG left the area in December 1847 before he could be caught, but FÉNG was arrested and put on trial.[21] FÉNG was finally released in the summer of 1848, but banished to his home town in Guǎngdōng.[22] He and HÓNG later returned to Purple Metal Mountain, where crowds of embittered paupers continued to swell the ranks of the Bài Shàngdì Huì.[23] When the Manchu authorities attempted to suppress the movement, the Tàipíngs took up arms against them.[24] After a successful campaign in Guǎngxī, Húnán, Húběi and Ānhuī, HÓNG established his capital in Nánjīng (Jiāngsū) in 1853 (**Box 5.19**). Once installed in Nánjīng, he left the daily running of his kingdom to his deputy, YÁNG Xiùqīng (YANG Hsiu-ch'ing), dividing his time between his harem and rectifying certain passages in the Bible.[25] When provoked, he could become physically violent, as a number of his concubines discovered.[26] When YÁNG openly challenged his authority in 1856, HÓNG had him, his family and thousands of his adherents put to death.[27] Since he could no longer trust others, he appointed his two elder brothers (HÓNG Réndá and Rénfā) to run his kingdom.[28] The brothers, fearing for their own power, scuppered various changes proposed by HÓNG's gifted cousin, Réngān ("Shield King", 1822 – 1864), who arrived in Nánjīng on 15 April 1859.[30] HÓNG became increasingly reclusive, never leaving his palace.[31] When his old friend Rev. Issachar Roberts, who had been attempting to reach Nánjīng for years, finally arrived on 13 October 1860, he was only given a single interview.[32] Towards the end of his life, HÓNG seems to have lost all sense of reality.[33] When food became scarce as the forces of ZĒNG Guóquán (1824 – 1890) tightened their grip on Nánjīng, HÓNG said that everybody should eat manna.[34] In order to show what he meant, he collected a number of weeds which he steamed and ate.[35] Some of the weeds may have been poisonous, for he subsequently became ill and died.[36] After the Manchus retook Nánjīng on 19 July 1864, they finally discovered his corpse in a sewer (30 July 1864).[37] HÓNG's bald head was sent to Běijīng and his body cremated.[38] To ensure that his remains should never find a resting place, his ashes were shot from a cannon.[39]

As Níngbō (Ningpo, "Peaceful Waves") was not yet controlled by the Tàipíngs, Fortune returned to this city to organize his next field campaign.[402] In May 1853 he hired a small covered boat with three compartments, the middle of which he occupied.[403] His first goal seems to have been Língfēng Shān ("Fairy Peak Hill"), about 20 km east of Níngbō. The crew sculled all night, reaching the end of the canal at Ayùwáng before dawn.[404] There Fortune was woken by the sounds of hundreds of voices.[405] Still half-asleep he imagined they had encountered part of the Tàipíng army.[406] Luckily the voices stemmed from pilgrims on their way to Ayùwáng Sì (Ah-yuh-Wang Temple, "The Educated King Temple", Ayuka's Temple; **Figs. 5.119 – 5.121**).[407] As the pilgrims disembarked from their boats, Fortune decided to follow the crowd to the temple.[408]

Fig. 5.119 Ayù Wáng Sì (The Educated King Temple), where Fortune stayed in May 1853, while visiting some nearby tea estates.

Fig. 5.120 A group of pilgrims at Ayù Wáng Sì. As in Fortune's day, most of the pilgrims are nowadays women.

Having entered the ancient gateway, he crossed the lotus pond by the ornamental bridge.[409] To the west of the lake there are a number of camphor trees, which were there in Fortune's day (**Fig. 5.121**). One of these had been blown down during a typhoon, but as Fortune noted, it was still being lovingly cared for by the monks.[410] The various buildings in the temple complex were crowded with worshippers, most of whom belonged to the fair sex.[411] This phenomenon still holds true today. Having paid his respects to the chief monk, and examined a precious relic belonging to the temple, Fortune went for a walk.[412] On the low-lying land the paddy fields presented a fresh green appearance.[413] On the better-drained ground at the base of the hills, the farmers were picking the immature tea leaves in the shadow of their ancestors' tombs.[414] In the hedges the yellow of *Forsythia viridissima* was making way for the pastel shades of several species of rose, *Spiraea cantoniensis* (Syn. *S. reevesiana*), clematises and *Wisteria sinensis* (Syn. *Glycine sinensis*).[415] On higher ground, where the soil had been prepared for sweet potatoes and maize, *Exochorda racemosa* (Syn. *Amelanchier racemosa*) and a number of deciduous rhododendron species including the yellow *Rhododendron molle* (Syn. *Azalea sinensis*) were in flower.[416] Towards evening Fortune returned to the now silent temple, which was to be his headquarters for the next few days.[417] From here he was able to make arrangements for seed with the tea-planters in the neighbourhood.[418] Once this job was completed, he took the boat back to Níngbō and on to the walled city of Cíchéng (Tse-kee, "Kind City") on the floodplain of the

Fig. 5.121 One of the camphor trees that Fortune saw at Ayù Wáng Sì in May 1853.

Yúyáo Jiāng, some 15 km northwest of Níngbō.[419] Although Cíchéng itself is situated on flat ground, the surrounding area is dotted with hills covered with *Pinus massoniana* (Syn. *P. sinensis*), *Cinnamomum camphora*, deciduous and evergreen oaks, two species of sweet chestnut, and one of Fortune's favourite fruit trees, the yángméi (yangmae or red bayberry, *Myrica rubra*).[420] He took some cuttings of yángméi for grafting, and decided to send someone back for the chestnuts in the autumn, as he felt they ought to be introduced into the Himalayas.[421] Some of the most beautiful spots on the hillsides had been chosen using the principles of geomancy (fēngshuǐ, i.e. wind and water) for the ancestral graves.[422] In general, evergreen conifers (arbor-vitae, cypress, juniper, pine) or *Photinia glabra* (Thunberg) Maximowicz / *P. serratifolia* (Desfontaines) Kalkman (Syn. *P. serrulata* Lindley) were planted in a half-circle round the graves, but Fortune found the occasional use of weeping willow particularly pleasing.[423] The tops of the hills were sometimes crowned by temples from which an excellent view of the surrounding countryside could be obtained.[424] Fortune particularly liked the lake to the north of Cíchéng with its causeway leading to some temples at the base of the hills (**Fig. 5.122**).[425] It was here that he decided to moor his houseboat for the duration of his stay.[426]

It would be wrong to think that Fortune was only interested in plants. While he was staying at Cíchéng, he made a large collection of insects for the leading entomologists in Europe.[427] Some of his collections were intended for the British Museum in London.[428] The idea of collecting insects for scientific study completely baffled the Chinese with whom he came into contact. They knew they could be used as food or in medicine, so they did not understand why Fortune wanted uncrushed specimens.[429] Nonetheless, once it became clear what he wanted, the children and old women willingly helped Fortune to search for specimens in exchange for some pocket-money.[430] In a short time he amassed a large number of species, many of which the local inhabitants had never seen before.[431]

Fig. 5.122 Lake to the north of Cíchéng (Kind City), close to the spot where Fortune moored his houseboat in 1853. "Between the north gate and the hills there is a pretty lake, which is crossed by a causeway with arches and alcoves." (Fortune, 1857: 47).

Fortune was impressed by the clean and well-dressed appearance of the citizens of Cíchéng.[432] He was also able to confirm with his own eyes that the women in this part of Zhèjiāng were attractive.[433] He got a rare opportunity to see some upper-class ladies when he visited a mandarin in the city.[434] In his own words he "was willing to look upon their pretty faces as long as possible".[435] However, he was less impressed by the narrow streets, the shabby shops, and the practice of skinning frogs alive![436]

When Fortune was in China, he was in the habit of collecting *objets d'art*, which were auctioned by Christie & Manson's once he got back to London.[437] So by 1853 he had become something of an expert on Míng porcelain. He therefore jumped at an opportunity to visit the house of a wealthy Chinese art collector while he was in Cíchéng.[438] He admitted that he had never seen a finer collection of ancient vases, bronzes, enamels, carved laquerware, and stones (agate and jade).[439]

While at Cíchéng, Fortune lived on board the boat.[440] However, by July, with daytime temperatures climbing to 38℃ and occasionally reaching 43°C, the weather was becoming too hot to live in the boat.[441] Fortune decided to return to Tiāntóng Temple, 23 km ESE of Níngbō. There he was kept busy making arrangements with the tea-planters to supply him with seed in the autumn, and supervising the collection of insects and snails by the local children.[442] As a result, he was not always able to take a siesta during the heat of the day.[443] Possibly as a result of over-exertion, Fortune succumbed to a violet attack of fever in August 1853.[444] As he was too far from western practitioners, a local doctor was called in.[445] A combination of pills, herbal tea, and a pinching of his taught muscles caused him to perspire, and in a matter of three days Fortune was better again.[446] Once he was cured, Fortune set about completing the arrangements for supplies of tea plants and seeds.[447] This was finished by the end of August.[448] Since the seed would not be ripe until October or November, Fortune decided to return to Shànghǎi to see what he could do about hiring tea-manufacturers from the black tea districts.[449] Quite by accident he chose to visit the Chinese part of Shànghǎi early on 7 September 1853, and thus witnessed the Small Sword Uprising (**Box 5.21**; **Fig. 5.123**) at first hand.[450]

Fig. 5.123 Small Sword Uprising in Shànghǎi (1853). "The morning of the 7th of September, being the day on which the mandarins usually pay their visit to sacrifice in the temple of Confucius, was chosen by the rebels for the attack upon the city. Without knowing anything about their plans, I happened to pay a visit to the city soon after daybreak." (Fortune, 1857: 118/119).

Box 5.21

Small Sword Uprising (7 September 1853 – 17 February 1855) After the First Opium War (**Box 2.2**) Shànghǎi became the major treaty port with an enormous turnover of trade. This prosperity attracted not only *bone fide* merchants, but also criminal elements involved in the opium trade.[1] These gangsters were often members of secret societies, such as the Small Sword Society founded in 1850.[2] About 500 men belonging to seven gangs of Guǎngdōng, Fújiàn and Zhèjiāng extraction were involved in the attack on the Chinese walled city of Shànghǎi in the early morning of 7 September 1853.[3] With the exception of a gatekeeper, who tried to resist the insurgents, the coup was largely bloodless.[4] However, the Shànghǎi magistrate YUÁN Zūdé (1811 – 1853) was killed as retribution for arresting 15 secret society members at the end of August.[5] On the other hand, the Dàotái (Taotai = Governor) WÚ Jiànzhāng (1791 – 1866) was spared because of his Guǎngdōng extraction.[6] However, to secure his safety, he turned over his treasury.[7] Foreigners like Fortune were unmolested.[8] Although the imperial troops tried to starve the rebels out by laying siege to Shànghǎi, and the ring leaders often quarreled about their policies, the uprising lasted for 17 months.[9] This was only possible through complicity by Chinese and foreigners, who supplied the gangsters with food and weapons.[10] A high wall was eventually built between the foreign settlement and the Chinese city to prevent this clandestine trade.[11] The end came when the Fújiàn leader CHÉN Ālín (CH'EN Alin) shot the secretary of LIÚ Lìchuān (1820 – 1855) from Guǎngdōng.[12] Fearing reprisals, the Fujianese fled.[13] LIÚ and more than a hundred of his henchmen followed.[14] Once the imperial troops were sure that most of the rebels had gone, they plundered and burned the city, beheading, drowning, torturing and raping any "suspicious persons" they encountered.[15] Those that survived the onslaught were dazed at the wholesale destruction of property.[16] In the process, some of the nurseries, which had supplied Fortune with ornamental plants were destroyed.[17]

At the beginning of October, having done all he could in Shànghǎi, Fortune returned to Níngbō.[451] He once more established himself at Tiāntóng Temple (**Fig. 3.25**).[452] He let it be known that he was interested in buying 500 – 600 catty (ca. 300 – 360 kg) of good tea seeds.[453] In the following days he and one of the monks vetted the tea seed from dawn to dusk.[454] At the same time he sent a dependable accomplice to the Huángshān area (SE Ānhuī) to bring back seed of the famous green tea and seed of the weeping cypress.[455] Another man was dispatched to Píngshuǐ ("Calm Water"), 14 km SSE of Shàoxīng (Zhèjiāng), to collect seed from the good-quality tea grown there.[456] On the way back to Níngbō, the latter collected large quantities of sweet chestnuts at Cíchéng (**Fig. 5.124**) for the Agricultural and Horticultural Society of India.[457]

Fig. 5.124 *Castanea mollissima*, one of two chestnut species found by Fortune near Cíchéng. "Amongst these woods I met with the chesnut [sic] for the first time in China. This discovery was of great importance, as I was most anxious to introduce this to the Himalayan mountains in India." (Fortune, 1857: 51).

The seed of other useful species, which Fortune intended to introduce into India, included the conifer *Cryptomeria japonica*, the palm *Trachycarpus fortunei*, both collected near Tiāntóng Temple, along with the highly poisonous Chinese varnish tree (*Toxicodendron vernicifluum*), the soap-bean tree (*Gleditsia sinensis*), and the wax-insect tree (*Fraxinus chinensis*).[458] In order to obtain plants and seeds of the Chinese green indigo (*Rhamnus utilis*), Fortune went all the way from Níngbō to the hilly country west of Hángzhōu ("Hang Prefecture").[459]

5.9 A Jaunt with Friends

Before returning to Shànghǎi with his plant collections, Fortune decided to visit Sìmíng Shān ("Four Distinct Mountains") to the southwest of Níngbō. He and a party of fellow countrymen hired some small boats for the occasion.[460] They took advantage of the incoming tide to make good progress up the Fènghuà Jiāng ("Revere Change River"), presumably so named because of its strongly meandering course.[461] The very next morning the party found itself at the base of the volcanic hills near Jiāngkǒu Zhèn ("Rivermouth Town") in the Fènghuà (Fung-hwa) District.[462] The group set off to climb the Shòu Fēng Tǎ ("Longevity Summit Pagoda", Kong-k'how Pagoda; **Fig. 5.125**) at the top of Shānhòu ("Rear Hill"), to obtain a good view across the Níngbō Plain as far as the Zhōushān ("Boat Mountain", Chusan) Archipelago (**Fig. 5.126**).[463]

Fig. 5.125 Shòu Fēng Tǎ (Longevity Summit Pagoda) at the top of Shānhòu (Rear Hill), which Fortune and his friends climbed to obtain a view of the Níngbō Plain.

Fortune, as usual, was so preoccupied with the plants, that he got separated from the rest of the group and could not resist the temptation to enter a garden with a number of interesting trees and bushes.[464] This garden presumably belonged to the Qīng Shuǐ Ān ("Clear Water Nunnery", **Fig. 5.127**), which would explain why it was surrounded by a fence.[465] The guard dogs soon drew the custodian's attention to the trespasser.[466] However, when the man discovered that Fortune was interested in the plants and not the nuns, he showed the plant hunter round the garden before serving him tea.[467] In the meantime, Fortune's friends were no doubt kicking their heels waiting for the botanist to put in an appearance. Once Fortune caught up with them at the pagoda, the party returned to their boats and proceeded a few kilometres upstream to Dàbùtóu ("Great Wharfside", **Fig. 5.128**), which is as far as the river was navigable.[468] As it was already late, the group spent the night in the village.[469]

Before breakfast the next morning the party walked less than a kilometre to the delightful Xiāo Wáng Miào ("King Xiao Temple", 1042 AD; **Fig. 5.129**) to see the elaborately carved stone altar and the "rude pictures" on the walls.[470] There they met a knowledgeable mandarin called A'chang, who guided them for the rest of their trip.[471]

Once coolies and chair-bearers had been organized, the group headed for the 40 km distant Shā Xī ("Sand Rivulet") where iron ore was said to be extracted from the black sand in a tributary of the Yúzé River (**Fig. 5.130**).[472] Fortune noted that the weeping cypress (*Cupressus funebris*; **Fig. 5.15**) was abundant in this area.[473] They returned to Dàbùtóu the next day.

Fig. 5.126 View over the Níngbō Plain from Shòu Fēng Tǎ. "When we reached the summit of this hill ... we were rewarded with one of those splendid views which are, perhaps, more striking in the fertile districts of China than in any other country." (Fortune, 1857: 172).

Fig. 5.127 Qīng Shuǐ Ān (Clear Water Nunnery), where Fortune jumped a hedge to examine some plants. "Inside of this fence there were a number of trees and bushes which seemed worth looking at, ..." (Fortune, 1857: 173).

Fig. 5.128 Dàbùtoú (Great Wharfside), where Fortune and his friends disembarked. This is as far as the Fènghuà Jiāng (Revere Change River) was navigable.

Fig. 5.129 Xiāo Wáng Miào (King Xiao Temple), "a pretty, small temple called the Sieu-Wang-Meou, which the people told us was well worth visiting." (Fortune, 1857: 176). It was here that the expats made the acquaintance of an old mandarin called A'chang.

Fig. 5.130 Tributary of the Yúzé River, said to yield iron ore. This was a ruse to cover up the real source in the nearby hills.

Fig. 5.131 Ornamental gate of Xuèdòu Sì (Snow Hole Temple).

On the next day A'chang led the group along the stream to Xīkǒu Zhèn ("Streamlet Mouth

Town"), where they started to ascend the hills in the direction of Sìmíng Shān.[474] They decided to stay the night at Xuèdòu Sì ("Snow Hole Temple"; **Fig. 5.132**).[475] However, as the coolies had not yet arrived with the baggage, A'chang took the group to see the high waterfall (Qiānzhàngyán Pùbù, "Very High Cliff Cataract"; **Fig. 5.133**) in the vicinity of the temple.[476] By the time the group got back to the temple, the cook was already preparing dinner.[477] Not to be outdone by his foreign friends, A'chang ate with knife and fork, drank beer and wine, followed by brandy and a cigar.[478] Before retiring to bed, A'chang and the British group had a sing-song.[479] The temple rang to the sound of "Rule Brittania" and "God Save the Queen"![480]

Fig. 5.132 Xuèdòu Sì, where Fortune and his friends had a singsong. "We had some difficulty in inducing our mandarin friend to leave us, as he was evidently prepared to "make a night of it ... " (Fortune, 1857: 183).

The party rose early the next morning to go and look up at the waterfall from the valley bottom.[481] On the way A'chang noticed Fortune's interest in the vegetation, so he started expounding on the medicinal virtues of the different plants they encountered.[482] When Fortune commented that he seemed to know everything, he was given an exposé on fēngshuǐ, another of A'chang's many interests.[483] After breakfast the group went to thank the chief monk for his hospitality.[484] As the monk was in voluntary confinement, they passed their present through a hole in the wall.[485]

Instead of going all the way to Sìmíngshān Village, the group turned off at Dōngshāncūn ("East Mountain Village"), and wound its way up the mountain. Turning a corner they would have spied Lántián ("Orchid Field") lying snug in a hollow beneath them. From Lántián a path led to Lǐcūn ("Inner Village"; **Figs. 5.134, 5.135**) where a large amount of *Strobilanthes cusia* (Syn. *Ruellia indigotica*, Acanthaceae) used to be grown for its blue dye.[486] Since there is still only a path linking these two villages, one is literally treading in Fortune's footsteps. On the way, one still encounters stands of young conifers (*Cryptomeria japonica*, *Cunninghamia lanceolata*, *Pinus massoniana*), tea plantations,

Fig. 5.133 High waterfall (Qiānzhàng Yán Pùbù) near Xuèdòu Sì, which A'chang showed to Fortune and his friends. "All at once we arrived at the edge of a precipice, which made us quite giddy as we looked over it. The water rolled out of the valley over the precipice, and long before it reached the bottom it was converted into showers of spray." (Fortune, 1857: 181).

numerous specimens of the multi-purpose hemp-palm *Trachycarpus fortunei* and clumps of the useful bamboo *Phyllostachys edulis* (mow-chok, maou-chok; **Figs. 5.136, 5.137**) mentioned by Fortune.[487]

Fig. 5.134 Looking down on Lǐcūn (Inner Village) from the path linking Lántián to Lǐcūn.

Fig. 5.135 Lǐcūn. "This little highland village is situated at the head of a glen which opens by various windings to the plain of Ningpo." (Fortune, 1857: 192/193).

Fig. 5.136 A grove of *Phyllostachys edulis* near Lǐcūn. Fortune considered this "the most beautiful bamboo in the world" (Fortune, 1857: 189).

Fig. 5.137 Sawing and splitting *Phyllostachys edulis*, the multifunctional bamboo *par excellence*. "It is used in the making of sieves for the manipulation of tea, rolling-tables for the same purpose, baskets of all kinds, ornamental inlaid works, and for hundreds of other purposes..." (Fortune, 1857: 190).

The mountains were and are thinly populated, with villages being confined to those areas where the soil is fertile enough to grow crops.[488] In the winter the villagers grew wheat, barley and some "greens", while in the summer they cultivated buckwheat, sweet potatoes, maize and two kinds of millet (presumably *Panicum miliaceum* and *Setaria italica*).[489]

Fig. 5.138 Xìng Cí Sì (Do Kindness Temple), Lǐcūn. "As we had another thirty *le* to go before we reached our boats, we rested ourselves in an old joss-house in order to allow our baggage to come up with us. Here the natives crowded round us in hundreds ..." (Fortune, 1857: 193).

Fig. 5.139 Zhāng Xī (Camphor Rivulet) near Yínjiāng Zhèn (Yin River Town), where boats were waiting in the evening to take Fortune and his friends back to Níngbō.

While the group waited for the baggage to catch up, they visited the Xìng Cí Sì ("Do Kindness Temple"; **Fig. 5.138**) in the centre of Lǐcūn, where they were surrounded by hundreds of onlookers.[490] From Lǐcūn it was downhill all the way to the Zhāng Xī ("Camphor Rivulet"; **Fig. 5.139**), so the visitors climbed into their bamboo chairs and enjoyed the beauty of the valley down which they weaved their way (**Fig. 5.140**).[491] The scenery reminded Fortune of the Bohea Mountains between Jiāngxī and Fújiàn.[492] Near the small town of Yínjiāng Zhèn ("Yin River Town") the boats were waiting to take the group back to Níngbō.[493] As parting presents they gave A'chang an English umbrella, a pencil-case, and a few other foreign articles he had shown an interest in.[494]

Fig. 5.140 View down the valley from Lǐcūn to Zhāng Xī. "The scenery in this glen is more strikingly beautiful than that in any part of the province which has come under my observation..." (Fortune, 1857: 193). The reservoir postdates Fortune's visit.

Once Fortune returned to Níngbō, he had to supervise the transfer of the plant collections he had left in the gardens of the British Vice-consul, John A. T. Meadows (**Box 5.22**) and Mr. Wadman to Thomas C. Beale's garden in Shànghǎi.[495]

Box 5.22

John A. T. Meadows (ca. 1810 – 1875) Son of a merchant, who appears to have traded with Germany and elder brother of Thomas Taylor Meadows (1815 – 1868).[1] He arrived in Guǎngzhōu in January 1845 and immediately started learning Mandarin.[2] In 1846 he began to study Cantonese, becoming public translator for Guǎngzhōu from September 1846 until November 1849.[3] During this time he acted as interpreter for the consular representatives of Belgium, France, the Netherlands and Prussia, and translated from the "Peking Gazette" for the "Hong Kong Register".[4] In the first half of 1850 he took over the job of interpreter at the British Consulate from his brother Thomas.[5] On 1 August 1850 he replaced Charles A. Sinclair (**Box 3.6**) as interpreter at the consulate in Níngbō.[6] He must have struck an impressive figure with his tall build, light blue eyes, and beard.[7] He was in charge of the British Consulate at Níngbō from April 1853 until March 1855.[8] His work involved dealing with mandarins, river police, smugglers, pirates, and collecting customs duties.[9] An incident in which a Portuguese subject was unlawfully given 35 lashes with a rope on Christmas Day 1854 led to his suspension, and he returned to his job as interpreter.[10] He took over as Consular Interpreter in Shànghǎi when Charles A. Sinclair was appointed Vice-consul at Níngbō in April 1856.[11] After more than two years in Shànghǎi, he returned to Guǎngzhōu, where his career had begun.[12] In 1861, after marrying a Chinese woman, he was obliged to resign from the consular service, and went into business.[13] He was to become an important member of the business community in Tiānjīn, acting as US Vice-Consul and consul for Denmark and the Netherlands.[14] His dinners were considered the best to be had.[15] After his death his illiterate Chinese wife led such a footloose life that their 3 daughters were sent to Europe "to safeguard their moral welfare".[16]

There he packed everything into Wardian Cases for shipment to Hong Kong and the subsequent voyage to India.[496] To make sure that nothing happened to his valuable cargo, he accompanied the cases to Hong Kong, where he divided the freight into four parts, which he sent by separate ships to Kolkata.[497] Luckily, none of the ships or their Wardian Cases was lost on the way.[498]

Before returning north, Fortune made a visit to Guǎngzhōu ("Extensive Prefecture") to see how scented- and caper teas were produced.[499] He observed that the dried tea leaves became scented after they had been mixed with fragrant flowers for 24 hours.[500] After this the flowers were removed, and any moisture taken up by the tea leaves expelled by drying them once more.[501] The caper teas were manufactured by rewetting the dried leaves, stuffing them into canvas bags, and expressing the moisture by treading on the bags.[502] While in Guǎngzhōu, he visited the Huādì ("Flowery field") nurseries in the company of a Mr. M[a]cDonald, a Chinese scholar.[503] Together they enjoyed the often weird wording of the notices intended for the Chinese visitors.[504]

At the beginning of April 1854 Fortune got a berth on a schooner bound for Fúzhōu ("Good Fortune Prefecture").[505] By visiting Fúzhōu, he hoped to be able to make arrangements to obtain plenty of tea seed and some tea manufacturers from the black tea district of Wǔyíshān, which lay near the source of the Mǐn Jiāng ("Fújiàn River").[506] As only a few foreign ships ever called at Fúzhōu, he was lucky to get a passage on SS "Confucius", which had been hired by the Manchu Government to deliver soldiers and money to Táiwān to help quell rebellions in Jiayi and Fengshan (near Kaohsiung, Gāoxióng), before returning to Shànghǎi.[507] The "Confucius" would have been anchored at Mǎwěi ("Horsetail") near Pagoda Island, 21 km downriver from Fúzhōu, where the Mǐn Jiāng is deeper.[508] This "island" is famous for its lighthouse, the 31.5 m high Luóxīng Tǎ (usually known as the "Falling Star Pagoda", but more correctly translated as the "Catching Star Pagoda", **Fig. 5.141**), erected in the Southern Sòng Dynasty (1127 – 1279 AD) by Lady LÍU Qīniáng in memory of her husband and rebuilt in 1621 – 1627 after it was destroyed by an earthquake or typhoon.[509]

Fig. 5.141 Luóxīng Tǎ, Mǎwěi, on Pagoda "Island". The pagoda was erected by Lady LIÚ Qīniáng in memory of her husband. It was originally known as the Móxīn Tǎ or "Tormented Heart Pagoda".

Fortune no doubt watched this landmark recede into the distance, as the "Confucius" set course for Táiwān in April 1854. By the time they reached the mouth of the Mǐn Jiāng, it was too late to thread their way between the treacherous sandbanks in the estuary.[510] They hoved to. Because the ship with its bullion was now a sitting target for pirates, Captain Dearborn considered it advisable to set a watch.[511] In the event, nothing happened, so they were able to head for northern Táiwān at daybreak.[512] With a strong northerly wind and heavy seas, it took all day to cross the 200+ km to Dànshuǐ ("Freshwater").[513] This meant they once more had to ride at anchor.[514] By the next morning Fortune was desperate to disembark to see what the large white flowers were, which he had spotted through a telescope.[515] They turned out to be none other than the endemic *Lilium formosanum*, which was common throughout Táiwān (**Fig. 5.142**).[516] Nearby, Fortune caught sight of the attractive shrub, *Tetrapanax papyrifer* (Syn. *Aralia papyrifera*), from which edible rice-paper is produced (**Fig. 5.143**).[517] Fortune got a Chinese soldier to dig up a few small plants for him to take to Shànghǎi.[518]

Fig. 5.142 *Lilium formosanum*, spotted by Fortune from the deck of the SS Confucius. "Before leaving the vessel I had been examining with a spy-glass some large white flowers which grew on the banks and on the hill-sides, and I now went in that direction, in order to ascertain what they were." (Fortune, 1857: 231/232).

Fig. 5.143 The rice-paper plant, *Tetrapanax papyrifer*, which Fortune took back to Shànghǎi with him. "The stems, usually bare all the way up, were crowned at the top with a number of noble-looking palmate leaves, on long footstalks, which gave to the plant a very ornamental appeaance." (Fortune, 1857: 232). The edible paper is prepared from the pith, which can also be used as fish bate as it floats.

Fortune spent the day exploring the countryside near Dànshuǐ.[519] He was suitably impressed by the fertile valleys and the hospitality of the population.[520] Most of those living in the coastal regions would have been Han Chinese or Hakkas, who immigrated to Táiwān after the island was annexed by the Manchus in 1683.[521] The original population, which is a Polynesian race, was a different kettle of fish, being given to headhunting, as many shipwrecked sailors discovered to their dismay.[522]

After dropping the man who had been responsible for the bullion at the mouth of the Mǐn Jiāng, so he could report its safe delivery, the "Confucius" steamed on to Shànghǎi.[523] By the time Fortune returned to Shànghǎi, Spring had arrived. The trees were covered with fresh green leaves and the birds were singing in every bush and tree.[524] Fortune enjoyed the riot of colour that announced the end of winter, namely the yellow trumpets of *Forsythia viridissima*, the pure white blossoms of *Spiraea prunifolia*, the pink papilionate flowers of *Cercis*, and the lilac blooms of *Daphne genkwa* (Syn. *D. fortunei*).[525]

Fortune soon set off for the tea districts of Zhèjiāng to order more seeds and plants.[526] He concentrated much of his effort on the Sìmíng Shān area, which he had visited the previous autumn. He once more travelled up the meandering Fènghuà Jiāng as far as Yínjiāng Zhèn.[527] He went to bed early and was up and about by 4 a.m. the next morning. Even at this early hour the roads were full of people converging on Yínjiāng Zhèn because a fair was to be held.[528] By the time Fortune got back to the town at 8 a.m. the main temple was the scene of enormous activity, which reminded him of the coming and going of bees at a hive.[529] The fortune tellers, who one still encounters near temples, were doing a brisk trade.[530] At the fair Fortune purchased a small porcelain bottle, which he thought resembled those found in ancient Egyptian tombs, and a porcelain seal such as those unearthed in Irish bogs.[531] In the afternoon he attended an opera performance, which went on until well after dark.[532]

To the northwest of the town there was a dilapidated Buddhist convent, the Léngshuǐ Ān ("Cold Water Nunnery") dating back to the Míng Dynasty, which Fortune visited in the summer when the heat became unbearable (**Fig. 5.144**).[533] Under the temple is a cellar carved out of the living rock from which a current of cold air still issues (**Fig. 5.145**).[534]

Fig. 5. 144 Léngshuǐ Ān (Cold Water Nunnery), which is now a monastery. "When complaining of the excessive heat to some of my visitors, I was recommended to go to a place called by them the *Lang-shuy-ain*, or "cold water temple," situated in the vicinity of the town in which I was staying." (Fortune, 1857: 262).

Fig. 5. 145 A corner of the cool cellar of Léngshuǐ Ān, where "a strong current of cold air was coming out of the earth at this particular point ... Beggars, sick persons, and others who had taken refuge from the heat of the sun were lolling about, evidently enjoying the cool air which filled the place." (Fortune, 1857: 263).

Having spent several days in the Yínjiāng Zhèn area, Fortune decided to visit Tiānjǐng Sì ("Heavenly Well Temple") in the hills about 8 km WNW of Yínjiāng Zhèn.[535] At first the narrow, well-paved road followed the Zhāng Xī, which by October had been reduced to a trickle.[536] In order to get their produce to market, the villagers were making use of rafts and flat-bottomed boats.[537] Beyond Hòulōngcūn the road left the river and entered a wide and highly cultivated valley.[538] The European in his mountain chair proved to be quite a sensation in the small towns he passed through, and he was not infrequently invited for tea.[539] Groves of the multi-purpose bamboo, máo zhú (*Phyllostachys edulis*) at the western end of the valley signalled that he had reached his destination.[540] Tiānjǐng Temple itself was not very imposing, but it was surrounded by impressive mountains covered by a mixed coniferous and hardwood forest.[541] Apart from the oaks, chestnuts and the widespread *Pinus massoniana*, there grew the golden larch.[542] These imposing conifers reached a height of almost 40 m.[543]

When Fortune entered the temple, the Buddhist monks, who had probably never seen a European, were dumbfounded.[544] One little boy even asked "If I go near him, will he bite me?"[545] The little boy was not the only person who was afraid of Fortune. The chief monk, whose room was behind Fortune's, could not be persuaded to pass the "foreign devil", so ended up sleeping in another room.[546]

Fortune was disappointed that the golden larch was not coning near the Tiānjǐng Temple in 1854, so gladly accepted the offer of WÁNG Ā'nào to show him some other trees near Guāndǐng Sì (Quanting Temple) about 3 km away.[547] The next day Fortune set off with a guide for his appointment with Mr. WÁNG. They ascended a steep pass, through mixed temperate forest and stands of bamboo before reaching Pǔxī Valley.[548] The inhabitants of the village viewed the stranger with a mixture of curiosity and fear.[549] In order to allay their fears, Fortune sat down beside an old man and talked to him.[550] This broke the ice, and Fortune soon found himself and his garments being examined in minute detail.[551] From Pǔxī it was another uphill slog until Fortune and his guide reached the summit.[552] After winding along the ridge in a westerly direction for some 600 m, they turned south and dropped down into the Guāndǐng Valley, where the temple was located (**Figs. 5.146, 5.147**).[553] Mr. WÁNG was busy in the kitchen preparing a meal for a party of visitors.[554]

Fig. 5.146 Guāndǐng Reservoir (1958), the site of Guāndǐng Sì, which is now under water. "The temple of Quan-ting has no pretensions as regards size, and appeared to be in a most dilapidated condition." (Fortune, 1857: 283).

Fig. 5.147 The corner of Guāndǐng Reservoir where the temple used to be.

While he was waiting for Mr. WÁNG to finish cooking, Fortune was taken to a small pond behind the dilapidated temple to see the golden bell in the water.[555] Unfortunately, as a non-Buddhist, he only saw some aquatic plants![556] The meal over, WÁNG, Fortune and Fortune's guide set off in a westerly direction for the golden larches.[557] It proved to be a long trek, involving a protracted climb at the head of Guǎndǐng Valley, followed by a gradual descent to the stand of golden larches.[558] As these trees were not coning either, Fortune dug up some seedlings to send to England.[559] Leaving WÁNG to return to the Guǎndǐng Temple, Fortune and his guide headed back over the mountains to Tiānjǐng Temple by another route, arriving as it was getting dark.[560] The next day Fortune set off for Níngbō, where he had to prepare the material he had collected during the summer and autumn for shipment to Shànghǎi.[561]

Box 5.23

Alexander Perceval Jr. (1821 – 1866) Third son of the Irish politician Colonel Alexander Perceval Sr. (1787 – 1858) and his wife Jane Anne L'Estrange (1790 – 1847).[1] He was presumably educated by private tutors. Through his relative Mary Jane Perceval, the wife of the opium magnate Sir James Matheson (1796 – 1878), he learned of the stupendous fortunes to be made in the China trade. He joined Jardine, Matheson & Co. in 1846, and by 1855 was the managing partner in Shànghǎi.[2] On 22 July 1858 he married Annie De Bois.[3] When Robert Jardine (1825 – 1905) returned to London in 1860, Perceval succeeded him as CEO in Hong Kong.[4] There he was able to add to his already large fortune. He was a member of the Legislative Council of Hong Kong from 1860 to 1864.[5] In 1860 he bought back his ancestral home, which his eldest brother Philip had been forced to sell on the death of their father in 1858.[6] In 1862, while he was still in China, he started to expand the already large mansion in preparation for his return to Ireland.[7] He retired to "Temple House" in 1864, but died two years later, leaving a widow and five children.[8]

As the coastal waters were infested by pirates, he was glad when Alexander Perceval (**Box 5.23**) gave him permission to travel on the "Erin", one of Jardine, Matheson & Co.'s well-armed sailing ships.[562] He had the company of two members of the Church Missionary Society, Rev. John Shaw Burdon (**Box 5.24; Fig. 5.148**) and Rev. John Hobson (Trinity Church, Shànghǎi) and his family during the passage.[563]

Fig. 5. 148 Bishop John Shaw Burdon (1826–1907), who travelled on the "Erin", one of Jardine, Matheson & Co's well-armed sailing ships.

Box 5.24

Bishop John Shaw Burdon (1826 – 1907) Only son of James Burdon and his second wife Isabella.[1] After his father's premature death, he was brought up by an uncle in Liverpool.[2] In 1850 he was accepted as a missionary by the Church Missionary Society (CMS) and spent two years training at Islington College (London).[3] He was ordained deacon on 19 December 1852 and sailed for Shànghǎi on 20 July 1853.[4] Burdon travelled widely in China, being the first CMS missionary to visit Běijīng, Hángzhōu, Shàoxīng and Yúyáo.[5] Fortune met him in 1854 on the "Erin" when he was on his way from Níngbō to Shànghǎi.[6] Three years later on 11 November 1857 Burdon married Burella Hunter Dyer (1835 – 1858), one of the daughters of the missionary Samuel Dyer (1804 – 1843).[7] However, on 16 August 1858 she died of cholera.[8] After the Second Opium War, Burdon became vicar to the British Legation in Běijīng during 1865 – 1872.[9] While there he contributed to the translation of the New Testament from Greek into Mandarin, and with the Lithuanian missionary Samuel Isaac Joseph Schereschewsky (1831 – 1906) completed a translation of the Prayer Book in 1872.[10] Thomas Francis Wade (**Box 10. 10**) recommended him for the job of English teacher at the Translation Bureau of the Chinese Foreign Ministry.[11] When Bishop Charles Richard Alford resigned in 1873, Burdon became Bishop of Victoria and Principal of St. Paul's College in Hong Kong during 1874 – 1897.[12] After resigning both posts, he retired to Běihǎi ("North [of the] Sea", Guǎngxī, Pakhoi), but in 1901 was forced by failing health to return to England, where he spent his final years with his youngest son, Edward Russell Burdon.[13]

Box 5.25

Captain Daniel Patridge (1820s – 1876+) Parents unknown. He was already involved in the opium trade as a young boy.[1] As third mate on board Jardine, Matheson & Co.'s brig "Ann" he took part in the storming and looting of Zhènhǎi on 10 October 1841.[2] Five months later, on 8 March 1842, the ship loaded with loot from Zhōushān headed for Macao, but was blown off course and wrecked off N Táiwān on 10 March 1842.[3] The crew were taken to the capital Táinán (Táiwānfǔ) where they were shackled, interrogated and suffered imprisonment among myriads of fleas, bugs, lice, ants, mosquitoes, cockroaches, centipedes and rats.[4] A trip undertaken by Rev. David Abeel (**Box 3.1**) to obtain the prisoners' release proved unsuccessful.[5] As the First Opium War drew to a close it was probably considered expedient to eradicate all trace of the mistreatment of the prisoners ("dead men tell no tales").[6] At the beginning of August many of the survivors were beheaded.[7] Three months after the signing of the Treaty of Nánjīng, and three days after Sir Henry Pottinger branded the executions as a war crime, Patridge was finally released.[8] He then dropped one of the "r"s from his surname, either to avoid being traced or because he thought it sounded better.[9] As captain of the "Erin" he was responsible for taking the consular mail from Níngbō to Shànghǎi, and remitting fees to Hong Kong.[10] In 1854 Patridge shared a house in Níngbō with the young Robert Hart (1835 – 1911), who described him as "a jolly fellow and a capital messmate".[11] Patridge was an outspoken man, with somewhat radical ideas, such as dispensing with the House of Lords and the right to sit in the presence of Queen Victoria.[12] He worked for Jardine, Matheson & Co. until 1872, when deteriorating health forced him to retire to London, where he wrote an account of his enforced stay on Táiwān.[13]

Taking advantage of ebb, Captain Daniel Patridge (**Box 5. 25**) sailed down the Yǒng Jiāng ("Níngbō River") to Zhènhǎi ("Calm Sea"; **Fig. 5.149**), where the river debouches into the sea.[564] As they headed for Jīntáng Island, they realized that pirate ships were seizing those vessels trying to leave Zhènhǎi.[565] However, when the pirates saw that it was the "Erin", they signalled that they did not intend to attack.[566] The rest of the trip to Shànghǎi passed without incident.

In the Winter of 1854/1855 and Spring 1855 Fortune was kept busy packing and dispatching his collections to India and Europe.[567] In April 1855 he paid another visit to Cíchéng.[568] He again moored the boat on the north side of the old city.[569] While he was there, he had a visit from two of his friends from Guǎngzhōu (Smith and Walkinshaw).[570] As William Walkinshaw also collected old vases, Fortune took him to an antiquarian to show him a vase he had just been offered for $80.[571] Walkinshaw also admired the vase, but the dealer could not be persuaded to part with it at a reasonable price.[572]

Fig. 5.149 Zhènhǎi (Calm Sea), the port of Níngbō, with the Áozhù Pagoda in the background.

The trio returned to Níngbō, and while Fortune got ready to visit the Húzhōu ("Lake Prefecture") area, Walkinshaw and Smith went to see Xuědòu Shān and the waterfall which Fortune had visited in the autumn of 1853.[573] The three friends then hired a boat each to take them part of the way to Shànghǎi by the inland route pioneered by Fortune in May 1845 (see Chapter 3).[574] It took them a day to reach Yúyáo, where they must have passed under the Tōngjì Bridge (**Fig. 5.39**), which linked the ruinous southern suburbs with the northern part of town.[575] On the northern bank, close to the bridge stands the Shùnjiāng Tower.[576] Having passed the bridge, they reached Lóngquán Shān ("Dragon Spring Hill"; **Fig. 5.38**) with its temples and Zhōngtiān ("Zenith") Pavilion, where the philosopher WÁNG Shǒurén (**Box 5.26**; **Fig. 5.150**) once taught.[577]

Fig. 5.150 WÁNG Shǒurén (1472–1529), an important philosopher who taught in Yúyáo.

Box 5.26

WÁNG Shǒurén (1472 – 1529) is probably the most famous philosopher that Yúyáo has ever produced.[1] He was the son of an earl and minister.[2] At the age of ten he already believed the most important thing in life was to become a sage or a worthy.[3] Having graduated at 21, Shǒurén held some minor governmental posts before becoming a provincial judge at the age of 30.[4] He resigned a year later to devote himself to Daoism and Buddhism, both of which he subsequently rejected.[5] At 33 he became a successful general, known for his strict discipline.[6] A year later he offended a corrupt eunuch, XÍNG Liè, by urging the profligate teenage Emperor Zhèngdé (1491 – 1521) to release two innocent officials from prison.[7] XÍNG Liè had him beaten and banished to Guìzhōu, where he had to look after the horses for dispatch riders during 1507 – 1510.[8] In this job he had no access to libraries, so he developed his philosophical concepts simply by thinking about various aspects of human life.[9] He argued that from birth everybody has an innate knowledge of the difference between good and evil.[10] Although WÁNG realized that the mind, being incorporated into the body, cannot act entirely independently, he stressed the need to free the mind of selfish desires, which cloud the purity of thought.[11] In 1516 WÁNG became Governor of South Jiāngxī.[12] As a Neo-Confucian he believed in the feudal system.[13] As a result, he faced a peasant revolt during 1517 – 1518, which he suppressed by sowing discord among the insurgents and tightening his control on the households.[14] After the province was pacified, he rehabilitated the rebels and started a programme of reconstruction.[15] Schools were also built.[16] In 1521 he was appointed Minister of War.[17] Later in life he returned to Yúyáo where as ZǏ Yángmíng ("Master Yangming", hence he is often referred to as WÁNG Yángmíng) he taught philosophy at Zhōngtiān ("Zenith") Pavilion on Lóngquán (Língxù) Hill.[18] He criticized the practice of learning texts by rote, as promoted by another great Neo-Confucian philosopher, ZHŪ Xī (1130 – 1200).[19] As a consequence, he was boycotted by ZHŪ Xī's followers.[20]

As they still had four hours of daylight, Fortune and his friends forced the crew to press on westwards.[578] Some distance from Yúyáo, the boats had to be winched up an inclined slope from one canal into another.[579] Unfortunately, recent rainfall had caused this N-S canal to be so full, that the boats needed to have their roofs removed in order to pass under the bridges.[580] The largest boat was therefore sent back to Níngbō.[581] Unfortunately, the further the friends went, the worse the situation became. Near Sǎmén ("Spill Gate") Fortune had to transfer his luggage to a small sand-pan and send his boat back too.[582] Rather than crowd into one boat, the friends decided to walk to Hángzhōu Wān ("Hángzhōu Bay").[583] While they were waiting for their ferry to Gǎnpǔ to arrive, the trio investigated the different methods that were employed to produce salt.[584]

The cabins on the ferry were so dirty and smelly that Fortune, Smith and Walkinshaw decided to sleep on deck.[585] Although they encountered some rough water in the middle of the bay, the ferry only took about three hours to cross the 23 km to Gǎnpǔ.[586] To save Fortune and his friends getting their clothes dirty, they were given a piggyback across the mudflats to *terra firma*.[587]

Fig. 5.151 Mulberries grown for silk production at Gǎnpǔ. "It is thought by some, and with pretty good reason, that this place is the same as that mentioned in Marco Polo's travels under the name of Kan-foo. In his day it was the seaport of Hang-chow-foo, and was frequented by ships from India and other parts of the world." (Fortune, 1857: 322).

As Fortune pointed out, Gǎnpǔ was probably the port referred to by Marco Polo (1254 – 1324) as Kan-p'u.[588] It faced out over the East China Sea and used to do a roaring trade as the seaport of Hángzhōu, the capital of Zhèjiāng.[589] Unfortunately, the large quantities of sediments being washed down the Cháng Jiāng caused siltation on a massive scale.[590] As a result, the city became cut off from the sea and sank into oblivion. When Fortune visited the place on 1 June 1855 it was no more than an insignificant town kept alive by some silk production and a few junks ferrying passengers and pigs across Hángzhōu Bay.[591] Today little remains of this once flourishing city. However, mulberry is still grown for silk manufacture (**Fig. 5.151**), as it was in Marco Polo's day.[592]

From Gǎnpǔ the friends walked some 3 – 4 km to Liùlǐ Cūn ("Six Li Village") to hire boats to take them to Shànghǎi.[593] They passed through Yuánhuā ("Yuán flower") on their way to Pínghú ("Calm Lake").[594] Here they had a break to buy some ancient porcelain and visit some nurseries.[595] Wherever they went, they were followed by a boisterous crowd, so the tradesmen were probably relieved when the foreigners left their premises.[596] Although Fortune had told his friends to try not to lose their tempers, he himself made the capital error of remonstrating with the crowd.[597] He soon regretted his error, as stones started to rain down from the city ramparts.[598] The friends were glad to get to Shànghǎi on 3 June 1855 without further incident.[599]

5.10 A Visit to the Silk District

The restless plant hunter only stayed in Shànghǎi for four full days, before setting off up the Sùzhōu Hé (Suzhou Creek) for the silk district of Húzhōu.[600] Fortune was sitting on deck in the evening of 8 June 1855 enjoying a cigar when the stench of decomposition reached his nostrils.[601] He then realized he was passing the headquarters of the Imperial Army, where thousands of rebels and suspected supporters of the Small Sword Uprising (**Box 5.21**) had been executed a few months earlier.[602] They had gone less than 20 km along the Sùzhōu Hé, when the canal leading to Nánxiáng (Nanziang) and Jiādìng (Cading) came into sight on the right. Fortune spent a few days at Jiādìng to examine a variety of cultivated plants, such as cotton, dyestuffs (safflower, woad), fruit (apples and melons) and bamboos, and collect ground beetles.[603] He employed children and old women to look for the Carabidae under stones, but had difficulty getting them to stop once he had 40 – 50 specimens of each species.[604]

Once he had completed this work, Fortune took a canal leading back to the Sùzhōu Hé. Continuing in a southerly direction, he eventually reached Qīngpǔ ("Green Riverside"). It was dark by the time his boat reached Diànshān Hú ("Silted Mountain Lake").[605] As this lake resembles a maze, Fortune's crew rightly refused to proceed any further in the darkness.[606] The boat was already under way, when Fortune woke at daybreak, and it was then that the plantsman realized the folly of attempting to travel at night.[607] Diànshān and the other lakes they crossed were only a few metres deep, so water-lilies (*Nymphaea tetragona*, *Euryale ferox*, **Fig. 5.152**) and water-chestnut (*Trapa natans*) were common.[608]

Fig. 5.152 ***Euryale ferox*, a large waterlily Fortune encountered in Diànshān Lake.**

The party finally found their way out of the maze of lakes and onto the Dà Yùnhé (Grand Canal) just north of the bustling town of Píngwàng.[609] At Píngwàng the canals fork. Fortune's men chose the righthand canal leading to Nánxún (Nan-tsin) and Húzhōu. Today the provincial road G318 runs parallel to this canal. In Fortune's day there was a paved pathway for pedestrians, so the plant hunter was able to leave the boat and look at the mulberry and other plants growing on the embankment.[610] Having crossed from Jiāngsū ("River revival") into Zhèjiāng Province, they arrived at Nánxún.

When Fortune visited Nánxún, this city of canals was one of the major centres of the silk trade, which in those days was still very much a cottage industry.[611] Fortune noted how the country people would arrive in the city soon after daylight with little parcels of silk to sell to the merchants.[612] The contrast between the animated peasants in front of the counter and the stoic silk inspectors behind it amused him greatly.[613] Judging from the amount of money which some merchants were able to invest in their houses and gardens (**Fig. 5.153**), the silk trade was a very lucrative business.[614]

Fig. 5.153 A rich merchant's house and garden in Nánxún, reflecting the profits that were to be made in the silk trade.

Fortune was impressed by the clean, healthy and contented appearance of Nánxún's inhabitants.[615] He spent a few days in the environs of Nánxún investigating the cultivation, grafting techniques and pruning of the mulberry bushes, which supplied the leaves for the larvae of the Silkworm Moth, *Bombyx mori*.[616] Once this inspection was complete, he sailed along the same canal to the other important silk centre of Húzhōu on 17 June 1855.[617]

Fig. 5.154 Fēiyīng Pagoda (1234 – 1236), Húzhōu, described by Fortune as "a pretty pagoda" (Fortune, 1857: 352). It represents the last extant pagoda-in-pagoda in China.

Fig. 5.155 The inner stone pagoda (884 – 894 AD), built to house one of Buddha's bones. It is covered in miniature images of Buddha.

Leaving the boat on the rundown south side of town, Fortune proceeded to walk in a northerly direction until he almost reached the North Gate.[618] He was surrounded by a large crowd, most of whom had probably never seen a European before.[619] With the stoning at Pínghú only two weeks previously, Fortune tried to remain good-humoured, although it cannot have been easy to examine the works of art in the antiquarian shops with so many people milling around.[620] One of the crowd offered to take him to another curio shop close by, but actually led him to a temple near the North Gate where a play was being performed.[621] Word spread that there was a foreigner in the audience, and Fortune soon found that he rather than the actors was the focus of attention.[622] As the play ground to a halt, Fortune decided he had better leave.[623] At the North Gate he admired the 55 m high Fēiyīng Pagoda (**Figs. 5.154, 5.155**), which was first built in 884 – 894 AD to house one of Buddha's bones.[624]

What Fortune actually saw was the outer shell, which was built in 1234 – 1236 after the original structure dating from 968 – 976 AD had been destroyed by lightning in 1150/1152.[625] Leaving the "pretty pagoda" behind, he walked round the ramparts to the East Gate, and then crossed the town from east to west.[626] It was quite late by the time Fortune got back to the boat. Having been on the go since morning, he was tired and just wanted to get some peace.[627] In order to shake off the inquisitive crowd, he asked the crew to moor the boat in a small creek on the east side of the Dōngtiáo Xī ("Eastern Chinese Trumpet Creeper Rivulet") Canal.[628]

Fortune spent the next few days visiting the neighbouring silk villages and adding to his entomological collections.[629] Having completed this work, he decided to visit the nearby temple of Wànshòu Sì ("Long-live Temple", **Fig. 5.157**) founded in 881 – 884 AD.[630] Starting from Dàochǎng Bāng ("Buddhist Rites Creek"), Fortune followed the pine-lined road to the temple (**Fig. 5.156**).[631] After some refreshment at the temple, he climbed the steps to the Dàochǎngshān Tǎ ("Buddhist Rites Mountain Pagoda"; **Fig. 5.158**).[632]

Fig. 5. 156 Tree-lined road leading to Wànshòu Sì (Long-live Temple).

Fig. 5.157 Wànshoù Sì, founded 881 – 884 AD, "a large and imposing building, or rather collection of buildings, founded about a thousand years ago by a certain Fuh-hu-shan-si—the 'Tamer of the Tiger.'" (Fortune, 1857: 362).

Fig. 5.158 Dàochǎngshān Tǎ (Buddhist Rites Mountain Pagoda), 1078–1085 AD (rebuilt 1987 – 1988), from which Fortune got a good view of Húzhōu. It is no longer possible to climb the tower.

Fig. 5.159 Wúshān Sì, founded 581 – 618 AD. Fortune was conducted "over all the halls and temples of the monastery, which, although very extensive, were in a most delapidated condition." (Fortune, 1857: 367). The present, much smaller, temple was erected in 2006.

To the north lay Húzhōu and Tàihú, to the east the flat agricultural land he had started from, while in the west the landscape was filled with mountains.[633] Fortune "gazed long with rapture upon the wonderful scene which lay beneath and around me."[634]

While Fortune was in the Húzhōu area, he learned of a river (Xītiáo Xī) with a source in the Tiānmù Shān to the SW of Húzhōu.[635] He determined to follow the river as far as it was navigable.[636] About 30 km west of Húzhōu Fortune spied a patch of rich vegetation with a large monastery to the south of the river.[637] This was Wúshān Sì (**Fig. 5.159**) founded in the Suí Dynasty (581 – 618 AD).[638] WÚ Chéng'ēn (1501 – 1582) is said to have written part of his epic novel "Journey to the West" here in 1568.[639] During its prime this temple had many courtyards and an almost perfect number of rooms, i.e. 99.5![640] However, by 1855 it was already in a very dilapidated condition.[641] It was abandoned shortly after Fortune's visit, during the Tàipíng Rebellion (**Box 5.19**), and completely destroyed some time prior to 1963.[642] When Fortune arrived at the temple, he was surprised to find that the dimly lit halls and outhouses were filled with silkworms.[643] After staying for a few days in the vicinity of the temple, Fortune continued on to Jīnjiātáng ("Jin Family Pond"), where he spent two days collecting insects and making notes on the local produce.[644] He finally arrived at the town of Méi Xī ("Plum Rivulet", **Fig. 5.160**), which was as far as the Xītiáo Xī was navigable for boats of any size.[645] Although Méi Xī is about 180 km from Shànghǎi, the low gradient means that the river is still tidal at this point.[646] During the rainy season (June and July) the tides prevent adequate runoff, with the result that the tributaries tend to overflow their banks, causing widespread destruction.[647] This explains the numerous embankments encountered in this area.[648]

Fig. 5.160 Méi Xī (Plum Rivulet). Although 180 km from Shànghǎi the Xītiáo Xī is still tidal. "Mei-che is a long town on the banks of the stream, and as the river is no longer navigable for the low-country boats a considerable business is done here in hill productions, which are brought down for sale." (Fortune, 1857: 371).

Fig. 5.161 The dissected leaves of paper mulberry, *Broussonetia papyrifera*, used in papermaking. Trees are planted on the embankments near Méi Xī.

These embankments are used for growing the paper mulberry (*Broussonetia papyrifera* , **Fig. 5.161**), which supplies the bark used in papermaking.[649] The low-lying land close to the rivers is occupied by mulberry plantations, while the tea estates are found on the lower slopes of the hills.[650]

When Fortune made a call at Wúshān Sì on his way back to Húzhōu, the monks were busy reeling and sorting the silk.[651] Fortune spent a day with the head monk, who was exempt from this work.[652] In the evening the monk accompanied Fortune back to the boat.[653] As a parting present, Fortune gave the monk copies of the magazines "Punch" and "Illustrated London News", which he had been reading.[654] Further downstream Fortune observed some villagers fishing from canoes. On one side of the canoe was a broad strip of white canvas, which dipped into the water when the man in the stern of the canoe lent over to starboard.[655] The fish finding themselves in shallow water tried to leap over the canoe, only to be caught by a net placed on the port side.[656] Fortune and his men watched the proceedings for more than an hour before continuing on their journey to Húzhōu.[657] The crew worked through the night, and by morning Fortune found himself back in Húzhōu.[658]

Fortune spent the next few days near the southern shore of Tàihú and in Nánxún, before heading back to Jiādìng to collect some seed which was still unripe when he was there in the second week of June 1855.[659] Pleased with the success of the mission to the "Silk Country", Fortune had almost reached Jiādìng in mid July (1855) when his boat grounded in the shallow water of the canal near Nánxiáng.[660] There was nothing to do but wait for the flood tide at 2 – 3 a. m. the next day.[661] When Fortune retired to bed between 9 and 10 p.m., it was still stuffy, so he left the small windows in the side of the boat open.[662] About 2 a.m. he was woken by a loud yell, which indicated that they had been robbed.[663] Although one of his servants and two of the crew plunged into the canal in pursuit of the thieves, the latter got away with Fortune's trunk containing his journal, irreplaceable drawings, accounts for the EIC, more than 100 Shànghǎi dollars, knives, pencils, and even the clothes he had been wearing.[664] Fortune swam to the bank of the canal to join the search for his property, but the only things they found were a few Manila cheroots, which the thieves had dropped during their get-away.[665] Leaving his men posted in the long grass in case the thieves should return to pick up some of the booty, a sopping Fortune returned to the boat to brood over his loss.[666] About an hour later a voice called out in the darkness from the other bank "come over here and receive the 'white devil's' trunk and clothes."[667] By the time Fortune's men had reached the opposite bank, the thieves had vanished.[668] However, to Fortune's relief, they had returned the trunk with everything except the money.[669] The thieves' motive for returning the trunk was probably to cover their tracks.[670] However, as a grateful Fortune realized, they could just as easily have destroyed anything they did not need.[671]

At daylight Fortune got dressed and went to report the loss to the chief mandarin in Nánxiáng.[672] Although two thieves were rapidly apprehended and given repeated beatings with a bamboo cane, Fortune's money still had not surfaced after two days.[673] Rather than hang about in Nánxiáng indefinitely, Fortune returned to Shànghǎi and reported the matter to the British Consul, Daniel Brooke Robertson (**Box 5.27**).[674] A few weeks later the Vice-consul, Frederick Harvey (**Box 5.28**) handed Fortune a handkerchief containing $35 and a few miscellaneous objects.[675] Fortune rightly or wrongly assumed that the mandarin had pocketed the rest of the money for services rendered.[676]

Box 5.27

Sir Daniel Brooke Robertson FRGS (1810–1881) Son of Daniel Robertson (ca. 1770–1849) and his wife Amelia Helen Clarke (ca. 1792–1869).[1] At the age of 16 he went out to India with the East India Company's merchantile naval service.[2] Fearing that he had less chance of promotion after one of his influential grandfathers died, he studied law, and was called to the Bar in 1840.[3] In 1842 he was employed by the British-Portuguese Commission for the settlement of the British Legion's claims stemming from the Portuguese Succession crisis (**Box 5.1**).[4] He was given a position at the British Consulate at Shànghǎi in December 1843.[5] When he sailed from Portsmouth in January 1844, his wife had to stay behind because their baby was ill.[6] On 20 July 1850 he took charge of the British Consulate at Níngbō, when Vice-consul George G. Sullivan moved to Xiàmén.[7] He allowed Fortune to use the consular garden for stockpiling his plants.[8] Although he was officially unable to intervene in disputes between Chinese and Portuguese citizens, he sent evidence of the cases to Hong Kong for transmission to Macao.[9] In a dispatch he declared "Ningpo people are polite to foreigners, and I am sorry to find foreigners not polite to them."[10] Robertson's efforts on the diplomatic front seem to have been recognized by his superiors, for he was made Vice-consul in Shànghǎi on 20 February 1851.[11] Two years later he was appointed Consul in Xiàmén.[12] 1853 was a turbulent year in Xiàmén's history. Rebels drove the Manchu troops out of the city on 18 May.[13] When the imperial forces recaptured the city on 11 November 1853 a blood-bath among the civilian population ensued.[14] After maintaining neutrality for four hours, Robertson stepped in to put an end to the bloodshed.[15] In May 1854 he was appointed Consul in Shànghǎi, but because he also had to run the Guǎngzhōu Consulate, he did not take charge in Shànghǎi until 9 March 1855.[16] It was here that Fortune met him once again in July 1855 at the end of his trip to the "Silk Country".[17] In the same year Robertson added a further feather to his cap by becoming Honorary Danish Consul in Shànghǎi.[18] During his time in Shànghǎi, he often had to deal with drunken seamen who went on the rampage.[19] Chinese persons and property were frequently the target of their excesses.[20] In one case in which a Chinaman was murdered, Robertson gave temporary assistance to the destitute and pregnant widow.[21] He also saw to it that pirates were rapidly apprehended and executed.[22] However, he was helpless to prevent the traffic in human beings (coolies), who were often kidnapped or enticed aboard what were in effect slave ships.[23] On 21 December 1858, while still at Shànghǎi, he was appointed British Consul in Guǎngzhōu.[24] He only held this position for a short period of time, for in 1859 he took his first home leave since arriving in China.[25] In January 1861 he became British Commissioner for the Kowloon Peninsula, which was ceded to Britain after the Second Opium War.[26] Having lived in China for most of his life, during which time he had made a point of cultivating the friendship of Chinese officials, Robertson had the ability to see things from a Chinese perspective.[27] The wise old widower, living in his residence in the heart of Guǎngzhōu surrounded by his cats and dogs, willingly gave advice on matters as diverse as an accidental shooting, the coolie traffic, the destruction of Prussian mission chapels, and a Christian convert accused of mutilating dying children.[28] He had little sympathy for the missionaries' complaints, pointing out that if Buddhist or Daoist priests built temples in England and went round denouncing Christianity as a farce, they would be in grave trouble.[29] Since he supported China's right to search its national vessels within Chinese territorial waters in an effort to curb opium smuggling, he was accused by the business community in Hong Kong of not representing British interests.[30] There were acrimonious exchanges with Sir Richard Graves Macdonnell (1814 – 1881) and Sir Arthur Edward Kennedy (ca. 1809 – 1883), successive Governors of Hong Kong from March 1866 until April 1877.[31] Robertson stood firm: "I care very little for the opinion of the governor of Hong Kong or his subjects. I know the position better than they do. I am not going to jeopardise friendly relations with the Chinese to suit their book."[32] He was clearly a man of principles. He probably also had a sense of humour, as he chose to retire on 1 April 1879.[33] He had just two years to live.

Box 5.28

Frederick E. B. Harvey (ca. 1828 – 1884) Son of James Vigers Harvey, British Consul at Bayonne (SW France).[1] When his father died in the early 1840s, he was appointed to an assistantship in China at 17.[2] On the voyage to China he taught himself the basic elements of Chinese writing.[3] He proved to be an able and efficient administrator.[4] In 1854 he was sent by Sir John Bowring (1792 – 1872) to investigate a riot at Xiàmén in which two British citizens were assaulted and a number of rioters and innocent bystanders shot and killed.[5] He replaced Thomas Francis Wade (**Box 10.10**) as Vice-consul in Shànghǎi in December 1854.[6] Four years later he became Consul in Níngbō.[7] He was in this Treaty Port, when it fell to the Tàipíngs on 9 December 1861.[8] Although he was treated courteously during the five months that the Tàipíngs held Níngbō, Harvey did not believe that their feelings were genuine.[9] He harboured an intense hatred for the Tàipíngs and their doctrines, and was partly responsible for the bad press they got in the UK.[10] When the ousted governor returned with a rabble to retake Níngbō he gave his permission for the British to shell the city, while his servant ZHÈNG Tóngchūn played an active role in the ensuing blood-bath on 6 May 1862.[11] In 1865 he was transferred to Zhènjiāng, which he referred to as an "insignificant hole".[12] Within two years he became ill and retired.[13] He lost all his money in the collapse of Dent & Co. (**Box 2.4**) in 1867, and when he died he left his wife without any means to support herself.[14] In 1902 and again in 1910 his widow applied unsuccessfully to the Foreign Office for a pension.[15]

When Fortune went to the British Consulate in Shànghǎi to report the theft, Consul Robertson no doubt told him that he had been able to find nine tea manufacturers from the Póyáng Lake area who were willing to go to India.[677] There was also good news from Fúzhōu. David Oakes Clark (1826 – 1883) of the American firm of Russell and Co. informed him that eight Fújiàn black tea manufacturers were already on their way to Hong Kong.[678] Fortune's old friend Charles A. Sinclair (**Box 3.6**), who had just moved from Xiàmén to Shànghǎi in March 1855, drew up a contract for Fortune and the Jiāngxī men to sign.[679] The tea manufacturers were paid an advance of $100 to support their families while they were away.[680] When the day of departure arrived only eight of the Jiāngxī men turned up.[681] However, a substitute was found at the last minute, so on 10 August 1855 Fortune and the nine men boarded a P&O steamer bound for Hong Kong.[682] In Hong Kong they met Clark's Fújiàn tea manufacturers.[683] A few days later both groups of tea manufacturers were on their way to Kolkata on board the P&O steamer "Chusan" (**Box 5.29; Fig. 5.162**).[684]

Fig. 5.162 SS Chusan (1852), which took Fortune's tea manufacturers to India, where "after having numerous adventures, which they related to me afterwards with great glee, they all arrived in safety and good health at their destination in the Himalayas." (Fortune, 1857: 397).

Box 5.29

SS Chusan (1852 – 1861) This propeller-driven, iron-hulled steamship was the first ship built by Miller, Ravenhill & Co. in their new yard at Walker-on-Tyne.[1] It was named after the Chusan, now Zhōushān, Archipelago in Zhèjiāng, China.[2] This ship was chosen by P&O to inaugurate the Royal Mail service between the UK and Australia.[3] The ship's arrival in Sydney on 3 August 1852 was an occasion for great celebration.[4] Captain Henry Downs and his crew were lionized and a ball held on 26 August at which the Chusan Waltz composed by Henry Marsh was performed.[5] Between November 1852 and April 1854 the ship ran between Australia and Singapore.[6] From May 1854 until August 1856 she linked Hong Kong with Kolkata.[7] As from 1858 she plied between Hong Kong and Shànghǎi, and Hong Kong and Manila.[8] Then in 1859 she was transferred to the Shànghǎi—Japan run.[9] She acted as a troopship on a number of occasions, most notably at the beginning of the Second Opium War.[10] On 3 June 1861 P&O sold her to R. D. Sassoon of Hong Kong.[11] After that she had various owners. Adrian & Co. (Shànghǎi) sold her in 1867 to the Matsuyama Clan (Japan), who renamed her "Kofuyo".[12] The "Kofuyo" was captured in 1868 by Chosen-han, who had her fitted out as a warship called the "Kayo".[13] In 1872 the "Kayo" was sold to C. Farnham of Shànghǎi and reduced to a hulk.[14]

5.11 Taking Leave of China

In the autumn of 1855 Fortune returned north by another P&O steamer commanded by Captain Jamieson.[685] As the ship neared Shípǔ, it was stopped by HMS "Bittern" and SS "Paou-shan", that had just waged a successful campaign against a pirate fleet.[686] As a few of his men had been wounded in the action, Commander Edward Vansittart (**Box 5.30**) requested Captain Jamieson to take the injured to Shànghǎi for treatment.[687]

Box 5.30

Vice-admiral Edward Westby Vansittart CB (1818 – 1904) Third son of Vice-admiral Henry Vansittart (1777 – 1843) and his wife Mary Charity Pennefather (1793 – 1834).[1] He entered the Royal Navy in June 1831, became midshipman on the brig HMS "Jaseur" in 1813 during the Carlist War in 1833 – 1839 (**Box 5.1**), and was mate in the 74-gun sailing ship "Wellesley" in 1815 at the Fall of Karachi in February 1839.[2] In December 1841 he was assigned to HMS "Cornwallis", the flagship of Sir William Parker (1781 – 1866), and saw action during the First Opium War (**Box 2.2**).[3] He was mentioned in dispatches and was promoted lieutenant on 16 September 1842.[4] After three years on the 16-gun brig HMS "Serpent", and a short stint on board the new paddle steamer "Gladiator", he joined Sir William Parker's 110-gun flagship the "Hibernia" in the Mediterranean.[5] He became Sir William's aide-de-camp.[6] Further promotion followed. On 1 January 1849 he became first lieutenant on the royal yacht "Victoria and Albert" commanded by Lord Adolphus FitzClarence (1802 – 1856).[7] The tall, strongly-built officer, must have made a good impression, as he was promoted to Commander on 23 October 1849.[8] On 25 August 1852 he was given charge of the ironclad brig, HMS "Bittern" charged with combating piracy in Chinese waters.[9] In September and October of 1855 he destroyed some 40 pirate junks at Shípǔ (Shie-poo) and out-manouvered the pirates waiting to destroy his landing party.[10] As thanks, a handsome subscription was raised by the British merchants, but the self-effacing commander turned this down, suggesting it be used to erect a stained-glass window in the church of his native Bisham (Berkshire).[11] His real reward came in the form of promotion to Captain on 9 January 1856.[12] In 1857 in another encounter with 30 – 40 pirate junks, their commander, the renegrade American sailor Eli Boggs, was captured.[13] On 18 November 1859 he was put in charge of the newly commissioned HMS "Ariadne".[14] This 26-gun, propeller-driven frigate was part of the squadron which accompanied the Prince of Wales (afterwards King Edward VII) in the "Hero" to the USA and Canada in 1860.[15] After more than four years as captain of the "Ariadne", he took command (1864 – 1868) of HMS "Achilles", the first ironclad battleship to be built in a Royal Navy dockyard. The last ship he commanded before retiring was the brand-new, propeller-driven, ironclad battleship, HMS "Sultan".[16] After retiral on 20 July 1873, he was promoted Rear-admiral on 19 January 1874 and finally Vice-admiral on 1 February 1879.[17] He died at home on 19 October 1904.[18] In his will he bequeathed £ 300 for the benefit of the poor of Bisham.[19]

As Fortune intended to leave China for good at the end of the year, he needed to settle a number of outstanding accounts and inform the people who had been supplying him with plants, seeds and invertebrates that he would no longer be requiring their services.[688] So, having arrived in

Shànghǎi, he immediately transferred to his favourite stamping-ground, Níngbō. At the end of October 1855 he was back at the temple of Tiānjǐng in the hope of finding seed of the golden larch.[689] At the village of Pǔxī he saw two fine specimens of Chinese Torreya (*Torreya grandis* Fortune ex Lindley 1857) growing in a garden.[690] Although these had come from Máohuò Cūn [" Thatched (*Imperata*) Cookingpot Village "] some 7 km due west of Pǔxī as the crow flies, Fortune was not to be put off.[691] He set off with the garden's owner for this locality. The journey was not easy, as it involved crossing a number of subtropical valleys and climbing a seemingly never-ending series of ridges with a plant cover of only wiry grass, gentians and spiraeas.[692] It was about 4 p. m. by the time they finally reached their destination (**Fig. 5.163**).[693]

Fig. 5.163 Máohuò Cūn (Thatched Cookingpot Village), "a pretty little town situated on the banks of a small stream which takes a winding course through the mountains to the eastward, and eventually falls into one of the branches of the Ningpo river." (Fortune, 1857: 414). It was here that Fortune bought seed of *Torreya grandis* and *Pseudolarix amabilis*.

Fortune was rewarded by the sight of numerous torreyas reaching 18 – 24 m in height.[694] As the seeds had already been collected, he had to order these and seeds of golden larch from a hill farmer.[695] By the time they had struck a bargain it was nearly 5 p.m. and almost dark.[696] The mist started to roll down the mountains and blotted out the landmarks, which the travellers needed to find their way back to Pǔxī.[697] To add to their misery, it had started to rain.[698] The sodden companions frequently had to retrace their steps.[699] Finally the guide admitted that he was lost.[700] For more than an hour they shivered in the lee of a granite outcrop, before the guide thought "he discerned a light at no great distance".[701] It turned out to be one of the shielings used during the maize harvest.[702] At first Fortune's appearance scared the inmates, an old woman and a boy, who retreated into a corner, but eventually the ice was broken.[703] The visitors roasted themselves and some corn cobs at the fire before falling sound asleep on the straw-covered floor.[704]

Fig. 5.164 Early morning scene near Máohuò Cūn. "A misty cloud hung here and there lazily on the sides of the hills, which only had the effect of making the sky look more clear and the scene around and below us more grand and lovely." (Fortune, 1857: 419/420).

When they woke in the morning, the sun was shining and the only sign of the previous night's downpour were some clouds of condensation rising from the forest vegetation (**Fig. 5. 164**).[705] On the clear autumnal day, it was no longer any problem to find their way back to the temple of Tiānjìng. Fortune spent another night there before returning to Níngbō.[706] One last visit to a nursery in Níngbō yielded a variegated form of *Farfugium japonicum* (Syn. *F. grande*), which was added to the long list of ornamental plants which Fortune introduced into Britain.[707]

Fortune thought that this was to be his last trip to this part of the world, so it cannot have been easy to say farewell to Zhèjiāng and his friends in Shànghǎi. From Shànghǎi he sailed to Hong Kong, and from there to Guǎngzhōu, where he engaged some scented-tea manufacturers and lead-box makers for India.[708] Fortune and the men caught Jardine's brand-new SS "Lancefield" with destination Kolkata.[709]

5.12 Second Tour of Inspection in Northern India

When they arrived in Kolkata on 10 February 1856 Fortune must have been disappointed to find that his friend Hugh Falconer (**Box 5.7**) was no longer there. The next day Fortune informed the Governor-General of India, Lord Dalhousie (**Box 5.32; Fig. 5.165**) that he had returned. By this time the worn-out governor-general was preparing to leave office, but he got his secretary Cecil Beadon (**Box 5.31**) to write to Fortune complimenting him on his work in China.[710] This pleased the plantsman no end.

Fig. 5.165 James Andrew Broun Ramsay, 10th Earl of Dalhousie (1812–1860), who complimented Fortune on his mission to China.

Box 5.31

Sir Cecil Beadon (1816 – 1880) Youngest of five children of Richard Beadon (1779 – 1858) and his wife, Annabella À Court (1781–1866).[1] After an education at Eton College and Shrewsbury School, he joined the Bengal Civil Service in 1836. In Bengal he met and married Harriet Sneyd (1818 – 1855), whose father Ralph Henry Sneyd (1784 – 1840) was a major in the Bengal Cavalry.[2] They had five children.[3] His administrative qualities and good sense were appreciated by successive governor-generals.[4] Under Lord Dalhousie (**Box 5.32**), he was responsible for the introduction of a uniform rate of postage in British India.[5] As Secretary to the Indian Government, he sent a letter to Robert Fortune in which he expressed Lord Dalhousie's approval at the success of Fortune's second mission on behalf of the East India Company.[6] Shortly after Fortune's visit to India, the Indian Mutiny (**Box 6.4**) broke out. Having been accused of underestimating the severity of the Indian Mutiny, Beadon subsequently cracked down on the native population.[7] During his Lieutenant-Governorship of Bengal from 1862 to 1866, he showed a real interest in Fortune's endeavours by visiting the tea plantations in Assam.[8] While he was at Darjeeling recovering from overwork, famine hit Orissa from 1866 to 1867. Although he could hardly be blamed for the famine, he played down its severity and refused any increase in government spending to the stricken area.[9] A subsequent inquiry criticized his handling of the crisis and he returned to England a broken man.[10]

Box 5.32

James Andrew Broun Ramsay (1812 – 1860) The third and youngest son of George Ramsay (1770 – 1838), 9th Earl of Dalhousie, and his plant-collecting wife Christian Broun (1786 – 1839), and cousin of Henry Ramsay (**Box 5.13**).[1] He spent his boyhood in Canada, where his father was Governor of Nova Scotia from 1816 – 1820 and then Governor-General of British North America from 1820 to 1828.[2] When he turned ten, he was sent back to the UK to study.[3] He attended Harrow School for a number of years before studying at Oxford University from 1829 to 1833.[4] After that he did some travelling in Italy and Switzerland.[5] Although he was a good orator, his attempt to become the MP for Edinburgh at the general election in 1835 was unsuccessful.[6] On 21 January 1836 he married Lady Susan Georgina Hay (1817 – 1853), the eldest daughter of his neighbour, the Eighth Marquess of Tweeddale.[7] The devoted couple had two daughters, Susan Georgiana (1837 – 1898) and Edith Christian (1839 – 1871).[8] In 1837 he was elected Conservative MP for Haddingtonshire, but when his father died in 1838, he became the 10th Earl of Dalhousie (both his elder brothers died young).[9] As a member of the House of Lords, he devoted his time and energy to Scottish issues.[10] He became Vice-president of the Board of Trade in 1843 and its President in 1845.[11] In this capacity, he was responsible for vetting proposed railway lines.[12] In 1846 he turned down an offer of a position in the Cabinet of Lord John Russell (1792 – 1878), but when the Prime Minister offered him the Governor-Generalship of India in 1847, he accepted on condition that he would be given a free hand.[13] Leaving their daughters in the UK, Lord and Lady Dalhousie sailed for Kolkata in November 1847.[14] On 12 January 1848 he was sworn in as the youngest Governor-General of India.[15] Notwithstanding the hot and humid climate, he worked on average eight hours a day, eating his lunch while at his desk.[16] He expected the same devotion to duty of his subordinates, in whom he took a personal interest.[17] Although he was obliged to entertain with the magnificence that became his high position, he much preferred the simple family dinners with his musical wife.[18] During his eight years as Governor-General, he took a special interest in the education and emancipation of women, promoting a bill to allow widows who were no longer obliged to commit sati (suttee, ritual burning), to remarry (Hindu Widows Re-marriage Act, 1856).[19] During his governorship British India grew by about 40% by applying the doctrine of lapsing successions, the Second Sikh War in Punjab during 1848 – 1849, the annexation of Sikkim in 1850 following the kidnap of the botanist Sir Joseph Hooker, and the conquest of Lower Burma in 1852.[20] In order to speed up communications between these far-flung parts of British India, he promoted railways, introduced a uniform postal rate, and established a telegraph system.[21] The latter was no mean feat, as the telegraph posts were removed by the natives for firewood and rafters, undermined by porcupines and bandicoot rats, and reduced to powder by termites.[22] They had to be encased in cast-iron sockets.[23] However, these innovations in turn encouraged commerce, communication, and thus welded India together.[24] Overwork and the death of his beloved wife undermined his health, and by the time he handed over responsibilities to Charles John Canning (1812 – 1862) on 29 February 1856 he was only a shadow of his former self.[25] After he arrived in the UK on 13 May 1856 his health continued to deteriorate.[26] He became a cripple, and was only able to walk with the help of crutches.[27] He was saddened to learn of the Indian Mutiny (**Box 6.4**), which could have been avoided if his plans to reduce the number of native soldiers and increase the British troops had been implemented.[28] Nonetheless, he has frequently been given the blame for the rebellion.[29] He was only 48 when he died of kidney failure.[30]

The tea plantations which Fortune visited in the North-West Provinces and Punjab in the next few months were found to be in an excellent state.[711] He submitted his official report to the Government of British India in October 1856, and having nothing more to keep him in India, left Kolkata on board the paddle-steamer "Bentinck" (**Box 5.33; Fig. 5.166**) on 9 November 1856.[712]

Fig. 5.166 SS Bentinck (1843), which Fortune travelled on from Kolkata to Suez in 1856.

Box 5.33

SS Bentinck (1843 – 1860) This wooden paddle-steamer, built for the Peninsular and Oriental Steam Navigation Company (P&O) by Thomas Wilson & Co. (Liverpool) was named after Lord William Bentinck (1774 – 1839), the cost-cutting Governor-General of India between 1833 and 1835, who was also known for suppressing sati, female infanticide and ritual murder.[1] Her hold was divided into a number of water-tight compartments, aimed at strengthening the vessel and preventing her from sinking.[2] She was originally designed to carry 102 first class passengers, 50 passengers' servants, and a crew of 173.[3] The well-ventilated cabins and the smoke-burners invented by the Irish shipowner and P&O Director, Charles Wye Williams (ca. 1780 – 1866), were aimed at making the passage Kolkata-Madras-Ceylon-Aden-Suez as comfortable as possible.[4] In 1850 the "Bentinck" returned to Britain for an extensive refit, including more and better cabins and the replacement of fixed paddle wheels by feathering paddles, which remained in a vertical position while in the water.[5] Due to the increased drive, the "Bentinck" was then able to achieve a top speed of more than 10 knots.[6] In this way the ship could cover the distance from Galle (Ceylon) to Suez in 17 days, as Fanny Loviot, the Frenchwoman kidnapped by pirates in 1854, recounted.[7] Ten years after her refit, the ship was finally sold to the Indian Government on 11 May 1860.[8]

The ship, commanded by Captain Caldbeck, called at Madras and Galle (Ceylon, now Sri Lanka) before crossing the Indian Ocean to Aden, and up the Red Sea to Suez.[713] After the trek through the desert, Fortune and the other passengers caught a train from Cairo to Alexandria.[714] There they boarded another P&O ship at the beginning of December. Fortune arrived in Southampton on 20 December 1856, just in time to celebrate Christmas with his family.[715]

Chapter 6 Intermezzo: Moving Up in the World

6.1 Back to London

Fortune had been away for exactly four years when his ship docked in Southampton on 20 December 1856.[1] He got back to London just a few days before Christmas. Helen (16) and John (13) would have remembered their father, but Mary, who was only a few months old when he left, would have no memory of her dad.

If he had not already learned about it before he left India, Fortune would now have heard about the "Arrow Affair" (**Box 6.1**), which was causing a furore in parliament. The Torys under Edward George Smith-Stanley, the 14th Earl of Derby sensed an opportunity to topple the Whig government.[2] Bowring and Parkes came in for a great deal of criticism for their handling of the incident.[3] Even some of Lord Palmerston's own party including his rival Lord John Russell (1792 –1878), the reformers Richard Cobden (1804 – 1865) and John Bright (1811 – 1889), and Sir Joseph Paxton (**Box 5.14**; **Fig. 5.91**) attempted to block the PM's call for war.[4] Paxton's friend William Gladstone (1809 – 1898), then a Tory, delivered a forthright speech in which he accused the government of unleashing "the whole might of England against the lives of a defenceless people".[5] When the vote went Gladstone's way, Palmerston simply dissolved the government and held new elections at the end of March/beginning of April 1857 ("Chinese elections").[6] As Benjamin Disraeli (1804 – 1881) had prophesied, many British including Fortune felt that it was high time to teach the Chinese a lesson, so the result was a landslide victory for the pro-war faction.[7] Having been discredited, in March 1857 Bowring was stripped of all his important functions, and replaced by James Bruce, the 8th Lord Elgin (1811 – 1863; **Fig. 6.1**) as Plenipotentiary.[8] In the next four years, as a result of the Second Opium War and the "Treaty of Yeddo" in 1858, Elgin went on to open up China and Japan to the western world. Without his intervention Fortune would have been unable to visit N China or Japan (see Chapters 8 and 9).

Fig. 6.1 James Bruce, the 8th Lord Elgin (1811–1863).

Fig. 6.2 YÈ Míngchēn (1807–1859), Governor of Guǎngdōng.

Box 6.1

Arrow Affair According to the Treaty of Nánjīng Guǎngzhōu was to be one of the five Treaty Ports open to foreigners. In fact, it remained largely closed with the exception of the foreign factories on the Pearl River. When Sir George Bonham (**Box 5.18**) was replaced by Sir John Bowring (1792 – 1872) in April 1854, the latter tried to meet the Governor of Guǎngdōng, YÈ Míngchēn (1807 – 1859; **Fig. 6.2**), but YÈ continually used his involvement in the suppression of the Tàipíng Rebellion (**Box 5.19**) as an excuse for not seeing Bowring.[1] On 8 October 1856 a minor incident triggered a chain of events that finally led to the unequal Treaty of Tiānjīn (**Box 9.1**), and the invasion of N China in 1860. While the lorcha "Arrow", a Chinese-owned vessel captained by Thomas Kennedy, was lying at anchor at Guǎngzhōu, the shipowner HUÁNG Liánkāi, spotted a pirate on board.[2] He reported this to the harbour authorities, who sent a squad of marines to investigate the claim.[3] They arrested the Chinese crew on suspicion of piracy and smuggling.[4] Kennedy complained to the Acting British Consul, the hot-headed Harry Parkes (1828 – 1885), who assuming the lorcha to be British, demanded an apology and the release of the crew within 48 hours.[5] When this did not happen, a Chinese junk was seized.[6] Parkes apparently had not bothered to check that the "Arrow" was no longer registered in Hong Kong.[7] Although YÈ released the crew of the ship on 22 October, Bowring had already made up his mind to teach him a lesson.[8] The Plenipotentiary sent the somewhat reluctant Admiral Sir Michael Seymour (1802 – 1887) with a squadron to Guǎngzhōu.[9] In the course of a week between 23 and 29 October, Seymour captured some of the Chinese forts, bombarded Guǎngzhōu, and sunk 23 Chinese war junks.[10] On 30 October YÈ placed a bounty on every English head, and refused to meet Bowring when the Plenipotentiary travelled to Guǎngzhōu in mid-November to see him.[11] On December 5 when a British sailor was killed, the village implicated was raised by UK troops.[12] The Chinese retaliated by burning the foreign factories on 14 – 15 December and killing any foreigners they encountered.[13] In a tit for tat the British troops set fire to thousands of houses in Guǎngzhōu on 12 January 1857.[14] This did not go unavenged. On 15 January 1857 many foreigners in Hong Kong became violently ill after breakfast, having eaten bread laced with arsenic.[15] While most recovered, Lady Maria Bowring never regained her health and died on 27 September 1858.[16]

Since Seymour's force was inadequate to capture and hold the spreading metropolis of Guǎngzhōu (with its population of more than one million), he withdrew but not before requesting reinforcements.[17] These were slow in coming, because of the political debate at home, new elections, and the fact that the first expeditionary force from the UK had to be diverted to India to deal with the Indian Mutiny (**Box 6.4**).[18] James Bruce (Lord Elgin) had to wait until the beginning of December for reinforcements to replace those that had stayed in India.[19] On 12 December 1857 he and his French counterpart Baron Jean-Baptiste Gros (1793 – 1870), who was enraged by the cruel death of the missionary Auguste Chapdelaine (1814 – 1856) in Guǎngxī, demanded direct negotiations, the payment of an indemnity, and a tract of land as compensation.[20] Since YÈ was in no position to grant these, he gave an evasive answer.[21] On 15 – 16 December Hénán Island, south of the Pearl River was occupied, and although Elgin had his misgivings, Guǎngzhōu was bombarded on the 28th, the day on which the "Massacre of the Innocents" was observed.[22] After shelling the city for 27 hours all resistance was over.[23] YÈ was captured on 5 January 1858 and taken to Kolkata where he died the next year on hunger strike.[24] In his place a provisional government ostensibly led by the Chinese but controlled by the French and British allies administered Guǎngzhōu for the next three years.[25]

One of Fortune's immediate concerns would have been to see how the plants collected during his second EIC trip were faring (**Fig. 6.3**). This time he had entrusted his material to Robert Glendinning (**Box 6.2**), who had a nursery close to the Horticultural Society's garden in Chiswick, some 7 km west of Thistle Grove Lane, where Fortune lived. On a fine day he could have walked there and back. He seems to have been satisfied by what he saw, as he refers to the material as "carefully cultivated" in his article "Notes on some Chinese plants recently introduced to England" in "The Gardeners' Chronicle".[9] He would have been particularly pleased by the progress of the golden larch, *Pseudolarix amabilis*, which Fortune considered his most important Chinese introduction.[10] The seedlings dug up in 1854 were thriving, and Glendinning had managed to germinate the seed, which Fortune had sent him in October 1855.[11]

Fig. 6.3A ***Amydalus triloba*** **var.** ***plena*****, a commonly grown flowering almond introduced to the UK by Fortune in 1855;**

Fig. 6.3B ***Ligustrum sinense*****, a deciduous ornamental shrub. The small fruits are eaten by birds, which disseminate the plant. It has become an invasive species in the eastern USA;**

Fig. 6.3C ***Pinus bungeana*****, a slow-growing pine, generally with multiple upright stems;**

Fig. 6.3D ***Pinus bungeana*** **is grown for its attractive scaly bark;**

Fig. 6.3E ***Pseudolarix amabilis*****, the golden larch, which Fortune considered "as the most important of all my Chinese introductions." (Fortune, 1857: 415). The asymmetrical whorls of needles are characteristic of this species;**

Fig. 6.3F ***Syringa oblata*****, an early-flowering lilac, commonly grown in China. It was introduced to the UK by Fortune in 1856.**

Box 6.2

Robert Glendinning (1805 – 1862) He took up gardening at an early age.[1] In 1824, while still a teenager, he went to England and worked for three years in Somerset.[2] He was then employed by the gardening enthusiasts Lord John Rolle (1750 – 1842) and his second wife Lady Louisa Rolle (née Trefusis, 1794 – 1885) at Bicton House, 17 km SE of Exeter (Devonshire).[3] He would seem to have satisfied his employers, as he remained their Head Gardener for 11 years.[4] During this time he created a lake in the grounds, and initiated the rockery and adjacent American Garden.[5] He summarized his experience of growing tropical fruits for his lordship's table in a treatise on the pineapple.[6] During his time at Bicton he collaborated with the Scottish landscape gardener John Claudius Loudon (1783 – 1843) in planting an arboretum.[7] This was considered one of the most complete arboreta of its day.[8] As the outspoken Loudon himself commented "It is a great mistake to suppose that the state of the gardens and the botanical riches of a country residence depend on the taste of the proprietor or his family; it depends much more on the knowledge and the tact of the head gardener."[9] In 1840 Glendinning became a partner of Lucombe and Pince in Exeter, hence his son born in 1840 was christened Robert Pince Glendinning.[10] In the Spring of 1843 he bought the rather neglected "Chiswick Nurseries" at Turnham Green from John Graham.[11] In their heyday these nurseries had belonged to Richard Williams, who specialized in plants from South Africa and Australia, and was responsible for propagating the pear sent to him from Aldermaston (Berkshire) in the late 18th Century, which subsequently became known as "Williams' Bon Chrétien".[12] Glendinning set about restoring the nurseries to their former glory by extending and remodelling them and building new glasshouses.[13] An advertisement appeared on the front page of "The Gardeners' Chronicle" on 16 December 1843 announcing the changes.[14] While he built up a reputation as a cultivator of heaths, conifers and fruit trees, he continued to seek employment as a landscape gardener.[15] For instance, in the 1850s he redesigned the Horticultural Society's arboretum at Chiswick.[16] This involved thinning the wild undergrowth, the creation of new mounds for evergreens and yuccas, and the laying down of new gravel paths.[17] As a member of the Horticultural Society's Council[18], he undoubtedly knew

Fortune. No doubt, he suggested that Fortune send him some seed from China. During his first expedition on behalf of the East India Company (EIC) in 1848 – 1851, Fortune sent him fruits of a new holly (*Ilex fortunei*) to cultivate.[19] Fortune was obviously pleased with the results, because he sent Glendinning most of the plants from his next EIC expedition during 1853 – 1856.[20] According to Fortune these were "carefully cultivated".[21] That this collaboration was thereafter terminated may have something to do with Glendinning's "severe and long-continued illness".[22] At the time of Glendinning's death, the firm was well-established.[23] His wife and sons continued the business for another nine years before selling the nursery to Henry Ewing.[24]

Box 6.3

Fortune's Plant Introductions from China 1853 – 1856

The name/names in brackets is/are the name(s) under which Fortune's material was once known. Bold type indicates that this was the first time the taxon was grown outside China. A question mark indicates that Fortune may not have been the first person to introduce the plant.

? *Amygdalus persica* Linnaeus 1753 'Camelliaeflora' (*Prunus persica* (Linnaeus 1753) Batsch 1801 'Camelliaeflora')[1]

? *Amygdalus persica* Linnaeus 1753 'Dianthiflora' (*Prunus persica* (Linnaeus 1753) Batsch 1801 'Dianthiflora')[2]

***Amygdalus triloba* (Lindley 1857) Ricker 1917 var. *plena* (Dippel 1893) S. Q. Nie 1982** (*Amygdalus pedunculata* Pallas 1789 var. *multiplex* Bunge 1833, *Prunus triloba* Lindley 1857 var. *plena* Dippel 1893, *Prunus triloba* Lindley 1857 forma *multiplex* (Bunge 1833) Rehder 1924, *Prunus triloba* Lindley 1857 'Multiplex')[3]

Callicarpa dichotoma (Loureiro 1790) K. Koch 1872 (*Porphyra dichotoma* Loureiro 1790)[4]

***Camellia japonica* Linnaeus 1753 'Cup of Beauty'**[5]

***Camellia japonica* Linnaeus 1753 'Princess Frederick William'**[6]

***Castanea mollissima* Bunge 1850**[7]

***Castanea seguinii* Dode 1908**[8]

***Cyclobalanopsis myrsinifolia* (Blume 1850) Oersted 1871** (*Quercus myrsinifolia* Blume 1850)[9]

? *Farfugium japonicum* (Linnaeus 1767) Kitamura 1939 (*Farfugium grande* Lindley 1857)[10]

Ilex cornuta Lindley & Paxton 1850 (*Ilex fortunei* Lindley 1857)[11]

***Ligustrum sinense* Loureiro 1790** (*Ligustrum fortunei* C. K. Schneider 1911)[12]

Lilium formosanum Wallace 1891 (*Lilium japonicum* auct. non Thunberg 1780)[13]

***Pinus bungeana* [Zuccarini ex] Endlicher 1847** (*Pinus excorticata* Lindley & Gordon 1850)[14]

***Pseudolarix amabilis* (J. Nelson 1866) Rehder 1919** [*Abies kaempferi* (Lambert 1824) Lindley 1833, *Chrysolarix amabilis* (J. Nelson 1866) H. E. Moore 1965, *Laricopsis fortunei* (Mayr 1890) Mayr 1906, *Laricopsis kaempferi* (Lambert 1824) A. H. Kent 1900, *Larix amabilis* J. Nelson 1866, *Larix kaempferi* (Lambert 1824) Carrière 1856, *Pinus kaempferi* Lambert 1824, *Pseudolarix fortunei* Mayr 1890, *Pseudolarix kaempferi* (Lambert 1824) Gordon 1858][15]

***Rhododendron fortunei* Lindley 1859**[16]

***Strobilanthes cusia* (Nees 1832) Kuntze 1891** [*Goldfussia cusia* Nees 1832, *Ruellia indigotica* Fortune 1857 nomen (Fortune, 1857: 158)][17]

***Syringa oblata* Lindley 1859** [*Syringa vulgaris* Linnaeus 1753 var. *oblata* (Lindley 1859) Franchet 1891][18]

***Syringa oblata* Lindley 1859 ‘Alba’**[19]

? *Taxus cuspidata* Siebold & Zuccarini 1846 [*Taxus baccata* Linnaeus 1753 var. *cuspidata* (Siebold & Zuccarini 1846) Carrière 1867, *Taxus baccata* Linnaeus 1753 subsp. *cuspidata* (Siebold & Zuccarini 1846) Pilger 1903][20]

Tetrapanax papyrifer (W. J. Hooker 1852) K. Koch 1859 (*Aralia papyrifera* W. J. Hooker 1852)[21]

***Torreya grandis* Fortune ex Lindley 1857**[22]

? *Toxicodendron vernicifluum* (Stokes 1812) F. A. Barkley 1940 (*Rhus verniciflua* Stokes 1812)[23]

6.2 Working on the Book

Once he was satisfied that the plants were in safe hands, Fortune settled down to work on his latest book *A Residence among the Chinese*. By April 1857 it was largely completed. On 4 April John Murray was writing to him[12]:

> My dear Sir
>
> I think your new Work on China likely to succeed as well as the last. I shall be happy to give you for the Copyright £ 200 by note at Six Months from the day of Publication together with 25 Copies of the Book. If you require any more you shall have them at the trade price—The sooner you can let me hear in reply to this the better as the Season is passing away.
>
> My dear Sir,
Yours very sincerely
John Murray

Fortune replied four days later (**Fig. 6.4**)[13]:

> My dear Murray
>
> It will give me great pleasure to come to you tomorrow at 7-The MS. will require, probably to be looked over by some carefull [sic] hand before it is sent to press. My last two works were carefully done by some practised hand whose alterations I fully appreciated. We must also have a consultation about the sketches—some of them I think you said might be improved.
>
> Yours truly
R. Fortune

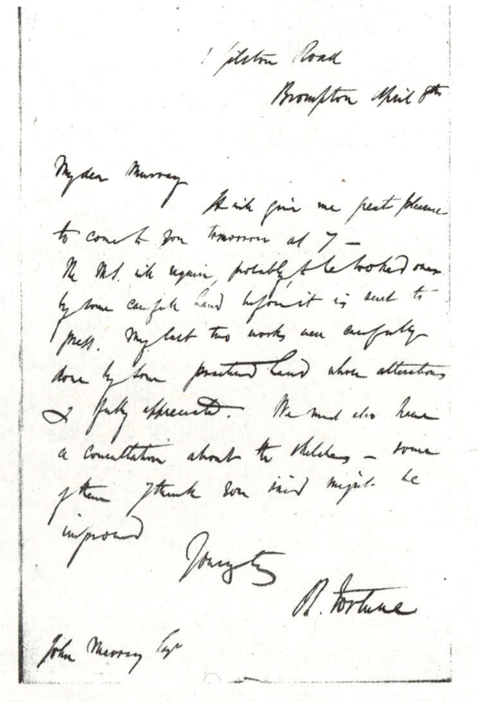

1 Gilston Road
Brompton April 8th

My dear Murray

It will give me great pleasure to come to you tomorrow at 7—
The MS. will require, probably to be looked over by some careful hand before it is sent to press. My last two works were carefully done by some practised hand whose alterations I fully appreciated. We must also have a consultation about the sketches — some of them I think you said might be improved

Yours truly
R. Fortune

John Murray Esq

Fig. 6.4 Fortune's letter to John Murray, 8 April 1857.

By 28 May 1857 all the problems had been resolved.[14] Murray's Ledger E records that 1500 copies were printed on June 30 and another 500 copies on December 31.[15]

In the meantime, news was coming out of India of a revolt of native soldiers (**Box 6.4; Fig. 6.5**). Fortune must have found it difficult to imagine that this mutiny could have started in the sleepy town of Meerut, which he had passed through at the beginning of August 1851.[16] As the uprising spread, Fortune must have followed the newspaper accounts with interest and no doubt with mounting horror. He was probably particularly shocked by the events at Kanpur (Cawnpore). The starving British garrison and their families had been offered a safe passage if they capitulated.[17] This they did on 27 June 1857.[18] However, no sooner had they boarded the boats that were to take them to Allahabad, than shots rang out (**Fig. 6.6**).[19]

Fig. 6.5 The death of Colonel John Finnis, the first British officer to be killed during the Sepoy Uprising (1857).

Fig. 6.6 The massacre at Kanpur (Cawnpore), which must have shocked Fortune.

The male survivors were executed, and the women and children imprisoned.[20] Finally, when it looked as though they were about to be rescued, on 15 July the 200 captives were all brutally murdered by five ruffians, as the sepoys were unwilling to do the job.[21] When Henry Havelock's relief force arrived on the scene, they found the floor caked with blood, blood-stained clothes scattered about, and a well choked by mangled human remains.[22] Understandably, the backlash was draconic. Brigadier-General James Neill (1810 – 1857) in particular was noted for hanging almost every Indian who crossed his path.[23] This earned him the nickname of "The Butcher of Cawnpore".[24] Like many of his contemporaries, who had a poor opinion of the Indians, Fortune probably considered the retributions just.

Box 6.4

Indian Mutiny also known as **the First Indian War of Independence** (1857 – 1859) Both terms are misnomers, as the fighting was largely confined to the Ganges Valley in N India and involved only a small proportion of the population.[1] The Sepoy or Sipahi Rebellion might be a better appellation (sepoy = native soldier; sipahi = Hindi for soldier). A number of factors led to the uprising. To start with the East India Company's (EIC's) policy of annexation which Lord Dalhousie (**Box 5.32**) implemented led to discontent particularly in the Province of Awadh (Oudh, now Uttar Pradesh), where many of the EIC's Indian soldiers were recruited.[2] These sepoys were commanded by British officers, some of whom had little or no interest in Hindu or Muslim customs or languages. Only thirty years previously these officers would have taken Indian mistresses ("bibis"), and thus become acquainted with the local customs and learned the language of their subordinates.[3] In this way any misunderstandings such as the rumour that the cartridges of the new Enfield rifles were greased by cow or pig fat [4] could have been defused. However, as the EIC's influence expanded, and more and more officers brought their families to India with them, bibis became socially unacceptable, and a barrier developed between officers and their men. The officers looked down on the superstitious "niggers", and the sepoys had little respect for the often young and inexperienced officers ("griffins").[5] To prevent draconian sentences being carried out for minor misdemeanours, most disciplinary matters had to be referred to army headquarters, which often overturned the officer's sentence and thus undermined his authority.[6] Moreover, the setbacks suffered by the British during the Afghan and Crimean Wars, demonstrated that the British were far from invincible.[7] Was it not time to rid the country of these foreigners? After all, the sepoys in most places outnumbered the British soldiers 6–8 to 1.[8]

Although discontent had been brewing for some time, the mutiny finally erupted in Meerut, which Fortune had visited in 1851. On 23 April 1857 Colonel George Munro Carmichael-Smyth (1803–1890), in an attempt to test the loyalty of his troops, urged them to accept the new cartridges.[9] When all but five of the sepoys refused, he had them shackled and imprisoned.[10] On 10 May 1857 their comrades released them along with more than 700 common criminals, who went on a killing spree before heading for Delhi.[11] Although the King of Delhi was not keen to get involved with the rebels who had committed this crime, he was eventually persuaded to act as their figurehead.[12] The foreigners in Delhi were hunted down and killed.[13] The British force on the ridge overlooking Delhi was initially too small to recapture and hold Delhi against the increasing numbers of rebels who converged on the Moghul capital.[14] In fact, for a long time the British were more like the besieged than besiegers.[15] Eventually when more reinforcements arrived General Archdale Wilson (1803–1874) was finally persuaded to launch an attack on 14 September.[16] After six days of street fighting with a large loss of life on both sides the British eventually prevailed.[17] The King was captured and two of his sons and a grandson, who were blamed for the killing of women and children in May, shot.[18] A few days later, on 24 September, Colonel Edward Harris Greathed (1812–1881) set out from Delhi in pursuit of the rebels fleeing to Awadh.[19] On the way he made a detour to relieve the siege of Agra Fort, but arrived to find the rebels had retired.[20] After thwarting a surprise rebel attack[21], he headed for Lucknow, which was still under siege.[22]

In the meantime, the chain-smoking Sir James Outram (1803–1863) and his Baptist friend, Brigadier Henry "Holy" Havelock (1795–1857) had battled their way through the streets of Lucknow to the besieged at the Residency on 25 September, only to find themselves unable to leave with all their wounded.[23] It took another force under Sir Colin Campbell ("Sir Crawling Camel", 1792–1863), a veteran of the Peninsular-and First Opium Wars, to break the 140 day siege and evacuate the women and children on 19 November.[24] Not until 21 March 1858 were the last of the rebels dislodged from Lucknow.[25] On that day the tireless Sir Hugh Henry Rose (1801–1885), diplomat turned soldier, with his small force arrived at Jhansi, where all but two of the 56 men, women and children who had been guaranteed their lives in exchange for the fort had been massacred by rebels on 8 June 1857.[26] Once he had conquered a relief force under Ram(a)chandra Pandurang Tope ("Tatya Tope" "Tantia Topi", 1814–1859) on 1 April[27], the siege of Jhansi was resumed. After some fierce fighting, the town was captured after a two-day battle on 3–4 April 1858, but Lakshmi Bai (1828–1858), the Rani of Jhansi, managed to escape dressed as a man.[28] When this intrepid woman was killed in battle on 17 June 1858[29], the mutiny all but collapsed. For eight months Tatya Tope's fast-moving army managed to evade the various British columns sent against him.[30] However, with their numbers severely depleted, in February 1859 the last of the rebels gradually dispersed.[31] Sir Colin Campbell and Brigadier-General James Hope Grant (**Box 9.9**) drove the remaining rebels in Awadh into the fever-ridden forests of S

Nepal where some died and others were captured and handed over to the British authorities.[32] Some leaders managed to evade capture. Disguised as a pilgrim Prince Firoz Shah escaped to Mecca where he died in poverty in 1877[33], while Nana Govind Dhondu Pant ("Nana Sahib") disappeared without trace.[34] The King of Delhi was found guilty of collaboration with the rebels and exiled to Burma, where he died on 7 November 1862.[35] In July 1859 a "State of Peace" was officially declared throughout India.[36]

The Indian Mutiny led to a number of changes in the administration of India. The EIC's rule came to an end on 1 November 1858, and its last Governor-General Charles John Canning (1812 – 1862) was appointed the first Viceroy of India, who was directly responsible to the British government.[37] "Clemency" Canning's policy was one of reconciliation, with a general amnesty promised to all those not guilty of murder.[38] The number of British soldiers was increased and that of the sepoys reduced.[39] Henceforth the artillery was to remain in British hands.[40]

While Fortune was finalizing his book, he was also preparing for an auction due to take place at Christie & Manson's on 13 – 14 May 1857.[25] This auction of "A very choice collection of ancient Chinese works of art" proved very lucrative. The 409 items, which consisted of old crackle (180 items), porcelain (78 items), ancient enamel on metal (34 items), old bronzes (32 items), jades (30 items), curiosities etc. (22 items), lacquerwork (19 items), seals etc. (9 items), glass (4 items), and marqueterie (1 item), sold for the princely sum of £ 2,566 – 3 – 6[26], which was equivalent to about £ 220,800 in 2015 (based on average earnings, the sum would be £ 1, 787,000). By now Fortune knew what the dealers and individual buyers were desirous to obtain. Since old crackle was in vogue, it constituted 44% of the items on sale.[27] The most expensive item was a small gourd-shaped bottle of turquoise crackle, which went to Mr. Webb for the not inconsiderable sum of £ 57 (£ 4,900 in 2015).[28] Mr. Webb also bought a square bottle of turquoise crackle for £ 49.[29] One globular crackly bottle, no more than 6.5 inches (16.5 cm) high went for as much as £ 48.[30]

As a result of this and other sales, the down payments on his books, and the 7+ years he had worked for the East India Company, Fortune was now able to afford a larger house for his wife and 3 children. They moved from 8 Thistle Grove to 1 Gilston Road (**Fig. 6.7**), a few blocks away in the first quarter of 1857.[31]

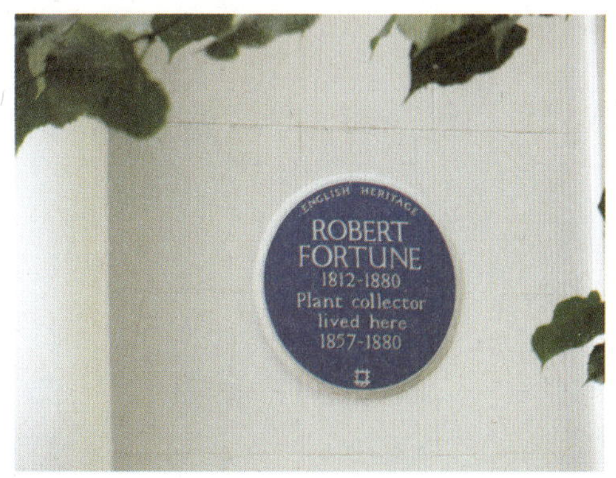

Fig. 6.7 (1–2)　Fortune's house at 9 Gilston Road, London SW10.

This was to be Fortune's home for the rest of his life. There was a small plot in front of the house, and a garden behind it. It seems likely that Fortune would have cultivated a number of his favourite introductions in his own garden. It would be nice to think that the *Anemone hupehensis* still growing in the next-door neighbour's garden (**Fig. 6.8**) originated from material cultivated by Fortune himself.

The Fortunes had not long been installed in their new home when a personal tragedy occurred. Little Mary, not yet 5 years old, died at the very end of June. She was buried in the family grave in Brompton Cemetery on Saturday, 4 July 1857.[32] Rev. N. Liberty officiated. Mary's death cast a shadow over family life. In order to escape the gloom, Fortune and presumably the rest of his family travelled to Scotland to visit family and friends.[33] They stayed there until mid-August.[34] By then Fortune knew there was a possibility of returning to China on behalf of the US Government.

Fig. 6.8 ***Anemone hupehensis* growing at 11 Gilston Road.**

Chapter 7 Working for the Americans

7.1 The History of Tea in the USA

In the course of the past 240 years various attempts have been made to grow tea in the southern states of the USA (The Carolinas, Florida, Georgia, Louisiana, but also Texas and California).[1] These private enterprises and government-sponsored initiatives were aimed at making the home market independent of external forces, which controlled the price of the imported product. The Americans knew from experience what it was to be dependent on the world market. After the costly war with France (1756 – 1763), the British Government raised taxes on a number of goods including tea in 1767.[2] Angered by what was seen as a violation of their independence, some colonists dressed up as Mohawks entered the cargo ships "Dartmouth", "Eleanor" and "Beaver", and threw some 340 chests of tea into Boston Harbour in what became known as "The Boston Tea Party" (**Fig. 7.1**).[3]

Fig. 7.1 The Boston Tea Party, 16 December 1773.

The port of Charleston (South Carolina) followed suit in 1774.[4] The tea tax was thus one of the factors which sparked America's Revolutionary War during 1775 – 1783.[5] During the Revolutionary War the Americans were supported by Britain's archrivals, the French.[6] This explains the strong French influence in post-war times.

At first tea was just grown as a novelty. When the French botanist François André Michaux (1770 – 1855) visited his friend Henry Middleton (1770 – 1846) at Middleton Barony near Summerville (South Carolina) in 1799, he brought with him tea plants and seeds.[7] From 1813 the rose breeder Philippe Noisette (1775 – 1835) was growing tea in the botanical garden of the South Carolina Medical Society on the outskirts of Charleston.[8] Tea was still being grown in the Charleston area in 1828, and by 1825 was widely cultivated in the former French colony of Louisiana.[9]

Fig. 7.2 Dr. Junius Smith (1780 – 1853), who established the Golden Grove Tea Plantation in Greenville, South Carolina, in 1848.

After the loss of the transatlantic steamship

"President" in March 1841, and the collapse of the British and American Steam Navigation Company, its chairman Dr. Junius Smith (**Box 7.1; Fig. 7.2**) turned to tea.[10]

Box 7.1

Dr. Junius Smith LL.D (1780 – 1853) Son of Major David Smith and his wife Ruth.[1] After graduating from Yale in 1802, he studied at Litchfield Law School, and subsequently practised at the New Haven bar.[2] In 1805 he travelled to London to argue for the release of his brother's ship "Mohawk", which had been seized by the British.[3] After winning large damages for wrongful seizure, he began a successful transatlantic partnership with his brother David during the Napoleonic Wars.[4] He met and on 9 April 1812 married the attractive Sarah Allen from Huddersfield.[5] They had a daughter, Lucinda, born in 1814.[6] Realizing how steamships could speed up transatlantic travel, Smith was finally able to interest the London business community in establishing the British and American Steam Navigation Company.[7] An order was placed with Macgregor Laird (1808 – 1861) for the construction of a large paddle steamer ("British Queen"), but when the engine-builders Claude Girdwood & Co. (Glasgow) went bankrupt, and it looked as though the Great Western Steamship Company would beat them to New York, Smith chartered the SS "Sirius" for the first transatlantic crossing under steam power.[8] Loaded with coal and with its normal quota of 40 passengers, the "Sirius" pushed its engines to the limit.[9] As the coal ran out, the crew were forced to burn cabin furniture, spare yards, and one of the two masts![10] The paddle steamer arrived in New York on 22 April 1838, just a few hours before its rival the "Great Western".[11] The "British Queen" came into service in 1839, and was followed the next year by an even larger paddle steamer, the SS "President".[12] For his contribution to bringing the Old and New Worlds closer together, Smith was awarded a LL.D by Yale in 1840.[13] However, just when the prospects for transatlantic travel looked so rosy, disaster struck. In March 1841 the "President" was lost in a gale between New York and Liverpool.[14] The company collapsed and Smith's shipping career came to an end.[15] With his wife dead and his daughter in India, Smith returned to the USA in 1843, where he worked in his nephew's garden, and penned diverse articles for magazines.[16] When his hopes of winning a contract to deliver mail between the USA and England were dashed in 1845, he bought a farm near Greenville (South Carolina), and established the Golden Grove Tea Plantation in 1848.[17] The enterprise succeeded and by July 1851 the shrubs were producing their first crop of commercial tea.[18] Seven months later he was attacked and badly wounded at his home, and never fully recovered from his skull fracture.[19]

He had always been a keen gardener, but the idea for a tea plantation would seem to have stemmed from his daughter Lucinda, who was living with her husband, the army chaplain Rev. Edward K. Maddock, in India.[11] In 1848 Smith bought a 200-acre farm near Piedmont, Greenville (South Carolina) and imported 500 tea plants from London in 1849 as well as seed from India (**Figs. 7.3, 7.4**).[12]

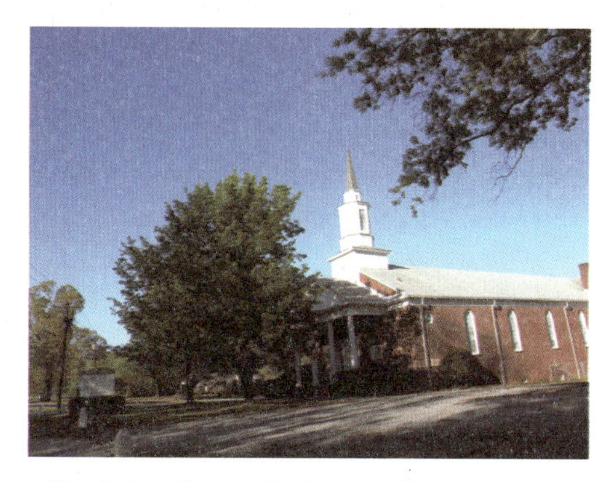

Fig. 7.3 Grove Station Baptist Church in the centre of Greenville, South Carolina, USA.

Fig. 7.4 Signposts indicating proximity to the former Golden Grove Plantation in Greenville.

However, this private enterprise was endangered from the start. On 11 April 1850 Smith felt obliged to complain to British-born Thomas Ewbank (1792 – 1870), the US Commissioner of Patents, when the Patent Office started to distribute tea seed free.[13] Then, just before Christmas 1851, he was beaten up at his house.[14] He went to live with his nephew, Henry Smith, in Astoria (Long Island), and it was there that he died on 22/23 January 1853.[15] Without protection his plants at Piedmont soon withered.[16]

Fig. 7.5 Hon. Charles Mason (1804–1882), who made enquiries about obtaining tea seed from China.

In May 1851 the Department of the Interior requested the US Navy to procure plants and seeds of useful Asiatic plants including those of tea.[17] Although the commander of the East India Squadron, Captain John H. Aulick (1787 – 1873) contacted Paul S. Forbes, the US Consul in Guǎngzhōu in 1852 nothing came of this effort.[18]

A few years later the US Government showed renewed interest in growing tea. On 21 July 1857 Charles Mason (**Box 7.2**; **Fig. 7.5**), at the end of his term as US Commissioner of Patents, wrote to the seed merchants Charlwood & Cummins in London to inquire about the probable cost of about ten bushels (just over 350 litres) of tea seed and the expenses of sending someone to collect it.[19]

Box 7.2

Hon. Charles Mason (1804 – 1882) Youngest son of Chauncey Mason (1770 – 1855) and Esther Dodge (1772 – 1836).[1] He graduated from the US Military Academy at West Point at the head of his class.[2] Two years later he resigned from the army, studied law, travelled and prospected in the western US before settling down in Burlington, Iowa.[3] On 1 August 1837 he married Angelica Gear (1804 – 1873), by whom he had three daughters.[4] He was First Chief Justice of Iowa from 4 July 1838 to 16 May 1847 and US Commissioner of Patents from 16 May 1853 to 4 August 1857.[5] He created a sensation when he appointed a female clerk, something that was unheard of in government circles at the time.[6] Having lost two of his three children, he tended to be melancholic.[7] In order to forget, he worked hard.[8] Travel also acted as a diversion.[9] He was twice the Democratic nominee for the Governorship of Iowa in 1861 and 1867.[10] Although he was opposed to secession of the confederate states during the American Civil War, he maintained that the USA "can never be perpetuated by force of arms and that a republican government held together by the sword becomes a military Despotism".[11] After losing a second time a philosophical Mason wrote in his diary: "I played the game of life at a great crisis and lost. I must be satisfied."[12]

The firm promised to consult Professor John Forbes Royle (**Box 4.1**) and Robert Fortune. Although he was visiting family and friends in Scotland in the aftermath of his daughter Mary's death at the very end of June[20], Fortune replied forthwith.

> Tea seeds are in great demand in India at the present time, and I doubt if a large supply could be obtained from that country. The finer varieties introduced by me from China, certainly could not be spared, and I would not advise the American Government to introduce and propagate inferior kinds. The best way would be to follow the example of the East India Company and introduce the *best kind* from China direct.
>
> The plan proposed in your letter, viz. "to send the seeds in tin cases" would not succeed. From long experience I have found that these seeds, like acorns, chesnuts etc., retain their vitality for a very short time when out of the ground; certainly not one in a thousand would vegetate on reaching America. Any money spent on an experiment of this kind would only be thrown away.
>
> You will find in my *Journey to the Tea Countries of China and India*, the plan I adopted with good success whilst engaged in introducing the Tea plant to India. If the American Government is determined to give the matter a fair trial and wishes to spend a reasonable sum to insure success I would have no objection to take the business in hand, and from the experience I have here, would most likely bring it to a successful issue.[21]

Fig. 7.6 Joseph Holt (1807–1894), US Commissioner of Patents 1857 – 1859, who engaged Fortune to collect tea seeds on behalf of the USA.

Although Charlwood & Cummins forwarded this letter to Washington on the 17 August 1857[22] Charles Mason was no longer in office, having resigned on 4 August, disillusioned by the newly elected President James Buchanan (1791 – 1868).[23] He was succeeded by Joseph Holt (US Commissioner of Patents 1857 – 1859; **Box 7.3**; **Fig. 7.6**) on 10 September. When Fortune returned to London he talked to the seed merchants. On 20 August 1857 they were able to report that:

> We have had an interview with Mr. Fortune, and he informs us that he would accept the same terms from you, that he had from the East India Company, which was £ 500 per annum and all

expenses paid, which would amount to about £ 700 additional: for this *he* would procure the best varieties of Teas. It would be too late this season as they ripen in October. He should leave this [country] in March and he would be able to get the seeds down from the North of China to the Port of shipment in November or December, and he would arrive in America (several shipments being made by various vessels[24]) during April or May. Thus 20 or 30 (or more) Ward's cases could be sent each containing seeds enough to produce say 2,000 plants. This could be effected during the year and Mr. Fortune assures us this way only (that is by the seeds being placed in soil in Ward's cases) is there any chance of success... Should you therefore entertain the project of sending an agent—you can let us know whether we might engage Mr. Fortune, as we know of no other man so capable or experienced to carry out your views in this matter.[25]

Box 7.3

General Joseph Holt (1807 – 1894) Second son of John W. Holt (1772 – 1838) and Eleanor Jennings Stephens (1782 – 1871).[1] He was educated in Kentucky at St. Joseph's College (Bardstown) and the newly-opened Centre College in Danville before training to become a lawyer under Robert Wickliffe.[2] He then went to work with Wickliffe's famous cousin, Benjamin Hardin (1784 – 1852) in Elizabethtown in 1828.[3] In January 1832 he set up his own practice in Louisville, where he also became assistant editor of one of the leading newspapers, the "Public Advertiser".[4] In 1835 Holt moved to the even more lucrative State of Mississippi.[5] While he was there, he started corresponding with Mary Louisa Harrison (1811 – 1846) of Bardstown.[6] After three year's courtship, they finally married on 24 April 1839.[7] Mary could not stand the humid climate in Mississippi, so while Holt's practice in Vicksburg thrived, she continued to live for the most part with her family in Bardstown.[8] After three years of occasional visits, the couple finally set up house in Louisville.[9] With the wealth he had accumulated in Mississippi, the couple could afford trips to Cuba, New Orleans and Charleston (South Carolina).[10] Unfortunately, these did not help Mary's ailing health.[11] When his wife died in 1846, Holt found some solace in the garden they had created together.[12] After two years of mourning, he decided to make the trip to Europe and the Middle East, from May 1848 to October 1849 that he and Mary had considered doing together.[13] After his return he opened another law office and got married to Margaret Wickliffe (1821 – 1860), a niece of Robert Wickliffe.[14] During their extended honeymoon Joseph showed Margaret some of the places he had visited in 1848 – 1849.[15] When the Democrat James Buchanan (1791 – 1868) was elected President in 1857, Holt decided to move to Washington D.C. in the hope of getting a paid position in the new administration.[16] That same year President Buchanan appointed him as Commissioner of Patents fron 10 September 1857 to 12 March 1859.[17] As he could not persuade his examiners to be more liberal in granting patents, he was inundated by appeals. To take the pressure of work off his shoulders, he established a Board of Appeals.[18] As a result, he was probably quite happy to accept the position of Postmaster General from 14 March 1859 to 31 December 1860 when Aaron Venable Brown (1795 – 1859) died.[19] In his new function Holt strove to wipe out fraud and streamline the postal service.[20] At the outbreak of the American Civil War, many cabinet members resigned. Holt, a staunch Unionist, became Secretary of War from 31 December 1860 to 6 March 1861.[21] The Republican President Abraham Lincoln (1809 – 1865) appointed him Judge Advocate General of the US Army from 3 September 1862 to 1 December 1875.[22] This position involved advising the president about the outcome of court-martials and evaluating the evidence against civilians accused of disloyalty.[23] Following the assassination of President Lincoln on 14 April 1865 by John Wilkes Booth (1838 – 1865), eight of Booth's accomplices were rounded up.[24] Holt, who was the presiding judge at their trial, came in for a good deal of criticism for using evidence supplied by a con-man, and the suppression of certain facts favourable to the defendants.[25] President Andrew Johnson (1808 – 1875) tried on numerous occasions to give Holt the blame for the hanging of Mary Surratt.[26] Although Holt published a pamphlet in 1873 vindicating his position, some questions still remain.[27] His belief that Jefferson Davis (1808 – 1889) was behind the assassination was never proven.[28] After Holt retired on 1 December 1875, he stayed in Washington, where he led a quiet life, reading, walking, chatting with the neighbours, writing to friends, and only occasionally entertaining.[29]

7.2 Fortune Employed and Dismissed

Fortune was engaged and started planning his fourth expedition. He departed from England on 4 March 1858.[26] He took with him the good wishes of the editor of the Gardeners' Chronicle on 13 March 1858.[27]

> A few days since Mr. FORTUNE sailed once more for China; but this time in the service of the United States Government. His object is to procure a very large supply of Tea plants for trial in some part of the American Union, and such other Chinese productions as it may appear desirable to introduce. We cannot but look upon this as an arrangement not only most honourable to our energetic and distinguished countryman, but reflecting much credit upon the government of Washington itself, which has shown how well it knows how to appreciate merit among strangers as well as its own people. Mr. FORTUNE carries with him the good wishes of all who know —as who does not? —the very great services he has already performed in carrying all that is most precious in the vegetation of the extreme east to the more civilized countries of the west.

Having taken the "overland route", he arrived in China at the end of April in time for the early tea picking.[28] What with the Anglo-French attack on Guǎngzhōu (Canton) to avenge the "Arrow Incident" (**Box 6.1**) in December 1857 and the capture of the Dàgū Forts on 20 May 1858 by the Western powers, China was still in a very unsettled state.[29] Travel for a British subject could be dangerous. However, writing from Shànghǎi on 18 August 1858 he indicated that:

> I have visited various great tea districts, and made my arrangements with the natives for large supplies of Tea and other seeds and plants at the proper season. I am now doing the same in the country about Shanghae, & if my health does not fail me, I hope to send you abundant supplies of interesting things during the autumn and winter. I am also employed in getting Wards cases made for their transmission.[30]

Because he had already won the trust of the farmers in Zhèjiāng ("Everywhere the people receive me kindly and wellcome [sic] me back amongst them." [31]), and knew where to find the best tea, he was able to complete his assignment in record time. All Fortune needed to do was tell the farmers what he wanted and then come to pick up the seed once it had been harvested. In this way he was able to obtain in one season what had taken three years to collect for the East India Company.[32] In early

December he was back in Shànghǎi to prepare the Wardian cases for shipment. Since he did not want the seeds to germinate until late May or early June 1859, when he expected to be in Washington to supervise their lifting, he covered the seeds of the first shipments with more soil to keep them dormant for longer.[33] Fortune completed the last shipment on 19 February 1859 and left Shànghǎi for England early in March 1859, a well-pleased man.[34] He was looking forward to visiting the USA to complete his horticultural mission.[35] The Patents Office had already bought a five-acre site in central Washington in 1858 where Fortune's acquisitions were to be raised.[36] But it was not to be. When he arrived in London the US Government agent handed him a letter from the Chief Clerk at the US Patents Office, Samuel T. Shugert, who was temporarily in charge after Joseph Holt was appointed Postmaster General on 14 March 1859.[37] Fortune was informed that his contract was terminated.[38] The deeply disappointed plant collector replied:

> As I have taken a deep interest in the success of this great experiment, it would have afforded me much pleasure to have given you the benefit of my experience in rearing and transporting to proper sites the Tea and other useful productions I have sent you from China. The most difficult part of this mission (namely the procuring and introducing these seeds and plants) being successfully accomplished, it will be a source of deep regret if the experiment should fail from want of that experience which can only be acquired in the country to which these plants are indigenous.[39]

It is not clear why the final leg of Fortune's journey on behalf of the US Patents Office was cancelled. The Secretary of the United States Agricultural Society had some scathing comments to make:

> It is to be hoped that this dismissal really arose from a desire on the part of the government to economize, and not from the jealous fears of any subordinate official that Mr. Fortune would receive the honors attendant on the successful introduction of the tea plant.[40]

He indicated that 26,000 tea plants had already been distributed by the US Patents Office and that "this valuable production will doubtless be fully and fairly tried." [41] The dissemination of tea plants by the Patent Office continued until the Civil War broke out in 1861.[42] Although there was no communication during the war with most of the areas where the tea plants had been sent, these survived the hostilities, and were used after 1867 to propagate more material for distribution.[43] Although further attempts were made to establish a viable tea industry in the USA, most of these failed because of the unreliability of the climate, and the high labour costs.[44]

7.3 Other Plants Introduced to the States by Fortune

As part of his contract Fortune agreed to collect seed of other economic and ornamental plants as well.[45] Of those plants which could be utilized as foodstuffs, there are only a few novelties. The fruits of buckwheat (*Fagopyrum esculentum*) would have been known as a source of flour to those who farmed poor, sandy soils, while the radish (*Raphanus sativus*) had already been introduced from Europe. Likewise, people would have been acquainted with *Amaranthus blitum* as a substitute for spinach.[46] On the other hand, only a few people would have been aware of the fruits of the yáng méi (*Myrica rubra*; **Fig. 7.7A**). The US Department of Agriculture introduced it at least 19 times between 1898 and 1927[47], which would suggest that Fortune's introduction was unsuccessful. However, even under the name "yumberry", this fruit did not catch on in the USA. It is not clear whether Fortune intended the maidenhair tree (*Ginkgo biloba*) and the Chinese nutmeg yew (*Torreya grandis*) as economic or ornamental plants. Both grow into magnificent trees. In China the resinous seeds of the nutmeg yew are eaten for their vermicidal properties[48], while the kernels ("silver apricots") of the maidenhair tree (**Fig. 7.7B**) are often served boiled or roasted.[49] The seeds are considered to relieve colds and asthma, improve blood circulation, alleviate high blood pressure, and ease urination.[50]

Fig. 7.7A ***Myrica rubra*** **(yáng méi). The taste of the fruits must have reminded Fortune of the brambles of his native Scotland. It is called "yumberry" in the USA.**

Fig. 7.7B ***Ginkgo biloba*****. Fortune would have been well acquainted with the maidenhair tree, as it is frequently grown near temples. The female plant produces "silver apricots", which are eaten in China. Unfortunately, these have a fleshy covering which smells like rancid butter when ripe.**

Fig. 7.7C ***Cinnamomum camphora*****, an evergreen tree, is often grown near temples, where its resinous wood acted as a source of incense. It requires mild winters, so it was intended for the southern states of the USA.**

Fig. 7.7D ***Boehmeria nivea*****. The ramie stems contain fibres, which for thousands of years have been used to produce grass cloth in China. "Fabrics of various degrees of fineness are made from this fibre, and sold in these provinces; but I have not seen any so fine as that made about Canton." (Fortune, 1857: 260).**

Fig. 7.7E ***Gleditsia sinensis*****. This tree is well-protected against herbivory by spines on the branches, and saponins, particularly in the pods. The pods were a source of detergent, while the thorns were used in traditional medicine.**

Fig. 7.7F ***Thuja orientalis*****, the Chinese Arbor-vitae. As a long-living evergreen tree, it is often planted near temples and graves. Fortune probably intended it for US cemeteries.**

Other plants were clearly meant to be grown for the oil that could be extracted from their seeds by crushing and steaming. The oil from rapeseed (*Brassica rapa* var. *chinensis*) was considered excellent for cooking and oil lamps[51], while oil from the fast-growing tung-oil tree (*Vernicia fordii*, Syn. *Aleurites fordii*; **Fig. 5.16**) could be used to waterproof paper and fabrics, and was a suitable varnish for boats and furniture, not the least because it is insecticidal.[52] Moreover, the residual oil-cake could be spread on the fields as a manure.[53] It would seem that the Americans were unaware of its excellent properties, because it had to be reintroduced by David Fairchild (1869 – 1954) in 1904.[54] In Fortune's time oil lamps and candles were the only sources of lighting in the houses[55], so it was natural for him to include plants that supplied suitable oils and waxes. The seeds of the tallow tree (*Triadica sebifera*; **Figs. 3.14**, **5.13**) produce an oil which not only burns brightly, but could be used to water-proof paper, and was taken internally as a purgative.[56] The seeds are covered by a white wax, which was removed by a combination of steaming and rubbing.[57] This wax was used to produce candles. Unfortunately, the wax tends to melt in warm weather[58], so it required to be coated with the harder wax produced by glands on the male wax scale-insect *Ericerus pela* (Syn. *Coccus pela*), which lives on two members of the olive family, Chinese ash (*Fraxinus chinensis*) and Chinese tree privet (*Ligustrum lucidum*).[59] Besides its use in candles, this wax was also employed to impart a gloss to high-quality paper, as a coating for pills, to make cloth lustrous, and as a polish on jade, soapstone and furniture, as well as in folk medicine.[60]

Fortune also sent fruits of *Cinnamomum camphora* (**Fig. 7.7C**), as the terpenoid present in the plant was essential for protecting clothing, dried plants, and stuffed animals from insect attack, as well as relieving sprains and rheumatic pains, and producing celluloid and fireworks.[61] The wild populations were being decimated in China, by the practice of chopping wood chips from the basal 3 m of the trunk and discarding the less productive upper parts.[62] So the idea of starting camphor plantations was of considerable interest to the US Government.[63] Moreover the fragrant wood was valued in ship-building and in cabinet-making not only because of its beauty but also its insect-repellant properties.[64] Fortune also included seeds of another fine timber tree, the Japan Cedar *Cryptomeria japonica* (**Fig. 3.51**).[65] These had already germinated in the US Patent Office's Garden by 1859.[66]

Naturally, Fortune forwarded fruits of his hardy hemp-palm, *Trachycarpus fortunei* to the USA. As the English name suggests, this palm was valuable for the fibres which it produces on its stems (**Figs. 3.15, 3.16**). These fibres have been employed to make durable ropes, mats, mattresses, and water-proof hats and capes.[67] Strips of the leaves can also be used as garden twine. Moreover, the flower buds are sometimes cooked as a vegetable, while the astringent seeds can be used both internally and externally to clot blood.[68] The introduction would appear to have been a success, as this palm is now widely grown in the USA.[69] Fortune included seeds of three more plants, which were harvested for their fibres, namely Chinese jute (*Abutilon theophrasti*), the phoenix tree (*Firmiana simplex*) and ramie (*Boehmeria nivea*; **Fig. 7.7D**).[70] While Chinese jute was employed in caulking boats and rope-making, the phoenix tree and ramie were also used to make clothes.[71] Ramie was exported to Europe and America in the form of "grass-cloth", which was cool in summer, hence its alternative name "summer cloth".[72] By sending the seeds to the USA, Fortune probably aimed at making America independent of Chinese imports. The phoenix tree, which used to be planted near houses in China as a source of inspiration to scholars, was successfully introduced into the USA, and is now grown in gardens and parks in Louisiana, Virginia and South Carolina for the shade that its large lobed leaves offer.[73]

Fortune did not forget to include seeds of a number of plants, which were used as dyestuffs in China, namely *Polygonum tinctorium* known for its blue dye, and a buckthorn species (*Rhamnus utilis*) which could be used to dye clothes green.[74] Unfortunately for Fortune, these natural dyestuffs were soon relegated to the past, as they were replaced by synthetic aniline dyes, that were cheaper to produce and lightfast into the bargain.[75]

Clothes require to be washed, so Fortune included seeds of the soap-bean tree (*Gleditsia sinensis*; **Fig. 7.7E**) in his shipment. The leaves, bark, and more particularly the pods of this tree contain saponins, which the plant uses to protect itself against attack by cold-blooded animals.[76] These saponins produce a good lather in hot and cold water, and as such were used for washing clothes, preparing hides, and cleaning furniture.[77]

Some of Fortune's plants simply had ornamental value. He noted that the Chinese weeping cypress (*Cupressus funebris*; **Fig. 5.15**) and the Chinese arbor-vitae (*Thuja orientalis*; **Fig. 7.7F**) were often grown on or around graves.[78] The evergreen habit and long lifespan of these trees symbolized eternal life. The actual species selected depended on the rank of the deceased.[79] Thus the arbor-vitae ("tree of life") was used for princes' graves.[80] Fortune always complained how forelorn British graveyards were[81], so these seeds would have been included in the package in the hope of changing this age-long tradition. Unfortunately, the weeping cypress proved too tender for all but the mildest areas in the British Isles[82], so Fortune's hopes would have been pinned on the warm temperate parts of the USA. Although it is not evergreen, Fortune could not resist the temptation to include what he considered his finest introduction, namely the golden larch (*Pseudolarix amabilis*; **Fig. 6.3E**).[83]

Chapter 8 Two Trips to Japan

8.1 The Race for Japan's Botanical Riches

During his active career, Fortune was no longer home, than he was planning his next move. It was said that when a friend greeted him just four days after he got back from China with "Well, Fortune, surely home for good now?" He replied, "No, I have just learned the Japanese ports are open to Europeans, and I go there on Thursday."[1] Although a bit of an exaggeration, when his US trip was cancelled in mid-1859, he was left to lick his wounded pride, and consider his options.

Fig. 8.1 John Gould Veitch (1839–1870). The Japanese ladies did not like his long whiskers.

After the forceful opening up of Japan in 1853 – 1854 by the American fleet under Commodore Matthew Perry (1794 – 1858) and the signing of the "Treaty of Yeddo" by Lord Elgin on 26 August 1858, the Land of the Rising Sun certainly offered the prospect of a rich botanical harvest, as only a fraction of its plants had been collected during the Dutch diplomatic missions (hofreizen) from Nagasaki to Edo (now Tokyo). The few plants that were known were certainly enticing to a plantsman interested in ornamental and economic plants.[2] However, if he wanted to corner the market before others did, he would have to act quickly. He probably knew that Veitch's Royal Exotic Nursery just 600 m from his house was intending to send John Gould Veitch (**Box 8.1**; **Fig. 8.1**) on a similar mission.

Box 8.1

Mr. John Gould Veitch (1839 – 1870) Eldest son of the nurseryman James Veitch Jr. (1815 – 1869) and his wife Harriott Reynolds Gould (ca. 1817 – 1879).[1] He was an intelligent child with a gift for languages.[2] Having studied botany at London University under Professor John Lindley (**Box 2.1**), he begged his father to allow him to travel.[3] With the signing of the Treaty of Yeddo, British citizens had the opportunity of visiting the botanical paradise of Japan for the first time. Nurseries were therefore keen to send their collectors there to skim off the novelties.[4] His assignment got off to a bad start. Veitch lost all of his equipment when the "Malabar" in which he and many other notables were travelling was wrecked off Ceylon.[5] However, he arrived just in time to join the party led by Rutherford Alcock (**Box 3. 10**), which successfully climbed Mt. Fuji in September 1860, the first foreigners to do so.[6] He later provided Alcock with various lists of plants to be found in Japan for the latter's book about that country.[7] On 13 November 1860, Alcock introduced him to Robert Fortune.[8]

Like Fortune, he used his charm and good manners to gain access to nurseries and private gardens.[9] In this way he managed to collect 17 new species of conifers and a number of flowering plants including the Star Magnolia (*Yulania stellata*), Boston Ivy (*Parthenocissus tricuspidata*) and the Japanese Maple (*Acer palmatum*) in a matter of four months.[10] In December 1860 the poop of the SS "England" from Yokohama to Shànghǎi was filled with his and Fortune's Wardian cases.[11] Like Fortune and Hugh Cuming (**Box 3.8**) before him, he went to the Philippines to collect *Phalaenopsis* orchids, returning to Japan in the autumn of 1861 in time for the seed harvest.[12] In early December he sailed for home, arriving in England at the beginning of 1862.[13] In the course of his short life, Veitch also visited eastern Australia during 1864 – 1866 from Melbourne to Cape York in search of ornamental plants.[14] He was responsible for reintroducing the Norfolk Island Pine (*Araucaria heterophylla*) into Europe.[15] While he was disappointed by the parched landscape, he was pleased to see "greenhouse plants" growing outdoors in the gardens.[16] During his visit to Australia he spent four months cruising in Polynesia on board HMS "Curacao".[17] He was frustrated by the fact that he was not always allowed to explore the islands properly, because the natives were considered unfriendly.[18] On one occasion a local chief had to be held hostage on board the "Curacao", while Veitch was collecting plants.[19] A palm he brought back from Fiji was named after him (*Veitchia joannis*) in 1870 by Hermann A. Wendland (1825 – 1903).[20] He died of tuberculosis aged 31, leaving his wife Jane Hodge (ca. 1845 – 1889+), whom he had married in the summer of 1867, to bring up their two infant sons (James Herbert, 1868 – 1907 and John Gould Jr., 1869 – 1914).[21]

Rather than delay his departure until he had found a suitable sponsor, by which time he would have been too late, he probably decided to take a calculated risk that he would be able to recoup the outlay once his plants were successfully introduced into cultivation.[3]

In order to raise the capital for the voyage, he put a large quantity of ancient porcelain, and a few carvings in jade and rock crystal up for sale at Christie & Manson's auction house in 1859. A total of 544 items were sold at three consecutive sales on 26 May, 23 June, and 22 July for £ 2,459.[4] One bottle of pale turquoise crackle only 17 inches (43 cm) high was bought by Mr. Webb for £ 131.[5] To make sure he did not run short of cash on his journey another 700 items were sold from Tuesday 28 February until Thursday 1 March 1860. These auctions raised an additional £ 1871 – 13 – 0.[6]

Fortune set off in the summer of 1860 by the "overland route" for the Land of the Rising Sun.[7] Once in Shànghǎi he boarded the barque "Marmora" for the 800 km journey to Nagasaki (**Fig. 8.2**).[8] On the 12 October they passed south of Fukue-jima (Goto-Retto or Gotto Islands, where Japanese Catholics were long able to practise their religion), rounded the steep-sided cliffs of Takahoko Island (Papenberg, "Priest's Mountain"; **Fig. 8.3**), from which many Christian converts were said to have been hurled to their deaths in 1622[9] and finally anchored in the sheltered harbour of Nagasaki (**Fig. 8.4**).[10]

Fortune stayed at the Myogyo Temple (**Figs. 8.5, 8.6**) with the Scottish businessman, Kenneth Ross Mackenzie (1801 – 1873), an agent for Jardine, Matheson & Co., who doubled as French Consul at Nagasaki.[11] Fortune probably met him when Mackenzie was working in Shànghǎi with his brother, Charles D. Mackenzie, in the 1850s, because he refers to Messrs. Mackenzie in a number of his publications.[12] Mackenzie had a reputation for shady business transactions which included smuggling.[13] Being a port, some of the largest buildings were brothels, but as Fortune noted, prostitution in Japan was not stigmatized (**Fig. 8.7**).[14]

Fig. 8.2 Map of Japan showing the chief localities visited by Robert Fortune.

Fig. 8.3 Takahoko Island, Fortune's "pretty little island of Papenberg", guarding the entrance to Nagasaki Harbour. From the steep cliffs many Christian converts were said to have been hurled to their deaths.

Fig. 8.4 Nagasaki Harbour, "one of the most beautiful in the world" (Fortune, 1863: 5), with a replica of an American "black ship".

Fig. 8.5 The Myogyo Temple, where Fortune stayed while in Nagasaki.

Fig. 8.6 Inner sanctuary of Myogyo Temple.

Fig. 8.7 Lantarns bearing girls' names outside a former brothel in Nagasaki.

Like other visitors, Fortune was impressed by the cleanliness of Nagasaki, but was disappointed by the poor quality of the products in the shops.[15] As he walked round the town, he was unable to resist the urge of poking his nose into the formal gardens at the rear of the houses.[16] In one belonging to Mr. Motoski (Matotski) he saw a nice collection of potted plants, including the conifers *Chamaecyparis* (Syn. *Retinospora* spp.), *Sciadopitys verticillata*, *Thujopsis dolabrata*, and variegated cultivars of bamboo, laurel (*Aucuba japonica*), orontium (*Rohdea japonica*), and specimens of *Hoya motoskei*, which had been called after their proud owner.[17] As Fortune was presented with a few of the owner's rare plants, it seems likely that some of these eventually found their way into cultivation in England.[18] One day he witnessed the Kunchi Procession, which had been established in the late 16th Century as a harvest festival, but soon became a means to assert the preeminence of Shintoism as the national religion.[19] When they reached the Suwa Shrine (**Fig. 8.8**), some Japanese boys dressed up in Dutch clothes, skilfully performed a number of military manoeuvres.[20]

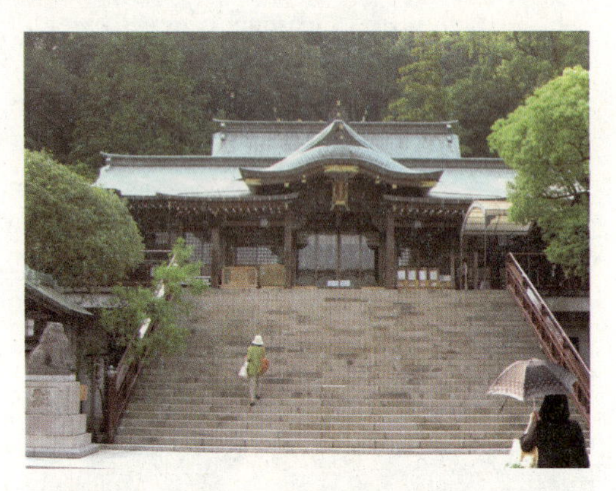

Fig. 8. 8 The Suwa Shrine, Nagasaki, where the annual Kunchi Festival, that Fortune witnessed, is held. Here the Japanese children get "dressed in the Dutch military costume" and are "put through various military manoeuvres, which were executed in a most creditable manner." (Fortune, 1863: 15).

Fig. 8.9 Artificial island of Dejima, built in 1635-1636 to house the Portuguese business community, but subsequently used to restrict the movements of Dutch traders. The Dutch required special permission to leave the island, and all the Japanese were obliged to quit the island at sunset with the exception of "women who had forfeited the first claim of their sex to respect or esteem ... " (Fortune, 1863: 7). An archaeological excavation is in progress on the left.

Fortune visited the artificial island of Dejima (Decima, Desima, "Moon[-shaped] Island", built 1635-1636; **Fig. 8.9**), where the Dutch colony had been cooped up prior to the "Treaty of Yeddo".[21] He was pleased to discover a large memorial stone dedicated to the pioneer naturalists, Engelbert Kaempfer (**Box 8.2**) and Carl Peter Thunberg (**Box 8.3**; **Fig. 8.10**), which Philipp von Siebold (**Box 8.4**; **Fig. 8.11**) had erected in the botanical garden on Dejima in 1826.[22]

Box 8.2

Engelbert Kaempfer (1651-1716) Second son of the Lutheran vicar Johannes Kemper (1610-1682) and his first wife Christine Drepper (ca. 1619-ca. 1654).[1] Between 1665 and 1673 he attended a succession of schools in Lemgo, Hameln, Lüneburg, Hamburg, Lübeck and Gdansk.[2] He went on to study philosophy and medicine in Torun, Kraków and Kaliningrad.[3] In 1681 he transferred to Uppsala (Sweden), where he studied under the polymath Olof Rudbeck Sr. (1630 – 1702).[4] In Sweden Kaempfer met his countryman, Samuel Von Pufendorf (1632 – 1694), then court historiographer.[5] Through Von Pufendorf and his brother, the diplomat Esaias Von Pufendorf (1628 – 1689), Kaempfer was appointed Secretary and Physician to a trade delegation to the Russian and Persian courts.[6] The delegation left Stockholm on 20 March 1683, crossed the Baltic Sea to Turku, continued on to Helsinki, before heading for Narva in Estonia.[7] From there they travelled overland via Novgorod and Tver to Moscow.[8] After almost two months in Moscow the group headed for Saratov and down the Volga to the Caspian Sea.[9] Kaempfer's assistant drowned in the Volga, and the hounds meant as a gift for Suleiman I, the Shah of Persia, were lost overboard during the stormy crossing of the Caspian Sea.[10] The party was detained at the border while they awaited a summons from the Shah of Persia.[11] This gave Kaempfer the opportunity to investigate the oil springs at Baku.[12] After crossing the snow-covered Elburz Mountains in January 1684, the delegation arrived in Qazvin.[13] From there they travelled to the holy city of Qom and finally reached the Persian capital of Isfahan on 29 March 1684.[14] Here Kaempfer learned a great deal about Persia from Father Raphael du Mans (1613 – 1696), who had been interpreter at the Persian court for 30 years.[15] At the end of months of unsuccessful negotiation in Isfahan, when the delegation was about to return to Sweden, Kaempfer decided to explore more of Asia, rather than return to an easy life at the Swedish Court.[16]

Fig. 8.10 Dr. Carl Peter Thunberg (1743 – 1828), who started his career as a pupil of Linnaeus. The latter encouraged him to visit the Netherlands. There he was engaged by the Dutch East India Company. He spent three years in South Africa, followed by 15 months in Japan (1775 – 1776). His "Flora Japonica" (1784) is still a basic reference.

He joined the Verenigde Oost-Indische Compagnie (VOC, Dutch East India Company) and travelled to Bandar Abbas on the Persian Gulf in December 1685.[17] On the way he visited Persepolis and Shiraz, studied cuneiform writing and the local method of preparing coffee.[18] In Bandar Abbas he endured what he described as "hell on earth" for 30 months, and only managed to survive by escaping to the mountains on two occasions.[19] It was here that he wrote his hundred-page report on the date palm.[20] Finally on 30 June 1688 he was given a job as doctor on board a cargo ship bound for southern India, Ceylon and Java.[21] During the 15 months he studied a number of tropical diseases including elephantiasis.[22] Once in Batavia (Jakarta) he tried unsuccessfully to obtain a job in the local hospital.[23] Failing this, he decided to visit Siam from 6 June to 11 July 1690 and Japan from 26 September 1690 to 31 October 1692.[24] With the exception of two annual trips to Edo to pay homage to Shogun TOKUGAWA Tsunayoshi (1646 – 1709), he was confined to the artificial island of Dejima at Nagasaki.[25] He was nonetheless able to amass numerous objects, books and information through the Japanese guards and translators.[26] These willingly supplied information, even on prohibited subjects, when they were plied with drink.[27] His servant IMAMURA Gen'emon Eisei (1671 – 1731) risked his own life obtaining maps for Kaempfer.[28]

Towards the end of 1692 Kaempfer returned to Batavia and from there sailed back to Amsterdam, which he reached on 6 October 1693.[29] After defending his dissertation at the University of Leiden on 22 April 1694, he returned to Lemgo in August 1694, where he became physician to Count Friedrich Adolf zur Lippe-Detmold (1667 – 1718) on 7 December 1698.[30] As the salary was meagre, he had to open a private practice, which prevented him from working on his manuscripts.[31] In the hope of gaining financial independence, on 18 December 1700 he married the only daughter of a rich merchant, the 16-year-old Maria Sophia Wilstach (1684 – 1761) by whom he had two daughters and a son named after his employer.[32] Unfortunately, his young wife proved to be self-indulgent, neglecting the housekeeping and taking no interest in his scientific work, so they finally decided to separate.[33] In order to prevent Sophia from inheriting anything, the embittered Kaempfer willed most of his estate to his nephew, Johann Hermann Kaempfer (1691 – 1736).[34] Philip Heinrich Zollmann (1680s – 1748) obtained Kaempfer's unpublished manuscript "Berichte auf heutiges Japan" (Reports on present-day Japan) with the intention of publishing an English translation.[35] However, his posts as Secretary to the British Ambassadors in Paris and Stockholm forced him to abandon this project, so in 1724 he sold the manuscript to Sir Hans Sloane (1660 – 1753), who had it translated into English by his (Sloane's) Swiss protégé Johann Gaspar Scheuchzer (1702 – 1729).[36] Fortune frequently referred to this book in his "Yedo and Peking" in 1863. On Sloane's death, Kaempfer's herbarium landed in the British Museum in London[37], where it was consulted by Carl Peter Thunberg (**Box 8.3**).

Box 8.3

Carl Peter Thunberg (1743 – 1828) Son of Johan Thunberg and his wife Margaretha Starkman.[1] He attended school in Jönköping, before enrolling at Uppsala University in 1761.[2] There he became a pupil of Carolus Linnaeus Sr. (1707 – 1778).[3] Linnaeus recommended him for the Kåhre Scholarship which enabled him to study abroad.[4] He set out for Paris in 1770 to complete his studies in medicine and natural history.[5] There he met the plant systematist Bernard de Jussieu (1699 – 1777).[6] In 1771 he visited Linnaeus' friend Johannes Burman (1707 – 1780) and his son Nicolaas (1733 –1793) in Amsterdam.[7] The Burmans were suitably impressed by the amicable Thunberg, and got him a position as a ship's surgeon with the Verenigde Oost-Indische Compagnie.[8] In this role he was expected to collect plants for the Hortus Medicus in Amsterdam, and some rich sponsors.[9] On 30 December 1771 Thunberg sailed for Cape Town.[10] He was lucky to arrive, as the ship's cook inadvertently used white lead instead of flour in the pancakes.[11] Thunberg stayed for three years in Cape Town from April 1772 to March 1775 in order to become fluent enough in the Dutch language to pass as a Dutchman in Japan.[12] During this time he made three expeditions—two with the Aberdonian Francis Masson (1741 – 1805) —into the interior of the Dutch Colony to collect plants and animals.[13] As this time was marked by an intense struggle between San hunter-gatherers and Afrikaans farmers[14], it was not without its dangers. An encounter with a wild buffalo (*Syncerus caffer*) in a narrow pass, which left a couple of his horses dead, must have also been a harrowing experience.[15] Pressed plants, seeds, bulbs and living plants were sent by him to the Netherlands and Sweden before he sailed for Java in March 1775.[16] He stayed from 18 May 1775 to 20 June 1775 in Batavia, before setting out for Japan on 20 June 1775.[17] He arrived in Nagasaki in the middle of August 1775.[18] Here he was confined to the artificial island of Dejima in Nagasaki Bay, but was able to obtain plants and seeds through a network of Japanese well-wishers, who were keen to learn about Western medicine, and by rummaging through the animal fodder.[19] He also collected plants during the 1776 hofreis from Nagasaki to Edo to pay homage to Shogun TOKUGAWA Ieharu (1737 – 1786).[20] Thunberg left Japan on 30 November 1776, and after six months on Java, arrived at Ceylon in July 1777.[21] In February 1778 he sailed for Europe via the Cape of Good Hope, arriving in Amsterdam on 1 October

1778 after a stormy passage through the English Channel during which the living plants were lost.[22] He made a short trip to London to meet Sir Joseph Banks (1743 – 1820) and examine the plant collection of Engelbert Kaempfer (**Box 8.2**), before returning to Sweden in March 1779.[23] In 1781 he was appointed supernumerary professor of botany at the University of Uppsala, becoming Full Professor (1784 – 1828) after the death of Linnaeus' son (1741 – 1783).[24] He made little effort to prepare his lectures, and neglected the Hortus Botanicus.[25] However, Thunberg was a kind-hearted man, who was popular with students.[26] In fact, his Uppsala years were highly productive. His Flora Japonica in 1784 was followed by the Prodromus Plantarum Capensium in 1794 and 1800, Icones plantarum japonicarum during 1794 – 1805, Flora Capensis during 1807 – 1823, Fauna Japonica during 1822 – 1823, Florula Javanica in 1825, and a number of smaller studies.[27] Many of the names he coined for the plants he described are still valid. [28]

Box 8.4

Dr. Philipp Franz Balthasar von Siebold alias *Shiboruto-san* (1796 – 1866) Second son and only surviving child of Professor Johann Georg Christoph von Siebold (1767 – 1798) and his wife Maria Apollonia Josephine Lotz (1768 – 1845).[1] Brought up by his mother and uncle (Rev. Dr. Franz Joseph Lotz, 1765 – 1839) after his father's untimely death from pneumonia, he attended the Gymnasium in Würzburg during 1809 – 1815 before going on to study medicine at Würzburg University during 1815 – 1820.[2] In 1817 he went to live with Professor Ignaz Döllinger during 1770 – 1841, one of his father's friends.[3] Here he read the travelogues of Alexander von Humboldt and Johann Reinhold Forster and dreamt of following suit.[4] One of his father's former students, Dr. Franz Joseph Harbauer (1776 – 1824), who had been Friedrich Schiller's doctor, was by then Inspector-General of the Medical Service of the Dutch Army and Navy. He suggested that by becoming a Dutch military doctor Siebold would have the opportunity of visiting the Dutch East Indies (now Indonesia) and thus realize his dream.[5] Siebold jumped at the opportunity. He was appointed medical officer on board the frigate "Jonge Adriana" which left Rotterdam for Java on 23 September 1822.[6] As a talisman he carried a strand of Döllinger's hair.[7] The five month journey gave him an opportunity to make a collection of marine animals, and improve his Dutch and learn some Malay.[8] On arrival in Batavia on 13 February 1823 he made a good impression on the Governor-general, Godert Alexander Gerard Philip, Baron van der Capellen (1778 – 1848) and the Director of the Botanic Gardens, Caspar Georg Carl Reinwardt (1773 – 1854).[9] In April 1823 he was appointed as scientist and physician to the Dutch trading post on Dejima in Nagasaki Bay.[10]

Fig. 8.11 Dr. Philipp Franz Balthasar von Siebold (1796–1866) as a young man. Fortune was keen to meet the senior author of the new "Flora Japonica", so he walked to Siebold's garden at Narutaki, where Japanese novelties were prepared for transportation to Europe. "I found him at home, and he received me most kindly." (Fortune, 1863: 17). This changed when Fortune paid a clandestine visit to Siebold's garden, and took cuttings of some of the plants.

Siebold sailed for Dejima on board the cargo ship "De Drie Gezusters" on 28 June 1823.[11] On 5 August the ship was caught in a typhoon off the north coast of Táiwān, and was lucky to reach

Dejima on 11 August 1823.[12]

Dejima was not much more than a state prison, visited only by custom officials, translators and prostitutes.[13] In this way Siebold met the courtesan Sonogi (KUSUMOTO Taki, 1807 – 1865, petname "Otakusa"), with whom he lived together—mixed marriages were in any case forbidden—and after whom he named a hydrangea (*Hydrangea otaksa* Siebold & Zuccarini).[14] They had a daughter (SHIMOTO Oine/Ine/Itoku, *later* KUSUMOTO, 1827 – 1903), who was to become the first Japanese female physician and finally gynaecologist to the emperor's household.[15] Although he was largely a prisoner on Dejima, after the successful treatment of an influential Japanese officer for cataract, Siebold acquired a piece of land in 1824 to start a school of western medicine (Narutaki-juku = Narutaki Cram School) and was granted a certain amount of freedom to treat patients in the Nagasaki area.[16] He never asked for payment, but did accept presents from his patients.[17] These formed the basis of his first ethnographic collection.[18] Through his contacts, Siebold was able to obtain specimens for a "Fauna Japonica" and plants for the garden in Dejima.[19] By 1825, the garden boasted a thousand Japanese plant species.[20] In 1825 he sent some of the tea seeds he obtained to Java packed in clay.[21] In this way they survived the sea journey and could be multiplied.[22] By 1833 tea plantations were well established on Java.[23] By then Siebold was back in the Netherlands, having been expelled from Japan in 1829 for obtaining a map of Japan from the court astronomer TAKAHASHI Sakuzaemon Kageyasu Jr. (1785 – 1829).[24] This was a capital offence. TAKAHASHI and some of Siebold's other collaborators were thrown into prison or executed.[25] The fact that Siebold was allowed to leave the country, indicates the high regard the Japanese had for his medical contributions.[26]

When he left Japan on 2 January 1830, he had to leave Sonogi and their daughter behind, but took with him to Java locks of their hair, and the remaining 2000 living plants in the appropriately-named frigate "Java".[27] On 5 March 1830 he headed back to the Netherlands, arriving in Vlissingen on 7 July 1830.[28] While the ethnographic material was stored in Antwerp, his dried plants were sent to the Herbarium in Brussels and the living material planted in the botanical garden of Ghent University.[29] Later that year the Belgian war of independence erupted.[30] He managed to transfer his ethnographic collection from Antwerp and the still unpacked herbarium specimens from Brussels to Leiden, but had to leave the living plants in Ghent.[31] This collection of Japanese plants was to play a role in the development of Ghent's horticultural industry (e.g. "Ghent" Azaleas of Louis Van Houtte, 1810 – 1876).[32] As a mark of recognition, the University of Ghent presented him in 1841 with specimens of plants from his original collection.[33]

In the Netherlands he was given a warm welcome.[34] King Willem I (1772 – 1843) freed him of official duties to enable him to study his collections.[35] As the State Herbarium and its director, Professor Carl Ludwig von Blume (1796 – 1862), with whom Siebold had cooperated on Java, had been moved from Brussels to Leiden, Siebold decided to settle there.[36] In the small university town, already accustomed to eccentric figures, Siebold in his Japanese kimono must have created quite a stir.[37] The fact that the Bavarian removed his shoes on entering his house, ate with chopsticks and slept on a mat on the floor only confirmed this impression of an idiosyncratic character.[38] More likely this was a ploy to draw attention to himself. Once in Leiden Siebold started writing up the results of his six years in Japan. In order to finance the various projects, he made a "Grand Tour" of Europe.[39] In 1832 the first of seven parts of his treatise "Nippon" (1832 – 1882) appeared, followed by "Bibliotheca Japonica" (1833 – 1841), an overview of Japanese literature, written in collaboration with KUO Chengchang and Johann Joseph Hoffmann (1805 – 1878).[40] The invertebrates and vertebrates in Siebold's collection were described by Dr. Coenraad Jacob Temminck (1778 – 1858), Hermann Schlegel (1804 – 1884) and Willem de Haan (1801 – 1855) as a series of monographs between 1833 and

1850.[41] Although Siebold only edited the series and wrote introductions to two parts (1838, 1850), he clearly felt that he deserved to be the first author of "Fauna Japonica".[42] In 1835 the first part of the "Flora Japonica" (1835 – 1870) appeared with Siebold as senior author.[43] In this undertaking Siebold collaborated with Professor Joseph Gerhard Zuccarini (1797 – 1848), whom he had met on his "Grand Tour".[44] Zuccarini's descriptions were accompanied by information on history, cultivation, use and Japanese names by Siebold.[45] Many of the new species they recognized found general acceptance in the botanical community.[46] Not to be outdone by the Belgian horticulturalists, Siebold established a Society for the Introduction and Cultivation of Japanese plants in 1842.[47] Within two years Siebold had bought out his partners and established Von Siebold & Co., which was specialized in the propagation and sale of Asiatic plants.[48] Although this firm was a financial flop, it was responsible for introducing a range of attractive ornamental plants into Europe.[49]

Because the Dutch climate was bad for his rheumatics, Siebold often travelled to spas in Germany.[50] In the winter of 1839 or 1840, he met the aristocratic Karoline Ida Helene Freiin von Gagern (1820 – 1877) in Bad Kissingen, 47 km NNE of Würzburg.[51] Though they were of different confessions, they married on 10 July 1845.[52] Although Siebold was almost 25 years older than his wife, it was a happy marriage, which was blessed in the next nine years by five children.[53]

Siebold always wanted to return to Japan, not the least to see Sonogi and their daughter Ine.[54] In 1852 he offered his services to the US Government, that was organizing an expedition to Japan under Commadore Matthew Perry (1794 – 1858).[55] However, his dealings with the Russians made him suspect.[56] Moreover, for the Japanese government, he was still "persona non grata". In the autumn of 1858, when his lifelong banishment was repealed through the intervention of the Dutch Consul Jan Hendrik Donker Curtius (1813 – 1879), he finally got his chance.[57] This time he took his eldest son with him.[58] As an official agent of the Dutch Trading Company Siebold took the "overland route" from Marseille on 13 April 1859, and finally arrived in Nagasaki on 4 August 1859.[59] News of Siebold's return spread like wildfire, and he was soon besieged by old friends and pupils.[60] Returning to Japan just before Robert Fortune arrived on the scene, he was able to introduce a number of plants to Europe before the Scottish plant hunter.[61] However, travel in the search of plants was not easy. Many Japanese loathed the Westerners for their gunboat diplomacy. In April 1860 a fire broke out in the Honren Temple in which the Siebolds were staying.[62] Philipp Siebold's face and his hands received burns.[63] To cover the scars, Siebold grew a beard.[64] In 1860 he bought an estate in Narutaki, where he established yet another botanical garden.[65] Fortune was suitably impressed.[66] The two plantsmen travelled on the same ship to Yokohama, when Siebold was invited for consultations in Edo. Warned of the dangers ahead by the Prussian envoy, Count Friedrich Albrecht zu Eulenburg (1815 – 1881), Siebold made his last will and testament before embarking on the "Scotland" on 13 April 1861.[67] On 6 July 1861, just 18 days after he and his son arrived in Edo, the nearby British Legation was attacked.[68] When he was informed of the attack, Siebold went to the scene to attend to the wounded.[69] This incident did not prevent Siebold from giving classes in his residence.[70] Siebold was an outspoken man, who did not always tow the official line. For instance, he advised the Japanese to encourage foreigners to use Nagasaki, rather than open a number of ports to foreign vessels.[71] This and his pro-Russian stance led to his recall to Yokohama on 18 November 1861.[72] Judging him to be a diplomatic risk, the Dutch government recalled him to Java in April 1862.[73] On 14 November 1862 he returned to Europe, and was reunited with his family on 10 January 1863.[74] On 7 October 1863 his request to be relieved of his functions was accepted.[75] Disillusioned at being marginalized, he offered his services to the Russian and French governments, but to no avail.[76] In April 1866 he travelled to Munich for an exhibition of his latest ethnographic collection of 4000 objects.[77] Weakened by consecutive colds, he contracted blood poisoning and died within a matter of days.[78] After a lot of haggling over the price with Karoline von Siebold, the Bavarian government finally bought the ethnographic collection in 1874.[79] Karoline died in 1877 and was buried next her husband in Munich's Old South Cemetery.[80]

Fig. 8.12 Hydrangeas in Siebold's garden at Narutaki. Siebold named a *Hydrangea* after his Japanese mistress. Note the bust of Siebold in old age on the right of the picture.

Fig. 8.13 Siebold's ecological garden. "On the hill-side above the house Dr. Siebold is clearing away the brushwood in order to extend his collections and to obtain suitable situations for the different species to thrive in." (Fortune, 1863: 18).

Fig. 8.14 Site of Siebold's ecological garden. This part of Siebold's garden has since been invaded by bamboos.

Fig. 8.15 Mount Inasa (Epunga), which, as Fortune pointed out, was "celebrated for the fine and extensive view to be obtained from its summit." (Fortune, 1863: 21/22).

One day Fortune went to visit the famous naturalist, Dr. von Siebold, who had a house with a commanding view at Narutaki ("Place of the murmuring waterfalls"), a few kilometres out of Nagasaki (**Fig. 8.13**).[23] The house was surrounded by nurseries for the propagation of plants prior to their shipment to Europe.[24] Fortune noted many of the plants described by Joseph Gerhard Zuccarini (1797 – 1848) in Siebold's "Flora Japonica", and was particularly interested in a number of variegated species that Siebold was growing.[25] On the hillside above the house Siebold was clearing the brushwood to make way for an ecological garden (**Fig. 8.13**).[26] Probably on Siebold's suggestion Fortune visited a nursery garden on Mount Inasa (Epunga), some 6 – 8 km NW of Nagasaki, which had a large collection of Japanese plants.[27] He purchased some of the species, before climbing to the top to admire the view (**Fig. 8.15**).[28] On the way back to Nagasaki he called in at a little well-kept garden with large azaleas belonging to an interpreter to the Japanese Government.[29]

8.2 Fortune Travels North

On 19 October the "Marmora" headed south to the Osumi Strait (Van Diemen's Strait) and out into the Pacific Ocean.[30] Unfortunately, once the barque reached the Pacific, it was caught in a gale and had to hove to for two days.[31] However, finally at 8 a.m. on 30 October the ship reached Yokohama (Yokuhama), where Commodore Perry had concluded his Treaty with Japan in 1854.[32] By the time Fortune arrived in Yokohama, the one-time fishing village had grown to a town with 18,000 – 20,000 inhabitants.[33] The town, which was surrounded by moats and had checkpoints to keep tabs on the movements of both foreigners and Japanese, was divided into a Japanese quarter in the northwest and a foreign settlement in the southeast (**Fig. 8.16**).

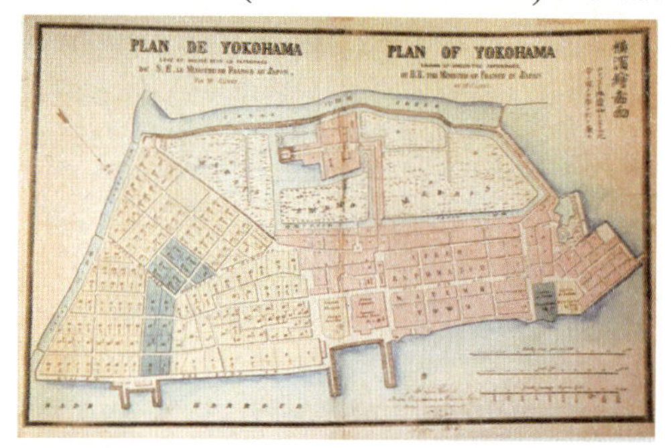

Fig. 8.16 Clipet's Map of Yokohama (1865) showing the foreign settlement (yellow) in the southeast, the Japanese quarter (pink) in the northwest, and the Miyozaki-cho, the red-light district, in the marshland to the west. The French were located in the blue area.

In the Japanese quarter Fortune discovered shops selling bronzes, ivory carvings, lacquerware, and porcelain.[34] Most of the objects were recently manufactured and of inferior quality, and although the lacquerware was good, he did not think it could compare with the beautiful old lacquerware from Kyōto (Miaco).[35] There were plenty of toys, including dolls which cried when their stomach was pressed![36] One wonders whether Fortune bought any of these toys for his growing family at home. In one of the streets Fortune found a shop selling maps, illustrated books, and woodprints, including some of the red-light district and other erotic prints, which he tactfully referred to as "very curious and instructive works".[37]

The head of Dent & Co. in Shànghǎi had provided Fortune with letters of introduction[38], so when he arrived at Yokohama, Fortune contacted the local representative, José da Silva Loureiro (**Box 8.5**).[39]

Box 8.5

Mr. José da Silva Loureiro (ca. 1835 – 1893) British subject of Portuguese extraction.[1] He worked for Dent & Co. and in this function could have met Fortune in Shànghǎi or London. Although Fortune spelt his name wrong (Loureira)[2], he nevertheless refers to him as "an old friend of mine".[3] When Fortune arrived at the port of Kanagawa at the end of October 1860, he stayed in the temple Loureiro rented.[4] Besides his tea business, Loureiro acted as Portuguese and French Consuls.[5] From 1861 to 1870 he lived and worked in Nagasaki, where he also served as the French and Portuguese Consuls.[6] While he was there he invested money in the newly established Shanghai Steam Navigation Company, which from 1867 until 1872 enjoyed a virtual monopoly of the steam traffic on the Cháng Jiāng (Yangtze) and played a leading role in trade with the ports on the Bóhǎi Sea.[7] From 1888 until 1892 he was Portuguese Ambassador in Tokyo.[8] When the Portuguese Embassy closed in 1892 as a consequence of the economic crisis, he left Japan for Hong Kong, where he died of pneumonia.[9]

Loureiro invited him to come and stay in the temple where he was living.[40] This suited Fortune admirably, as there was plenty of space in the temple and surrounding garden for his natural history collections.[41] Dr. James Curtis Hepburn (**Box 8.6; Fig. 8.17**) and Rev. Samuel Robbins Brown (**Box 8.7; Fig. 8.18**), who were living in nearby temples, provided Fortune with valuable information about the Yokohama area.[42] Fortune was particularly interested in the Buddhist temples, as it was here that he could expect to find fairly intact vegetation.[43]

Box 8.6

Dr. James Curtis Hepburn MD (1815 – 1911) Elder son of the judge Samuel Hepburn (1782 – 1865) and his wife Ann Clay (1788 – 1865).[1] His mother was interested in the foreign missions, and as his first teacher must have passed on some of her passion to her son.[2] James was a precocious scholar. He completed his studies at Princeton University in 1835 while still in his teens.[3] In 1836 he graduated MD at the University of Pennsylvania.[4] Then in 1838 he met Clarissa Mary Leete (1818 – 1906), while she was teaching at Norristown Academy.[5] Unlike her future father-in-law, "Clara" shared Hepburn's missionary dreams.[6] The couple married at Fayetteville on 27 October 1840 but having arrived in Boston too late to catch the "United States", finally set off on 15 March 1841 for Batavia in the Dutch East Indies in an old whaling ship, the "Potomac".[7] During the voyage, Hepburn marvelled at the flying fish, porpoises and the great variety of birds, distributed Bibles to the crew, bound their wounds and started studying Malay.[8] They had not been at sea for two months when the Hepburns suffered their first set-back. On 12 May 1841 Clara gave birth to a premature stillborn boy.[9] This traumatic event must have been still in the forefront of their minds when they finally disembarked at Batavia on 11 June 1841.[10] After almost a month on Java, the couple sailed on to Singapore during 7 – 12 July 1841, which was then the centre of American missionary work in the Far East.[11] The Hepburns worked at the mission station in Singapore for almost two years. During this time their second son was born, but only lived for a few hours.[12] Once the First Opium War (**Box 2.2**) was over and the Treaty Ports established, the Hepburns sailed for Macao. They arrived there on 9 June 1843 and stayed with Rev. Walter Macon Lowrie (1819 – 1847) and Dr. Samuel Wells Williams (1812 – 1884).[13] In the few months at the Portuguese enclave, the Hepburns enjoyed walking and bathing.[14] In October 1843 they went to join Rev. David Abeel (**Box 3.1**) and Dr. William H. Cumming at Xiàmén (Amoy) in southern Fújiàn.[15] Here their third son, Samuel David Hepburn (1844 – 1922) was born.[16] Unfortunately the Hepburns were often down with malarial fever, so in 1845 they decided to return to Macao to recuperate.[17] When this did not help, they took a passage on the "Panama" bound for New York.[18]

Fig. 8.17 Dr. James Curtis Hepburn (1815 – 1911), US missionary, who supplied Fortune with information about the climate and frequency of earthquakes. Tomi, one of his servants, helped Fortune to collect plants.

Over the next 13 years Dr. Hepburn built up a lucrative practice in Manhattan.[19] During this period three more children were born, but all died of scarlet fever and dysentry.[20] With only Samuel David Hepburn left, the Hepburns were once more free to travel. When the "Harris Treaty" was signed between the USA and Japan on 29 July 1858, the Hepburns finally got their opportunity. They sent Samuel to a boarding school and sailed on the "Sancho Panza" from New York to Shànghăi.[21] Detained in Shànghăi by sickness until 1 October 1859, the Hepburns did not arrive at Kanagawa, Japan until 18 October 1859.[22] The couple lived in Kanagawa from 1859 to 1863.[23] Hepburn opened a clinic there, but this was soon closed by the authorities.[24] He complained to the US Consul Townsend Harris (**Box 8.11**), but the latter retorted that the "Harris Treaty" was made for merchants and not for missionaries.[25] This being the case, Hepburn devoted himself to the study of the Japanese language and natural phenomena.[26] When Fortune was in Japan, Hepburn supplied him with information on the climate and frequency of earth tremors in the Kanagawa area.[27] Not only earthquakes, but the anti-foreign sentiment, made Japan a dangerous place to live in. Once when they were walking home, Mrs. Hepburn was attacked from the back.[28] Luckily, the club struck her shoulder and not her head. Hepburn was on the spot to dress the wounds of Mr. William Marshall (Yokohama merchant) and Mr. Woodthorpe C. Clarke (A. Heard & Co.) at the US Consulate in Kanagawa after the attack of 14 September 1862 in which Fortune's good acquaintance Charles Lenox Richardson was killed (**Box 10. 2**).[29] In 1863 the Hepburns moved to Yokohama, where James was allowed to open a clinic. With the help of five to ten Japanese assistants, he saw at least 20-70 patients each morning for the next 16 years.[30] Hepburn's afternoons were devoted to literary work, with the exception of an hour or so before dinner, when he went for a walk or visited patients or friends.[31] In the evenings he did only light work or visited friends.[32] Hepburn had a thirst for knowledge, and read all manner of books including popular Japanese literature.[33] This made him the ideal lexicographer.[34] In 1867 he finally published the Japanese-English dictionary he had been working on for a number of years.[35] Then, with the help of OKUNO Masatsuna (1823 – 1910) he started on the translation of the New Testament in earnest.[36] The Gospels of St. Mark and St. John appeared in the autumn of 1872, followed by the Gospel of St. Matthew in the following spring.[37] Not content with this achievement, he agreed with some colleagues in June 1874 to undertake a complete translation of the Bible into Japanese.[38] This magnum opus finally appeared in 1887.[39] That very year he was appointed to the Chair of Physiology and Hygiene at the Meiji Gakuin (Hall of Learning of the Era of Enlightened Government) in Tokyo.[40] In 1889 he became its first President.[41] By the time he retired from this job three years later, he was a fêted man.[42] On 22 October 1892 the Hepburns left Japan on the steamship "Gaelic", arriving at San Francisco on 10 November 1892.[43] He spent the last 19 years of his life at East Orange, just 15 km W of New York City.[44] Unfortunately, his wife, who had helped him all his life, became aggressive and had to be committed to a sanatorium near Paterson, New Jersey in 1904, where she died on 4 March 1906.[45] For his various contributions to Americo-Japanese understanding Hepburn was awarded the Imperial Order of the Rising Sun by the Emperor of Japan on his 90th birthday and an Honorary LL. D. from Princeton University three months later on 14 June 1905.[46] Finally on 21 September 1911, at the age of 96, he drifted into a coma and died.[47] By a strange coincidence, Hepburn Hall in the grounds of Meiji Gakuin went up in flames on that very same day.[48]

Box 8.7

Rev. Samuel Robbins Brown D. D. (1810 – 1880) Son of a carpenter Timothy H. Brown (1780 –1853) and his wife Phoebe Allen Hinsdale (1783 – 1861), noted for her hymn-writing.[1] He attended Amherst College before studying at Yale University.[2] While he was at Yale (1828 – 1832) he waited on the tables and gave music lessons at a boys' school in New Haven in order to pay for his fees.[3] From 1832 to 1835 he taught at the Institution for the Deaf and Dumb in New York City to pay his father's debts.[4] He then enrolled at the Theological Seminary at Columbia, South Carolina, supporting himself by teaching vocal and instrumental music at Barhamville Young Ladies' Seminary.[5] In October 1838 Brown was caught up in a whirlwind of events which were to change his life. On 4 October he was offered the post of teacher at the Morrison Chinese School, a bilingual school in Macao set up by the Morrison Education Society in memory of the Northumbrian missionary Robert Morrison (1782 –1834).[6] However, the aptly-named "Morrison" was due to sail from New York to Macao in just two weeks' time.[7] He had to act fast. He had to find out whether his childhood heart-throb Elizabeth Goodwin Bartlett would go with him.[8] She would, so the couple were married on 10 October, just a day before his ordination in New York.[9] The newly-weds stayed at the Manhattan residence of the American merchant David Washington Cincinnatus Olyphant (1789 – 1851), one of the founders of the Morrison Education Society, before setting sail for China with Rev. David Abeel (**Box 3.1**) on 17 October 1838.[10] Abeel found the young couple amiable, intelligent and pious, and thought they would make very valuable missionaries.[11] He was also impressed by Brown's fine tenor voice and his ability to play the violin, double bass, flute and seraphine organ.[12] It took the "Morrison" 125 days to reach Macao via the Cape of Good Hope.[13] Once in Macao Samuel spent his mornings studying Mandarin, in order to understand the Chinese psyche.[14] In the afternoons and evenings he taught English at the Morrison School.[15] He was a fine teacher with the ability to explain things clearly.[16] He endeared himself to his pupils, because he sympathized with their efforts to master his subject .[17] While the First Opium War (**Box 2.2**) was raging, the Browns visited Singapore in 1841, and there met Dr. James Hepburn (**Box 8.6**).[18] After the First Opium War, the Morrison School was moved from Macao to Hong Kong.[19] On 1 November 1842 the school was attacked and ransacked by pirates.[20] Brown's right leg was wounded in the attack, and he and the other inmates had to flee to the henhouse, where they stayed until the danger had passed.

Fig. 8.18 Rev. Samuel Robbins Brown (1810 – 1880), another US missionary, who accompanied Fortune to Buken-ji.

In December 1846 he was obliged to return to Monson (Massachusetts) because of his wife's failing health.[21] The three Chinese pupils he took with him caused a sensation.[22] In 1848 he became Principal at Rome Academy (New York State), where he spent his spare time searching for rocks and fossils.[23] Unfortunately, the school proved to be unprofitable, so he resigned on 31 March 1851.[24] In April 1851 the family moved to Owasco Outlet (New York State), where he was not only the pastor of Sand Beach Church, but farmed and established springside Boarding School for boys.[25] After his father's death in 1853, his mother came to live with them.[26] No doubt her presence reminded him of the importance of giving girls a good education. He thus became one of the founders of Elmira College in 1855.[27] Now that his wife's health was restored, Brown applied to the Dutch Reformed Church of America to go to China or Japan as a missionary on 11 December 1858.[28] His wish was granted.

On 7 May 1859 the Browns sailed from New York on the "Surprise" with two other missionaries and Francis Hall (1822 – 1902), who had worked with Brown to establish Elmira College.[29] During the voyage Brown devoted some of his time to learning the Japanese language.[30] After running aground in the Straits of Banka, the family finally arrived in Hong Kong on 23 August 1859.[31] From there they sailed to Shànghǎi, where Brown boarded the "Mary Louisa" bound for Yokohama.[32] On arrival at Yokohama on 3 November 1859 he and Hall were met by the Consul, General Eben M. Dorr.[33] Because Christianity was still illegal in Japan, Townsend Harris (**Box 8.11**) appointed Brown Chaplain to the US Legation.[34] He went to stay with Harris in Edo on three occasions between his arrival in Japan and the spring of 1861.[35] On his first sightseeing tour he saw Edo Castle, the Asakusa Temple, visited the Tokugawa Mausoleums, sailed down the Sumida River, and went for a ride with Henry Heusken (**Box 8.12**) to what remained of the Gohyaku Rakan-ji (500 Rakan Temple), which had been destroyed by the Ansei Earthquake in December 1854.[36] Back in Yokohama Brown got involved in the Seamen's Friend Society which aimed at curbing drinking among the foreign sailors by offering them alternatives like a reading room and a temperance refreshment house.[37] During the week he taught a variety of subjects (English grammar and pronunciation, arithmetic, physics) in a governmental school (Shubunkwan) belonging to the Customhouse.[38] Saturday was spent writing sermons in preparation for the Sunday service.[39] In his spare time he took up photography.[40] In spite of this busy schedule, Brown found time to accompany Fortune to Buken-ji.[41] The Browns first lived in a converted temple (Joryu-ji) before moving on 30 June 1864 into a new house built specially for them by subscription.[42] Unfortunately, the house with Brown's valuable collection of books, manuscripts and notes burnt down in May 1867.[43]

He decided to take time off to visit his son Robert at Rutgers College (Newark, New Jersey) and find a place for his daughter Hattie in a good school.[44] In June 1867 an Hon. DD was conferred on him by the University of the City of New York for his work in Japan.[45] While he was in the USA he had an offer of the Principalship of a new governmental school to be opened at the port of Niigata.[46] He could not wait to return to Japan. He arrived back in Yokohama on 26 August 1869 by which time Edo had become the new capital of Tokyo.[47] It took him 16 days to cross the mountains and reach Niigata on the west coast of Honshu.[48] Although his 30 pupils were intelligent and well-behaved, in Niigata he felt cut off from his old friend and fellow-translator, Dr. James Hepburn (**Box 8.6**).[49] So, when he was offered the headmaster's job at a new school in Yokohama, he gladly accepted.[50] The new school opened on 11 September 1870.[51] On 1 August 1872 Brown left the Japanese educational service to devote himself wholly to translation.[52] In the autumn of 1872 the Japanese version of the Gospel of Mark was published, followed by the Gospel of St. Luke in 1876.[53] While he pushed ahead with the work of translation, Brown also felt the need to train a group of Japanese Christians to take over the role of spreading the gospel once he was gone.[54] His health suffered, and he was in some pain as a result of neuralgia and angina pectoris.[55] In 1877 he was forced to take a well-deserved rest.[56] Brown's wife and some of his friends encouraged him to join a trip to the South Seas, which had been mounted to locate some castaways.[57] The castaways were never found, but Brown enjoyed the cruise to Guam, the Mariana Islands, New Guinea, Easter Island, the Spice Islands, the Philippines, and Hong Kong.[58] In Hong Kong some of his former pupils came to see him.[59] He visited the site of his former home and discovered that the tree which he had planted on Morrison Hill in 1843 had grown into a magnificent specimen with a trunk 1.5 m in diameter.[60] At Macao he found his old house in ruins, but brought away a tile as a memento.[61] He sailed up the Chinese coast, where he was lionized by former pupils in Xiàmén and Shànghǎi.[62] In the Chinese capital he stayed for 8 days with the American missionary, Dr. William Alexander Parsons Martin (1827 – 1916), a specialist in international law and author of a number of books about China.[63] He finally got back to Japan, and was able to note with satisfaction the great strides that the Japanese church had made.[64] On 26 June 1879 two surgeons ordered a complete rest and suggested he return to Philadelphia for treatment.[65] Once back in the USA he travelled to see various people.[66] There he learned of the completion of the translation of the New Testament into Japanese.[67] Samuel Brown passed away peacefully on his way to a reunion of the 1832 Yale class in the summer of 1880.[68] He was survived by his wife, two sons and two daughters.[69]

One day at the beginning of November Rev. Brown walked with him to the picturesque Buken-ji (Bokengee) Valley with its temple complex some 3 – 5 km from Yokohama.[44] The road led along a stream between paddy fields in which the rice was ripening.[45] Near the temple gateway stood a variety of ornamental plants, while a little further on, in front of one of the principle temples, Fortune was delighted to see a magnificent specimen of *Sciadopitys verticillata*, the umbrella pine, so called because its needles are arranged like the ribs of a parasol or umbrella (**Fig. 8.19**).[46]

Fig. 8.19 The umbrella pine, *Sciadopitys verticillata*, which Fortune saw at the temple of Buken-ji. Fortune considered it "a tree of great beauty and interest ... it will be a very great acquisition to our list of ornamental pines." (Fortune, 1863: 48). The needles are arranged in whorls, like the stretchers of an umbrella, hence the English name for this conifer.

Fortune remarked that the buildings were beautifully thatched and the surrounding gardens well-kept, the hedges neatly trimmed, and the paths swept daily.[47] Brown and Fortune took one of the paths leading up a hill, where they obtained a magnificent view of Mt. Fuji (Fusi-yama), which by then was half-covered by snow.[48] On the way back to Yokohama, the companions visited a number of farmhouses.[49] In one of the gardens Fortune found a fine collection of chrysanthemums, and managed to purchase some of them.[50]

Fortune was keen to obtain seed of a number of evergreen conifers (*Pinus* spp., *Sciadopitys*, *Taxus*, *Thuja*, and *Thujopsis*)[51], so Dr. Hepburn lent him one of his male servants to act as his guide.[52] Tomi discovered that there were some fine trees of *Thujopsis dolabrata* (**Fig. 8.22**) at the temple of Torin-ji (To-rin-gee, **Figs. 8.21**, **8.23**, **8.24**), close to Uraga Harbour (**Fig. 8.20**), where Commodore Perry had first landed.[53] On 10 November they located the stately, 24 – 30 m trees, in the cemetery to the west of the little temple.[54] Although the trees were coning, the cones were restricted to the higher branches, so with the Buddhist monk looking on in awe, Fortune and Tomi climbed the trees to obtain the seed.[55] Not one of these magnificent trees is now to be found in the cemetery (**Fig. 8.23**). They were presumably felled to make way for more graves. Likewise, the paddy fields, which used to line Uraga Harbour, have been replaced by countless houses (**Fig. 8.25**).

Fig. 8.20 Uraga Harbour, which gained notoriety as the place where US Commadore Matthew Perry (1794–1858) and his fleet of smoke-belching ships docked on 8 July 1853. This event heralded a radical change in Japanese society.

Fig. 8.21 Temple of Torin-ji, Uraga Harbour, which Fortune visited to collect seed of *Thujopsis*. "The temple of To-rin-gee is a small one, and has only one priest and priestess to minister at its altars." (Fortune, 1863: 54).

Fig. 8.22 *Thujopsis dolabrata*, an attractive conifer sought after by Fortune. He considered it "a beautiful tree ... having leaves of a fine dark-green colour ... Beneath they are of a silvery hue, which gives them a somewhat remarkable appearance when blown about by the wind." (Fortune, 1863: 55/56).

Fig. 8.23 Cemetery of Torin-ji with a diminutive *Cephalotaxus* peeping from the graves. *Thujopsis* no longer grows in the cemetery.

Fig. 8.24 The "pretty garden filled with flowers, and kept in the most perfect order", mentioned by Fortune (1863: 54), is still to be found next the priests' quarters at Torin-ji.

Fig. 8.25 Torin-ji no longer overlooks "a quiet and secluded rice valley" (Fortune, 1863: 55), but is hemmed in by a housing estate.

On their return journey Fortune and Tomi passed through an agricultural area with the thatched farmhouses occupying the lower slopes of the hills between the yellow paddy fields and the evergreen woodland.[56] Fruit trees abounded and some tea was cultivated for personal use.[57] Fortune also encountered a grape-vine, which he thought would be suitable for cultivation in the USA.[58] Once he was back in Yokohama, he suggested this to the American merchant, George Rogers Hall (**Box 8.8; Fig. 8.26**), who had already introduced a number of plants into the USA from China.[59]

Box 8.8

Dr. George Rogers Hall MD (1821–1899) Son of the landowner Benjamin Hall (1795–1873) and his wife Ruth Miller Rogers (1795–1851).[1] After attending Washington College (now Trinity College), Hartford (Connecticut), he studied at Harvard Medical School during 1842–1846.[2] In the wake of the Treaty of Nánjīng, there was a need for doctors to attend to the American and European seamen, who called in at the Treaty Ports.[3] In 1850 he opened a practice with the Scottish doctor John Ivor Murray.[4] It is possible that Fortune first met the frank and pleasant American through his Scottish partner, but if so, he failed to mention him.[5] In 1852 Hall and Murray opened the "Seaman's Hospital" at Shànghǎi.[6] Unfortunately, the hospital was not a financial success, so Hall decided to go into business in order to support his growing family.[7] However, after the Small Sword Uprising (**Box 5. 21**) during which Hall helped the Governor of Shànghǎi to escape, and with the Tàipíng Rebels in nearby Nánjīng (**Box 5.19**), they decided in 1854 that it would be safer for their three young boys if Mrs. Hall returned to the USA.[8] George Hall stayed on in China to earn a fortune by buying and selling porcelain, lacquer work, bronzes, jade and ivory, and speculating in gold and silver.[9] In 1859, following the "Harris Treaty" signed between the USA and Japan (**Box 8.11**), he moved to Japan, where he became the Yokohama representative of Walsh & Co.[10] In his garden in Yokohama he started to collect ornamental plants to take back with him to the States.[11] Some of these were obtained from Philipp von Siebold (**Box 8.4**).[12] Fortune used his garden on the waterfront as a depot for his collections, so presumably a certain amount of swapping went on.[13] It cannot be denied that a number of species were introduced almost simultaneously into the USA, the UK and the Netherlands.[14] Fortune was particularly grateful to the American for allowing him to dig up a male plant of *Aucuba japonica* growing in Hall's garden, for this offered the prospect of producing fruits on the exclusively female plants, which had been introduced into Britain.[15] Fortune was probably responsible for persuading him to use Wardian Cases for transporting plants.[16] In 1861 Hall entrusted several Wardian Cases filled with Japanese plants to the shipping magnate Franklin Gordon Dexter (1824–1903), who delivered them to Francis Lowell Lee (1823–1886) of Chestnut Hill in Boston.[17] As

Fig. 8.26 Dr. George Rogers Hall (1821–1899), a doctor turned businessman, who was responsible for introducing a number of Chinese and Japanese plants into the USA. He gave Fortune a male plant of *Aucuba japonica* with which to pollinate the female plants already introduced into the UK.

Lee was about to enlist in the Massachusetts Volunteer Infantry to fight the Confederate Army, he passed the plants on to his friend, the historian and horticulturalist Francis Parkman (1823 – 1893) in Boston.[18] On 3 January 1862, when Hall himself returned to the USA, he took with him a second consignment of living plants.[19] These were propagated by Samuel Bowne Parsons (1819 – 1906) in Flushing (New York) for the American market.[20] Hall was keen to show his wife the country from which these plants had come, but the trip had to be abandoned when Helen fell ill.[21] Following her death in 1864 Hall made his first visit to the "Confederate" South, which was just recovering from the American Civil War during 1861 – 1865.[22] As a convinced unionist, Hall hung the American flag from his house near Rome (Georgia), only to have it torn down repeatedly.[23] He then built a house in Jacksonville in NE Florida.[24] Except for a visit to Japan in 1875 – 1876, his life alternated between Bristol (Rhode Island) and his Florida winter home, where he had an orange grove.[25] After a few winters in Florida, Hall and some associates built a hotel at Fort George Island, some 40 km from Jacksonville.[26] Unfortunately, this was destroyed by a fire just after the insurance had lapsed, his shares in a Georgian gold mine plummeted, and his orange grove succumbed in the "Great Freeze" of 1880 – 1881.[27] As a result, Hall had to sell most of his Florida property and mortgage the farm at Bristol.[28] He managed to eke out a living by selling his Asiatic works of art.[29]

It was in Hall's garden on the Yokohama waterfront that Fortune obtained a male *Aucuba japonica* as a pollinator for the female plants which were invariably grown in the UK at the time.[60] In this way Fortune hoped to enliven Victorian gardens with their crimson drupes borne in the winter and spring.[61]

Consul Rutherford Alcock (**Box 3.10**) got word that Fortune was in Japan, and invited him to come to Edo (Yedo).[62] On 13 November 1860 the plantsman set off on horseback for the almost two-million metropole accompanied by an armed guard of mounted police or yakunins (yakoneens).[63] The party soon joined the Tōkaidō, the Imperial Highway between Kyōto and Edo, thronged by streams of coolies, pedestrians, palanquins, and packhorses fitted with grass shoes, transporting whole families in their panniers.[64] There were no carts to be seen, although they were used extensively in Edo.[65] The roadside was lined by houses, small shops, refreshment booths, beggars and Buddhist monks and nuns soliciting alms, travelling musicians and flower-sellers armed with sprays of aniseed (*Illicium anisatum*) for the graves.[66] In the few places where there were no buildings, trees provided the travellers with a certain amount of shade (**Fig. 8.27**).[67]

Fig. 8.27 Tōkaidō, the imperial highway between Kyōto and Edo. Photograph taken by Felice Beato near Kanagawa in 1865.

At the busy market town of Kawasaki (Kawasaky) the party was punted across the Tama River (Loga River) in flat-bottomed boats.[68] After going for another 3 km, the riders stopped at the Ume-yashiki ("Mansion of Plum Trees") Tea-house in Omori (Omora), where Fortune found himself surrounded by a number of pretty, good-humoured girls, who fed him with cakes, sweetmeats and hard-boiled eggs, as he sipped his tea.[69] While his mounted guard finished their meal in another room, Fortune strolled round the formal garden with its rockeries, ponds and rustic bridges.[70] He imagined how charming it must be in spring and summer when the large numbers of plum trees were in full flower.[71]

It was only a short ride from Omori to Shinagawa (Sinagawa) on the southern outskirts of Edo. Turning west off the Tōkaidō, Fortune's party finally arrived at the British Legation, which was housed next the temple of Tozen-ji (**Fig. 8.29**).[72]

Fig. 8.28 The present-day Tōkaidō in Shinagawa, on the outskirts of Tokyo.

Fig. 8.29 Yard in front of Tozen Temple. The former British Legation was situated in the lefthand corner of the temple complex.

Fig. 8.30 Atagoyama, a small hill with a look-out post for fires. Fortune climbed the 85 steps to the top, to get the "Grande Vue".

Fig. 8.31 When Fortune and Alcock visited Atagoyama in 1860 they "gazed upon the vast and beautiful city which lay below us spread out like a vast panorama." (Fortune, 1863: 78). The "Grande Vue" is now blocked by apartments and offices.

Alcock was there to greet him.[73] The next day was sunny, so Alcock and Fortune went sightseeing.[74] In order to give Fortune an impression of the size of Edo, Alcock first took Fortune to Atagoyama (Atango-yama, 29 m), which was used as a look-out post for the fires that often caused extensive damage to the wooden buildings of the metropole.[75] Leaving their horses at the foot of the hill, they climbed the 85 steps to the Atago Jinja, the shrine dedicated to Atago Dai-gongen, the god of fire-prevention, which Shogun TOKUGAWA Ieyasu (1542 – 1616) had had erected in 1603 (**Fig. 8.30**).[76] While they sipped their tea, they took in the panorama, which was referred to as the "Grande Vue" by the foreign community.[77] Today the "Grande Vue" is blocked by apartment blocks and office buildings (**Fig. 8.31**).

To the south of Atagoyama lay the burial ground of six of the fifteen TOKUGAWA Shoguns, which Fortune wrongly referred to as the Imperial Cemetery.[78] Alcock and Fortune rode round the walled compound and past the Sanmon (Main Gate; **Fig. 8.32**) before doubling back to Edo Castle where the shogun lived.

Fig. 8.32 Sanmon (Main Gate) of Zoji-ji, where six shoguns were buried. The deodar in the foreground was planted in 1879 by US President Ulysses S. Grant (1822 – 1885) during his world tour.

Fig. 8.33 Massive front gate of IKEDA Inshu's residence in Tokyo. "All we saw of the houses of the Daimios was the outer walls, the grated windows, and the massive-looking doors, many of them decorated with the armorial bearings of their owners." (Fortune, 1863: 80).

As they approached the castle from the south, they entered the area where the feudal princes (daimyos) resided. These lived with their family and samurai retainers in walled residences with massive doors (**Fig. 8.33**).[79] When the daimyo returned to his fiefdom, he had to leave his family and some of the samurai behind to ensure his loyalty to the shogun.[80] Alcock and Fortune passed more daimyo residences to the east of Edo Castle as they circled it in anticlockwise direction. Once they left the daimyo houses behind, they soon found themselves on rising ground to the north of the Shogun's Palace.[81] Here at a point that the foreigners had aptly called "Belle Vue" they got another splendid view of the city and Edo Bay, before the setting sun forced them to return to the British Legation.[82]

It was unfortunate for Fortune that the British Legation was located in Shinagawa, as it meant he had to cross Edo each time he wanted to visit the nurseries in the northern suburbs of the metropolis.[83] To the chagrain of his armed guard, who would have preferred to relax at one of the many tea-houses in the valley, he spent an entire day going from one nursery to another in Dango-saka (Dang-o-zaka).[84] While he was there, he was surprised to meet female dummies made from thousands of chysanthemum flowers.[85] These were part of the once-famous Dango-saka Chrysanthemum Doll Festival, which was discontinued at the end of the Meiji Era, and only recently re-established.[86] On another occasion he headed for the village of Somei (Su-mae-yah), which was renowned for its extensive gardens and nurseries.[87] It was here that the most popular flowering cherry in Japan, *Cerasus* x *yedoensis* (Somei-yoshino; **Fig. 8.34**) is thought to have originated in the mid 19th Century.[88]

Fortune was in his element, for a 2-km long street was lined with innumerable nurseries with a great variety of bonsai, potted- and bare-root plants.[89] Fortune was particularly impressed by the large number of variegated plants.[90] Here he was able to select a great number of ornamental shrubs and trees, which he hoped would "produce a striking and novel effect upon our English parks and pleasure-grounds".[91] Unfortunately for plant-lovers the nurseries in Dango-saka and Somei have long since been built over, as the population of Tokyo expanded.

Fig. 8.34 ***Cerasus* x *yedoensis*, the popular Yoshino cherry, was first put on the market in Somei, where Fortune obtained some of his Japanese plants.**

On 28 November 1860 Fortune had an appointment with Abbé Girard (**Box 8.9**; **Fig. 8.35**), who was temporarily employed as Interpreter at the French Legation.[92]

Box 8.9

Abbé Prudence Séraphin Barthélémy Girard (1821 – 1867) When he was only four he dreamed of becoming a priest.[1] After receiving a basic education from the Curate Bias of Châteaumeillant (Cher, C France), he entered the Seminary of Bourges at the age of ten.[2] In October 1838 he was enrolled in the Foreign Missionary College in Paris.[3] On 17 May 1845 he was ordained, and served for some time as a priest in the parish of St. Pierre-le-Guillard in Bourges.[4] While he was there he decided to become a missionary in the Far East. He first worked in Hong Kong, before transferring to Guǎngzhōu, where he had a floating chapel.[5] During this time he studied English, Portuguese, Chinese and Japanese.[6] In 1854 he sailed for the Ryūkyū Islands, where he spent four difficult years living among malevolant people and coping with a hostile administration.[7] He was one of three French Catholic missionaries who stayed on in the Ryūkyūs after a treaty had been signed between France and the Ryūkyū Kingdom in November 1855.[8] He left the Ryūkyū Islands in October 1858, and eventually arrived in Edo on 6 September 1859 where he acted as secretary/interpreter to Gustave Duchesne de Bellecourt (1817 – 1881), the French Chargé d'Affaires in Japan.[9] He was enthusiastic about the country, and tried to encourage a group of French nuns living in Singapore to come to Japan.[10] When Fortune visited Edo in November 1860, Girard took him to see the Asakusa Temple and the left bank of the Sumida River.[11] He was with Henry Heusken (**Box 8.12**) when the latter died.[12] That same year Girard moved to Yokohama, where he supervised the construction of the Church of the Sacred Heart (Tenshu-do, "House of God") on the edge of

Fig. 8. 35 Abbé Prudence Séraphin Barthélémy Girard (1821 – 1867), took Fortune to see the temples of Asakusa Kan'non and Eko-in on 28 November 1860.

Chinatown.[13] It was completed by the end of 1861 and consecrated on 12 January 1862.[14] Its mixture of Neoclassical and Gothic styles[15] attracted not only foreigners, but Japanese as well.[16] Girard took advantage of their curiosity to preach to his captive audience.[17] For some of the Japanese visitors, their tour of the building ended in prison, as Christianity was still a forbidden religion for them.[18] It took interventions by Girard and the French Consul-General to obtain their release.[19] In 1863 he acted as interpreter for Admiral Constant Louis Jean Benjamin Jaurès (1823 – 1889) during the Bombardment of Shimonoseki.[20] Girard died of a fever at the age of 46.[21]

Fortune had heard that the Asakusa Kan'non Temple (Ah-sax-saw Temple) was well known for the variety and beauty of its chrysanthemums, so he wanted to pay it a visit.[93] However, it was almost a two-hour ride to get to the temple, which is close to the Sumida River in northern Edo.[94]

Between the outer gate (Kaminari-mon), guarded by the Gods of Wind (Fujin) and Thunder (Raijin) with their bulging eyes (**Fig. 8.36**), and the inner gate (Hozo-mon, Treasure House Gate), there was a shopping mall (Nakamise-dori, **Fig. 8.37**), where a large variety of products, such as toys, lacquerware, looking-glasses, pipes, pictures and porcelain were for sale.[95] It is easy to imagine how Fortune felt, as the appearance of the temple complex seems to have changed little since Fortune's day. However, one should remember that with the exception of the simple Asakusa Shrine in 1649, all the buildings were destroyed in the American air raids at the beginning of 1945.[96]

Fig. 8.36 The Outer Gate (Kaminari-mon) of the Asakusa Kan'non Temple, guarded by the Gods of Wind and Thunder.

Fig. 8.37 The Nakamise-dori, between the outer and inner gates of the Asakusa Kan'non Temple. "Each side of the avenue was lined with shops and stalls, open in front like a bazaar, in which all sorts of Japanese things were exposed for sale." (Fortune, 1863: 125). Fortune may have bought some toys for his children here.

The Buddhist temple of Eko-in (Eco-ying), on the left (east) bank of the Sumida River, where Girard and Fortune now headed, suffered the same fate in 1945. Today there is little more than the cemetery left (**Fig. 8.38**).

This temple was erected in memory of the (more than) 100,000 persons who lost their lives in the great fire of Meireki in 1657, which destroyed 60% of Edo.[97] The two weeks in Edo were soon over, but before he left on 28 November 1860, Fortune had his plants packed into baskets and asked Alcock's amiable assistant Mr. John MacDonald (ca. 1838 – 1866) to send them by boat to Yokohama.[98]

Fig. 8. 38 Cemetery of Eko-in Temple. The temple itself was destroyed in 1945.

Fortune and his armed guard returned to Yokohama by the Tōkaidō.[99] When they got to Omorikaigan, one of his yakunins pointed out the Suzugamori execution ground, which was in operation between 1651 and 1871.[100] A well for washing the severed head, a round base with a hole for an iron post for burning arsonists (a serious crime at a time when the houses were all wooden structures), and a square base for erecting a crucifix for the worst criminal offences—the criminals were slowly speared to death[101]—are still on display (**Figs. 8.39, 8.40**).

Fig. 8.39 The Suzugamori execution ground, which was still in operation during Fortune's visits to Japan. The round base held an iron post used for burning arsonists, a serious crime at a time when most of the buildings were made of wood.

Fig. 8.40 The square stone base was used to hold a crucifix for the worst criminals. These were slowly speared to death.

A greengrocer's daughter, OSHICHI Yaoya (ca. 1667 – 1683), who attempted arson in the hope of meeting a temple page again[102], and the Japanese Robin Hood, NAKAMURA Jirokichi (1797 – 1831), were executed here. NAKAMURA's remains were subsequently buried under a maidenhair tree at Eko-in (**Fig. 8.41**).

Once the party had crossed the Tama River, they stabled their horses at the Man'nen-ya Inn ("Ten Thousand Years") and proceeded on foot to the famous Buddhist temple of Kawasaki Daishi.[103] Shops lined the approach to the main gateway on the east side of the complex (**Fig. 8.42**).[104]

Before ascending the steps to the main temple (**Fig. 8.43**), which was decorated with huge paper lanterns, the pilgrims first cleansed themselves by sprinkling themselves with holy water from a tank on their right.[105] Although he was impressed by its massive structure, Fortune considered the temple far inferior to the major temples he knew from China.[106] Could this have been the reason why he did not explore more of the temple complex (he made no mention of a Shinto Shrine, the bell tower or the pagoda), or was he simply looking forward to being fed by the "pretty damsels" at the inn? [107] However, as Fortune wryly wrote "the best of friends must part at last, so I was obliged to bid adieu to mine host and his fair waiting-maids of the 'Ten Thousand Centuries', and pursue my way to Kanagawa".[108]

Fig. 8.41 Tombstone of the Japanese Robin Hood ("rat boy"), NAKAMURA Jirokichi (1797–1831), at Eko-in. Pilgrims still rub the white stone at the front with one of the pebbles, hence its unweathered appearance.

When the plants arrived in Yokohama, Fortune had them transferred to George Hall's garden on the Yokohama waterfront.[109] There were problems completing the Wardian Cases, as the carpenter would not do the glazing, and another carpenter, who agreed to fit the glass panes, broke his diamond cutter.[110] However, everything was finally resolved and the collection stowed on board the SS "England" with destination Shànghǎi.[111] As John Gould Veitch (**Box 8.1**) was also returning on this vessel, the whole poop was crammed with Wardian Cases.[112]

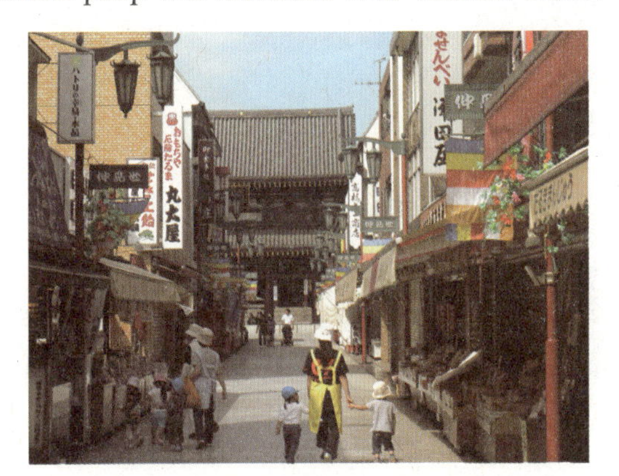

Fig. 8.42 The Kawasaki Daishi. Shops line the approach to the main gateway on the east side of the temple complex. "On the roadside there are many little shops in which tea and dried fruits were exposed to tempt the weary pilgrim on his way to worship at the temple." (Fortune, 1863: 140/141).

Fig. 8.43 The Main Hall of Kawasaki Daishi, where Charles Lenox Richardson (1834–1862) was heading when he met his death.

The steamship left Yokohama on 17 December 1860.[113] Captain Dundas steered his ship along the Pacific coast of Honshu (Nipon) as far as the Kii-suido (Kino Channel) between Honshu and Shikoku (Sikok).[114] Since he had presents from the Shogun for Queen Victoria, he got special permission to enter the calmer Inland Sea.[115] Unfortunately the Japanese pilots misjudged the draft of the "England", and grounded it on two occasions.[116] At the southwestern extremity of the Inland Sea, where it is connected to the Pacific Ocean by the Bungo Channel, the ship encountered a gale and made slow progress.[117] However, once the wind had died down Captain Dundas steered his ship along the east coast of Kyushu, passed through the Osumi Strait and headed north to Nagasaki. The steamship docked at Nagasaki on Christmas Day 1860. Fortune took advantage of the three days that the "England" stayed at Nagasaki to add some novelties to his collection.[118] As the Wardian Cases were full, these plants were placed in baskets and stowed away in the long-boat on the starboard side.[119] This was a lucky choice, as the long-boat on the port side was swept into the sea between Nagasaki and Shànghǎi.[120] Once the ship arrived at Shànghǎi on 2 January 1861, Fortune had to repack the plants for their journey to the UK.[121] Most of the material was shipped back to Port Suez on the "Tung-yu" (Captain Taylor), and arrived in England in mint condition.[122] Fortune stayed with Edward Webb, who had become head of Dent & Co. in Shànghǎi, following the death of his friend Thomas Chaye Beale (**Box 5.5**).[123]

8.3 Fortune's Second Season in Japan

At the beginning of April 1861 Fortune returned with his Chinese servant Tunga by the SS "Scotland" to Japan.[124] He arrived in Nagasaki just in time for the Hata-age Kite Festival (**Fig. 8.44**).[125]

As the ship stayed at Nagasaki for two days, Fortune was able to make an excursion into the countryside.[126] In the gardens numerous varieties of azaleas, camellias, flowering cherries, and *Kerria japonica* were now in full bloom, while the cabbage-oil plant (*Brassica rapa* var. *chinensis*, Syn. *B.* "*sinensis*") enlivened the fields with their fragrant yellow flowers.[127] By this time the winter wheat and barley were almost ready for harvesting.[128] As Siebold (**Box 8.4**) travelled on the "Scotland" to Yokohama[129], the two men must have met. However, by this time the two plantsmen were no longer on speaking terms, as Fortune had paid a clandestine visit to Siebold's garden while the latter was out, and helped himself to sprays of all the new plants.[130]

Fig. 8.44 Hata-age Kite Festival. "In the air above the town, and all over the country, there was a swarm of paper kites, which I at first sight mistook for a flock of seagulls." (Fortune, 1863: 171). Finely crushed glass on the lines of the fighter kites is used to cut the lines of rival kites.

All went well until the steamship got close to the Osumi Strait, where it was buffeted for two days by gale-force winds.[131] The ship made no headway, even though Captain Bell tried to find some shelter by staying close to the cliffs at Sata-Misaki (Cape Chichakoff).[132] On the evening of the second day the storm moderated, and the ship was able to continue on its journey, reaching

Yokohama on 19 April 1861.[133]

By the time Fortune returned to Yokohama, spring was in the air. The insects were busy pollinating the camellias and cowslips (*Primula sieboldii*, Syn. *P. cortusoides*) in the woods, and the gaudy azaleas lining the streams.[134] On the grassy banks the red flowers of *Chaenomeles japonica* contrasted with the pastel shades of the violets.[135] At the Buddhist temples and in cottage gardens double-flowering cherries and flowering peaches were covered in blossom.[136] In the fields the beans, peas and cabbage-oil plant were already blooming, while the winter cereals were coming into ear.[137] The farmers were mulching the paddy fields with freshly cut grass and herbs, prior to sowing the rice.[138] When they were not too busy Fortune was able to enlist the local people to collect the endemic ground beetle *Damaster blaptoides* and other insects for his London agent Mr. Samuel Stevens (**Box 8.10**; **Fig. 8.45**).[139]

Fig. 8.45 Mr. Samuel Stevens (1817 – 1899), Fortune's London agent.

Box 8.10

Mr. Samuel Stevens FLS (1817 – 1899) Son of John Stevens Sr. and his wife Augusta.[1] As a child he showed considerable artistic flair, winning a Royal Society of Arts medal at the age of thirteen.[2] In order to improve his poor health, he started collecting natural objects outdoors.[3] He became intrigued by insect diversity, a fascination which never left him. He joined the Entomological Society of London on 6 November 1837 and rarely missed a meeting.[4] He served on the Council during 1841 – 1847 and 1873 – 1887, was Treasurer from 1853 to 1873, and Vice-President in 1885.[5] After working as a junior partner in his brother John's auction rooms in Covent Garden during 1840 – 1848, he established his own Natural History Agency at 24 Bloomsbury Street in London, selling animals and plants to musea and private fanciers.[6] In this way explorers like Henry Walter Bates during 1825 – 1892 and Alfred Russel Wallace during 1823 – 1913 were able to finance their expeditions.[7] He not only sold the naturalists' objects, but invested the proceeds on their behalf, and actively attracted sponsors by displaying their specimens at scientific meetings, and writing short accounts of their ongoing work for journals.[8] When the shipwrecked Alfred Wallace returned to London in 1852 with nothing more than the clothes he was wearing, Stevens took him to a clothes shop, and had him measured for a new suit.[9] He moreover invited Wallace to stay at his house, where Stevens' mother Augusta nursed the explorer back to health.[10] In 1867 Stevens sold his agency to look after his mother, who died the following year.[11] In 1874 he married Frances Wood (61) and went to live in Croydon (Surrey).[12] Here he devoted the rest of his life to gardening, fishing, painting, and further insect-collecting.[13]

On 7 May 1861 Fortune and Tunga moved to the temple which José Loureiro (**Box 8.5**) had occupied.[140] Here Fortune had room for his growing collections, and a focal point for nurserymen who had heard of his interest in purchasing ornamental plants.[141] He was particularly pleased with a basket of *Primula japonica* with magenta flowers arranged in tiers, which was brought to him one morning.[142] By mid-May he was able to admire the beauty of *Paulownia tomentosa* (Syn. *P. imperialis*; **Fig. 8.46**) in full bloom in the grounds of an adjacent temple.[143] What more could a plantsman want?

Fig. 8.46 ***Paulownia tomentosa*** **(Syn.** ***Paulownia imperialis*****), a fast-growing tree from China, which was and is cultivated elsewhere for its showy flowers. It was named by Siebold & Zuccarini after Grand Duchess Anna Pavlovna of Russia (1795–1865), the wife of Willem II of the Netherlands.**

Having scoured the Yokohama area for interesting plants, insects and shells, Fortune felt he ought to pay another visit to the nurseries in Somei and Dango-saka to see if there were any spring flowers worth adding to his collection.[144] He wrote to Rutherford Alcock (**Box 3.10**) at the end of April telling him of his intention. Unfortunately, Alcock was then in China and was not expected to return to Edo until the end of June, by which time the spring flowers would be past.[145] In a letter to Sir William Hooker (Kew) Alcock explained that:

> Fortune is, or was, at Kanagawa; & would give his ears no doubt to be put on my suite—but there is no time to communicate with him. He is sighing to get back to Yeddo, now that the vegetation & flowers are at their best, & wrote a month ago to ask for a Map, which I sent him.[146]

When he did not hear from Alcock, Fortune contacted the US Minister Resident, Townsend Harris (**Box 8.11**; **Fig. 8.47**), who invited him to come to Edo for as long as he liked.[147]

Fig. 8.47 Mr. Townsend Harris (1804–1878), US Minister Resident in Japan from 1856 to 1862. He showed Fortune round Edo in 1861.

Box 8.11

Mr. Townsend Harris (1804 – 1878) Youngest of five sons of Jonathan Harris (1757 – 1816) and his wife Eleanor Watson (ca. 1760 – 1847).[1] Although he was a good pupil, his father as a hatter could not afford to give him a full education.[2] When he was 13 he was apprenticed to one of his father's friends in New York City before working in the crockery import business with his brother John.[3] In his spare time he read widely and taught himself French, Spanish and Italian.[4] Wishing to give those in a similar situation a good education, he used his Democratic Party connections to lobby for a "Free Academy of the City of New York".[5] A bill was passed by the New York State Legislature on 7 May 1847, a building designed by the architect James Renwick (1818 – 1895) erected, and in January 1849 Harris witnessed the admission of the first pupils.[6] Harris was very fond of his mother and rarely left her while she was alive.[7] Once she died in November 1847, and with his "Free Academy" up and running, Harris felt free to indulge in his wanderlust. He bought a ship, travelled to California via Cape Horn, and in 1849 began trading with China, India, Ceylon, Malaya, the Philippines, Dutch East Indies (now Indonesia), and New Zealand.[8] While he was in China he learned of Commodore Matthew Perry's treaty with the Japanese.[9] With the help of his old friend, the Secretary of State William Learned Marcy (1786 – 1857), the approval of Commodore Matthew Perry, and possibly backed by William Henry Seward (1801 – 1872), Harris was appointed Consul General to the Empire of Japan by President Franklin Pierce (1804 – 1869) on 4 August 1855.[10] He employed Henry Heusken (**Box 8.12**) as his secretary/interpreter.[11]

On 17 October 1855 Harris left New York for Europe.[12] In Paris he visited the Louvre and went to the opera[13], and had some flashy clothes made with which to impress the Siamese and Japanese.[14] On the way to Japan Harris successfully concluded a trade agreement with the Siamese king on 29 May 1856.[15] Harris finally arrived in Shimoda (Japan) on 21 August 1856[16], less than 20 months after the town had been devastated by a 8.3 magnitude earthquake on 23 December 1854.[17] With no news from the USA and confronted with initially antagonistic Japanese officials, Harris and Heusken led a secluded life, broken only by the arrival of the odd ship.[18] Harris was subject to depressions and frequently fell ill.[19] Once a month he would drink himself stupid, and be incapacitated for three days.[20] After 13 months he was finally invited to Edo for an audience with the shogun, TOKUGAWA Iesada (1824 – 1858).[21] On 23 November 1857 Harris left for Edo, where US President Frankin Pierce's letter to the Emperor of Japan was mistakenly presented to the shogun on 7 December 1857.[22] This audience was followed by prolonged negotiations, during which Harris tried to obtain a maximum number of concessions, and the Japanese attempted to limit these to an absolute minimum.[23] A Treaty of Amity and Commerce ("Harris Treaty") was finally signed between the USA and Japan on 29 July 1858.[24] The treaty was ratified by the US Senate in December 1858, but not signed by the Japanese Emperor until 1868.[25] Under this agreement Harris was able to open the US Legation in the temple of Zenpuku-ji in Edo on 7 July 1859.[26] However, with the majority of the daimyos (feudal lords) violently opposed to the treaty, Edo was a dangerous place for foreigners to live.[27] Harris' secretary (**Box 8.12**) was assassinated and the British Legation attacked on two occasions. After Heusken's death, most diplomats retreated to Yokohama. However, as a sign that he could not be intimidated, Harris stayed on in Edo.[28] While this won him the respect of the merchants[29], his self-importance and disdain for lesser mortals alienated him from the US citizens in Japan.[30] On Christmas Day 1860, when he went to Yokohama to spend the day with his countrymen, he found that they were not willing to spend the day with him![31] On 10 July 1861, just 7 weeks after Fortune's visit, Harris wrote his letter of resignation to President Abraham Lincoln (1809 – 1865), quoting age and ill-health as the reasons for his decision.[32] A formal request by the Japanese to the US Government that he be kept in office and the desire of Lincoln's Secretary of State (William H. Seward) for him to continue at his post could not make him change his mind.[33] On 11 May 1862 Harris left Japan for the USA, where he was to become a favourite with New York society.[34] However, with the American Civil War during 1861 – 1865 in full swing, people soon lost interest in an old man reminiscing about Japan.[35] Towards the end of his life he became a recluse.[36]

On 20 May 1861 Fortune was met at the Tama River by Harris' interpreter Anton L. C.

Portman (**Fig. 8.48**).[148] After the obligatory stop at the "Mansion of Plum Trees", where the garden was now delightfully shaded by the trees, and the waitresses as pretty as ever ("Pleasant, very pleasant", Fortune sighed), Fortune, Portman and their armed guard left the Tōkaidō and headed inland.[149] On the way to Edo they stopped at another tea-house with an enormous *Wisteria floribunda* (Fuji) said to be 600 years old.[150] Although Francis Hall (1822 – 1902) described this tea-house as small and dingy with tea of the worst description[151], Fortune seems to have enjoyed sipping his tea and smoking a cigar at a table under the wisteria.[152]

Fig. 8.48 Japanese caricature of Harris' second Dutch interpreter, Anton L. C. Portman, who accompanied Fortune to Edo. Portman stayed on in Japan after Townsend Harris left, and played a role in the modernization of Japan. He drowned at sea.

As they entered the suburbs of Edo, they met a party of young men from the British Legation out for a ride.[153] A few pleasantries were exchanged, before they parted company.[154] When they arrived at the American Legation, which was housed in the temple of Zenpuku-ji (Jempuku-ji; **Fig. 8.49**), Fortune was impressed by two enormous maidenhair trees (*Ginkgo biloba*, Syn. *Salisburia adiantifolia*) guarding the entrance.[155]

Although the now 750-year male tree lost most of its trunk during an air raid in 1945, it still stands.[156] However, the temple in which Fortune stayed has long since vanished. It was burnt to the ground in an attack in 1863[157], and its replacement destroyed on 25 May 1945. The present Main Hall was brought from Yao near Osaka in 1970.[158]

Fig. 8.49 Temple of Zenpuku-ji, the US Legation.

While Townsend Harris and Fortune were dining that evening, a letter arrived for Fortune from the British Legation.[159] It was from Dr. Francis Gerhard Myburgh (1837 – 1868), who was acting as the Chargé d'Affaires while Alcock was in China.[160] Myburgh was piqued that Fortune, as a British subject, had not asked him for permission to visit Edo, and requested him to leave the city without delay.[161] In a reply early the next morning Fortune explained the motives for his action, and then without waiting for a reply, headed off for the nurseries in Somei and Dango-saka.[162] It took all day to visit the numerous nurseries and make a number of purchases, so it was evening by the time he got back to the American Legation.[163] There was a reply from Myburgh, in which Fortune was told that his reasons for visiting Edo were irrelevant, that Harris had no right to invite him to Edo, and that he must leave at once.[164] The next day, while the plants were being packed into baskets for the journey to Yokohama, Fortune and Harris went for a ride.[165] They first visited the grave of Harris' first Dutch Interpreter, Henry

Heusken (**Box 8.12**), behind the temple of Korin-ji (**Fig. 8.50**).[166]

Fig. 8. 50 Henry Heusken's grave at Korin-ji. "The tomb is placed in a quiet and beautiful spot on the hill-side amongst some lofty trees. A neat and substantial monument, with a simple inscription, has been placed on the grave by Mr. Harris, and a hedge of evergreen oak and camellias has been planted around it by his orders." (Fortune, 1863: 198). These plants have since been replaced by a juniper on either side of the tombstone. The two stone jars of Dutch gin on the base are a reminder of Heusken's country of origin.

Fig. 8.51 Henry Heusken on horseback, a contemporary illustration. The metal shoes on Heusken's horse would have betrayed him to his assassins.

Box 8.12

Henricus Conradus Joannes Heusken (1832 – 1861) Son of an Amsterdam merchant, Johannes Franciscus Heusken and his wife, T. F. Heusken-Smit.[1] He was educated at a boarding school in Breda in the south of the Netherlands.[2] In 1847 he returned to Amsterdam to help in his father's business.[3] Unfortunately, his father died shortly thereafter, so the inexperienced young man was left to run the family business single-handed.[4] When the business failed in 1853, Henry decided to emigrate to the USA.[5] Life was hard in New York, as he was unable to find permanent employment.[6] Through Rev. Thomas De Witt of the Collegiate Reformed Church of New York he learned that Townsend Harris (**Box 8.11**) was looking for a secretary who knew Dutch, the *lingua franca* in Japan.[7] Even though his English was not perfect he was given the job, and early on 25 October 1855 he left New York on board the cockroach-infested US SS "San Jacinto".[8] After stops in Siam, China, Macao and Hong Kong, Harris and Heusken finally arrived in Shimoda (Japan) on 21 August 1856.[9] They were housed in the Gyokusen-ji Temple, which doubled as US Consulate.[10] The fifteen months they stayed in Shimoda proved pretty tedious. To pass the time, Heusken made ink drawings, went for walks, and learned to ride a horse.[11] Although he was once threatened by a samurai, in general he got on well with the Japanese, who referred to him as Hiyofusukuwan.[12] He is said to have had three Japanese girlfriends while in Shimoda.[13] Although everybody who met Heusken found him friendly and very obliging, he had a forgetful streak.[14] Harris complained that even though Heusken was sitting by the fireside in mid-winter, he would forget to tend the fire.[15] Nevertheless, Harris must have respected Heusken's other qualities, for when he was suffering a period of ill-health, he appointed Heusken as his Vice-Consul.[16] Heusken became highly valued as an interpreter by the British, French and Prussian envoys to Japan.[17] For his assistance to the British delegation, Lord Elgin (**Fig. 6.1**) presented him with his valuable watch, while Queen Victoria sent him a costly gold snuff box with her monogram in diamonds.[18] Since September 1860 he was in the habit of visiting the Prussian envoy, Count Friedrich Albrecht zu Eulenburg (1815 – 1881), who was anxious to sign a similar treaty with the Japanese Government.[19] By returning to the US Legation at night, he exposed himself to a possible attack by some of the more xenophobic Japanese.[20] He had been warned by Harris (**Box 8.11**), his pregnant girlfriend Otsuru, and the Japanese Minister of Foreign Affairs, but all to no avail.[21] On the cold, rainy evening of the 15 January 1861, as he was riding back from the Prussian Legation (**Fig. 8.51**), he was set on by about seven masked men and mortally wounded.[22] Ninety critical minutes elapsed between the attack (9 p.m.) and his receiving proper medical attention at the American Legation, by which time he had lost a lot of blood.[23] Although Dr. Robert Lucius (1835 – 1914) from the Prussian Legation stitched his abdominal wound, he died between midnight and 0:30 a.m. the next day aged 28.[24] Abbé Prudence Girard (**Box 8.9**) was there to comfort Heusken in his final hour.[25] Heusken was buried with much pomp in the graveyard behind Korin-ji Temple on 18 January 1861.[26] His death no doubt hastened the signing of the "Treaty of Amity, Commerce and Navigation" between Japan and Prussia on 24 January 1861, as the Japanese Government was anxious to get Count Eulenburg out of harm's way.[27] Although the identity of the assailants would appear to have been known to the Japanese authorities, Realpolitik prevented them from being brought to justice.[28] By way of compensation the Japanese Government paid $10,000 to Heusken's mother.[29]

Fig. 8. 52 The Juniso Shrine, a Shinto shrine founded in the 14th Century, which was very popular in Fortune's day. "Report says that many of the visitors are particularly fond of composing and reciting poetry in one of the avenues near the temple ..." (Fortune, 1863: 199). Today most of the visitors are from the nearby business centre. The torii in the foreground indicates that this is not a Buddhist temple.

Harris' simple inscription read: "Sacred to the memory of Henry C. J. Heusken Interpreter to the American Legation in Japan. Born at Amsterdam January 20, 1832. Died at Yedo January 16, 1861." Because he made no mention of the circumstances of Heusken's death, Harris came in for some criticism.[167] The tombstone of the Interpreter to the British Legation, only 3 m away, is more explicit: "Dan-Kutch, Japanese Linguist to the British

Legation murdered by Japanese assassins 21 January 1860".[168] Having paid their respects, Harris and Fortune rode for two hours to the Juniso Shrine (Joo-ne-shoo, "Temple of the Twelve Altars"; **Fig. 8.52**), set in Shinjuku Central Park in the western suburbs of Edo.[169]

After admiring the waterfall (**Fig. 8. 53**), the companions sat beside one of the two lakes, and drank some tea.[170] In Fortune's day people used to swim in these lakes, but having become so polluted that even the carp died, these were finally infilled in 1968.[171]

Fig. 8. 53 Waterfall in Shinjuku Central Park, which Harris and Fortune admired.

On the 23 May 1861 Fortune headed back to Yokohama.[172] The flowering shrubs (*Deutzia scabra*, *Weigela hortensis*) and small trees (*Styrax japonicus*, **Fig. 8. 54**) on the hillsides and in the hedgerows competed for his attention with the blooms of scrambling roses and the sweetly-scented honeysuckle (*Lonicera japonica*, **Fig. 8.55**).[173]

Fig. 8. 54 *Styrax japonicus* with its star-shaped, pendulous flowers, which prove attractive to insects and gardening connoisseurs alike.

Fig. 8. 55 *Lonicera japonica*. Its semi-evergreen leaves gives this scrambling vine an edge over plants with a shorter growing season. As a result, it has become an invasive plant in many parts of the world. The white flowers turn yellow after they have been pollinated.

In the gardens the spring flowers had made way for irises, clematises, herbaceous paeonies (*Paeonia obovata*), Banksian roses (*Rosa banksiae*, **Fig. 8.56**), Reeves' spiraea (*Spiraea cantoniensis*), and early-flowering chrysanthemums.[174] In the fields the barley was yellow and the rape-seed was being harvested.[175] Once the seed had been collected by treading on the fruiting capsules, the uprooted plants were burnt to fertilize the soil in preparation for a variety of summer crops.[176]

Once the summer crops had been planted, there was more time for other activities. At the end of June/beginning of July Fortune became aware of a marked increase in the number of pilgrims at an adjoining temple.[177] Curious to know what was going on, he visited the temple.[178] As in China most of the pilgrims were women, but they were nonetheless partial to a draft of sake between their devotions.[179] Fortune must have caused quite a sensation, as he was followed back to his temple by a crowd of people, who proceeded to investigate his clothes, books and specimens.[180] The beetles, butterflies and shells, which he and Tunga had been collecting with the help of the local inhabitants[181], aroused their particular curiosity. "What was he going to do with them? Eat them, use them as medicine?"[182] They clearly considered him insane to have paid money for them.[183]

Fig. 8.56 ***Rosa banksiae*, a sweet-scented rose, introduced in 1807 by Kew's collector, William Kerr, and named after his employer's wife, Dorothea Banks.**

At the beginning of July 1861 Dr. Walter Dickson, who had a medical practice in Guǎngzhōu, arrived in Yokohama after an epic journey from Guǎngzhōu to Hànkǒu ("Han River Mouth", Greater Wǔhàn) between 11 April and 20 May 1861.[184] He had been accompanied by the American missionary Samuel William Bonney (1815 – 1864), Rev. William R. Beach of the Church Missionary Society, and the Scottish merchant Mr. Thorburn.[185] No doubt Fortune was curious to know what this part of China was like. An excursion to the ancient capital of Kamakura would offer an opportunity to discover from Dickson more about the provinces Guǎngdōng and Húnán, through which the group had passed. Dickson and Fortune set off on horseback on 4 July accompanied by two Yokohama merchants, J. B. Ross and a Mr. Hope.[186] They spent the first night in an inn at Kanasawa, some 8 km from Kanagawa, where they were amused by the sometimes risqué questions posed by the Japanese.[187] The next morning, while he was waiting for his breakfast after a bathe in Sagami Bay, Fortune discovered some fine examples of *Podocarpus macrophyllus* (**Fig. 8.57**) and *Pinus thunbergii* (Syn. *P. massoniana*), and what he assumed was a new arbor-vitae, which he referred to as *Thuja falcata*.[188]

Fig. 8.57 ***Podocarpus macrophyllus*, an evergreen conifer with strap-shaped leaves, often cultivated at Buddhist temples, hence its common name "Buddhist pine". As they mature, the cone-scales become fleshy and attractive to birds, that disperse the enclosed seeds.**

Breakfast over, the party took the high road to Kamakura, in order to enjoy the panoramic views of Mt. Fuji and the island of Enoshima (Ino-sima).[189] It was a hot (ca. 38℃), cloudless day, so they were glad to find an inn in the centre of Kamakura, close to the avenue (Wakamiya Oji, "Young Prince Avenue") leading from the beach to the most important Shinto Shrine, the Hachiman-gu (**Fig. 8.58**).[190]

After resting in the inn for a short time, they set off to see "the sight" in Kamakura, the giant bronze Buddha (**Fig. 8.59**).[191] This had been a "must" for British tourists, ever since Captain

John Saris (ca. 1581 – 1643) and Richard Cocks (ca. 1566 – 1624) visited it in 1613 and 1616 respectively.[192] The Buddha had been cast in 8 sections, which were subsequently welded together.[193] They were invited to go inside the hollow statue, which was lit by two windows at the back.[194] Everybody was suitably impressed, and their only regret was that their poor knowledge of Japanese prevented them learning more about the origin and history of the edifice.[195] Had they understood Japanese, they would have discovered that the Buddha (1252-ca. 1263) had once been indoors, but that the temple had collapsed during a typhoon in August 1334/1335 killing most of the 500-odd samurai who had sought refuge there.[196] The temple was rebuilt, but was destroyed during the Meio Earthquake and subsequent tsunami[197], which inundated large areas of Kamakura on 20 September 1498.[198] The Buddha had been exposed to the elements ever since. Before returning to their inn for lunch and a siesta, the quartet visited the Buddhist temple of Hasedera nearby (**Figs. 8.60, 8.61**).[199]

Fig. 8.58 The Hachiman-gu (1180 AD), the most important Shinto Shrine in Kamakura, is "approached by an avenue terminating in a broad flight of stone steps ..." (Fortune, 1863: 234). To the left of the steps is a *Ginkgo* tree behind which MINAMOTO Kugyo (1200–1219) was said to have hidden before stabbing his uncle, the Shogun MINAMOTO Sanemoto (1192–1219).

Fig. 8.59 Giant Bronze Buddha (1252-ca. 1263) in Kamakura. On 20 September 1498 the temple which housed it was destroyed by the Meio Earthquake and subsequent tsunami.

Fig. 8.60 Main Hall of the Hasedera-ji in Kamakura, famous for its 9 m high gilded statue of Kan'non, the Goddess of Mercy.

Later in the afternoon, once it had become a little cooler, the colleagues continued their cultural programme.[200] From Fortune's descriptions they would appear to have visited the Hongaku-ji (**Fig. 8.62**), where people with eye-diseases prayed for a cure, and the Daigyo-ji with its "curiously formed" stone, which was thought to help childless women to conceive (**Fig. 8.63**).[201] Before leaving Kamakura for Kanasawa by the coastal road, a visit was paid to the simple grave of the first Shogun, MINAMOTO

Yoritomo (**Fig. 8.64**).[202] On 6 July 1861, when they reached Yokohama, they learned that the British Legation in Edo had been attacked by a group of samurai on the previous night.[203] That Consul Alcock had just returned from an overland journey from Nagasaki on 4 July, would appear to suggest that the attack had been planned. Sword nicks and bullet holes can still be seen in the wooden pillars on either side of the entrance to the former legation (**Fig. 8.65**).

Fig. 8.61 Cave at Hasedera with the sea goddess, Benzaiten, Fortune's goddess "in a dark place". (Fortune, 1863:232).

Fig. 8.62 The Hongaku-ji in Kamakura, which was visited by those suffering from eye diseases.

Fig. 8.63 Phallus-like stone at the entrance to Daigyo-ji, Kamakura. "This stone had the remarkable property, we are told, of rendering barren women fruitful." (Fortune, 1863: 235) It has become somewhat damaged in the course of time.

Fig. 8.64 The simple grave of the first shogun, MINAMOTO Yoritomo (1147–1199). "As we were leaving Kamakura I rode up to the foot of a hill on our left to see the tomb of Yuritomo, a celebrated general, the founder of the race of Japanese Temporal Emperors, and a man who is remembered among the people as William Wallace or Robert Bruce is in Scotland." (Fortune, 1863:236).

In the wake of the attack the Japanese government took measures to protect the foreigners.[204] A few holes in the fences round the temple in which Fortune was staying were blocked, and the gate to the adjoining cemetery nailed up.[205] Of course, Fortune realized that these steps would be useless against a fanatical ronin (unemployed samurai) intent on murdering him.[206] So with his work in Japan almost finished, Fortune went to stay with William Aspinall (**Box 8.13**; **Fig. 8.66**) in Yokohama's foreign settlement towards the end of July.[207]

Fig. 8.65 Bullet holes in a wooden pillar at the entrance to the British Legation, Tozen-ji.

Fig. 8.66 Mr. William Gregson Aspinall (1822 – 1879), Fortune's Liverpudlian friend.

Box 8.13

Mr. William Gregson Aspinall(1822 – 1879) Son of a Liverpudlian broker Richard Aspinall Sr. (1777 – 1856) and his wife Martha Goulburn.[1] For some time he may have worked for his elder brother Richard Jr. (1816 – 1884), who was a tea-broker in London.[2] In 1845 he married Caroline Hudson.[3] They had two children before the family moved to Shànghǎi in 1853 where he became a partner of Mackenzie Brothers and Co.[4] This is where he would have met Fortune and his future partner Frederick Cornes (1837 – 1927) for the first time.[5] With Shànghǎi under threat from the Tàipíng rebels (**Box 5.19**) and Japan opening up after the Treaty of Yeddo, he decided to move to Yokohama, where he set himself up as a tea inspector and general commission agent on 1 May 1860.[6] Frederick Cornes (1837 – 1927), who had been a silk buyer in Shànghǎi for the Manchester-based firm Holliday Wise and Company from 1858 to 1861, followed him to Japan.[7] They joined forces and founded Aspinall, Cornes and Co. in 1861 in Yokohama to export silk and green tea.[8] The business quickly expanded to include the import of Lancashire cotton, metals, consumer goods, kerosene, coal and other raw materials.[9] In the 1860s Aspinall-Cornes became agents for a number of important companies including the Peninsular & Oriental Steam Navigation Co. in 1863, Universal Marine, London & Oriental, Commercial Union and Queen Insurance, and Lloyds of London in 1868.[10] While Cornes was in England during 1867 – 1871, Aspinall was left to run the Japanese side of the business.[11] By the time Cornes returned, Aspinall was estranged from his wife, in poor health and financially pressed.[12] In 1873 Cornes bought out his partner and paid him £ 1,300 per annum for his share of the company's Yokohama property, plus £ 500 a year for 3 years on condition that he did not begin trading on his own account.[13] This should have been enough for him to live in style and support his wife and four children.[14] For some unknown reason (alcoholism? gambling debts?) he continued to be short of money.[15] In 1875 he was declared bankrupt.[16] When he died he had less than £ 200.[17] A simple gravestone marks his last resting place in the Foreigners' Cemetery in Yokohama.[18]

There he met Kew's plant collector, Richard Oldham (1837 – 1864), on the 24th of July.[208] A number of Wardian Cases were hurredly prepared and filled with examples of Japanese trees and shrubs.[209] The British Consul in Yokohama, Captain F. Howard-Vyse (**Box 8.14**) supplied a note, exempting the collection from duty and inspection.[210]

Box 8.14

Captain Francis Howard-Vyse (1828 – 1891) Youngest of ten children of the notorious egyptologist Major-General Sir Richard William Howard-Vyse MP (1784 – 1853), who used gunpowder to investigate the Cheops Pyramid, and his wife Frances Hesketh (1784 – 1841).[1] Nothing is known about his early life. On 11 March 1859 "Punch" Vyse left Paris with Frederick Bruce (**Box 9.4**) and George Wyndham (**Box 9.6**) to take up his position as British Vice-Consul in Edo.[2] He was promoted Consul in Yokohama towards the end of November 1860.[3] In this function he was able to waive the duty Fortune should have paid for shipping his collections out of the country.[4] One of the first problems he faced in Yokohama was to obtain the release of Michael Moss from prison, after the poacher had accidentally shot one of the Japanese policemen during his arrest.[5] This he did by threatening to kidnap the Japanese governor![6] When news of Charles Lenox Richardson's murder (**Box 10.2**) reached him on 14 September 1862, the impetuous Howard-Vyse immediately set off with a mounted guard in the direction of the scene.[7] In this way he contravened the orders of the Chargé d'Affaires, Lieut.-Colonel Edward St. John Neale (1812 – 1866), who no doubt realized that the apprehension of the perpetrators would aggravate an already explosive situation. For his insubordination Vyse was transferred in December 1862 to the British Consulate in Hakodate (S Ezo), where it was considered he could do less harm.[8] In 1865 he was moved to Nagasaki, but was forced to tender his resignation in December 1866, when it became clear that he had been involved in robbing Ainu graves for skeletons of the primitive inhabitants of Ezo (later renamed Hokkaidō).[9]

On 29 July 1861 Fortune bade farewell to the "lovely scenery of Japan", as the SS "Fiery Cross" (Captain Crockett) left Edo Bay and moved out into the Pacific.[211] Fortune arrived in Shànghǎi on 4 August 1861, and stowed his collection in Webb's garden until he was ready to return to the UK.[212]

Fortune probably had mixed feelings about his trip to Japan. On the one hand, he had been able to collect a number of interesting plants (**Fig. 8.67**) and many variegated cultivars (**Box 8.15; Fig. 8.68**).

Fig. 8.67 ***Lychnis senno*. Although Fortune concentrated on introducing woody plants, he could not resist this "ragged robin".**

Fig. 8.68 ***Osmanthus heterophyllus* 'Variegatus'. Variegated plants were in vogue in the Victorian Era, so Fortune obtained as many of these as possible while he was in Japan.**

Box 8.15

Fortune's Plant Introductions from Japan

The name(s) in brackets is/are the name(s) under which Fortune's material was once known. Bold type indicates that this was the first time the taxon was grown outside Japan. A question mark indicates that Fortune may not have been the first person to introduce the plant.

Acer spp.[1]

Aquilegia flabellata Siebold & Zuccarini 1846[2]

***Arachniodes standishii* (Moore 1863) Ohwi 1962** (*Lastrea standishii* Moore 1863)[3]

? *Aralia elata* (Miquel 1840) Seemann 1868 'Variegata' [*Aralia elata* (Miquel 1840) Seemann 1868 forma *variegata* (Rehder 1900) Nakai 1924, *Aralia chinensis* Linnaeus 1753 forma *variegata* Rehder 1900, "*Aralia variegata*"][4]

***Aucuba japonica* Thunberg 1783 (male shrub)**[5]

***Aucuba japonica* Thunberg 1783 'Limbata'** (*Aucuba japonica* Thunberg 1783 forma *limbata* [Bull ex] J. Dix 1864)[6]

Buxus microphylla Siebold & Zuccarini 1846[7]

***Camellia sinensis* (Linnaeus 1753) Kuntze 1887 'Variegata'** (*Thea viridis* Linnaeus 1762 *variegata*)[8]

Chamaecyparis obtusa (Siebold & Zuccarini 1844) Endlicher 1847 (*Retinospora obtusa* Siebold & Zuccarini 1844)[9]

***Chamaecyparis obtusa* (Siebold & Zuccarini 1844) Endlicher 1847 'Argentea'** (*Retinospora obtusa* Siebold & Zuccarini 1844 *variegata* Standish 1861 nomen)[10]

Chamaecyparis pisifera (Siebold & Zuccarini 1844) Endlicher 1847[11]

Chamaecyparis pisifera (Siebold & Zuccarini 1844) Endlicher 1847 'Argentea' (*Retinospora pisifera* Siebold & Zuccarini 1844 *variegata* Standish 1861 nomen)[12]

Chamaecyparis pisifera (Siebold & Zuccarini 1844) Endlicher 1847 'Aurea' (*Retinospora pisifera* Siebold & Zuccarini 1844 var. *aurea* [Fortune ex] Gordon 1862)[13]

***Chamaecyparis pisifera* (Siebold & Zuccarini 1844) Endlicher 1847 'Filifera'**[14]

***Chamaecyparis pisifera* (Siebold & Zuccarini 1844) Endlicher 1847 'Plumosa Argentea'**[15]

***Chamaecyparis pisifera* (Siebold & Zuccarini 1844) Endlicher 1847 'Plumosa Aurea'**[16]

Chrysanthemum indicum Linnaeus 1753 cultivars [*Dendranthema indica* (Linnaeus 1753) Des Moulins 1855 cultivars][17]

Chrysanthemum morifolium Ramatuelle 1792 'Bronze Dragon' (*Dendranthema morifolia* (Ramatuelle 1792) Tzvelev 1961 'Bronze Dragon')[18]

Chrysanthemum morifolium Ramatuelle 1792 'Yellow Dragon' (*Dendranthema morifolia* (Ramatuelle 1792) Tzvelev 1961 'Yellow Dragon')[19]

***Clematis patens* Morren & Decaisne 1836 'Fortunei'** [*Clematis fortunei* Moore 1863][20]

***Clematis patens* Morren & Decaisne 1836 'John Gould Veitch'** [*Clematis coerulea* Lindley 1837, *Clematis fortunei* Moore 1863 *coerulea* (Lindley 1837) Standish nomen?][21]

***Clematis patens* Morren & Decaisne 1836 'Standishii'** (*Clematis standishii* Van Houtte 1865)[22]

***Cleyera japonica* Thunberg 1783 'Fortunei'** (*Cleyera fortunei* J. D. Hooker 1895, *Cleyera japonica* Thunberg 1783 'Tricolor', *Eurya japonica* Thunberg 1783 forma *variegata* Hayashi 1963, *Eurya latifolia* K. Koch 1869 var. *variegata* Carrière 1869)[23]

Convallaria majalis Linnaeus 1753 'Variegata' ("*Convallaria variegata*")[24]

? *Corylopsis pauciflora* Siebold & Zuccarini 1835[25]

Corylopsis spicata Siebold & Zuccarini 1835[26]

? *Daphne odora* Thunberg 1784 'Aureomarginata' ("*Daphne variegata*")[27]

***Deutzia scabra* Thunberg 1781 'Plena'** (*Deutzia fortunei* Carrière 1866)[28]

***Elaeagnus pungens* Thunberg 1784 'Variegata'** ("*Elaeagnus variegata*")[29]

Euonymus fortunei (Turczaninov 1863) Handel-Mazzetti 1933 var. *radicans* (Siebold ex Miquel 1865) Rehder 1938, (*Euonymus radicans* Siebold ex Miquel 1865)[30]

? *Filipendula multijuga* Maximowicz 1879 (*Spiraea palmata* Pallas 1776)[31]

Hosta sieboldiana (Loddiges 1869) Engler 1887 var. *fortunei* (Baker 1876) Ascherson & Graebner 1905 (*Funkia fortunei* Baker 1876)[32]

***Kerria japonica* (Linnaeus 1771) De Candolle 1818 'Aureovariegata'** [*Kerria japonica* (Linnaeus 1771) De Candolle 1818 var. *aureovariegata* Rehder 1927, *Kerria japonica* (Linnaeus 1771) De Candolle 1818 forma *aureovariegata* (Rehder 1927) Rehder 1949][33]

***Ligustrum japonicum* Thunberg 1784 'Rotundifolium'** (*Ligustrum coriaceum* Carrière 1874)[34]

***Ligustrum japonicum* Thunberg 1784 'Variegatum'** (*Ligustrum japonicum* Thunberg 1784 forma *variegatum* (Nicholson 1885) Hara 1949)[35]

Lilium auratum Lindley 1862[36]

Lilium x *maculatum* Thunberg 1794 (*Lilium fortunei* Lindley 1862)[37]

? *Lonicera japonica* Thunberg 1784 'Aureoreticulata' (*Lonicera aureo-reticulata* Moore 1863)[38]

***Lychnis senno* Siebold & Zuccarini 1839**[39]

***Lychnis senno* Siebold & Zuccarini 1839 'Variegata'**[40]

***Nageia nagi* (Thunberg 1784) Kuntze 1891 'Variegata'** (*Myrica nagi* Thunberg 1784, *Nageia ovata* Gordon 1862, *Podocarpus nageia* [R. Brown ex] Endlicher 1825)[41]

***Osmanthus* x *fortunei* Carrière 1864** (*Osmanthus fragrans* x *O. heterophyllus*; *Olea aquifolia* Siebold & Zuccarini 1846)[42]

Osmanthus heterophyllus (G. Don 1832) P. S. Green 1958 (*Olea ilicifolia* Hasskarl 1844, *Osmanthus ilicifolius* (Hasskarl 1844) Carrière 1885)[43]

***Osmanthus heterophyllus* (G. Don 1832) P. S. Green 1958 'Rotundifolius'** (*Osmanthus variegatus nanus*)[44]

Osmanthus heterophyllus (G. Don 1832) P. S. Green 1958 'Variegatus' (*Osmanthus variegatus*)[45]

Parthenocissus tricuspidata (Siebold & Zuccarini 1845) Planchon 1887 (*Ampelopsis tricuspidata* Siebold & Zuccarini 1845)[46]

***Pittosporum tobira* (Thunberg 1780) Aiton 1811 'Variegatum'** ("*Pittosporum variegatum*")[47]

Pleioblastus fortunei (Van Houtte 1863) Nakai 1933 (*Arundinaria fortunei* (Van Houtte 1863) Rivière 1878, *Bambusa fortunei* Van Houtte 1863, Bambusa variegata Siebold ex Miquel 1866, *Pleioblastus variegatus* (Siebold ex Miquel 1866) Makino 1926, *Sasa variegata* (Siebold ex Miquel 1866) Camus 1913)[48]

***Pleioblastus viridistriatus* (Regel 1866) Makino 1926** [*Arundinaria auricoma* Mitford 1896, *Arundinaria fortunei* (Van Houtte 1863) Rivière 1878 var. *aurea* (Carrière 1887) Bean 1894, *Bambusa fortunei* Van Houtte 1863 var. *aurea* Carrière 1887, *Bambusa viridistriata* Regel 1866, *Pleioblastus auricomus* (Mitford 1896) D. C. McClintock 1991][49]

***Primula japonica* A. Gray 1858**[50]

Rhaphiolepis umbellata (Thunberg 1784) Makino 1902 (*Rhaphiolepis ovata* Briot 1870/1871)[51]

***Rhapis excelsa* (Thunberg 1784) Rehder 1930 'Variegata'** (*Rhapis flabelliformis* Aiton 1789 var. *variegata* B. S. Williams 1870)[52]

***Rhododendron degronianum* Carrière 1869 var. *heptamerum* (Maximowicz 1870) Sealy ex Davidian 1992** [*Rhododendron heptamerum* (Maximowicz 1870) Balfour fil. 1920, *Rhododendron metternichii* Siebold & Zuccarini 1835 nom. illegit., *Rhododendron metternichii* Siebold & Zuccarini 1835 var. *heptamerum* Maximowicz 1870][53]

***Saxifraga fortunei* J. D. Hooker 1863** [*Saxifraga cortusifolia* Siebold & Zuccarini 1843 var. *fortunei* (J. D. Hooker 1863) Maximowicz 1871][54]

***Saxifraga stolonifera* Curtis 1774 'Tricolor'** (*Saxifraga fortunei* J. D. Hooker 1863 var. *tricolor* Lemaire 1864, *Saxifraga sarmentosa* Linnaeus fil. 1782 var. *tricolor* (Lemaire 1864) Maximowicz 1871)[55]

Sciadopitys verticillata (Thunberg 1784) Siebold & Zuccarini 1842 (*Taxus verticillata* Thunberg 1784)[56]

Skimmia japonica Thunberg 1783[57]

***Skimmia japonica* Thunberg 1783 cultivars**[58]

Thuja orientalis* Linnaeus 1753 var. *falcata Lindley 1862 [*Biota fortunei* (Hort. ex) Carrière 1867, *Platycladus orientalis* (Linnaeus 1753) Franco 1949 'Falcata', *Thuja falcata* (Hort. ex) Carrière 1867][59]

***Thuja standishii* (Gordon 1862) Carrière 1867** (*Thujopsis standishii* Gordon 1862)[60]

Thujopsis dolabrata (Thunberg ex Linnaeus fil. 1782) Siebold & Zuccarini 1844 (*Libocedrus dolabrata* (Thunberg ex Linnaeus fil. 1782) J. Nelson 1866, *Platycladus dolabrata* (Thunberg ex Linnaeus fil. 1782) Spach 1841, *Thuja dolabrata* Thunberg ex Linnaeus fil. 1782, *Thujopsis laetevirens* Lindley 1861)[61]

Thujopsis dolabrata (Thunberg ex Linnaeus fil. 1782) Siebold & Zuccarini 1844 'Variegata'[62]

Tricyrtis hirta (Thunberg 1784) W. J. Hooker 1863 (*Uvularia hirta* Thunberg 1784)[63]

? *Woodwardia japonica* (Linnaeus fil. 1782) Smith 1793[64]

***Woodwardia orientalis* Swartz 1801**[65]

In hindsight Fortune probably wished that he had gone to Japan immediately after the Treaty of Yeddo, instead of opting to collect on behalf of the US Government (Chapter 7). In this way he would have cornered the market before John Gould Veitch (**Box 8.1**) and Philipp von Siebold (**Box 8.4**) got there. This would certainly have assured the financial success of the mission. Now he would not know whether it had been a profitable venture until the plants had been grown and marketed by John Standish (**Box 5.15**). The plant hunter now considered his next move.

Chapter 9 A Final Farewell to China

9.1 En Route to the Chinese Capital

When Fortune embarked on his voyage to Japan in the summer of 1860, he cannot have intended to visit the capital of China, as this was still barred to foreigners. However, by October 1860 the Anglo-French Expeditionary Force, which had been sent to China to enforce the Treaty of Tiānjīn (**Box 9.1**; **Fig. 9.1**), had battled its way past the Dàgū Forts, captured Tiānjīn ("Heavenly Ford") and forced Běijīng to capitulate.

Fig. 9.1 Signing of the Treaty of Tiānjīn on 24 June 1858.

Box 9.1

Treaty of Tiānjīn. The main provisions of this unequal treaty signed by the Chinese under duress in June 1858 were:

1. Britain, France, Russia and the USA could establish legations in Běijīng or visit the Chinese court there.[1]
2. Ten more ports were to be opened to foreign trade.[2]
3. Warships could call for repairs and supplies at any Chinese port.[3]
4. Foreigners could travel in China if their country's passport was countersigned by the local Chinese authorities.[4]
5. Missionaries had the right to preach Christianity anywhere in China.[5]
6. Foreigners acting wrongfully were to be tried according to the laws of their own country.[6]
7. The use of "yí" (barbarian) to refer to foreigners was to be forbidden in all official documents.[7]
8. Four millions taels of silver were to be paid to Britain and two million to France as indemnity.[8]

The Chinese had no choice but to ratify the treaty, which stipulated, among others, that foreigners were allowed to travel in China with a passport signed by their consul and countersigned by the local Chinese authorities.[1] This was the opportunity that Fortune had been waiting for. He would at last get a chance to collect all the hardy ornamental plants from temperate China without any competition from other plant hunters. One wonders how his wife reacted when she heard that he was intending to postpone his return.

Just a week after arriving in Shànghǎi, Fortune sailed for the port of Yāntái ("Smoke Platform") in NE Shāndōng in HM dispatch boat "Attalante" (**Figs. 9.2–9.5**).[2]

Fig. 9.2 **Approaching Yāntái from Shànghǎi. This is roughly the panorama Fortune would have had from the "Attalante" on 16 August 1861. Yāntái is on the far right.**

Fig. 9.3 **The smoke platform (Yāntái) on the hill overlooking the harbour, which in the Míng Dynasty was used to warn for approaching pirates.**

Fig. 9.4 **Sea of masts at the newly opened Treaty Port of Yāntái. Photograph (1861) displayed in lighthouse on Yāntáishān.**

Fig. 9.5 **Yāntái was considered as the "Brighton of China". It still has the "fine beach for sea-bathing" mentioned by Fortune (1863: 306).**

On arrival on 16 August 1861 he met his old friend Martin C. Morrison (**Box 3.7**), who had become British Consul at Yāntái.[3] Fortune spent the second half of August exploring the environs of Yāntái. He noted that while the valleys and low-lying areas were very fertile with crops of peas, beans and several kinds of millet, the hills were extremely barren (**Fig. 9.6**).[4] Although the hills may have been worthless from an agricultural point of view, they were full of wild flowers.[5] At the time of Fortune's visit *Platycodon grandiflorus* (**Fig. 3.8**), several species of *Veronica*, *Potentilla*, and *Belamcanda chinensis* (Syn. *Pardanthus chinensis*)

Fig. 9.6 **Barren limestone hills near Yāntái, which Fortune heard were "covered with wild flowers in the spring of the year ..." (Fortune, 1863: 307).**

were in flower.[6] Fortune collected seed of pines and arbor-vitae in these hills.[7]

While he was at Yāntái, Fortune met Brigadier-General Charles Staveley (**Box 9.2**; **Fig. 9.7**), the Commander of the troops at Tiānjīn, an avid collector of Míng porcelain[8], and Dr. Charles A. Gordon (**Box 9.3**; **Fig. 9.8**), the Inspector-General of Hospitals and amateur botanist, who were recovering from ill-health brought on by the extreme weather in Tiānjīn.[9]

Box 9.2

General Sir Charles William Dunbar Staveley (1817 – 1896) Eldest of 12 children of Lieutenant-general William Staveley (1784 – 1854) and Sarah Mather (1797 – 1871).[1] After an education at the Scottish Military and Naval Academy in Edinburgh, he was commissioned as Second Lieutenant in the 87th Regiment (Royal Irish Fusiliers) on 6 March 1835.[2] He was made Lieutenant on 4 October 1839 before following his father to Mauritius.[3] He was aide-de-camp (ADC) to the Governor of Mauritius from July 1840 to June 1843.[4] When his father became Acting Governor for several months in 1842, he served directly under him.[5] After a short time in the UK, he became ADC to the Governor-General of Canada.[6] At that time the Canadian-USA boundary west of the Rocky Mountains was in dispute. It became an issue during the US Presidential election campaign of 1844, with the Democrats claiming all the territory between the Mexican province of California and Russian-owned Alaska.[7] However, at the outbreak of the Mexican-American War during 1846 – 1848, the Americans were in no position to wage war on Britain.[8] The British, on the other hand, had no desire for a conflict over a remote part of the Empire.[9] With the help of Staveley's sketches, a compromise in the form of the 49th Parallel was reached in 1846.[10] Staveley was then posted to Hong Kong where he again worked under his father as Assistant Military Secretary.[11] He fought in the Crimean War during 1854 – 1856, helped to suppress the Indian Mutiny (1857 – 1858), and led the First Infantry Brigade during Sir Hope Grant's Anglo-French expedition to Běijīng in 1860.[12] After the hostilities he became Commander of the British Occupational Force at Tiānjīn.[13]

Fig. 9. 7 General Sir Charles William Dunbar Staveley (1817 – 1896), a lover of Míng porcelain, whom Fortune met when he arrived in Yāntái in August 1861. At the time Staveley was Commander of the British Occupational Force at Tiānjīn. After Fortune left China, he became Head of all the British land forces in China. He retired from the army in 1883.

Staveley was a pretentious man with an unshakable confidence in his ability to train crack troops.[14] When the Tàipíng rebels threatened Shànghǎi in 1862, Staveley arrived in March with a force of some 2000 men to clear the country around the Treaty Port.[15] In April 1862 Staveley replaced Sir John Michel (1804 – 1886) as Head of the British land forces in China.[16] In this function he pestered the pacifist Frederick Bruce (**Box 9.4**) for more troops.[17] Unlike his predecessor he had no respect for the mercenary "Ever-Victorious Army" under Frederick Townsend Ward (1831 – 1862).[18] After Ward's death on 22 September 1862, he reached an agreement with Lǐ Hóngzhāng (Li-Staveley Agreement, on 14 January 1863) to place the "Ever-Victorious Army" under joint British-Chinese control.[19] He eventually got his way. After a short period in which the "Ever-Victorious Army" was led by the volatile Henry Burgevine (1836 – 1865), it was finally handed over to Staveley's close friend and brother-in-law, Charles George Gordon (1833 – 1885; "Chinese" Gordon or "Gordon of Khartoum") on 24 March 1863, thereby increasing Staveley's control over the region.[20]

In 1863 ill-health forced him to return to the UK.[21] While there, he married Susan Millicent Minet (1833 – 1918) on 31 October 1864.[22] The next year he was back in Asia, this time in command of the First Division of the Bengal Army.[23] When the Christian Emperor Tewodros II (= Theodore II, 1818 – 1868) of Magdala (Amba Mariam) in Ethiopia, in exasperation at not receiving a reply to his letter (October 1862) addressed to Queen Victoria, imprisoned six British subjects in 1867 an expeditionary force of 12,000 soldiers was assembled at great cost by Sir Robert Napier (1810 – 1890), the Commander-in-Chief of the Bombay Army.[24] Napier, who had become acquainted with Staveley during the Anglo-French expedition to Běijīng, appointed him as Commander of the 1st Division.[25] With his customary energy, Staveley made sure that the baggage trains were well organized, and the mules well-watered and fed.[26] Although he was suffering from rheumatism, Staveley nonetheless took part in the storming and looting of the deserted fortress of Magdala in April 1868.[27] On his return from the Ethiopian Campaign, Staveley took command of the troops at Plymouth from 1869 from 1874, before returning to India as Commander-in-Chief at Mumbai during 7 October 1874 – 7 October 1878.[28] He retired from the army on 8 October 1883 and was awarded the Grand Cross of the Bath (GCB) on 24 May 1884.[29] He died at Aban Court in Cheltenham, but was buried in Brompton Cemetery, not far from Robert Fortune.[30]

Box 9.3

Sir Charles Alexander Gordon MD (1821 – 1899) Illegitimate son of General William Alexander Gordon (1769 – 1856) and Elizabeth (Betty) Leys (1797 – 1851 +).[1] He studied medicine at the University of Edinburgh.[2] After receiving his Licentiate of the Royal College of Surgeons (Edinburgh) and a MD from the University of St. Andrews in April 1840, he joined the Army Medical Department in June 1841 as assistant-surgeon.[3] He was sent to India.[4] At first the wide-eyed Gordon was able to do some sightseeing. He visited the Taj Mahal on several occasions.[5] Later when the Maratha rose against British rule in December 1843, he was confronted with the bloody reality of his job.[6] Once the rebellion was crushed, Gordon returned with the troops to the British base at Meerut.[7] In the middle of March 1844 he was sent with a small body of native troops to police the ritual bathing on 11 April 1844 (Maha Mela Festival) at the holy town of Haridwar on the Ganges.[8] Once this was over he rejoined his regiment at Allahabad.[9] Late in September 1844 orders were given to proceed to Kolkata and embark for England.[10] On 29 April 1845, after almost 15 weeks at sea, the "Monarch" docked at Gravesend (Kent) on the south bank of the Thames, and the troops proceeded to their barracks at Chichester.[11] While at Chichester, Gordon received an offer of promotion conditional on his volunteering for service in West Africa.[12] In the first week of January 1847 he sailed in the overcrowded brig "Emily" from Gravesend to Ghana.[13] During his 15 months there he

Fig. 9.8 Sir Charles Alexander Gordon (1821–1899). As "an ardent lover of botanical pursuits", he accompanied Fortune when the plant hunter visited some nurseries near Tiānjīn.

took careful notes about the climate, flora, fauna, and the various tribes, their languages, customs and economic plants.[14] In April 1848 he was one of six white men to take part in the West African Coast Expedition against the bloodthirsty King of Apollonia, Koko- or Quako Acka.[15] When in May 1848 the "Baretto Junior" arrived at Cape Coast bound for Barbados, Gordon made his "getaway" from the "White Man's Grave".[16]

From Barbados he returned to Britain, where he served with the 57th at Enniskillen in Ireland.[17] On 14 March 1850 Gordon married Annie Mackintosh (1826 – 1910) of Torrich (Nairnshire).[18] They had four sons and a daughter.[19] In early June 1851 the Gordons with their infant boy (Henry King Gordon) boarded the "Bentinck" (**Box 5.33**) bound for India.[20] A daughter and two more sons were born in India.[21] In the course of 1855, Gordon became seriously ill and had to apply for sick leave.[22] After a disease-plagued journey from Kolkata, the Gordons arrived at Gravesend on 14 July 1856, and proceeded to Aberdeen, where they thought the bracing air would do them all good.[23] Having prepared his will, the incompletely recovered Gordon took leave of his family in the Spring of 1857 and returned to India.[24] During the Indian Mutiny (**Box 6.4**) he saw active service as Superintending Surgeon of the 10th Foot. At the time of the siege and capture of Lucknow (March 1858) he initiated a system by which one of the regimental surgeons was embedded in the fighting force to give immediate aid to those wounded in battle.[25] In the aftermath of the capture of Lucknow, the 10th Foot joined a field force under Sir Edward Lugard (1810 – 1898) to raise the siege of Azamgarh.[26] For his wholehearted commitment to duty during this field campaign he received the CB.[27] Early in 1859 the 10th Foot was ordered back to Headquarters.[28] On arrival in Gravesend on 13 July 1859, the troops were transferred to the "Himalayah" bound for Portsmouth, where Gordon was reunited with his family.[29]

In 1860 he was sent to China. From 21 June until 28 November 1860 he was Chief Medical Officer in Hong Kong responsible for the sick and wounded from the Anglo-French Expedition to Běijīng during the Second Opium War.[30] Once the campaign was over, he was transferred to Tiānjīn, where he became Inspector-General of Hospitals of the occupying force under Sir Charles Staveley (**Box 9.2**).[31] While there he compiled a hefty book about the flora, fauna, and the influence of the weather on the health of the soldiers.[32] A short stay at Nagasaki on his return journey to Britain in 1861 convinced him of the suitability of this Japanese port as a sanitorium for British soldiers stationed in China.[33] After his arrival in Britain in 1862 he had a well-deserved holiday with his wife in Paris and Rouen.[34] Shortly after their return, Gordon was ordered back to India.[35] He left Southampton on 4 September 1862.[36] From 1862 to 1867 he was in charge of the Benares and Presidency Divisions, which included the inspection of hygiene at military posts and on the troopships.[37] In the course of his tours of inspection he visited Darjeeling, Patna, Monghyr, Hazaribagh and Benares.[38] On his return from Hazaribagh, he became ill and was incapacitated for two months.[39] He was therefore glad when his term of foreign service finally came to an end.[40] On 12 January 1868 he was back in Portsmouth, where his duties involved inspections of military establishments and the embarkation and disembarkation of troops.[41] During his tours of inspection, he was able to indulge in a bit of geology.[42]

During the Franco-Prussian War of 1870 – 1871, he was sent to Paris as War Office Medical Commissioner to the French Army.[43] In the confusion he was arrested as a spy, and spent a few hours at a Paris police station until he was cleared.[44] He became a member of a committee appointed to distribute an English donation among the sick and wounded, and supervised the work of the Red Cross orderlies.[45] While he was in Paris he met Henry Dunant (1828 – 1910), the founder of the Red Cross.[46] As the winter set in and the fighting escalated, he and his men were forced to work long hours with little food to nourish them.[47] After the French capitulation on 26 January 1871, Gordon stayed on to help with the wounded.[48] For his services during the siege and bombardment of Paris he was made an Officier de la Légion d'Honneur on 21 February 1871.[49] On the 14th March 1871 he left Paris for England.[50] As his time in France was considered equivalent to a tour of foreign service, he was able to take things easily for the next three years.[51] On 1 April 1874 he was promoted to Surgeon-General.[52]

In early September 1874 Gordon, his wife and daughter embarked for India.[53] They landed in Mumbai (Bombay), and transferred to Chennai (Madras).[54] Sir Frederick Paul Haines (1819 – 1909), the Commander-in-Chief of the Madras Army, who was sent to test the feasibility of a trade route between Burma and Yúnnán, asked Gordon to accompany him.[55] On 31 December 1874 they arrived at Yangon (Rangoon).[56] After some sightseeing in Yangon, the party steamed up the Irrawaddy as far as Thayat Myo, where horses and elephants were hired for the overland journey to Yedashe on the Sittoung River.[57] At one stage Gordon fell from the charpoy, and was lucky not to have been tampled by the elephant.[58] At Tantabin the adventurers boarded dugouts which took them almost as far as the estuary of the Sittoung, before branching off to Bago Pegu.[59] They were back in Yangon on 9 February 1875.[60] Later that year Gordon published a very readable account of the round trip, with (as was his wont) extensive notes on palaeontology and economic geology, the climate, fauna, palms and forest trees, the various inhabitants and languages, their festivals, government, trade, weights and measures, the wars with Britain and the state of British Burma.[61] In December 1879 he embarked for England.[62] Having landed in Southampton in January 1880, he proceeded to Portsmouth to take up his position.[63] He retired on 25 May 1880, and was appointed Queen's Honorary Physician.[64] After his retirement he continued to play an active role in medical affairs. He was President of the Hamilton Association for Providing Trained Male Nurses in 1885, and President of the Army Medical Officers' Friendly (Widows') Society, which was established to supplement the poor army pensions.[65] He was knighted by Queen Victoria in 1897.[66]

Together they boarded the French SS "Feilung" ("Flying Dragon"), which delivered mail from Shànghǎi to Dàgū (Taku) every month.[10] The passage took less than a day. They arrived at Dàgū on the morning of the 2 September 1861.[11] The next day they were taken up the Hǎi Hé ("Sea River") to Tiānjīn by the French gunboat "L'Etoile" ("Star").[12] Just before they got to the city, they passed some salt-pans. On the raised ground between the salt-pans the crystallized salt had been bagged and piled into heaps of salt almost 10 m in height.[13] Although mats had been thrown over the piles of salt to prevent it being washed away by rain, the ground was covered by a crust of salt, which gave the landscape a wintry appearance (**Fig. 9.9**).[14]

Fig. 9.9 Salt-pans near Tiānjīn. As Fortune mentioned, "The whole place had a wintry aspect, the ground was whitened as if with hoar-frost, and as I walked over it a crisp crushing noise was heard as if one was walking on frosted snow." (Fortune, 1863: 309).

Tiānjīn was a walled city with four gates, but by 1861 its walls and ramparts were already in a ruinous state.[15] Although trade was obviously thriving, Fortune was surprised by the rundown appearance of the city.[16] The paving stones were broken and covered with mud, which made going difficult if it rained.[17] Although Fortune had commented adversely on Xiàmén and Shànghǎi, he didn't think Tiānjīn was very clean.[18] Although the curiosity shops were well-stocked with jade, quartz, porcelain, bronzes, and nicely modelled clay figures, some of the shops only contained the basic necessities of life (**Fig. 9.10**).[19] Even the tradesmen in second-hand clothes did not have many customers.[20] As the various shopkeepers and the street hawkers vied for trade, the streets were filled with a dissonant clamour[21], to which was added the noise of beggars beating the shop counters with sticks and stones, and howling for alms.[22] Fortune was no doubt glad to leave this cacophonous city behind.

Fig. 9. 10 Pedestrian precinct in Old Tiānjīn, with the Drum Tower in background. Fortune was impressed by the shops in Tiānjīn.

Although Fortune found the flat coastal plain round Tiānjīn very dreary, with its scattered trees and saline vegetation[23], he hoped to find some novelties in the nurseries to the west of the city (**Fig. 9.11**), which he visited with Charles Gordon (**Box 9.3**).[24] In this he was disappointed, as most of the nurserymen preferred to grow more exotic plants, such as *Jasminum sambac*, *Osmanthus fragrans*, limes, oranges and pomegranates.[25] The only hardy forms were such well-known plants as *Amygdalus triloba*, *Jasminum nudiflorum*, *Weigela florida*, roses and honeysuckles.[26]

Fig. 9.11 A nursery near Tiānjīn. Fortune was disappointed that he did not find any novelties here.

Fig. 9. 12 Walled enclosures ("winter-houses") to protect tender plants during the winter. In Fortune's day the opening facing south was covered with paper. Polythene is now used.

During the winters the tender plants were either protected by lowering them into straw-covered holes in the ground or by placing them in walled enclosures, which were only open to the south.[27] Good examples of these "winter-houses" can still be seen between Tiānjīn and Běijīng (**Fig. 9.12**). However, the translucent paper, which used to be used to cover the southern face, has been replaced by polythene sheeting. In the vineyards on the left bank of the Grand Canal yet another method was employed to protect the tender shoots. They were simply taken off the trelliswork and buried in the soil.[28] This method was presumably widespread in those regions experiencing a cold winter, and is still employed in NW China, where *Vitis vinifera* was first introduced into China at least 2,300 years ago.[29] In the days before refrigeration, the grapes and other fruit such as apples and pears were kept fresh by storing them in ice-houses.[30] Near the vineyards there were more nurseries, this time with a range of herbaceous plants, for which Fortune showed little interest[31], as well as pomegranate and poplar, a tree which is now very common on the coastal plain.

While he was in Tiānjīn, Fortune applied for a passport to travel within China. This was granted on 16 September 1861, his 49th birthday. So, after a fortnight spent in Tiānjīn, Fortune finally set off for the Chinese capital Běijīng on 17 September 1861.[32] In order to see something of the countryside and its produce, Fortune chose to travel by covered cart (**Fig. 9.13**).[33]

Fig. 9.13 The unsprung cart in which Fortune travelled from Tiānjīn to Běijīng was very uncomfortable. "Nearly shaken to pieces, and thoroughly tired with the day's journey, I retired early to rest." (Fortune, 1863: 348).

The journey became an ordeal before they had even left the suburbs of Tiānjīn, for the road was badly potholed and the cart without springs.[34] After being thrown from side to side for 45 km, the plantsman was happy when the cart finally stopped outside a simple inn at Náncài Cūn ("Southern Cài Village").[35] Forgetting that the occupation of Zhènjiāng ("Guard River") by the Tàipíng (**Box 5.19**) was preventing rice from reaching the north of China, Fortune ordered a simple meal of rice and eggs.[36] He had to make do with a mutton chop, hard-boiled eggs and millet bread![37] After a tiring day and the prospect of an early start the next morning, Fortune soon retired to bed.[38] At dawn they were rattling on their way again. The countryside was still fairly flat, so the giant millet (*Sorghum bicolor*, 3 – 5 m; **Fig. 9.14**) often blocked their view.[39]

Sometimes the carters got lost and had to retrace their steps.[40] However, occasionally they got a good view of the Běi Yùnhé ("Northern Canal", the northern branch of the Grand Canal; **Fig. 9.15**). The mud embankments were not always effective in preventing flooding. In places where the road was under water, they had to make a detour through the fields.[41] On the evening of the 18 September 1861 they stopped at an inn in Zhāngjiāwān ("Zhang Family Meander"; **Figs. 9.16, 9.17**).[42] After another early start, they headed west to Běijīng.

Fig. 9.14 Giant millet, *Sorghum bicolor*, made it difficult to find the way. "The carters themselves had frequently to halt, not knowing where they were; and on more than one occasion we had to retrace our steps and get into another by-road." (Fortune, 1863: 349).

Fig. 9.15 The Běi Yùnhé (Northern Canal), the northern branch of the Grand Canal, which Fortune saw on his way to Běijīng by cart. On his return journey to Tiānjīn, he hired a boat instead of a cart, "and we sailed rapidly and pleasantly down the stream." (Fortune, 1863: 387).

Fig. 9.16 A bridge connecting the old and new town of Zhāngjiāwān (Zhang Family Meander). Fortune arrived here on 18 September 1861, and stayed the night in a cheap hotel.

Fig. 9.17 Rutted road leading to the old town of Zhāngjiāwān.

Shortly after midday on 19 September 1861 they arrived at the Guǎngqúmén ("Wide Canal Gate"; **Fig. 9.18**), where they had to explain who they were, and where they had come from.[43] When Fortune's servant told the guard that his master was an Englishman, who was on his way to the British Legation, they were waived on without having to show Fortune's passport![44] The paved streets in the Chinese section of the city were in such a poor state of repair, that Fortune preferred to walk.[45] After walking in a westerly direction and then due north, they entered the Manchu ("Tartar") part of Běijīng and proceeded towards the Forbidden City.[46] Fortune arrived at the British Legation, which was housed in the renovated Liáng gōng fū ("Liáng's Mansion"; **Fig. 9.19**), that had belonged to the Guǎngdōng playright, historian and cartographer, LIÁNG Tíngnán (1796 – 1861) on the east side of Tiān'ānmén ("Heavenly Peace Gate") Square.[47] He was welcomed by the British Minister at the Chinese Court, Frederick Bruce (**Box 9.4; Fig. 9.20**).[48]

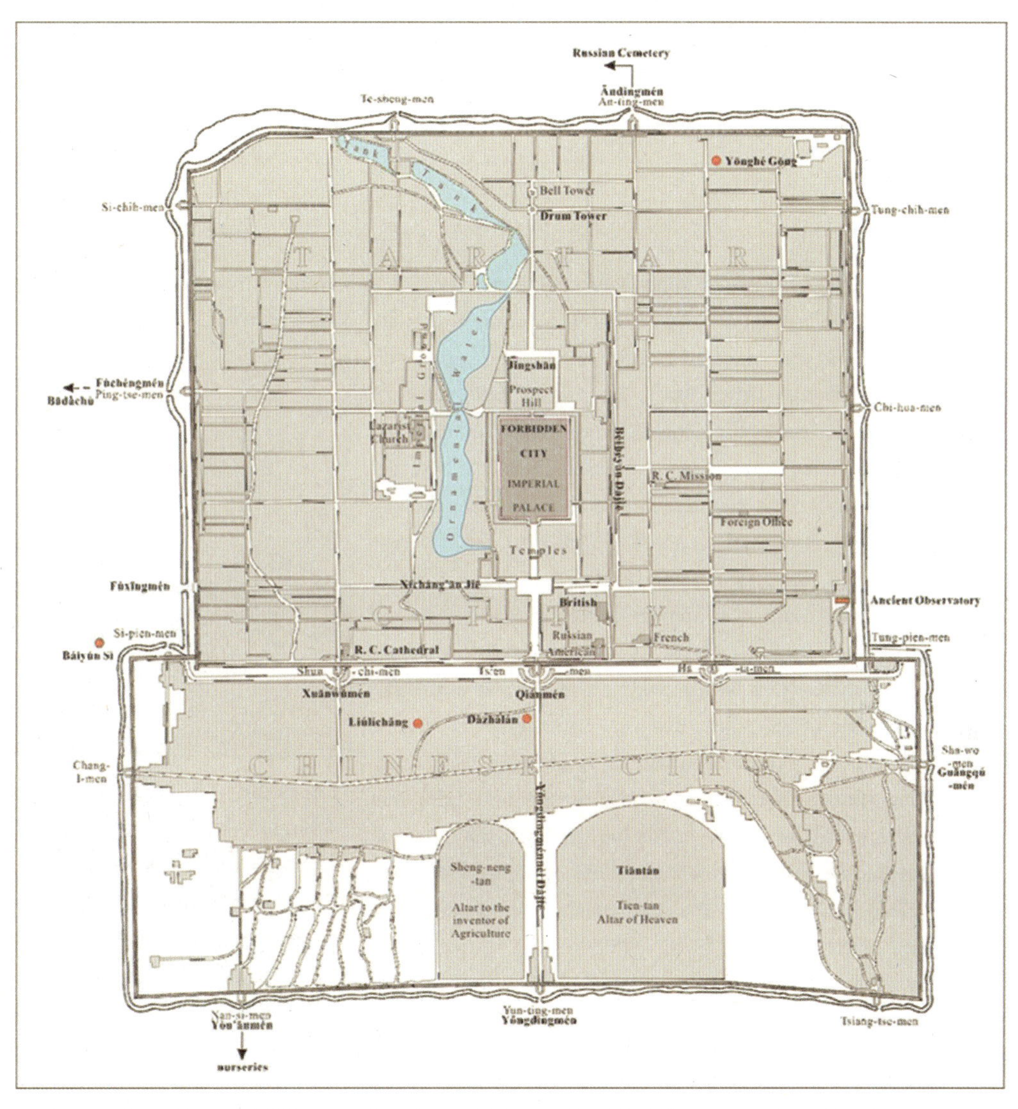

Fig. 9.18 Map of Běijīng at the time of Fortune's visit. Places mentioned by Fortune are indicated in bold.

Fig. 9.19 Liáng Gōngfǔ (Liang's Mansion), the British Legation, "a most gorgeous place", where Fortune stayed in 1861.A. Main entrance to legation; B. Inner courtyard.

Box 9.4

Sir Frederick William Adolphus Bruce (1814 – 1867) Third son of Thomas Bruce, Seventh Earl of Elgin (1766 – 1841) and his second wife Elizabeth Oswald (1790 – 1860).[1] He was educated at Eton.[2] As younger brother of the powerful politician James Bruce, 8th Earl of Elgin (1811 – 1863), he quickly rose through the diplomatic ranks. In 1842, as attaché to Alexander Baring, Lord Ashburton (1774 – 1848), he was involved in settling the disputed border between Maine (USA) and New Brunswick (Canada).[3] This had led to war between the two countries in 1812 (see **Ch. 1.1**), and had more recently been the cause of a number of ugly incidents, such as the burning of the US Steamboat "Caroline", which had been supplying Canadian rebels with money, provisions and arms, the retaliatry burning of the British steamer "Sir Robert Peel" on 29 May 1838, and the arrest in February 1839 of Rufus McIntire, a Maine land agent sent to expel Canadian lumberjacks from the Aroostook area.[4] Troops were called up on both sides, but a truce was hurredly arranged in March 1839 by General Winfield Scott (1786 – 1866) and the Lieutenant-Governor of New Brunswick, Sir John Harvey (1778 – 1852) before fighting broke out.[5] The deliberations were far from easy, due to the hard stance taken by the Maine delegates, who not only hated the British, but distrusted the US Government's involvement in their internal affairs.[6] However, the Webster-Ashburton Treaty was finally signed by Daniel Webster, US Secretary of State (1782 – 1852) and Lord Ashburton in Washington on 9 August 1842.[7] Two years later Bruce was appointed Colonial Secretary (Second-in-command) to Sir John Davis, the Second Governor of Hong Kong.[8] In 1846 he was promoted to Lieutenant-Governor of Newfoundland.[9] In 1847 he was posted to South America. He was British Consul (1847 – 1848) and then Chargé d'Affaires (1848 – 1851) in Bolivia and Chargé d'Affaires in Uruguay (1851 – 1853), before being sent to Egypt as Consul-general in 1853.[10]

Fig. 9.20 Sir Frederick William Adolphus Bruce (1814 – 1867), Britain's first Ambassador to China. Bruce was known for his modest manner and cordiality. "On my arrival at the residence of the English Ambassador, I was kindly received by His Excellency Mr. Bruce." (Fortune, 1863: 351). Bruce grew whiskers to hide his tendency to blush.

In April 1857 he accompanied his brother Lord Elgin to China.[11] A year later, after his brother had signed the Treaty of Tiānjīn (**Box 9.1**), he was left to conduct the affairs in China.[12] He became Envoy Extraordinary and Minister Plenipotentiary on 2 December 1858.[13] However, when it came to ratifying this unequal treaty in Běijīng in 1859, the Chinese tried to divert Frederick Bruce and his French counterpart, Count Alphonse de Bourboulon (1809 – 1877) to Běitáng (Peitang, "Northern Embankment"), a port used for the reception of envoys from tributary states.[14] When their fleet attempted to force an entry at Dàgū, the forts opened fire and many of the ships were sunk.[15] In 1860 a 20,000 strong expeditionary force under Sir James Hope Grant (**Box 9.9**) landed at Běitáng, took the Dàgū Forts and Tiānjīn, before homing in on Běijīng.[16] When it was discovered that four British subjects including "The Times" correspondent Thomas William Bowlby (**Box 9.8**) and some Sikh soldiers had been left to die in a Chinese (Chāngpíng) prison, Lord Elgin ordered the destruction of the Yuánmíngyuán (Emperor's Summer Palace).[17] After that the Treaty of Tiānjīn (**Box 9.1**) plus the supplementary Convention of Běijīng were quickly ratified on 24 October 1860.[18] Once the treaty was signed, Frederick proceeded to Běijīng, only to be driven back to Tiānjīn by the harsh winter.[19] It was not until 26 March 1861 that he was able to take up residence in Liáng's Mansion.[20] Frederick was more conciliatory towards the Qīng officials than his elder brother James had been. He considered gunboat diplomacy ultimately counterproductive and urged his consuls to cooperate with the Chinese, and only use force if it was absolutely necessary.[21] He was proud "to have established satisfactory relations at Peking and become in some degree the advisors of a government with which eighteen months since we were at war".[22] However, his good contacts in the Chinese capital, gave him a biased view of the unrest then sweeping China. Like many of his diplomatic contemporaries, he may have had his doubts about the ability of the Tàipíng to govern, but he was probably particularly worried that should the Tàipíng gain power, the indemnity repayments agreed in the Treaty of Tiānjīn (**Box 9.1**) would stop.[23] He was responsible for repeating the rumours that the Tàipíng (**Box 5.19**) had committed atrocities.[24] His reports thus influenced official British policy.[25] When Fortune arrived in China in September 1861, Bruce supplied him with a passport for the Chinese capital and gave him permission to stay at the British Legation in the heart of the city.[26] There Fortune admired the collection of Míng porcelain that Bruce had managed to acquire.[27] Bruce stayed for only four years in Běijīng. On 1 March 1865 he was appointed British Minister to the USA.[28] Having arrived just a week before Abraham Lincoln's assassination, the two men never met.[29] Bruce, who never married, died in office in 1867 and was remembered for his unpretentious manner and genial hospitality.[30]

9.2 Excursions In and Around Běijīng

At the British Legation Fortune met his old friend Dr. William Lockhart (**Box 3.11; Fig. 3.49**), who had only just arrived to establish a hospital in the Chinese capital.[49] Lockhart took Fortune to see the Ancient Observatory (**Figs. 9.21, 9.22**), some 2.6 km from the legation in the southeast corner of the Manchu City. Here they were able to admire a map of the world prepared under the direction of the Jesuit Matteo Ricci (**Box 9.5; Fig. 9.23**), and view a set of astronomical instruments cast for the Kāngxī Emperor by the Flemish Jesuit, Ferdinand Verbiest (1623 – 1688).[50]

Fig. 9.21 Ancient Observatory, Běijīng, which Fortune visited in the company of Dr. William Lockhart.

Fig. 9.22 A highly ornamental armilla (1744), used to determine solar time.

Box 9.5

Matteo Ricci (1552 – 1610) Eldest child of the nobleman Giovanni Battista Ricci and his wife Giovanna Angiolelli.[1] Being one of 13 or 14 children, he did not get much attention from his parents.[2] He received private tuition before attending a Jesuit school in Macerata.[3] He then studied law in Rome for two years (1568 – 1571), before joining the Jesuit order in 1540 against his father's wishes.[4] He received a thorough grounding in mathematics, cosmology and astronomy from Father Christopher Clavius (1538 – 1612) best known for the Gregorian Calendar.[5] In May 1577 Ricci left for Coimbra in Portugal without even taking leave of his parents.[6] On 24 March 1578, after a farewell audience with the young King of Portugal, he sailed in the "São Luiz" from Lisbon for Goa, the Portuguese enclave in western India.[7] The journey was not without its excitement. The ship was shadowed for days by two well-armed French vessels, and then ran aground in Mozambique.[8] The already cramped conditions became even worse when 300 – 400 slaves were taken on board.[9] After almost 6 months of poor sanitation and disease, Ricci was glad to bathe in clean water again.[10] In Goa during 13 September 1578 – 26 April 1582 Ricci continued his study of theology in preparation for his mission to the Far East.[11]

Fig. 9.23 A statue of Matteo Ricci (1552 – 1610) in front of "his" Roman Catholic Cathedral in Běijīng.

He arrived at Macao on 7 August 1582 and started learning the Chinese language and customs.[12] In 1583 he was allowed to settle in Zhàoqìng ("Clear Celebration", Guǎngdōng) by the local prefect, WÁNG Pàn, but was expelled in 1589 by WÁNG's successor, who required the site of Ricci's mission for his own shrine.[13] He moved to Sháoguān ("Beautiful Pass", North Guǎngdōng) and in 1595 to Nánchāng ("Southern Prosperity"), the capital of Jiāngxī.[14] He attempted to establish himself first in Nánjīng ("Southern Capital") in 1595 and then in Běijīng ("Northern Capital") in 1598, but as a possible spy he was not welcome in either of the cities during the invasion of Korea of 1592 – 1598 by the Japanese expansionist HIDEYOSHI Toyotomi (1536 – 1598).[15] When HIDEYOSHI died and his troops were withdrawn from Korea, Ricci was able to establish himself in Nánjīng.[16] From there he undertook another journey along the Grand Canal to Běijīng in 1600 in the hope of converting the Chinese Emperor Wànlì (1563 – 1620) and thus win over a large proportion of the population.[17] Wànlì was particularly fascinated by the clocks and clavichord which Ricci had included among the gifts for the emperor, but Ricci was never to meet the by then reclusive Wànlì.[18] However, by remaining humble, donning the robes of a Chinese scholar, adopting a Chinese name (LÌ Mǎdòu), learning to write classical Mandarin, and by combining the ethics of Kǒng zǐ (Confucius, 551 – 479 BC) with Christian beliefs, Ricci soon gained the confidence of the Chinese, who were fascinated by the breadth of his knowledge and his prodigious memory.[19] His books, such as "Treatise on Friendship" in 1595 and "Ten Discourses by a paradoxical man" in 1608, were well received.[20] By filling the mission house with all manner of curiosities (chiming clocks, Venetian prisms, musical instruments, European engravings, large tomes such as the Plantin Bible [1568 – 1572] etc.), he lured the Chinese to his abode, where the inquisitive visitors were confronted by a picture of the Madonna and Child.[21] In this way he was able to introduce his visitors to Christianity.[22] With a growing Christian community throughout China, Ricci spent an increasing amount of time corresponding with priests.[23] This bureaucracy, combined with the preparation of a new world map in 1602, the translation of Christopher Clavius' books into Chinese, a book about his experiences in China (Historia, 1608 – 1610) and a succession of courtesy visits wore him out prematurely.[24] He was so admired that when he died, instead of being taken back to Macao for burial, Emperor Wànlì granted special permission for him to be buried in the Zhàlán Mùdì ("Portuguese Cemetery") to the west of Běijīng's city walls.[25] Four centuries after his death his grave continues to draw crowds of Chinese admirers.[26]

From the platform Fortune looked back over the plain through which he had come, then across the Chinese City to the Temple of Heaven (**Fig. 9.33**).[51] To the northwest Jǐngshān ("Prospect Hill" "Coal Hill"), where the last Míng Emperor had hung himself, was clearly visible.[52] From the observatory it was even possible to see the Western Hills to the west and northwest of Běijīng.[53] Today the Ancient Observatory is surrounded by an agglomeration of high-rise buildings, which block the once magnificent panorama (**Fig. 9.24**).

Fig. 9.24 High-rise buildings now block the once magnificent panorama from the Ancient Observatory.

The next day Lockhart and Fortune skirted the eastern wall of the Forbidden City, and arrived at the foot of Jǐngshān, with its pavilions, temples and trees (**Figs. 9.25, 9.26**).[54] Fortune had been told that "Prospect Hill" was composed entirely of coal, which had been stockpiled in case of a siege.[55] In fact, this artificial hill was created from spoil when the North, Central and South Lakes to the west of the Forbidden City were excavated during the Míng Dynasty.[56] Since foreigners were not allowed to climb the hill and thus look down into the Forbidden City, the friends rode south between the Forbidden City and the three lakes, before turning west along the Xīcháng'ān Jiē ("West Cháng'ān Street") to the Fùxīngmén ("Revival Gate").[57] They made a short stop at Báiyún Sì ("White Cloud Temple") to get a glimpse of the pretty gardens, rockeries and artistic bridges (**Figs. 9.27, 9.28**), but as this temple was still out-of-bounds to foreigners, they continued north along the Sānlǐhé Dōnglù ("Three Li River East Road") to the Zhàlán Mùdì (Portuguese Cemetery).[58]

Fig. 9.25 The Jífāng (Collect fragrance) Pavilion (1750) on Prospect Hill, overlooking the Forbidden City. Close to this spot the last Míng emperor hung himself.

Fig. 9.26 View of the Forbidden City from Jǐng Shān (Prospect Hill). Note the renovation being undertaken prior to the Olympic Games in 2008.

Fig. 9.27 Entrance to Báiyún Sì (White Cloud Temple). "Although we could not enter the sacred enclosures, we got glimpses of pretty gardens with rock-work and artistic bridges, which gave us very favourable impressions of its internal beauties and made us long for a nearer view." (Fortune, 1863: 358).

Fig. 9.28 Temple of Founder QIŪ (1443), Báiyún Sì, which Fortune never managed to see.

Fig. 9.29 Zhàlán Mùdì (Portuguese Cemetery), where Matteo Ricci and other Jesuits lie buried. "Pines, junipers, and other trees grow all over the cemetery, and throw a pleasing shade over the last resting-places of the ancient fathers." (Fortune, 1863: 360).

Fig. 9.30 Roman Catholic Cathedral (1605), which Emperor Wànlì gave Matteo Ricci permission to build.

Here, in the shade of pines, junipers and other trees, Matteo Ricci (**Box 9.5**) and other Jesuits lay buried (**Fig. 9.29**).[59] On the way back to the British Legation, the companions passed the Roman Catholic Cathedral ("South Cathedral"; **Fig. 9.30**), which Emperor Wànlì gave Matteo Ricci permission to build just inside the Manchu City and close to the Xuānwǔmén ("Proclaimed Gallant Gate").[60]

Fig. 9.31 Dàzhàlan, "famous for its collections of works of art both ancient and modern." (Fortune, 1863: 364).

Fig. 9.32 Shop window in Dàzhàlan. "Specimens of carved jade-stone and rock crystal are plentiful in this street..." (Fortune, 1863: 364).

As Fortune had heard about some nurseries in the southern suburbs, he determined to see whether they had any novelties. He headed for the central gate (Qiánmén, "Front Gate") on the south side of the Manchu City. There he had to pick his way between creaking carts, camels, donkeys and pack-horses, as well as the inevitable beggars and strident street hawkers.[61] The Chinese City which he now entered, was packed with haberdashers selling silks and cotton, but also furs and padded quilts for the approaching winter, butchers with all sorts of meat (chicken, duck, beef, mutton and pork), greengrocers with Chinese cabbage, carrots, peas and beans, and fruiterers selling grapes, peaches and pears.[62] One of the side streets, Liúlichǎng ("Coloured glaze factory") was full of booksellers, while in Dàzhàlan ("Major railings") there were shops with jade (**Figs. 9.31, 9.32**), quartz, bronze, and porcelain.[63]

Fortune was unable to resist buying some porcelain to add to his collection.[64] As he continued down the Yǒngdìngménnèi Dàjiē ("Perpetually Calm Gate Inner Avenue") he noticed on his left a walled park with the Tiāntán (Temple of Heaven, **Fig. 9.33**).[65] As this area was still reserved for the Emperor, Fortune continued on his way.[66] He passed through the Yǒngdìngmén ("Perpetually Calm Gate"), and followed the southern wall of the Chinese City as far as the south-western gate (Yòu'ānmén, "Western Secure Gate").[67] Once he got there, he was told that the nurseries he was looking for were another 4 km further south.[68] There were some 10 - 12 nurseries altogether, but unfortunately not one of them seemed to have any novelties.[69] As in Tiānjīn, most of the potted plants were tender, and the few hardy plants such as *Amygdalus triloba*, *Cercis chinensis*, *Jasminum nudiflorum*, *Weigela florida*, honeysuckles and roses, had already been introduced into England.[70] However, a *Forsythia* with broader, darker leaves than *F. viridissima* caught his eye.[71] Intuitively he felt that this must be a new taxon, and indeed when it flowered in England it clearly was (**Fig. 9.34**).[72] Professor Lindley named it after its discoverer.[73]

Fig. 9.33 The Imperial Hall of Prayer for Good Harvests in the extensive grounds of Tiāntán (Temple of Heaven). The circular form of the building symbolizes heaven. As Tiāntán was officially out-of-bounds to foreigners in 1861, Fortune passed it by.

By the time he had gone round all the gardens, and had satisfied himself that there was nothing more to be had, it was quite late.[74] He had to ride rapidly to reach the gates of the Manchu City before these closed shortly after sunset.[75]

George Wyndham (**Box 9.6; Fig. 9.35**) of the British Legation had promised to show Fortune the northern part of the Manchu City the next day, the 23 September 1861.[76]

Fig. 9.34 *Forsythia suspensa*, which Fortune discovered in one of the nurseries south of Běijīng.

Box 9.6

Sir George Hugh Wyndham (1836 - 1916) Son of Charles Wyndham MP (1796 - 1866) and his wife Elizabeth Scott.[1] He was educated at Twyford School in Hampshire and the University of Oxford, before entering the diplomatic service.[2] Wyndham was first sent to Singapore as an unpaid attaché in 1859.[3] Two years later he became an attaché at the British Legation in Běijīng.[4] He arrived in Běijīng just ahead of Sir Frederick Bruce (**Box 9.4**) at the end of March 1861.[5] From the start Wyndham took a great interest in Běijīng and its environs. Within a month of his arrival he had visited the Temple of Sun, Temple of Earth, the Lama Temple, the Chinese Emperor's hunting ground south of the capital, and the Russian Cemetery where Thomas W. Bowlby

(**Box 9.8**) lay buried.[6] His sketch of the graves made on a hot and sultry day shows him to be an accomplished artist. He also took longer trips to different parts of the Western Hills, sketching as he went.[7] He was therefore ideally qualified to show Fortune round the Chinese capital when the plant hunter arrived in Běijīng in late September 1861. He also supplied Fortune with a drawing of the lacebark pine (*Pinus bungeana*) for his latest book.[8] After Fortune had gone and before the Chinese winter set in, he took a fortnight off at the beginning of November to visit the Great Wall at Luówényù Cūn ("Luó Wén's Valley Village, 140 km ENE of Běijīng) and Gǔběikǒu ("Ancient North Gateway", 110 km NE of Běijīng). As such he was the first Englishman to visit Gǔběikǒu since Lord Macartney's Embassy passed this way on 5 September 1793.[9] On his return to Běijīng he visited the Míng Tombs near Chāngpíng ("Prosperous level") and the hot springs at Xiǎotāng Shān ("Little Hotspring Hill").[10] By this time Wyndham understood colloquial Chinese, quite a feat even with the help of a teacher.[11] No doubt he would have become an accomplished sinologist had he stayed in China. However, the career diplomat soon left the Far East to become Second Secretary at the British Legation in Berlin.[12] He then became Secretary of the British Legation in Athens in 1875, before being transferred to the equivalent job in Madrid in 1878.[13] Three years later he became Secretary at the British Embassy in Constantinople (Istanbul), before being promoted as Envoy Extraordinary and Minister Plenipotentiary to the Kingdom of Serbia in Belgrade from 1886 to 1888.[14] On 1 February 1888 Queen Victoria appointed him Envoy Extraordinary and Minister Plenipotentiary to the Emperor of Brazil in Rio de Janeiro.[15] He ended his diplomatic career in the Rumanian capital Bucharest.[16]

Fig. 9.35 Sir George Hugh Wyndham (1836 – 1916), who showed Fortune round part of Běijīng on 23 September 1861.

However, they had to wait until Empress Quán, and some of the ladies of the Court had returned to the Forbidden City following the death of her son Emperor Xiánfēng in self-imposed exile in Chéngdé (Jehol) on 22 August 1861.[77] More were to follow. On a cool, but sunny November 1, the new Emperor Qíxiáng ("Auspicious") entered Běijīng.[78] As he was still only five years old, his mother Cí Xǐ (**Box 9.7**; **Fig. 9.36**) "temporarily" took over the reigns of power. She was to become one of the most powerful women in Chinese history, who moulded the course of the Qīng Dynasty for more than forty years.

Box 9.7

Cí Xǐ ("Motherly and auspicious"; 1835 – 1908) Eldest child of YEHENALA Huìzhēng (1805 – 1853), a minor Manchu official, and his wife.[1] As a child she felt mistreated, neglected and unloved.[2] She became a petite young lady (153 cm) with dainty, unbound feet, delicate hands and full lips with a winning smile and a sweet voice.[3] According to some sources she was betrothed to her cousin Rónglù (1836 – 1903).[4] However, in 1851 she was summoned to the Forbidden City for the selection of concubines for Emperor

Xiánfēng ("Universal Prosperity", 1831 – 1861).[5] Along with tens of other candidates, she was scrutinized by the Dowager Kāngcí (1812 – 1855), and finally appointed as a low-level concubine in 1852.[6] On bearing Xiánfēng's first and only surviving son (Zǎichún, "Completely Honest") on 27 April 1856, she was promoted, followed one year later by a further promotion, and referred to as Noble Consort Yì ("Exemplary").[7]

When the Anglo-French Invasion was on the verge of entering Běijīng in September 1860, the court decided that it was time for the Emperor to quit the capital for the relative safety of Chéngdé (Jehol) beyond the Great Wall. Even though his half-brother Prince Gōng ("Respectful", 1833 – 1898), whom he had left in Běijīng to negotiate with the foreigners, repeatedly requested Xiánfēng to return to Běijīng, the emperor did not want to see the ruins of the Yuánmíngyuán, face the people he had deserted, or suffer the humiliation of giving audiences to foreign diplomats, who no longer needed to kotow.[8] News of the destruction of the Yuánmíngyuán made Emperor Xiánfēng turn to drink and drugs.[9] As the Emperor's health declined, the power of his advisors led by Sùshùn (1816 – 1861) grew accordingly. They expected to become regents to Zàichún once the Emperor died.[10] They drew up an edict to this effect, but without the imperial seal were unable to validate it.[11] Sùshùn suggested murdering Cí Xǐ on the road back to Běijīng.[12] This was foiled at the last minute by Rónglù, who rode to her aid.[13] With the help of Prince Gōng, who had become Cí Xǐ's brother-in-law, Cí Xǐ had the advisors ("Gang of Eight") arrested and convicted of treason (Xīnyǒu Coup).[14] On 8 November 1861 Sùshùn was beheaded like a common criminal at Běijīng's vegetable market, and his place taken by Rónglù.[15]

Fig. 9.36 Empress Dowager Cí Xǐ (1835–1908). While Fortune was in Běijīng, Cí Xǐ was in Chéngdé, fighting for the right to act as regentess during the minority of her son. She was to become the most powerful woman in Qīng history.

As a child Emperor Qíxiáng ["Auspicious"; renamed Tóngzhì ("Joint rule") after the Xīnyǒu Coup] was a puppet in the hands of his ambitious mother. Although Cí Xǐ was nominally co-regent along with Emperor Xiánfēng's widow, Cí'ān ("Motherly and tranquil", 1837 – 1881), the latter with her preference for literary pursuits was only too pleased to leave the boring state business, such as policy decisions, to Cí Xǐ.[16] By the time Tóngzhì was 15 he was drinking heavily and consorting with prostitutes.[17] As his mother, Cí Xǐ may have tried to turn a blind eye to his misbehaviour.[18] After he died (ostensibly of smallpox, but possibly syphilis) on 12 January 1875, his loving wife Empress Alute (1854 – 1875), commited suicide, which was then considered the height of conjugal loyalty.[19] While Cí Xǐ is often given the blame for Alute's death[20], this disregards the fact that within 48 hours of Tóngzhì's death an attempt was made to poison Cí Xǐ herself.[21] Shortly before, her three-year-old nephew, Zǎitián ("Grow Calm", 1871 – 1908) had been chosen as the new Emperor Gūangxù ("Glorious Succession"), thus allowing Cí Xǐ to continue in power for another 18 years.[22]

On 26 February 1889, just a week before Gūangxù took over the reigns of power, Cí Xǐ arranged for him to marry her niece Xiàodìng ("Pious Calm", 1868 – 1913). Since Cí Xǐ probably meant well, she was disappointed that neither her niece nor the emperor were happy with the arrangement.[23] Be that as it may, their antagonism meant that Cí Xǐ had no fear that they would connive against her, while she was living in retirement at the Yíhéyuán (new Summer Palace, 1886 – 1891).[24] There she devoted her time to walking, poetry, painting, calligraphy, theatricals, breeding Pekinese dogs, cultivating chrysanthemums and gourds, and boating on Kūnmíng Lake.[25] Gūangxù's father (Prince Chún, 1840 – 1891), the newly appointed Chairman of the Admiralty Board, tried to ingratiate himself with Cí Xǐ by diverting funds designed for the navy into the Summer Palace.[26] This possibly contributed to China's defeat in the Sino-Japanese War during 1894 – 1895, which led to a loss of territory (Táiwān, Pénghú Islands and Liáodōng Peninsula) and a crippling indemnity.[27] To most Chinese it was a shock to be beaten by a small nation which they considered inferior.[28]

In the wake of the war it was clear that reforms were necessary. Apart from a vague suggestion to train the army Gūangxù did little to start with.[29] It was Cí Xǐ who initiated the reforms in the name of the emperor.[30] Together they decreed changes to streamline the bureaucracy and improve education, including the establishment of Peking University and provincial agricultural schools, renewed the idea of sending students abroad, offered special awards to authors and inventors, encouraged commerce by improving transport, and ordered the adoption of modern military training.[31] Their decrees were endorsed by the self-styled reformist, KĀNG Yǒuwèi (1858 – 1927), who suggested establishing an Advisory Board to help the emperor make decisions.[32] Although this particular idea was not taken up, KĀNG and his cronies managed to ingratiate themselves with the emperor.[33] At their suggestion Gūangxù dismissed numerous conservative Manchu officials who opposed the reforms.[34] The officials begged Cí Xǐ to intervene on their behalf.[35] At first Cí Xǐ did nothing.[36] However, when she was informed that there was a plot to surround the Summer Palace and possibly kill her, she had the guard at the Forbidden City replaced by Rónglù's trustworthy troops.[37] She then returned to the Forbidden City on 19 September 1898 to keep an eye on developments.[38] The next day Emperor Gūangxù had an interview with ITŌ Hirobumi (1841 – 1909), the very man who had initiated the Sino-Japanese War and dictated the humiliating terms of the Treaty of Shimonoseki.[39] This was enough to make any Chinese patriot wary. On 21 September 1898 Cí Xǐ put an end to the "Hundred Days' Reform", had six of Gūangxù's advisors executed a week later[40], and banished her submissive nephew to Yíng Tái ("Ocean Terrace") Island in Nánhǎi ("South Lake") to the west of the Forbidden City.[41] Since Gūangxù was impotent, and his mental and physical health were giving cause for alarm, the grandees decided it was time to designate an heir to the throne.[42] As the choice fell on the son of ambitious Prince Duān, the 12-year-old Pǔ Jùn ("Universal Handsome", 1886 – 1929), Cí Xǐ continued in her function as regent.[43] In this capacity she broke with tradition by inviting the wives of the foreign diplomats to a reception on 13 December 1898.[44] Expecting to meet a scheming monster, they were favourably impressed by Cí Xǐ.[45] Although it had been rumoured that Gūangxù had been assassinated after the Hundred Days' Reform, his presence at the reception proved otherwise.[46]

At this time China was faced by an uprising of Shāndōng peasants intent on wiping out the foreigners and their Christian converts, who were blamed for meddling with their way of life and causing the drought, flooding and famine in 1897 – 1898.[47] By early 1900 the Yìhétuán (Boxer) Rebellion had spread west into Héběi and Shānxī, where westerners were attacked and symbols of western technology such as railways and telegraph lines destroyed.[48] In Shānxī more than a hundred foreigners were killed.[49] This bloodshed reached a peak on 9 July 1900 when Governor Yùxiàn executed missionaries, their wives and children, 45 persons in all.[50] Such incidents were used by the foreign powers to make more demands on the Qīng government, which in turn disaffected even the moderates in the regime.[51] If Cí Xǐ had opted to put down the rebellion with the help of the foreign powers, she would have alienated her own subjects and endangered the dynasty.[52] Instead she considered using the Boxers to rid China of the foreigners.[53] Having heard of their invincibility, the superstitious Cí Xǐ invited the Boxers to Běijīng for a demonstration of their prowess in May 1900.[54] Unfortunately, when the impetuous German Minister, Baron Clemens Freiherr von Ketteler (1853 – 1900), attacked a man dressed as a Boxer and imprisoned a boy, whom he later shot, the Boxers went on a rampage attacking shops which traded with foreigners and set fire to two of the three Roman Catholic cathedrals where many converts burned to death.[55] Despite more western provocations, the legations were not stormed.[56] Had the well-armed Boxers wanted to, this could have easily been accomplished.[57] On the very day that the Dàgū Forts were overrun by western forces on 17 June 1900, Cí Xǐ and Gūangxù gave General Rónglù specific orders to protect the legations.[58] In order to placate the conservatives ("Ironhats"), Rónglù had his soldiers shoot into the air and set off firecrackers.[59] General Rónglù himself feared for the political consequences of overrunning the legations of such powerful nations.[60] Although it has sometimes been interpreted as a cynical move, Cí Xǐ twice arranged for fresh fruit and vegetables to be sent to the besieged legations as a sign of her goodwill.[61]

Shortly before an international relief force under the considerate Sir Alfred Gaselee (1844 – 1918) finally got through to Běijīng on 14 August 1900, Cí Xǐ's advisors decided it was time for her to flee.[62] However, to make sure she was not usurped, she took Gūangxù with her.[63] They set off in the direction of Inner Mongolia, where they thought they would be safe.[64] However, the Russian occupation of northern China, which had been triggered by the Boxers' destruction of the Trans-Siberian Railway, made them change their plans.[65] The party then traversed the province of Shānxī from north to south before entering Shǎnxī (Shaanxi) and heading for the ancient capital of Xī'ān ("Western Calm"), which they reached on 26 October.[66]

Once the allied forces captured Běijīng, it was looted, and many innocent Chinese citizens were tortured, raped or killed in revenge for the so-called Boxer atrocities.[67] Alute's father, her brother and many of the family committed suicide.[68] The foreign powers put Cí Xǐ on their list of war criminals, but she got Gūangxù to take the blame for the Boxer Rebellion.[69] In the Boxer Protocol (Xīnchǒu Treaty, signed on 7 September 1901), the foreign powers once more imposed a huge war indemnity, thus raising taxes and causing dissention among the population.[70] Only when she was sure that her name had been cleared, did Cí Xǐ return to Běijīng.[71] In contrast to the hardships experienced on the outbound journey, the return trip was made in style, with a luxurious train engaged for the last 260 km of her triumphal return.[72]

Thankful that she had not been humiliated on her return to Běijīng on 7 January 1902, Cí Xǐ held many receptions for foreign diplomats and their wives, as well as missionaries and tourists, and soon won numerous friends by her cordial and charming manner.[73] She enjoyed being in the limelight, had her portrait painted a number of times, and frequently posed for photographs.[74] In August 1907 Cí Xǐ had a mild stroke, and realized that she did not have long to live.[75] Although Rónglù's grandson Pǔ Yí ("Universal Gift", 1906 – 1967) had been designated to succeed her[76], there was always the chance that Gūangxù might make a bid for power when she died. In September 1908 Gūangxù started to suffer from severe stomach pains, which could suggest that he was being poisoned.[77] Analysis of his hair has confirmed that he died of arsenic poisoning.[78]

Cí Xǐ was an habitual early riser, and by 6 a.m. on the 15 November 1908, some 12 hours after Gūangxù's death, she was presiding over the Imperial Council to discuss the consequences of Gūangxù's demise.[79] Plans were made for Pǔ Yí's inauguration.[80] However, after lunch she herself was suddenly taken ill and was dead by 3 p.m. in what would seem to have been an act of reprisal by the pro-Gūangxù faction.[81] No doubt fearing for his own life, Cí Xǐ's Grand Eunuch Lǐ Liányīng ("Repeatedly Hero", 1848 – 1911) fled from the Forbidden City and went into hiding.[82] In less than three years the crumbling Qīng Dynasty was finally swept away by the Xīnhài Revolution.[83]

Once the cortege was safely inside the Forbidden City, Wyndham and Fortune rode along a wide street (presumably the Běihéyàn Dàjiē, "North Riverside Avenue") to Āndìngmén (Anting- or Yan-ting Gate, "Quiet Gate"; **Fig. 9.37**), leaving the Forbidden City and the Drum Tower (**Fig. 9.38**) on their left. Having passed through the Āndìng Gate and into the countryside, the compatriots turned westwards to reach the Russian Cemetery where "The Times" Correspondent, Thomas Bowlby (**Box 9.8**; **Fig. 9.40**) and some others, who had died in prison during the 1860 Anglo-French Campaign, had been buried.[79]

Fig. 9.37 Āndìngmén (Quiet Gate), which would have been bombarded by the Allied troops in 1860, had it not been surrendered at the last minute. Note the Union Jack and French Tricolor flying from the city wall. Fortune would have seen this illustration in his copy of "The Illustrated London News". The gate was demolished in the 1960s to improve traffic conditions.

Fig. 9.38 Drum Tower (1272 AD), Běijīng, which Fortune would have noticed as he headed towards the Āndìngmén.

Fig. 9.39 The beating of a drum during the Yuán, Míng and Qīng Dynasties (1271–1911) was one of the few possibilities the inhabitants had of telling the time. Since 2002 it is beaten four times a day as a tourist attraction.

Fig. 9.40 Thomas William Bowlby (1818–1860), who covered the Second Opium War for "The Times". He was captured by the Chinese, and died in prison.

Box 9.8

Thomas William Bowlby (1818 – 1860) Eldest of nine children born to Captain Thomas Bowlby (1790 – 1842) of the Royal Artillery and his first wife Williamina Martha Arnold Balfour (1798 – 1834).[1] When he was only a few days old the Bowlbys returned to Durham, where his father became a timber merchant.[2] Young Bowlby was educated at Grange School near Durham, before training to become a solicitor under his cousin Russell Bowlby (1792 – 1865).[3] Having completed his training he moved to London, where in 1846 he became a junior partner in the firm of Lawrence, Crowdy and Bowlby.[4] He continued to work for the firm until 1854, but was increasingly drawn into a career as a writer.[5] In March 1848 he went to Berlin to report on the riots for "The Times".[6] After his return, on the 23 September 1848, he married Frances Marion Mein (1824 – 1891), the younger sister of his father's second wife Margaret Matilda Mein (1804 – 1882).[7] On the death of his father-in-law, Pulteney Mein (1769 – 1853), his wife inherited a considerable fortune.[8] However, during the "railway mania" of the mid 1840s Bowlby overinvested in risky railway companies and landed in debt.[9] Between 1852 and 1853 he misappropriated more than £ 5,777 of a client's savings.[10] In 1854 the Bowlbys moved with their two boys and a girl to Belgium, where another four children were born between 1855 and 1858.[11] In April 1860 he accepted the job of war correspondent to "The Times" with the British Expeditionary Force to China.[12] The six foot tall correspondent was a great favourite with the troops.[13] On 18 September 1860, during the advance towards Běijīng, he was captured and thrown into prison, where he died of maggots forming in his tightly-bound wrists.[14] His body was then thrown over the prison wall to be eaten by dogs and pigs.[15] When what remained of his body was handed over by the Chinese on 16 October 1860, Lord Elgin (1811 – 1863) decided to avenge his death and those of others tortured by the Chinese by setting fire to the Yuánmíngyuán.[16] This act of desecration was begun on the 18 October 1860, the day after Bowlby, Lieutenant Robert B. Anderson, Attaché William de Norman, and the gallant Private John Phipps were buried in the Russian Cemetery.[17]

Fig. 9. 41 Yōnghé Gōng (Harmonious Peace Lamasery), which suffered some damage during its occupation by Sir James Hope Grant's army in 1860.

Fig. 9.42 Sir James Hope Grant (1808 – 1875). A painting by his artistic brother, Sir Francis Grant (1803–1878).

After paying their respects, the duo returned to the Manchu City, where they visited the extensive Yōnghé Gōng ("Harmonious Peace Lamasery"; **Fig. 9.41**), which the troops of James Hope Grant (**Box 9.9**; **Fig. 9.42**) had occupied in 1860.[80]

Box 9.9

Sir James Hope Grant GCB (1808 – 1875) Fifth and youngest son of Francis Grant (1746 – 1818) and his wife Anne Oliphant (1765 – 1837).[1] After two years at Edinburgh High School, he was sent to Baron Philipp Emanuel von Fallenberg's finishing school at Hofwil (= Hofwyl) near Berne.[2] Here he acquired a knowledge of foreign languages, which later came in useful.[3] As music played an important part at Fallenberg's school, Grant soon became an accomplished cellist as well as a fine horseman.[4] On returning to Britain in 1826, he joined the British Army as a cornet in the 9th Lancers.[5] He quickly rose through the ranks from lieutenant in 1828, to captain in 1835 and then brigade-major in 1841 under Alexander George Fraser (17th Lord Saltoun, 1785 – 1853).[6] He probably owed his appointment in part to his proficiency on the cello, as Lord Saltoun loved music and was himself a good guitar player.[7] With his cello, a small piano and his Newfoundland dog, Grant left for China on board HMS "Belleisle" on 13 December 1841.[8] During the 164-day voyage via Rio de Janeiro, the Cape of Good Hope, Java and Singapore, Grant befriended Lieutenant-Colonel Colin Campbell (1792 – 1863), the commander of the 98th Regiment, with whom he was to serve during the Indian Mutiny (**Box 6.4**).[9] The "Belleisle" finally docked at Hong Kong on 2 June 1842.[10] On 22 July 1842 Grant helped to co-ordinate troop movement during the attack on Zhènjiāng ("Guard River") during the First Opium War, and was present at the signing of the Treaty of Nánjīng on 27 August 1842.[11] After the war he accompanied Lord Saltoun to the Philippines but did not see much of Luzon, as he was prostrate with fever for most of the month.[12] After the First Opium War, the 9th Lancers were stationed in India and Grant appointed major.[13] He joined his regiment in Kanpur (Cawnpore) in March 1844.[14] There he was housed in style in a mansion overlooking the Ganges, where he had plenty of room for his cello, piano, some curiosities from China, and a number of paintings by his brother Francis.[15] However, this life of luxury came to an abrupt end, when fighting broke out with the Sikhs (First Anglo-Sikh War, from 18 December 1845 to 21 February 1846). On his way to the battlefront, he nevertheless found time to visit Agra, where he marvelled at the Taj Mahal.[16] During the First Anglo-Sikh War he got into trouble for reporting the colonel of the 9th Lancers for drunkenness during combat.[17] After the war Grant obtained leave to recuperate in the Himalayan foothills near Simla. He required 93 coolies to carry all the necessities including his piano and cello![18] However, he soon tired of the fashionable company, so he spent the next six weeks walking, shooting and collecting ferns in the hills of Koti-Ghur.[19] One day his dog "Wolf" was carried off by a leopard.[20]

In February 1847 he married Elizabeth Helen Tayler (1826 – 1891) at Agra.[21] It proved to be a harmonious marriage.[22] After a year of comparative calm, the murder and mutilation of two British political agents in May 1848 sparked off the Second Anglo-Sikh War.[23] After some intense fighting, the Sikh offensive was finally crushed on 12 March 1849 and the Punjab incorporated into British India.[24] Grant was promoted to the rank of lieutenant-colonel and obtained six-months' leave to visit his wife at Simla.[25] He rejoined his regiment at Wazirabad (= Wuzeerabad, Punjab) in October 1849.[26] However, he contracted malarial fever, which was to plague him on and off for the rest of his life.[27] In 1850 he was sent to Simla to recover from sunstroke.[28] In October 1850 he was finally sent back to England on sick-leave.[29] The Grants sailed from Mumbai (Bombay) in February 1851 and eventually arrived in London after a leisurely trip through Italy and Switzerland.[30] Once his health had improved, he and his wife returned to Ambala (Umballa) in March 1854.[31] In 1856 the improved Enfield rifle was introduced, that required grease to lubricate the barrel.[32] That this grease may have contained beef-fat was abhorrant to the Hindu soldiers.[33] The soldiers' revolt finally led to the bloody Indian Mutiny (**Box 6.4**), which was only put down with the greatest of difficulty by the numerically-inferior British and Sikh forces. Grant lost many of his comrades in the conflict and he himself was almost killed on a number of occasions.[34] He was promoted to brigadier-general by his old friend Sir Colin Campbell, who had been sent to India to replace General George Anson (1797 – 1857) as Commander-in-Chief.[35] After the mutiny he was given overall command of the mopping-up force.[36]

When in 1859 the Chinese prevented Sir Frederick Bruce (**Box 9.4**) and his French counterpart (Count Alphonse de Bourboulon) from establishing legations at Běijīng, as agreed in the Treaty of Tiānjīn (**Box 9.1**), Grant was appointed Commander of the British troops of an Anglo-French Expeditionary Force.[37] He and Mrs. Grant sailed in the "Fiery Cross" from Kolkata (Calcutta) to Hong Kong on 26 February 1860, arriving at their destination on 13 March 1860.[38] After a little sightseeing in Guǎngzhōu (Canton) with his wife, he sailed for Shànghǎi on 31 March 1860 to start coordinating the campaign.[39] Considering that approximately 20,000 men and a fleet of 113 vessels were involved, this was no mean feat.[40] After a number of delays, the troops were finally landed at Běitáng (Pei-tang) on 1 August 1860.[41] Once the forts at Dàgū (Taku), which guarded the entrance to the Hǎi River, fell on 21 August 1860, resistance all but evaporated.[42] The expeditionary force reached the Chinese capital Běijīng on 6 October 1860 and was billeted in the Lama Temple near Āndìngmén (Anting Gate) on the north side of the city.[43] In order to avenge the deaths of four British subjects, Grant agreed with Lord Elgin to torch the Yuánmíngyuán on 18 October 1860.[44] This precipitated the Chinese capitulation[45], and on 24 October 1860 the Treaty of Tiānjīn (**Box 9.1**) was duly ratified and the supplementary Convention of Běijīng signed. With winter setting in, Grant left Běijīng on 9 November 1860.[46] However, before returning to India, he and his staff spent three weeks in Japan.[47] They stayed at the British Legation in Edo and were taken to the tea gardens at Ogi (Ogee, Oh-gee) by Sir Rutherford Alcock (**Box 3.10**).[48] Alcock arranged for Grant to meet some of the Shogun's ministers before the group left Kanagawa for Nagasaki.[49] After a short stay in Hong Kong to organize the defence of Kowloon, and a visit to Europe to receive his knighthood and the Légion d'Honneur from Napoleon III (1808 – 1873), he was back in India as Commander-in-Chief of the Madras Army.[50] In 1865 he finally returned to headquarters in England, where he became involved in the reform of army training following the Franco-Prussian War of 1870.[51] He never got round to writing his memoirs, perhaps because his English grammer was far from perfect.[52] Luckily for posterity, this job was done for him by Sir Henry Knollys (1840 – 1930).[53]

Leaving the Manchu City by one of the northern gates, they galloped across the grassy plain to the northeast corner of the Manchu City, rode south, and entered one of the eastern gates, before returning to the legation.[81]

Since the Běijīng nurseries had hardly yielded any novelties, Fortune thought he would try his luck in the Western Hills.[82] As there were a number of Buddhist Temples at Bādàchù ("Eight Great Places"), where the vegetation was presumably fairly intact, Fortune decided to go there.[83]

He started one morning at daybreak.[84] Rather than get jolted in the cart, Fortune decided to walk.[85] So, even though there was a cold northwesterly blowing, he felt quite warm.[86] Fortune left the Manchu City via the Fùchéngmén ("Abundant Achievement Gate") in the middle of the western wall, and headed westwards along the Fùchéng Lù ("Fùchéng Road") to the long straggling town of Bālǐzhuāng ("Eight Li Village"; 12 km from the Legation), where he stopped for breakfast about 9 a.m.[87] While he was in Bālǐzhuāng he admired the highly ornamental Císhòu Pagoda ("Pagoda of Benevolence and Longevity"; **Fig. 9.43**).[88]

Fig. 9.43 Bālǐhuāng (Eight Li Village): The elaborate Císhòu Pagoda (Pagoda of Benevolence and Longevity), which Fortune passed on his way to Bādàchù. "Altogether it is one of the most remarkable specimens of Chinese architecture that have come under my observation." (Fortune, 1863: 375/376).

Fig. 9.44 The Zhāo Xiān Tǎ (Enlist Immortal Pagoda), Língguāng Sì, which contains one of Buddha's teeth. Fortune was lulled to sleep by the tinkling of the bells on the eaves of this pagoda.

The large temple in 1576 to which the pagoda belonged was in a ruinous condition[89], and has since vanished. A little further on in Bālǐzhuāng Fortune visited a large 200 – 300 year old cemetery with lofty junipers, arbor-vitae ("cypresses") and pines.[90] His attention was drawn to two stiff pines with multicoloured, flaky bark.[91] Thinking these might represent a new species, he went to inspect them more closely.[92] However, they turned out to be *Pinus bungeana* (**Fig. 6.3 C, D**), which he had already introduced from Shànghǎi.[93]

Fortune finally arrived at Bādàchù, where the chief monk of Língguāng Sì ("Effective Brightness Temple") offered him a room for the night.[94] Although he had often stayed in Buddhist temples, he had never had such good accomodation before.[95] In the courtyard was a pretty pagoda with bells suspended from the eaves (**Fig. 9.44**)[96], and an old *Ginkgo* covered in *Wisteria sinensis*.[97]

It is an uncanny experience to stand in this courtyard and view the same things that Fortune once saw (**Fig. 9.45**). Having walked some 26 km from the centre of Běijīng, Fortune had a short rest before setting off for the other temples with a young monk as his guide.[98] In Dàbēi Sì ("Great Compassion Temple"; **Fig. 9.46**), where the botanist Alexander Georg von Bunge (1803 – 1890) had stayed in May 1831[99], some Russians had defaced the temple by scratching their names on a wall in 1832.[100] On the way to the Xiāngjiè Sì ("Fragrant Boundary Temple"), Fortune and his guide crossed a marble bridge over a mountain stream (**Fig. 9.47**).[101]

Fig. 9.45 Courtyard of Línguāng Sì (Effective Brightness Temple), Bādàchù, where Fortune stayed. "Here I noticed some fine old specimens of the "Maidenhair tree" (*Salisburia adiantifolia*), one of which was covered all over with the well-known glycine. The creeper had taken complete possession of this forest king ... but the Salisburia evidently did not like its fond embraces, and was showing signs of rapid decay." (Fortune, 1863: 380–381). In order to save the ancient ginkgo in the centre of the picture, the wisteria which covered it in 1861, was removed.

Fig. 9.46 Dàbēi Sì (Great Compassion Temple), where the botanist Alexander Georg von Bunge (1803–1890) stayed in May 1831.

Fig. 9.47 Marble bridge "of great age" over stream at Bādàchù, which Fortune crossed on his way to Xiāngjiè Sì (Fragrant Boundary Temple).

Fig. 9.48 Xiāngjiè Sì is famous for its large *Pinus tabuliformis*, and the fact that Emperor Qiánlóng (1711 – 1799) stayed here.

The Xiāngjiè Sì (**Fig. 9.48**) was the largest of the temples[102] with a sizeable courtyard in front of the Hall of Heavenly Kings. In this courtyard was and is a large *Pinus tabuliformis*, and a stele with a command "Worship Buddha" written by Emperor Qiánlóng (1711 – 1799) who had stayed there.[103]

As they proceeded up the hill, Fortune noted a number of interesting plants, such as *Vitex negundo* var. *heterophylla*, *Aleuritopteris argentea* (Syn. *Pteris argentea*), *Thuja orientalis*, and what he took to be a new species of oak, which he referred to as *Quercus sinensis*.[104] This oak, which is still to be found growing on the hill, is none other than *Quercus variabilis* (**Fig. 9.55**). Once he reached the highest point, Fortune sat down on a cairn of stones to enjoy the panorama (**Fig. 9.49**).[105] From this vantage point, he could see the Tiānjīn Plain, Běijīng's walls and watchtowers, the yellow-roofed palaces, the ruins of the former Summer Palace Yuánmíngyuán ("Circular Míng Garden"), Kūnmíng Lake at the new Summer Palace, and the Lúgōu Qiáo ("Lú Channel Bridge")[106], which later gained notoriety as the place where the Japanese launched a surprise attack on 7 July 1937.[107]

Fig. 9.49 View from the top of Bādàchù in autumn. "The most charming views were obtained from this situation ..." (Fortune, 1863: 383). The leaves of *Cotinus coggygria* are responsible for the reddish hues.

Fig. 9.50 Wild jujube, *Ziziphus jujuba*, which Fortune encountered in the Western Hills.

To the north lay wild and barren hills with jujube (**Fig. 9.50**) and spotted bellflower (*Campanula punctata*), which reminded Fortune of Scotland.[108] As the sun started to set, Fortune and his guide picked their way down the steep steps to the Língguāng Sì.[109] All the exercise and the mountain air, had given Fortune "a tolerable appetite".[110] After dinner he was visited by some mandarins, who wanted to know why he had visited Bādàchù.[111] He humoured them with cigars and wine, and they did not even bother to ask to see his passport![112] After they had gone, Fortune retired to his room, and soon fell asleep to the sound of the leaves rustling in the wind and the tinkling of the bells on the pagoda.[113]

The next morning he was up before sunrise, and was able to watch as the sun's rays gradually lit up the whole plain round Běijīng.[114] Later that day he visited a number of temples and gardens on the other side of the valley.[115] He collected some plants for his herbarium, and picked seeds of those ornamental and economic plants worth introducing into Europe.[116] In the afternoon he returned to Běijīng.[117] On the way back he stopped at the cemetery in Bālǐzhuāng to obtain a fresh supply of seeds of *Pinus bungeana*.[118]

After packing his collections, Fortune left Běijīng on 28 September 1861.[119] Lockhart accompanied him part of the way to Tōngzhōu ("Connecting Prefecture"), where he caught a boat back to Tiānjīn.[120] Just a little more than 3 km before Tōngzhōu, he passed Bālǐqiáo ["Eight Li Bridge" (1446); **Fig. 9.51**], the scene of a fierce battle between French troops and Chinese forces on 21 September 1860(**Fig. 9.52**).[121]

Fig. 9. 51 Bālǐqiáo (Eight Li Bridge, 1446), the site of a fierce battle between French troops and Chinese forces on 21 September 1860. Fortune passed the bridge a year later, on 28 September 1861.

Fig. 9. 52 Battle of Bālǐqiáo. The French troops are on the right.

Fortune remained for a few days in Tiānjīn, until he obtained a passage on the dispatch boat "Contest" from Dàgū.[122] He arrived back in Shànghǎi on 20 October 1861, and was pleased to find his Japanese collection in excellent condition.[123] He spent the next fortnight packing the plants into Wardian Cases for the long voyage via the Cape of Good Hope.[124] As a precautionary measure, these were sent by two separate ships.[125] He himself took two little hand greenhouses containing special favourites, including a creeping saxifrage with variegated leaves, now known as *Saxifraga stolonifera* 'Tricolor' (**Fig. 9.53**).[126] Following the voyage to Port Suez and the overland train journey through Egypt, he boarded the P&O Steamer "Ceylon" (**Box 9.10**; **Fig. 9.54**; Captain Evans) in Alexandria in mid-December, and arrived back in Southampton on 2 January 1862.[127]

Fig. 9.53 *Saxifraga stolonifera* 'Tricolor'. This was one of Fortune's "especial favourites", which was "brought home by the overland route under my own care." (Fortune, 1863: 388).

Fig. 9.54 SS Ceylon in 1858, which carried Fortune from Alexandria to Southampton, on the last leg of his return journey from China.

Box 9.10

SS Ceylon (1858 – 1907) This propeller-driven iron steamship, built for the Peninsular and Oriental Steam Navigation Company (P&O) by London's leading shipyard (Samuda Brothers), was designed to carry 160 passengers in comfort.[1] Its innovative super-heated furnaces saved 500 tons of coal on the round voyage Southampton-Alexandria-Southampton.[2] On 17 February 1862, just over 6 weeks after delivering Fortune safely to Southampton, she ran aground on Portland Bill in thick fog, thereby losing her bowspit and mainmast.[3] In January 1863 she rode down the brig "Ridesdale", which was being towed by the "Aid" in Southampton Water.[4] The brig sank and the tug's funnel fell and killed the pilot.[5] In 1873 she was substantially refitted in Southampton.[6] When P&O sold her to John Clark of London in 1881, she was refurbished for cruises around the world.[7] Judging by the limited number of passengers between Southampton and San Francisco[8], these cruises were not a financial success. Clark sold her to the Ocean Steam Yachting Company (Limerick) in December 1883, who resold her to Michael Drury-Lavin (London) in 1885.[9] On 27 September 1889 while en route from Stockholm to Oslo the ship ran aground on the Kulla Grunden Shoal near Trelleborg on the southern tip of Sweden.[10] The ship was towed to nearby Copenhagen, where the damage was repaired so she could return to London.[11] She then had a succession of owners, before being sold to shipbreakers at Bo'ness in December 1907.[12]

When Fortune set out for Běijīng he had high hopes of collecting a large number of northern Chinese ornamental plants, which would be hardy in Britain.[128] In this he was to be disappointed, as the nurseries almost exclusively grew plants which he had already encountered further south. The only exception was the new *Forsythia*, which John Lindley (**Box 2.1**) named after him.[129] It was almost autumn by the time he arrived in Běijīng, and with just a limited amount of time at his disposal, he only managed to spend two days looking for suitable candidates in the wild. At Bādàchù he saw what he took to be a new species of oak, which he dubbed *Quercus sinensis* (**Fig. 9.55**)[130]

Fig. 9.55 ***Quercus variabilis*, which Fortune introduced to the UK from Bādàchù.**

Although this oak had already been described by Professor Carl Ludwig von Blume (1796 – 1862) as *Quercus variabilis* in 1850, the acorns which Fortune presented to John Standish (**Box 5.15**), represented the first introduction of this chestnut-leaved species into culture.[131]

Chapter 10 Out of the Limelight

10.1 Life in London

After an absence of a year and a half, it cannot have been easy for Fortune to re-establish his role as paterfamilias. His infant daughter Alice must have wondered who this strange man was, who walked into their house. Since his wife Jane had been solely responsible for the childrens' upbringing, they would naturally have had a stronger bond with her, although his older children and London acquaintances were no doubt eager to learn all about Asia. However, one can imagine that his wife must have got fed up hearing about his adventures in the Far East, since she had had to stay at home and look after the children.

One of Fortune's first tasks on his return was to make sure that John Standish (**Box 5.15**) got his material for propagation. Armed with his "especial favourites", he would probably have caught the train at Vauxhall Bridge Station (**Fig. 10.1**) to save the difficulty of finding the London & South Western train to Woking in the maze of platforms at Waterloo Bridge Station.[1]

Fig. 10.1 Vauxhall Bridge Station opened on 11 July 1848. Fortune would have caught the train to John Standish's nursery in Ascot here.

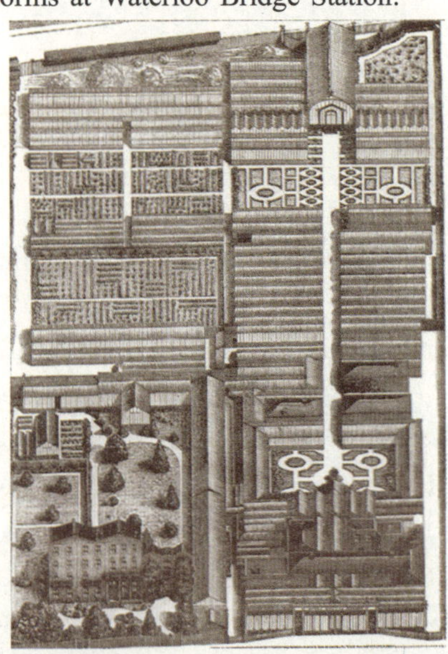

Fig. 10.2 Plan of Veitch's Royal Exotic Nursery at Gunter Grove. No doubt Fortune would have visited it to vet their Japanese collection.

At Woking the plantsman would have changed to a Staines, Wokingham and Woking train bound for Ascot.[2] After delivering the material for propagation, he probably went to have a peep at the Royal Exotic Nursery, just 600 m down Fulham Road at Gunter Grove (**Fig. 10.2**), to see how the plants brought back from Japan by John Gould Veitch (**Box 8.1**) were faring.

Once the Wardian Cases were safely in Standish's hands, Fortune had to start unpacking the works of art, which he had collected while in Japan and China. Less than 3 months after his return, on 27 – 29 March, Messrs. Christie, Manson & Woods put on view "a very choice collection of ANCIENT PORCELAIN, ENAMELS, AND LACQUER WORK", totalling 290 items.[3] The auction itself was held in Christie's Octagonal room at 8 King Street on Monday 31 March 1862 and Tuesday 1 April 1862. It seems likely that Fortune would have attended. As a respectable, middle-class, gentleman he most probably wore a dark, three-piece suit.[4] As he left the house, he would have picked up a top hat from the hat stand as an outward sign that he had "arrived".[5] Fortune would probably have taken a walking stick with him, as he had to be on his guard not only for pickpockets, but more aggressive criminals, who in 1862 carried out a spate of brutal muggings ("garottings") even in broad daylight.[6] If he was in a hurry, he could have hailed a horse-drawn omnibus to take him from Fulham Road to King Street. However, on a nice day he, like most of his contemporaries[7], would probably have walked the 6 km, as the auctions did not start until 1 p.m. As he walked along Fulham Road, he would have passed Elm Park, where Queen Elizabeth I (1533 – 1603) and her chief advisor William Cecil (Lord Burghley, 1520 – 1598) were said to have sheltered from the rain under a great elm tree (**Fig. 10.3**).[8] If he turned into Brompton Road he would have seen the partly completed South Kensington Museum (**Fig. 10.4**)[9], which was being built in the grounds of Brompton House for the International Exhibition of 1862.[10]

Fig. 10.3 Elm Park, where Queen Elizabeth (1533 – 1603) and her chief advisor, William Cecil (1520 – 1598), were said to have sheltered from the rain under a great elm tree. All the large trees are now London Plane, *Platanus* x *hispanica*.

Fig. 10.4 South Kensington Museum, built for the International Exhibition of 1862. It was renamed the Victoria and Albert Museum in 1899.

He would have passed Brompton Square (**Fig. 10.5**) where so many famous writers including Charles William Shirley Brooks (1816 – 1874), a lead writer for "The Illustrated London News" and contributor to "Punch" lived.[11] On the other hand, he may have preferred to walk to Knightsbridge along Walton Street/Basil Street, which was quieter. On the way he would have passed the neogothic St. Saviour's Church (**Fig. 10.6**).[12] Either way, he would have ended up at Hyde Park, and from there made his way to the auctioneers in King Street.

Fig. 10.5 Brompton Square, where so many famous writers lived.

Fig. 10.6 The neogothic St. Saviour's Church.

Fortune would have learned from earlier sales which objects were most coveted, and therefore worth importing to England. For instance, turquoise crackle vases were in vogue. There seems to have been considerable interest in this latest collection of works of art, as it fetched £ 977 - 14 - 6.[13] Even after Christie's had deducted their 8 – 10% commission, this was still a considerable sum. No doubt, Fortune came away satisfied. With this sort of income, the Fortunes could easily afford a number of servants.[14] Another auction of "a very choice collection of ANCIENT PORCELAIN AND JAPAN LACQUER WORK", which was held on Friday 13 June 1862, was less successful. The 157 items raised only £ 299-19-6.[15] Lord Southwark bought 6 white porcelain seals "same as found in the Irish bogs" for only £ 1-9-0 (**Fig. 10.7**).[16]

Fig. 10.7 Chinese white porcelain seals, which Fortune believed were identical to those found in Irish bogs.

Fortune must have been disappointed, as he mentioned that "The peculiar white or rather cream-coloured porcelain of which they are composed, has not been made in China for several hundred years ...They are very rare in China at the present day."[17] The last item "a beautiful specimen of the seaweed '*Hyalonema*' found on the coasts of Japan" remained unsold.[18] Fortune might have wondered whether there was some truth in the saying that Friday the 13th is an unlucky day.

Between the auctions, Fortune probably visited his successor, Thomas Moore (**Box 10.1**; **Fig. 10.8**) at the Chelsea Physic Garden to discuss the latest turn of events.

Box 10.1

Thomas Moore (1821 – 1887) Nothing seems to be known about his childhood. He started his horticultural career at Dickinson's nursery in Guildford before being employed by John Fraser at his Lea Bridge nurseries in Leyton, Essex in 1839.[1] From there he moved to Park Hill, Streatham as an under-gardener in 1842.[2] In 1844 he became assistant to the landscape gardener Robert Marnock

(1800 – 1889), who was responsible for developing the Royal Botanic Society's garden in Regent's Park.[3] As editorial assistant on the "Gardeners' Chronicle" he came to the attention of John Lindley (**Box 2.1**), who clearly thought highly of him.[4] When Fortune relinquished his post as Curator of the Chelsea Physic Garden in 1848, Lindley had his new protégé appointed.[5] He was an authority on ferns, publishing a number of new taxa, and stimulating the Victorian craze for these plants (pteridomania).[6] He was also interested in *Pelargonium*, hollies and along with George Jackman Jr. (1837 – 1887) was responsible for developing *Clematis* x *jackmanii* (Fortune's *C. lanuginosa* crossed with *C. viticella*), the first of the modern large-flowered hybrids.[7] He was also very active as an editor of various journals.[8] His curatorship was overshadowed by uncertainties regarding the future of the Physic Garden. Plans for the Chelsea Embankment meant that the garden would be cut off from its natural water supply, the Thames.[9] Moreover, in 1859 the West of London and Pimlico Railway Company had plans to build a railway on the river frontage, which would diminish the value of the property.[10] There was talk of moving elsewhere.[11] As a consequence, the Society of Apothecaries had no desire to invest in their garden. Moore struggled to cope with reduced financial resources and personnel.[12] As a result, he has wrongly been branded as a poor curator.[13] However, he was widely respected in the horticultural world, serving on various committees including the International Horticultural Congress in 1866.[14] Moreover, with the support of Dr. Nathaniel Ward (**Box 2.6**), he was able to develop and rearrange the garden to a certain extent.[15] During his curatorship, women were encouraged to study botany at the garden, and sit an examination set by Moore.[16]

Fig. 10.8 Thomas Moore (1821 – 1887), Fortune's successor as Curator of the Chelsea Physic Garden.

The garden had fallen on hard times since Fortune had left.[19] The budget had been cut, and the glasshouses with their tender plants sold.[20] However, the current question was how to react to the letter which had appeared in "The Times" on 7 April 1862 by its Special Correspondent to China in 1857, George Wingrove Cooke (1814 – 1865). Cooke, who lived nearby, was irritated by the exclusiveness of the garden, with its high wall designed to protect it from prying eyes (**Fig. 10.9**).

Fig. 10.9 The high wall, that, despite some public criticism, still surrounds the Chelsea Physic Garden. It helps to create a haven of peace in the heart of a busy city.

He suggested replacing the wall by a railing and allowing the general public access.[21] In his reply in the "Independence", Moore pointed out that the wall was necessary to protect the half-hardy plants, and provide the medical students with the necessary peace to study the plants.[22]

Moreover, although the garden belonged to the Society of Apothecaries, no one to his knowledge had ever been barred from entering.[23]

In a carefully worded report to the Garden Committee of the Chelsea Physic Garden, the long-suffering Moore pointed out on 30 May 1862[24]:

> The Committee is probably aware that within the last month or two, an attack has been made in the Times newspaper, upon the wall which surrounds the gardens, and to some extent, also upon the rights of the Society in respect thereof. The question having been further brought before the Chelsea vestry on the motion of Mr. Wingrove Cooke, who was the writer of the letters in the Times, I felt called upon to notice the subject in one of the local newspapers. The letter addressed by me to the "Independence" together with such other of the documents as I have been able to collect are placed on the table for the information of the Committee (papers marked A).
>
> I have also within the last few days received a letter from the Secretaries of the "National Association for the promotion of Social Science" in respect to the admission of foreign visitors to the garden. This letter, I have also placed on the table, having replied, that it should be submitted to the proper authorities at the earliest opportunity which presented itself (paper marked B).
>
> I have latterly refrained from troubling the Committee with the wants of the garden, but it is my duty to mention that if it is to be retained the buildings have become urgently in need of painting and repairs.
>
> I have the honor to be, Gentn.,
> Your obet. Sert.
> Thos. Moore

In the face of public criticism, the Society of Apothecaries finally acted. Fortune's "old friend" Nathaniel B. Ward (**Box 2.6**) headed a new Garden Sub-Committee to rejuvenate the garden.[25] The section of hardy herbaceous plants was enlarged, and a fresh collection of tender medicinal plants established.[26] A new lean-to glasshouse ("Cool fernery") on the western wall (**Fig. 10.10**) was erected and a third gardener hired.[27]

Fig. 10.10 Lean-to glasshouse erected against the west wall of the Chelsea Physic Garden. It was here that pteridologist Thomas Moore grew some of his beloved ferns, hence its designation as "Cool fernery".

By this time the cotton industry in Lancashire was suffering from a shortage of raw material caused by the blockade of the Confederate ports during the American Civil War during 1861 – 1865.[28] In Lancashire only 11% of the 355,000 people normally employed in spinning and weaving cotton were working full time.[29] On 8 April 1862 a soup kitchen was opened in Manchester to alleviate some of the suffering.[30] While most people were concerned about the possibility that the UK might be drawn into the conflict, Fortune must have wondered what had become of the tea plants he had so carefully selected for the American market (Chapter 7).

The outbreak of the American Civil War meant that the USA was not represented at the

International Exhibition of Fine Arts and Industry which opened on Cromwell Road on 1 May 1862.[31] This exhibition was the brainchild of Queen Victoria's husband Prince Albert (1819 – 1861), who intended it to surpass the highly profitable Great Exhibition of 1851.[32] A gigantic complex of buildings designed by Francis Fowke (1823 – 1865), the architect of, e. g., the Royal Scottish Museum, the National Gallery of Ireland, and the Royal Albert Hall[33], was erected between Cromwell Road and Kensington Road (**Fig. 10.11**) to house the 28,000 exhibits from 36 countries.[34]

Fig. 10. 11 Building designed by Francis Fowke (1823 – 1865) to house the International Exhibition of Fine Arts and Industry (1862). It was demolished after the exhibition.

Fortune would have been particularly interested in the exhibits from China and Japan, which Rutherford Alcock (**Box 3. 10**) had helped to organize.[35] These included exquisite carving, lacquerware, porcelain, bronzeware, silk, jade, books on natural history and chemistry, Japanese armour and instruments, and examples of Chinese medicine.[36] In the six months that it ran, the exhibition was visited by over 6,000,000 people.[37] Although the exhibition made a slight profit, it did not prove as successful as the Great Exhibition of 1851.[38] In the press the building was condemned as "a wretched shed", and when Parliament refused to sanction its purchase, it was demolished and the material used to build Alexandra Palace, a centre of recreation and education in the north of London.[39]

Fortune spent much of 1862 recording his experiences in Japan and China for publication. By the autumn the manuscript was so far advanced that he was able to show it to John Murray III (1808 – 1892; **Fig. 10.12**) for his approval.

The publisher and Robert Fortune continued to be on good terms. On 12 November 1862 Murray sent the following letter to the plantsman:[40]

Fig. 10.12 John Murray III (1808–1892) at his desk.

> My dear Fortune,
>
> I will give you £ 200 for the copyright of your new book on China—Pekin & Japan if you prefer that method of arrangement to sharing 1/2 profits—I bearing the risque [sic]. I shall be at home on Saturday 12 to 2 if you should be passing.
>
> My dear Sir
> Yours very sincerely
> John Murray

This was the same amount Murray had given him for his last two books. While the earnings in no way compared with those which accrued from the auctions, this new book would have the effect of bolstering Fortune's claim to fame. To reach the famous publishing house at 50 Albemarle Street, Fortune would have taken roughly the same route as he did when going to Christies, the only difference being that instead of turning right into St. James's Street, he would have turned left into Albemarle Street. Some twenty-three brisk paces would have brought him to the doorstep of John Murray's house (**Fig. 10.13**), where he would have been shown upstairs into the room that had witnessed the meeting of so many famous authors (**Fig. 10.14**).

Fig. 10.13 Facade of John Murray's house at 50 Albemarle Street in the heart of London.

Fig. 10.14 Reception Room/Library at 50 Albemarle Street, which witnessed many a gathering of great minds.

While Fortune was working on his book "Yedo and Peking", he kept in touch with events in Asia. We know this because he mentioned the Namamugi Incident in which one of his Shànghǎi acquaintances, Charles Lenox Richardson, was killed on 14 September 1862 (**Box 10.2**; **Fig. 10.15**).[41]

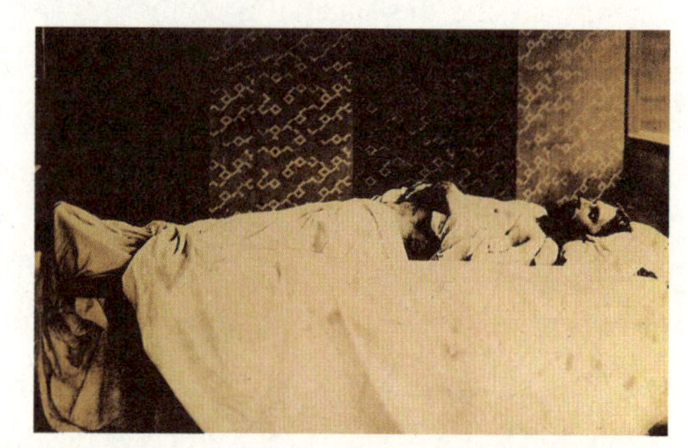

Fig. 10. 15 The Shànghǎi businessman, Charles Lenox Richardson (1834 – 1862) lying in state after he was killed by a samurai in the Namamugi Incident.

Box 10.2

Namamugi Incident On 14 September 1862 the merchant Charles Richardson (1834 – 1862) and his friends Woodthorpe Charles Clarke, William Marshall and Marshall's sister-in-law, Mrs. Margaret Watson Borradaile set out from Yokohama on horseback to visit the Kawasaki Daishi (see **Figs. 8.42–8.43**).[1] They were either unaware of or ignored the official warnings that daimyo processions would be passing along the Tōkaidō that day.[2] When they reached the village of Namamugi, the road narrowed, so the horsemen formed into pairs, with Richardson and Borradaile leading the way.[3] At this point they met a cortège belonging to Prince SHIMAZU Hisamutsu of Satsuma (1817 – 1887), returning from Edo after escorting an imperial messenger to the shogun.[4] Accounts differ as to what happened next. According to Mrs. Borradaile they obeyed the sign to move to the side of the road.[5] However, they neither dismounted nor prostrated themselves, which could have been interpreted as an affront to SHIMAZU Hisamutsu.[6] Marshall said he shouted to Richardson to stop and turn back, but the latter retorted "I know how to deal with these people".[7] According to a Japanese account, Richardson's horse grew frightened, reared up, and moved into the middle of the road, which made a clash inevitable.[8] Richardson received a sword blow to the left shoulder and another which slit his belly open.[9] He fell from his horse.[10] Once on the ground the unarmed man had little chance against the samurai. When he was found his right hand was merely hanging by a strip of flesh and the back of the left hand was cut through, wounds which may have been inflicted while he tried to parry further blows to his neck and heart.[11] According to one account, SHIMAZU ordered his samurai to give Richardson the "coup de grace".[12] While Marshall was wounded in his side and back, Clarke's left arm was almost severed at the shoulder during the attack[13], but he and Marshall managed to escape. Once they reached Kanagawa their wounds were treated by Dr. James Hepburn (**Box 8.6**).[14] By ducking a sword Mrs. Borradaile lost her hat and some hair, but not her head.[15] In a state of shock she rode for her life until she reached the first house in Kanagawa.[16] Unaware that Marshall and Clarke were safely back at the US Consulate, the British Consul Francis Howard-Vyse (**Box 8.14**) immediately set off with an escort to Namamugi.[17] News of the attack soon spread among the foreign community.[18] The merchants were outraged and felt that SHIMAZU should be brought to justice.[19] Realizing that this would be tantamount to declaring war on the Japanese Government, the Chargé d'Affaires St. John Neale resisted the cries for revenge, and opted for a diplomatic solution.[20]

Acting on orders from the Foreign Office, the British Chargé d'Affaires, Lieutenant-Colonel Edward St. John Neale (1812 – 1866) demanded an apology and a huge indemnity for the deed.[42] When this was not forthcoming, a British fleet bombarded the Satsuma capital Kagoshima on 15 – 16 August 1863 (**Fig. 10.16**).[43] The Satsumas were so impressed by British fire power, that they not only paid the indemnity but made enquiries about buying a warship![44] Following this incident, a power struggle developed between the pro-western shogunate and the Japanese emperor, who advocated expelling the foreigners.

Fig. 10.16 Kagoshima being bombarded in August 1863 as retribution for Richardson's death.

By the time Fortune's book was published in 1863, the shogunate had become increasingly discredited for giving in to the foreigners. Consequently, the power of the Sonnō-Jōi Movement to support the emperor and expel the barbarians increased. In April 1863 Emperor Kōmei (1831 – 1867) summoned the teenage Shogun TOKUGAWA Iemochi (1846 – 1866) to Kyōto and ordered him to close the ports to all foreigners by 25 June 1863 (10 May according to the lunar calendar).[45] As the shogun was in no position to carry out this order, the Chōshū Clan took the law into their own hands.[46] They fired on every foreign ship entering the Shimonoseki Straits after the emperor's deadline expired.[47]

The political situation in Japan does not seem to have had any influence on the auction of Japanese porcelain and lacquerware which Fortune put up for sale at Messrs. Christie, Manson and Wood's offices in London on Thursday, 3 March 1864 ("VERY CHOICE PRIVATE

COLLECTION OF ANCIENT PORCELAIN AND LACQUER WORK,"; **Fig. 10.17**).[48]

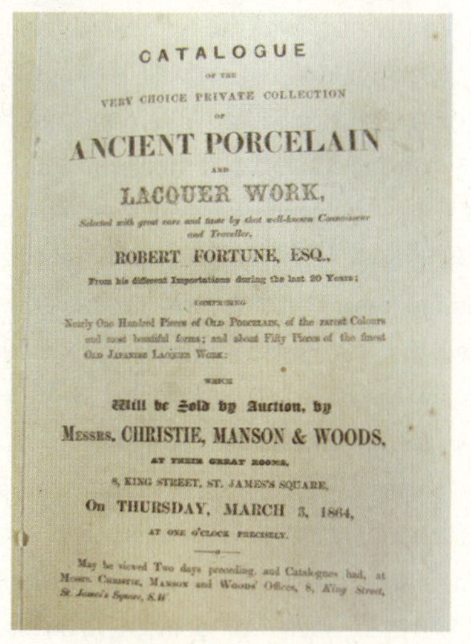

CATALOGUE

OF THE

VERY CHOICE PRIVATE COLLECTION

OF

ANCIENT PORCELAIN

AND

LACQUER WORK,

Selected with great care and taste by that well-known Connoisseur and Traveller,

ROBERT FORTUNE, ESQ.,

From his different Importations during the last 20 Years;

COMPRISING

Nearly One Hundred Pieces of Old Porcelain, of the rarest Colours and most beautiful forms; and about Fifty Pieces of the finest Old Japanese Lacquer Work:

WHICH

Will be Sold by Auction, by

Messrs. CHRISTIE, MANSON & WOODS,

AT THEIR GREAT ROOMS,

8, KING STREET, ST. JAMES'S SQUARE,

On THURSDAY, MARCH 3, 1864,

AT ONE O'CLOCK PRECISELY.

May be viewed Two days preceding, and Catalogues had, at Messrs. Christie, Manson and Woods' Offices, 8, *King Street, St. James's Square, S.W.*

Fig. 10.17 The title page of the last auction of Robert Fortune's *objets d'art* at Messrs. Christie, Manson & Woods on 3 March 1864.

Fig. 10.18 The well-known Victorian hostess, Lady Dorothy Nevill (1826 – 1913), who bought two tortoise-shell stands and two porcelain seals at the auction of Fortune's Japanese ornaments on 3 March 1864.

It was attended by many dealers and a number of prominent figures.[49] The 134 items raised a total of £ 680 – 3 – 0.[50] The highest price (£ 31 – 10 – 0) was fetched by a bottle enamelled with ornaments and foliage on a carved stand.[51] Two pilgrim bottles, one with plants and ornaments in blue and another with trees and birds in gold, were bought by Lord Lansdowne for £ 15-10-0.[52] The politician also purchased two oblong boxes decorated with landscapes for an additional £ 9-5-0.[53] The well-known Victorian hostess, Lady Dorothy Nevill (1826 – 1913; **Fig. 10.18**) bought two tortoise-shell stands for £ 9, and two ancient seals of white porcelain surmounted by figures of animals for as little as 11 shillings.[54]

Fig. 10. 19 The tree-loving Baron William Buller Fullerton-Elphinstone (1828 – 1893).

Box 10.3

Baron William Buller Fullerton-Elphinstone (1828 – 1893)

Son of James Drummond Fullerton-Elphinstone (1788 – 1857) and his wife Anna Maria Buller (1799 – 1845).[1] He entered the Royal Navy in 1841, serving in the West Indies on HMS "Queen" and "Illustrious" under his cousin, Admiral Charles Adam (1780 – 1853) of Blair Adam (Kinross-shire).[2] He recorded his impressions of the places they visited in the form of sketches.[3] He then served under the accomplished artist Captain Henry Byam Martin (1803 – 1865) on HMS "Grampus" in his unsuccessful attempt to prevent the French from annexing Tahiti.[4] He became Lieutenant in 1848 and sailed to China on the "Cleopatra" to combat piracy on the China Sea.[5] He may have met Fortune while in

Hong Kong. During the Crimean War, he was assigned to the "Dauntless" and "Royal Albert" in the Black Sea.[6] As Commander of HMS "Hornet", he was posted to the East Indies and China in 1860 – 1861 (Second Opium War).[7] Following the deaths of his cousins the 13th and 14th Lord Elphinstones in quick succession, he became 15th Lord Elphinstone on 13 January 1861.[8] He retired from the navy in 1863, and on 16 June 1864 married the 6th Earl of Dunmore's second daughter, Constance Euphemia Woronzow Murray (1838 – 1922).[9] They had five children.[10] Lord Elphinstone sat in the House of Lords as a Scottish Representative Peer from 27 November 1867 until 18 November 1885 and was Government Whip in the House of Lords in the Conservative administrations of Benjamin Disraeli and Lord Salisbury.[11] During his time in the upper house he travelled widely, e.g. Australia, New Zealand, and more particularly in western Canada where he purchased some land.[12] As a lover of exotic trees, he invested much time and money in the policies at Carberry Tower (**Fig. 10.20**), Haddingtonshire (now East Lothian).[13] No doubt, Robert Fortune and he met and discussed their joint passion for plants when Robert was visiting his son at Elphinstone Tower Farm (**Fig. 10.21**).

Fig. 10.20 Carberry Tower, the castle once owned by Lord Elphinstone. It is now a pleasant hotel.

Fig. 10. 21 Sitting Room in Carberry Tower, where Lord Elphinstone would have received Robert Fortune.

At some stage between 1862 and 1864 Fortune appears to have met or become reacquainted with Lord Elphinstone, a lover of exotic trees, who had been to China in the course of his naval career (**Box 10.3; Fig. 10.19**). If they had not already met while HMS "Cleopatra" was in Hong Kong in 1848—Fortune quotes Captain Thomas Lecke Massie (1802 – 1898) of the "Cleopatra"[55]—then it is possible that John Murray, with his large circle of friends and acquaintances, may have introduced the two Scots.[56] They seem to have become friends, as Robert's 21-year-old son, John Lindley Fortune (1842 – 1913), became a tenant on the Elphinstone Estate in 1864. In the "Haddingtonshire Courier" of 25 November 1864, he is described as "the new tenant", who was given some help by his neighbours to plough about 30 acres near Elphinstone Tower, where the Protestant reformer George Wishart (1513 – 1546) had been detained in January 1546 prior to his

Fig. 10. 22 Elphinstone Tower Farm (1865), where Fortune's eldest son, John Lindley Fortune, was tenant.

execution in St. Andrews by burning at the stake.[57] This "friendly ploughing" was repeated in the Spring of 1865, as the 22-year-old Fortune was "far back with his field work" as a result of all the cartage in connection with the demolition of the late 17th Century house adjoining the medieval tower, and the erection of the new steading and cottages at Elphinstone Tower Farm (**Fig. 10.22**).[58]

Even though it was hard work, in the course of time John turned Elphinstone Tower Farm into a "veritable paradise".[59] He stayed there with a housekeeper until the early 1880s.[60] Robert spent the autumn of each year with his eldest son on the farm[61], something that has led many biographers to believe that Robert himself took up farming during his retirement.[62] One author even went as far as to state that "He now settled down to farming with his son-in-law", a relative the plant hunter never had![63]

At the time John started farming at Elphinstone, his father was still very much involved in the horticultural world. In 1864 he was co-opted onto the Royal Horticultural Society's Committee on the Improved Education of Gardeners chaired by the RHS's Secretary,William Wilson Saunders (1809 – 1879).[64] The committee also included the nurserymen William Paul (1822 – 1905; rosarian etc.), Thomas Rivers (1798 – 1877; pomologist), and James Veitch Jr. (1815 – 1869; orchid-grower), the pteridologist Thomas Moore (**Box 10.1**), the legendary Sir Joseph Paxton (**Box 5.14**), and Sir Charles Wentworth Dilke (1810 – 1869), one of the driving forces behind the Great Exhibitions of 1851 and 1862.[65] One can imagine the often charged atmosphere at the committee meetings, as each of its members laid out his proposals and defended his stance. While Paxton would have remained his amiable self, Dilke could be bossy, while Veitch was an outspoken person, who brooked no contradiction.[66] Under these circumstances, the self-effacing Moore probably said little. However, in the end, many of the ideas bandied about were incorporated into the comprehensive report published in 1865.[67]

In their report the committee recommended that the Royal Horticultural Society's garden at Chiswick should include collections of all manner of ornamental plants, fruits and vegetables, both hardy and tender, and facilities for hybridization by which to produce novel cultivars.[68] Instruction should be given in the theory and practice of horticulture, and the students, who required to be recommended by a RHS Fellow, examined twice a year.[69] The final examination after three years should also include a test of general knowledge.[70] Most of these recommendations were gradually put into effect.[71]

The RHS's Committee would have allowed Fortune and Moore to get to know one another better. They were both protégés of John Lindley (**Box 2.1**), and seem to have got on reasonably well.[72] Since the Chelsea Physic Garden is just over 2 km from Gilston Road, it seems more than likely that Fortune would have visited his old haunt. However, as the population of London increased in the course of the 19th Century[73], so had the problem of sewage (not to forget the animal droppings from the horse-drawn omnibuses, and cattle being driven to market). At low tide the air in the Chelsea Physic Garden adjoining the Thames must have been filled with the stench of rotting/fermenting organic matter. As diseases such as cholera, typhus and malaria (literally "bad air") were then attributed to this "miasma", it was not considered a healthy place to be.[74] As London expanded and the transportation costs for night soil rose, the farmers turned to guano to fertilize their fields.[75] As a result the cesspits were not emptied regularly.[76] This problem was compounded by an increase in the number of flush toilets after the Great Exhibition of 1851, which caused the cesspits to overflow.[77] All the untreated sewage and industrial waste eventually found its way into the Thames, and floated up and down river on the successive tides.[78] Writing in 1840 the building contractor Thomas Cubitt (1788 – 1855) stated that "The Thames is now made a great cesspool instead of each person having one of his own."[79] In 1855 Michael Faraday (1791 – 1867) in a letter to "The Times" commented on the foul condition of the Thames, and warned of the dire consequences of a hot season.[80] He did not have to wait long to see his prophecy come true. In the warm summer of 1858 the stench of the rotting/fermenting mass became unbearable.[81] The "Great Stink" finally forced the government to act.[82] Joseph Bazalgette (1819 – 1891; **Fig. 10.23**), the chief engineer of the Metropolitan Board of Works, was asked to draw up a scheme to build large sewers to carry the waste further downriver.[83]

Fig. 10.23 Joseph Bazalgette (1819–1891), who designed London's sewers.

There was now talk of building the Chelsea Embankment. However, this would cut off the Chelsea Physic Garden from its river frontage, and lead to a loss of bargehouse letting fees.[84] In fact, the Chelsea Embankment was the last of the three Thames embankments to be built. Started in July 1871, it took three years to complete.[85] As it turned out, the garden gained land by the project, and was able to re-use the bricks from two of its bargehouses to build a storeyard wall.[86]

Fortune's contribution to the Royal Horticultural Society's Committee on the Improved Education of Gardeners (1864 – 1865) was apparently appreciated, because after a preliminary meeting attended, among others, by John Standish (**Box 5.15**) and Thomas Moore (**Box 10.1**), he and Sir Charles Wentworth Dilke, William Paul and James Veitch Jr. were called upon to serve on the Executive Committee of The International Horticultural Exhibition and Botanical Congress, held on the site of the 1862 Exhibition from 22 to 31 May 1866.[87] Fortune's admiration for the much-travelled geologist Sir Roderick Murchison (1792 – 1871), one of the congress vice-presidents, may have played a role in his acceptance.[88] Not to be outdone by Sir Roderick, Fortune also made a donation of 5 guineas.[89] Although the Veitch Nurseries creamed off many of the horticultural prizes, including one for Japanese novelties, Fortune would have been pleased with Standish's and Glendinning's awards.[90] He must have felt proud that a number of prizes were awarded for *Aucuba japonica* in berry (**Fig. 10.24**), which

Fig. 10.24 *Aucuba japonica* in fruit, made possible by the introduction of a male plant by Robert Fortune.

relied on the male plant he had introduced from Japan (see Chapter 8).[91]

There were also three separate watercolours of *Lamprocapnos spectabilis*, which Fortune had introduced from China.[92] Fortune was no doubt delighted that the profits of the exhibition were used by the Royal Horticultural Society to buy the library of his late mentor, John Lindley (**Box 2.1**).[93]

Fortune sometimes attended the Tuesday meetings of the Royal Horticultural Society, and was asked on one occasion by George Fergusson Wilson (1822 – 1902), Chairman of the Fruit Committee, to contribute an exposé on the kumquat.[94] This hardy orange was later to be called *Fortunella* in honour of the plant hunter.[95]

In the last ten years of Fortune's life, it proves difficult to trace his movements in any detail. As a colleague later remembered "... he occasionally came among his old friends and companions ..."[96] He may have made a conscious effort to spend more time with his family, which he had neglected for so long, spending part of each year with his eldest son at Elphinstone Tower Farm, and no doubt visiting some of his siblings in the Borders.[97] To get there he would have taken a Great Northern train from London King's Cross (**Fig. 5.112**) to York, where he had to change to the North Eastern line to Edinburgh.[98] In those days the only source of light in the compartments was a dim oil lamp, while the lack of toilets and heating necessitated frequent stops at stations where the portable foot-warmers could be swapped.[99]

When he was back in London, he probably took his youngest children on outings to places like Battersea Park (**Fig. 10.25**), that had been opened in 1858 by Queen Victoria.[100] They would have walked along Beaufort Street, where Sir Thomas More (1478 – 1535) had once lived, and crossed the picturesque Old Battersea Bridge (1771 – 1885; **Fig. 10.26**) immortalized by painters such as Walter Greaves, John Atkinson Grimshaw, Joseph M. W. Turner and James McNeill Whistler.[101]

Fig. 10. 25 Battersea Park, which was opened by Queen Victoria in 1858.

Fig. 10.26 Old Battersea Bridge, a painting by Walter Greaves (1874).

The park, which was located on a point bar of the Thames, was prone to flooding.[102] Before the marshland could be converted into a park, the Battersea Fields were covered by a layer of earth excavated from the London Docks.[103] John Gibson (1815 – 1875; **Fig. 10.27**), a protégé of Sir Joseph Paxton (**Box 5.14**), was responsible for landscaping the garden, that included a boating lake (**Fig. 10.28**) and a sub-tropical garden (**Fig. 10.29**), inspired by Gibson's visit to India in 1836 – 1837.[104] Today one of the sub-tropical elements in the park is the palm *Trachycarpus fortunei* (**Fig. 10.30**). Fortune was pleased to note that *Forsythia viridissima*, which he had introduced from China in 1844, had been extensively planted in the park.[105]

Fig. 10.27 John Gibson (1815–1875), a protégé of Sir Joseph Paxton, was responsible for landscaping Battersea Park.

Fig. 10.28 South shore of lake, Battersea Park.

Fig. 10.29 Banana plant in the Subtropical Garden, Battersea Park.

Fig. 10.30 *Trachycarpus fortunei* in Battersea Park.

The Cremorne Pleasure-gardens (1836 – 1877; **Fig. 10.31**) were even closer to Gilston Road, with a gate on King's Road.[106] They were set in the sumptuous park of the late Thomas Dawson, Viscount Cremorne (1725 – 1813), which could easily be reached from London by horse-drawn omnibus or steamboat.[107] However, they were run like a fun fair with acrobatics, marionettes, dancing, ballets, pageants, farces, fireworks, and balloon ascents to attract the masses, that milled about.[108] During the day they were reasonably sedate, but after 10 p.m., they became the stamping ground of women of doubtful character.[109] The Fortunes probably considered the gardens to be unsuitable for their children, as they had "acquired the reputation for being the resort of all the rowdies in the neighbourhood".[110] Under the influence of drink, fights would break out, so the gardens had to be

Fig. 10.31 Remaining fragment of Cremorne Pleasure-gardens (1836–1877).

policed.[111] It was not an uncommon sight to see a policeman in hot pursuit of a hooligan.[112] The din of the firework displays, which lasted until closing time, must have disturbed many an evening in Gilston Road.[113] After the gates closed at midnight, the neighbours were kept awake by the rabble in the streets.[114] Complaints were lodged.[115] When the lease expired in 1877, the pleasure gardens were finally closed down to the relief of the Chelsea inhabitants.[116] That part of Edith Grove south of King's Road was later built over the site.

10.2 The Emergence of Modern Japan

While his children were growing up, Fortune would almost certainly have kept abreast of events in the Far East, which by the mid 1860s were reaching boiling point. On 29 August 1866 the young shogun died—possibly poisoned—and was replaced by TOKUGAWA Yoshinobu (1837 – 1913).[117] Just 5 months later, on 30 January 1867, the Japanese emperor died under mysterious circumstances and power was transferred to the 14-year-old Prince Mutsuhito (**Box 10.4; Fig. 10.32**), who became Emperor Meiji on 3 February 1867.[118]

Fig. 10.32 Emperor Meiji (1852 – 1912), who heralded in the modern era in Japan.

Box 10.4

Emperor Meiji (1852 – 1912) Born as Sachinomiya ("Prince Sachi"), son of Emperor Kōmei (1831 – 1867) and NAKAYAMA Yoshiko (1835 – 1907), a daughter of Kōmei's major counsellor, NAKAYAMA Tadayasu (1809 – 1888).[1] Soon after birth, in line with court etiquette, he was adopted by ASAKO Nyogo (1833 – 1898), Kōmei's principal consort.[2] The two became very attached to one another.[3] Following his adoption he was then proclaimed Kotaishi ("Crown Prince") on 11 November 1860, and given his adult name Mutsuhito.[4] His formal education had already started in 1859 with calligraphy lessons and the study of the Confucian classics.[5] As he himself admitted, he found writing boring.[6] This might explain why he wrote hardly any letters and never kept a diary.[7] However, encouraged by his father, he loved to express his feelings about current events in tanka poems.[8] It is estimated that during his lifetime, he wrote more than 100,000 poems![9] For someone with a good education, it is perhaps surprising that he only occasionally read books or newspapers.[10] However, as a poem written in 1905 shows, he was irritated by the errors which appeared in print.[11] For information about world events, he tended to rely on his counsellors.[12] After his father's untimely death, and his ascension to the throne in 1867, Emperor Meiji ("Enlightened Rule", his posthumous name) continued his classical education, which did not include any foreign languages or information about world leaders until 1871.[13] While he was fascinated by the new knowledge, he never discarded traditional values.[14] For instance, although he realized that western medicine had its good points, he still favoured traditional medicine.[15]

He got his first glimpse of the wider world when he left the confines of his palace on 25 February 1868.[16] On 23 March 1868 he took a further step by granting his first audience to foreign diplomats, something his xenophobic father would never have considered.[17] He soon conquered his shyness, and learned to shake hands, smile and make small talk.[18] At the first of these meetings he appeared in Japanese fashion with painted eyebrows, rouged cheeks and blackened teeth.[19] His official coronation, which had been postponed because of unrest, took place on 12 September 1868.[20] Instead of the traditional "Chinese" robes, Meiji was dressed like a Shintō priest.[21] Once this obligation was out of the way, a visit to Tokyo ("Eastern Capital") was planned. Accompanied by over 3300 retainers the emperor left Kyōto on 6 November 1868.[22] On 14 November the emperor caught his first glimpse of the Pacific Ocean, followed by the sight of snow-capped Mt. Fuji on 20 November.[23] He arrived in Tokyo on 26 November 1868, where his visit was celebrated in December by a two-day long binge paid for by the emperor.[24] On 5 and 6 January 1869 he met the foreign diplomatic corps.[25] On 20 January the emperor started back to Kyōto, arriving there on 5 February 1869.[26] Unlike his predecessors, who were virtual prisoners in their Kyōto palace, Meiji travelled to different parts of the country to acquaint himself with the lives of his subjects.[27] It is characteristic of the man that when he was advised to protect himself with mosquito netting, he replied, "The whole purpose of this journey is to observe the suffering of the people. If I did not myself experience their pains, how could I understand their condition? I do not in the least mind the mosquitoes."[28] He also attended numerous military manoeuvres, often getting dirty in the process.[29] As he probably considered that he had to set an example, he stoically suffered the hardships of the soldiers without complaining.[30] Luckily he had a strong constitution, so could afford to pay little attention to his health.[31] He had an aversion for doctors and dentists, and particularly disliked being examined.[32] In 1877 and 1883 he suffered from beriberi, caused by a lack of vitamin B in his diet.[33] In 1888 and again in 1894/1895 he was *hors de combat* as a result of pneumonia.[34] In 1891 he was confined to bed for 40 days with influenza.[35] Since 1904 he suffered from diabetes and by 1906 he had chronic hepatitis.[36] He bore this suffering stoically.[37] Towards the end of his life his kidneys started to malfunction, causing a weakening of the heart with an irregular pulse, which made climbing exhausting, and a tendency to fall asleep during the day.[38]

He was married to the talented ICHIJŌ Haruko (Empress Shōken, 1849 – 1914), who despite her diminutive size was to prove a tower of strength in the changing times.[39] Starting in 1886 the empress fulfilled Meiji's appointments when he was indisposed.[40] In order to emphasize her new role, she wore western dress, a fashion which soon caught on.[41] Being open to western ideas, she probably helped the emperor to adapt to the numerous changes which occurred during his lifetime. The marriage remained childless, but Emperor Meiji had 15 children by five different concubines.[42] Many of the children died of meningitis at an early age.[43] Although it would have been natural for Meiji to be particularly protective of the five that survived, he showed little outward affection for them.[44] He repeatedly refused to give them an audience.[45] On the other hand, he treated the Korean Prince Yi Eun (1897 – 1970), who came to study in Japan, like a son.[46]

The emperor was a modest man, who detested having his photograph taken and was reluctant to spend money on his clothes or the private quarters in his palace.[47] He was devoted to duty, and regularly attended cabinet meetings up until two weeks before his death.[48] Although he rarely made any comments[49], his opinion was sought when the government was divided on an issue.[50] After all, a decision made by the respected emperor was more likely to be accepted by the Japanese people. Moreover, he personified stability at a time of innumerable cabinet changes. Furthermore, by cultivating contacts with monarchies throughout the world[51], he paved the way for the general acceptance of Japan. By the time of his death from heart failure, Japan had emerged as one of the great powers on the world stage.[52]

As his position had become untenable, the new shogun relinquished power in November 1867.[119] From then on Emperor Meiji not only wielded spiritual but also worldly power. However, fearing for their fiefs and a loss of prestige, the daimyos loyal to the shogun advocated civil war.[120] The shogun retracted his resignation and sent some of his followers to attack Kyōto (Battle of Toba-Fushimi, the start of the Boshin ["Earth dragon"] War).[121] After four days of fighting in the suburbs of the imperial capital they were soundly defeated by the emperor's less numerous but better equipped troops on 30 January 1868 (**Fig. 10.33**).[122]

Fig. 10.33 Battle of Toba-Fushimi, near Kyōto, in January 1868, a contemporary illustration.

Fig. 10.34 SAIGŌ Takamori (1828–1877), known as "the last samurai". He initially fought for the emperor, but later led the Satsuma Rebellion.

Fig. 10.35 Battle of Ueno on 4 July 1868.

Fig. 10.36 Monuments to the fallen at the Battle of Ueno.

The shogun's forces retreated northwards pursued by the imperial army under SAIGŌ

Takamori (1828 – 1877; **Fig. 10.34**).[123] In Edo the shogun sought refuge in the family temple of Kan'ei-ji, protected by some 2000 soldiers.[124] About 300 of his troops died in the subsequent Battle of Ueno on 4 July 1868 (15 May 1868 of the lunar calendar; **Fig. 10.35**).[125] A small stone erected by a Kan'ei-ji priest in 1869 and a larger monument in 1874 built by OGAWA Okisato, one of the survivors, mark the spot where the cremated remains of the fallen were buried (**Fig. 10.36**). The shogun retired to Mito, where (between siring tens of children) he devoted himself to archery, hunting, polo, photography and oil-painting.[126]

Part of the shogun's forces with their French military advisors fled to the island of Ezo ("Barbarian Land").[127] After eliminating a little local resistence, they established the Republic of Ezo modelled on US lines on 27 January 1869 according to the Gregorian calendar.[128] The winter over, imperial troops landed in mid-May 1869 and progressively overran the republican positions and attacked the rebel headquarters (Fortress of Goryōkaku).[129] On 27 June 1869 (lunar calendar: 17 May 1869) the Republic of Ezo became a thing of the past.[130] In order to eraze any memory of the republic, the island was renamed Hokkaidō ("Northern Sea District") on 20 September 1869 (lunar calendar: 15 August 1869).[131]

The transfer of power from the daimyos to the central government meant that almost two million samurai became redundant.[132] This caused much hardship, and led to a number of samurai insurrections between 1874 and 1877, the most important of which was the Satsuma Rebellion in 1877 led by SAIGŌ Takamori (**Fig. 10.34**).[133] What started out as an offensive campaign aimed at the centre of power in Tokyo, got bogged down at Kumamoto, and soon turned into a rearguard action.[134] The few remaining samurai rebels (300 – 500) were eventually cornered on a hill overlooking Kagoshima. Even after their numbers had been further reduced by a bombardment in the early hours of 24 September, they fought bravely against the overwhelming odds.[135] In a final suicidal attack the handful of swordsmen hurled themselves on the Imperial Army only to be mown down by imported US Gatling guns.[136]

The failure of the Satsuma Rebellion is in many ways symbolic of the far-reaching changes that had taken place in Japan since Fortune visited the country. Fortune must have been surprised to learn that commoners were encouraged to take family names and permitted to wear hakama (loose trousers) and haori (tunics), which were formerly reserved for the samurai.[137] These soon made way for western clothes.[138] After 12 November 1872 western dress was adopted for all court and official ceremonies.[139] When Satsuma samurai appeared in Tokyo in 1873 in their traditional dress, they were stared at by bypassers.[140] By 1870 commoners were also allowed to ride horses or make use of the recently-invented jin-riki-sha ("human-powered vehicle"; **Fig. 10.37**).[141]

Fig. 10.37 Jin-riki-sha (Human-powered vehicle) in Kamakura.

Two years later some 40,000 rickshaws were operating in Tokyo.[142] By then there was even a train connection between Yokohama and the Tokyo suburb of Shinagawa.[143] On 1 January 1873 the lunar calendar of Chinese origin was replaced by the Gregorian Calendar in order to facilitate communications with the outside world.[144] Communications were also improved by the establishment of a postal service in 1871, the brainchild of MAEJIMA Hisoka (1835 – 1919), who had been sent to the UK in 1870 with this in mind.[145] The Japanese also turned to the West for ideas on how to reorganize their educational system.[146] In 1872 all children over the age of six had to attend school for 16 months.[147] By the time of Fortune's death in 1880, this had been increased to three years.[148] Unfortunately, to start with there was a shortage of teachers with the necessary background, and the school texts, which were often translations of western books, proved unsuitable.[149]

During the Edo Period (1603 – 1868) citizens had to register annually with a Buddhist temple in order to prove that they were not Christians.[150] In return for the certificate, households (danka) were expected to pay for the upkeep of the temple and its Buddhist monks, and provide free labour.[151] Without the certificate it was impossible to get married, travel or apply for certain jobs.[152] Moreover as a hinin (non-person), one ran the risk of being executed as a Christian.[153] This exploitation created widespread resentment[154], so when the Edo Period gave way to the Meiji Restoration popular feeling against the oppressive shogunate and its foreign religion erupted in a movement referred to as Haibutsu Kishaku ("Abolish Buddhism and destroy Buddhist effigies").[155] In their enthusiasm for all things western, a number of Japanese turned to Christianity.[156] Although Fortune would have welcomed the fact that after 1873 Christians were no longer persecuted for their beliefs, he was probably dismayed at the widespread destruction of Buddhist temples, images and texts, and the concurrent rise of Shintoism as the state religion.[157]

However, he was probably more disturbed by the first signs of militarism as a means of enhancing Japan's resources, something the Japanese copied from the foreign powers.[158] As from 10 January 1873 all males over the age of 20 had to serve in the army for 3 years followed by 4 years as a reservist.[159] Moreover, in 1872 they annexed the Ryūkyū Islands[160], and two years later invaded Táiwān on the pretext that some of "their" Ryūkyū fishermen had been murdered in 1871 by the head-hunting aborigines.[161] They only withdrew temporarily after China paid for the cost of the Japanese expedition and compensated the relatives of the murdered men.[162] On 20 September 1875 the Japanese Navy provoked an incident by sending a small Japanese warship up the Han-gang River, 45 km from Seoul.[163] The outcome was a "Treaty of Friendship" on 27 February 1876 similar to that which Commodore Matthew Perry had exacted from Japan in 1854, i.e. right to free trade, permission to enter and survey coastal waters, and extraterritorial rights for Japanese citizens.[164] This represented the first step towards a complete takeover of Korea in 1910.[165]

10.3 China's Continuing Contentions

As Fortune would have noticed, change in China took much longer for a number of fundamental reasons:

1. After the Anglo-French Invasion in 1860 the country found itself saddled with a hefty indemnity as stipulated in the Treaty of Tīanjīn (**Box 9.1**).

2. This left less money to combat the revolutions which had broken out in many parts of the country (**Boxes 10.5–10.9**).

Box 10.5

Niǎn Rebellion (1851 – 1868) This rebellion was started by a handful of mafiosi referred to as "Red-beard Bandits".[1] Because they were local people, the bandits painted their faces black and beards red to conceal their identity.[2] In exchange for provisions or financial support members of a village were offered protection against raids by neighbouring mafiosi (Niǎn = "group" "unit" or "horde").[3] Life in these villages tended to be peaceful.[4] Those villages which did not pay protection money were repeatedly plundered, and the inhabitants raped, kidnapped or killed.[5] In some districts the Niǎn bribed their way into government service, and were able to warn other gang members of any plans to arrest them.[6] Prior to 1851 the Niǎn were active in parts of Ānhuī, Héběi, Hénán, Húběi, Jiāngsū and Shāndōng.[7] Until then the movement was of little significance and generally ignored by the central government.[8] However between 1851 and 1857 the Huáng Hé ("Yellow River") changed its course, causing flooding in some areas and drought in those parts of Jiāngsū deprived of water.[9] As the Qīng Government did nothing to repair the breach in the dyke, anti-government sentiments grew, which swelled the ranks of the Niǎn.[10] The Niǎn also won the support of the local population by sharing some of their booty.[11] Increasingly, the Niǎn targeted government granaries, treasuries and pawnshops, and held rich people to ransom.[12] Government forces, that had no way to distinguish between innocent farmers and Niǎn adherents, often over-reacted by massacring indiscriminately, which drove the inhabitants to side with the rebels.[13] The movement also seems to have been stimulated by the arrival of the Tàipíngs (**Box 5.19**) in the region in 1853, and there was a certain amount of collaboration between the two movements.[14] Like the Tàipíngs the Niǎn showed their opposition to the Qīng Government by wearing their hair long.[15] There were other similarities to the Tàipíngs, e.g. the allocation of titles, use of similar flags and uniforms.[16] However, unlike the Tàipíngs the Niǎn had no religious ideology, smoked opium, tried to avoid fighting pitched battles, preferring to use their cavalry for surprise attacks.[17]

Seeing how successful the well-organized Tàipíng were, the Niǎn decided to follow suit. At a meeting of the various Niǎn chiefs at Chíhé ("Ponded River", Ānhuī) in 1853, they agreed to coordinate their actions.[18] They elected the illiterate but none the less capable ZHĀNG Luòxíng (CHANG Lo-hsing, 1811 – 1863) to be their leader.[19] Under ZHĀNG's leadership more and larger units were established, which enabled the Niǎn to take on government forces successfully.[20] Moreover, the increasing use of horses after 1853, made the Niǎn much more mobile.[21] Although ZHĀNG was officially in charge, many of his chiefs had their own agenda.[22] Some like the ambitious MIÁO Pèilín (1798 – 1863) and the wily LǏ Zhāoshòu switched allegiances regularly, depending on which side was offering the best prospects for advancement.[23] This weakened the Niǎn Movement considerably. On the other hand, the Niǎn received additional support from a number of Tàipíng leaders like QĪU Yuāncái (CH'IU Yüan-ts'ai), FÀN Rǔzēng (1840 – 1867) and LÀI Wénguāng (1827 – 1868) after the fall of Nánjīng in 1864.[24] The Tàipíng influx not only reinforced Niǎn numbers, but also improved Niǎn strategy greatly.[25] Without them the Niǎn might have been suppressed much earlier.[26]

By luring the Qīng troops into following them until exhausted and then suddenly striking, the Niǎn cavalry successfully combatted the famous Mongolian general Sēnggélínqìn (1811 – 1865), who had been sent against them in 1860.[27] After Sēnggélínqìn was killed in an ambush, he was replaced by the Chinese general ZĒNG Guófān (1811 – 1872), who had crushed the Tàipíng Rebellion (**Box 5.19**).[28] By erecting a number of fortified towns around the Niǎn area and enacting a scorched-earth policy, ZĒNG attempted to quell the Niǎn by containment rather than pursuit.[29] These fortifications offered the population protection from the marauding Niǎn, so the population tended henceforth to side with the government.[30] However, as ZĒNG's tactics did not yield immediate results, he was criticized in government circles.[31] In October 1866, a month before ZĒNG handed over command to his pupil LǏ Hóngzhāng (1823 – 1901; **Fig. 10.46**), he managed to split the Niǎn into an eastern group (Dōngniǎn) and a western group (Xīniǎn).[32] As the Niǎn's freedom of movement became restricted, they were forced to fight more frequently against the better-armed government forces.[33] Although the eastern group broke across the Grand Canal and threatened the Treaty Port of Yāntái, they were finally pushed back to the mouth of the Mí Hé ("Overflowing River") where they were annihilated on 24 December 1867.[34] The western group under ZHĀNG Zōngyǔ ("Small King of Hell") went to the aid of Muslim rebels fighting government forces in northern Shǎnxī, but when they got there they were attacked from the rear by the troops of DǑNG Fùxiáng (1839 – 1908) (**Box 10.6**).[35] Then in December 1867 they received an appeal for assistance from the eastern Niǎn.[36] By launching an attack on Běijīng they attempted to draw off Qīng forces that were attacking their comrades.[37] They got as far as the Marco Polo Bridge on the western outskirts of Běijīng, causing panic in government circles, before they were driven further east.[38] Once they arrived in Tiānjīn, they discovered that the eastern group, with whom they intended to reunite, no longer existed.[39] It was only a matter of time before they themselves were cornered near Chípíng ("Chí Mountain Plain", NE of Dōngchāngfǔ ("East Prosperous Prefecture", Shāndōng) and wiped out on 16 August 1868.[40] To avoid capture ZHĀNG Zōngyǔ jumped into the Túhài River and was presumed drowned.[41]

Box 10.6

Xián-Tóng Rebellion ("Miáo Rebellion", 1854 – 1873) A series of local revolts in the impoverished province of Guìzhōu, which took place during the reigns of the Emperors Xiánfēng and Tóngzhì.[1] The appelation "Miáo Rebellion" is misleading, as other ethnic groups such as the Bùyī, Dòng, Hàn, Huí and Shuǐ were also involved.[2] The chief causes of these revolts were the confiscation of land by the government, high taxation and other official abuses, which pushed the people below the subsistence level.[3] Some of the Miáo became so destitute that they dug up their family graves to obtain silver funerary ornaments with which to pay their taxes.[4] Faced with starvation, the peasants had nothing to lose and much to gain by opposing the excessive demands of a government that provided little in return.[5] The early successes of the Tàipíngs (**Box 5.19**), which caused the government to raise taxes once more, acted as catalysts for the anti-Qīng feelings which had been simmering for some time.[6] However, unlike the Tàipíngs, the Guìzhōu tribesmen tended to employ guerrilla tactics.[7] While there was a certain amount of cooperation between the rebel bands, each rebel group had its own agenda.[8] Some groups like the Báihào ("White Signals") were only interested in obtaining booty, while the Miáo attempted to hold the towns which they had captured.[9] Because of these different attitudes, it was not unknown for the rebels to fight among themselves.[10] When a town was besieged by the rebels for any length of time, the population starved, and was often forced into cannibalism.[11] Children, who had died of starvation, and the corpses of criminals were consumed, while fresh graves were robbed of their contents.[12]

Although the government often granted financial aid to put down the revolts, only some of it reached its proper destination.[13] The local Qīng army was therefore too small and too badly equipped to deal with all the revolts.[14] Governor ZHĀNG Liàngjī (1807 – 1871) likened it to putting out a fire with a glass of water![15] Moreover, because the province was basically bankrupt, the soldiers went for a long time without pay, so they were less keen on fighting, and more interested in looting.[16] When a town was retaken, the inhabitants often suffered badly at the hands of the government forces.[17] This alienated the inhabitants, who raised their own militia as a means of protecting themselves and their property.[18] Once the army had left to deal with a revolt in another part of the province, the militia could rise against the ineffectual government with impunity.[19] This explains why the rebellion lasted so long.

Once the Tàipíng Rebellion (**Box 5.19**) was largely suppressed in 1864, more money became available to deal with the other uprisings. However, Guìzhōu was not high on the central government's list of priorities.[20] In 1868 better equipped and more disciplined armies from Sìchuān and Húnán were finally brought in to deal with the rebels.[21] This turned the tide of battle in the government's favour.[22]

When the fighting was over up to 70% of the total population of Guìzhōu was dead or missing.[23] For decades many towns remained no more than heaps of rubble, while large tracts of the province were depopulated.[24]

Box 10.7

Panthay Rebellion (1856 – 1874) In the latter part of the 18th Century Běijīng tried to consolidate its grip on the southwestern province of Yúnnán by encouraging Hàn from all over China to emigrate to Yúnnán.[1] They took over land belonging to the local population to grow mushrooms and other plant produce.[2] This generated a certain amount of animosity with the Yúnnánese Hàn, the largely Muslim Huí, and other ethnic minorities.[3] Between 1796 and 1856 there were 70+ local disturbances involving long-time residents and the new arrivals.[4] The fact that the Huí were successful miners and had a large stake in the caravan trade meant that they stood in the way of the newcomers' aspirations.[5] With the connivance of local officials, who doubted the loyalty of the native population, the immigrants started what can only be described as a campaign of genocide.[6] The inhabitants of whole Huí townships were wiped out. At first the Huí appealed to Běijīng to mediate, but when they discovered that this was in vain, they took the law into their own hands.[7]

The Hàn, used to having it their own way, were caught off guard when the Huí finally counterattacked in August 1856.[8] After capturing Dàlǐ ("Great truth"), the charismatic DÙ Wénxiù (1823 – 1872), with his knowledge of Chinese history, law and bureaucracy, was chosen to lead the Huí cause.[9] All attempts to dislodge DÙ from Dàlǐ failed.[10] From his base in Dàlǐ, DÙ extended his influence over most of western Yúnnán and brought peace to the region.[11] Trade is said to have flourished.[12] During his reign as sultan, DÙ managed to gain the support of numerous ethnic groups such as the Bái, Dǎi, Hānī, Jǐngpō, Lìsù, Yí, Zhuàng, who probably viewed the Huí as the lesser of two evils.[13] DÙ offered a position in his government to the fervent Muslim leader MǍ Rúlóng (1832 – 1891), who had had considerable success in southern Yúnnán.[14] Preferring to retain his independence, MǍ refused the post.[15] MǍ may have had other plans. In 1862, while he was besieging Kūnmíng ("Offspring Míng") for a second time, he was persuaded by CÉN Yùyīng (1829 – 1889) to switch sides in exchange for a position as regional commander of his birthplace Jiànshuǐ ("Establish Lake").[16] The Huí were not only deprived of an important tactician, but were split into two factions.[17] This sealed the outcome of the rebellion.[18] After MǍ Rúlóng set off for Jiànshuǐ in the winter of 1862, Kūnmíng was taken over by other Huí forces under MǍ Liánshēng and MǍ Róng in March 1863.[19] MǍ Déxīn (1794 – 1874), who had travelled widely in the Middle East, and was Yúnnán's most important Huí scholar and religious leader, became governor-general.[20] However, before the Huí had time to consolidate their success, MǍ Rúlóng returned, recaptured Kūnmíng and handed it back to Qīng officials.[21] While MǍ Liánshēng and MǍ Róng managed to escape, Rúlóng's teacher MǍ Déxīn was placed under house arrest.[22] Rúlóng then set off to capture MǍ Róng, who was brought back to Kūnmíng, where he was publically executed.[23] In 1863 and again in 1864 Rúlóng tried unsuccessfully to negotiate DÙ's surrender by sending first MǍ Fùtú and then MǍ Déxīn to Dàlǐ.[24] In 1867, in an effort to turn the tide in their favour, DÙ's forces marched on Kūnmíng, which was the only major Qīng stronghold left in Yúnnán.[25] Rúlóng's men started deserting him.[26] To make things worse, Rúlóng's troops were being decimated by plague.[27] In early March 1868 DÙ's army began to close in on Kūnmíng.[28] The city was finally saved by CÉN Yùyīng, who broke the blockade with his militia.[29] With more financial support from Běijīng, CÉN's protégé YÁNG Yùkē (1838 – 1885) managed to push DÙ's forces out of central Yúnnán.[30] However it took the government troops another four years of fierce fighting to reach DÙ's powerbase. On Christmas Day 1872, DÙ's top general advised him to surrender.[31] The next morning DÙ donned his ceremonial robes, said goodbye to his family and advisors, and proceeded to the Qīng camp in his sedan chair.[32] On the way he took a fatal dose of opium, and was dead before the Qīng commander had a chance to execute him.[33] In a semblance of respecting the sultan's last wish to spare the population, CÉN ordered a three-day pause to the hostilities.[34] Then on 29 December 1872 he held a banquet for DÙ's top generals.[35] Once seated all seventeen were decapitated.[36] In the next three days Dàlǐ's population was massacred.[37] The soldiers cut ten thousand pairs of ears from the dead, which were sent to Běijīng along with Sultan DÙ's head as evidence that CÉN's mission had been accomplished.[38] In the next few months most of the remaining centres of resistance were "mopped up".[39] However, it took until 4 May 1874 for the last Huí stronghold to fall into Qīng hands.[40] The rebellion, which had cost the lives of up to a million people, ended as it had begun with renewed violence against the Huí.[41] As a result those Huí who had somehow managed to survive sought safety in the highlands of Burma, Laos and Thailand.[42] The total population of Yúnnán, which had stood at between 8 – 10 million prior to the rebellion, fell to less than 50% .[43]

Box 10.8

Huí Rebellion in Shǎnxī and Gānsù (1862 – 1873) In the Spring of 1862, as the Tàipíng (**Box 5.19**) poured into Shǎnxī(Shaanxi), the Hàn Chinese organized militia to combat the rebels.[1] Fearing that the Chinese muslims might be aiding the Tàipíng, these militia started to slaughter the Huí.[2] As a reaction to these massacres, the Huí of Líntóng, 25 km NE of the capital Xī'ān, rose in June 1862.[3] The uprising soon spread throughout the Wèi Valley and before the month was out Xī'ān ("West Calm") itself was being besieged.[4] When General Shèngbǎo's campaign against the Huí proved ineffective, he was disgraced and replaced by the hardline Manchu general Duōlōng'ā (1817 – 1864).[5] Duōlōng'ā retook a number of rebel strongholds.[6] However, after Duōlōng'ā was fatally wounded in May 1864, the rebellion spread to various parts of Shǎnxī and Gānsù.[7] It was not until the Niǎn Rebellion (**Box 10.5**) had been suppressed that the Qīng government could pay more attention to the rebellion in Shǎnxī and Gānsù.[8] In 1866 they appointed ZUǑ Zōngtáng (1812 – 1885) as Governor-General of Shǎnxī and Gānsù, with the specific purpose of suppressing the Huí rebellion.[9] However, it was to take until September 1868 before ZUǑ's plans were finally approved.[10] The first goal was to flush out the rebels in N Shǎnxī .[11] The rebel leader DǑNG Fùxiáng (1839 – 1908) surrendered Suídé

(" Peaceful Mind") to General LIÚ Sōngshān (1833 – 1870) and thereafter fought for the Qīng.[12] By the middle of 1869 Shǎnxī and SE Gānsù had been pacified.[13] The Qīng forces could now concentrate on the rebel bases in other parts of Gānsù. Three Qīng army columns converged on Jīnjī Fortress, MǍ Huālóng's headquarters near Wúzhōng (now Níngxià) on the Huáng Hé (" Yellow River").[14] The initial attack at the beginning of 1870 was repulsed with heavy losses.[15] General LIÚ Sōngshān was killed in the attack and replaced by his nephew LIÚ Jǐntáng (1844 – 1894).[16] When food ran out and MǍ Huālóng was forced to surrender at the beginning of 1871, Jǐntáng revenged his uncle's death by razing the town, and submitting all male members of MǍ Huālóng's family and many officials to a lingering death by slicing.[17] BÁI Yànhǔ (" Big Tiger"), who had been opposed to surrendering escaped to the walled city Línxià [" Near Xià (River)"], 70 km SW of the capital Lánzhōu.[18] Although ZUǑ's initial assault on Línxià was repulsed in February 1872, MǍ Zhàn'áo (1830 – 1886) and his son Ānliáng (1855 – 1919) defected to the Qīng in order to spare the Huí population.[19] ZUǑ needed to clear the Héxī Corridor of any opposition before he could attack the rebel state of Kashgaria in Chinese Turkestan (**Box 10.9**). So before attacking Jiǔquán (" Alcohol Spring"), he had to eliminate rebel-held Xīníng (" Western Peace") on his left flank.[20] After a siege lasting three months, the city fell in December 1872.[21] The Qīng then marched on the last rebel stronghold of Jiǔquán, which on 4 November 1873 witnessed an orgy of bloodletting even worse than that which had occurred at Jīnjī.[22] Almost 7,000 Huí were executed, and the rest resettled in SE Gānsù to prevent them from collaborating with those Huí in Chinese Turkestan.[23]

Box 10.9

Muslim Rebellion in Chinese Turkestan (1864 – 1878) Chinese Central Asia was first annexed by the Qīng in the middle of the 18th Century.[1] Rather than aggravate the local population by direct rule, the Chinese acted through local muslim nobles (begs), who were responsible to the Chinese authorities.[2] However, as these officials only received small salaries, this extra layer of bureaucracy only led to more bribery, corruption and extortion.[3] Being a poor area Chinese Central Asia was initially supported by subsidies from the richer Chinese provinces.[4] However, when these subsidies were diverted to deal with other rebellions (**Boxes 5.19, 10.5, 10.6, 10.7, 10.8**), taxation and forced labour increased to breaking point.[5]

A rumour that the emperor had ordered the massacre of the local Huí sparked off a revolt in the town of Kùchē (" Storehouse Machine").[6] News of the capture of the Manchu fort and the killing of the Qīng officials in the night of 3/4 June 1864 spread like wildfire.[7] This had a domino effect with the uprising spreading spontaneously to other towns.[8] The leaders who were chosen to coordinate the rebellion were almost invariably religious men, who turned the rebellion into a jihād (holy war) against an infidel state.[9] This had the effect of mobilizing the Muslims regardless of origin and class.[10] Thus the Turkic Uyghur sooned joined in the fray.[11] The poorly-manned Qīng garrisons were no match for the fanatical Muslim fighters, as the soldiers were not regularly paid, and were consequently poorly motivated.[12] Moreover, they themselves often belonged to the same ethnic minority as the rebels.[13]

However, once the common enemy had been exterminated or driven away, fighting broke out among the Muslims themselves.[14] The Kùchēan leader Rāshidīn Khoja attempted to extend his domain throughout the Tǎlǐmù Péndì (Tarim Basin) by sending an expeditionary force eastwards to secure the towns north of the Tǎkèlāmǎgān (Taklamakan) Desert, and others westwards in the direction of Shāchē (Yarkant, "Vast land" in Wéi language).[15] However, because the Huí soldiers were poorly armed and ill-trained, they failed to subjugate cities like Shāchē, Kǎshí and Hétián ("Harmonious land").[16] With his more professional Uyghur army the ruthless Ya'qūb Beg (BǍI Ā'gǔ, ca. 1820 – 1877; **Fig. 10.38**) routed the much larger Kùchēan Army at Khan Ariq (Hǎn Nán Lì Kè), 45 km SE of Kǎshí in the summer of 1865.[17] With the help of 7,000 battle-seasoned troops from his native Kokand (now Uzbekistan) he was then able by June 1867 to subdue further resistance and create an independent Muslim state under his leadership.[18] With the fall of the Huí cities of Tùlǔfān (Turpan, "Plentiful place" in Huí language) and Wūlǔmùqí (Urumqi, "Beautiful pasture" in Mongolian) in November 1870, he managed to secure the eastern border of his territory for the time being.[19]

Fig. 10.38 Portrait of Ya'qūb Beg (ca. 1820 – 1877), who established an independent Muslim state (1867–1877) in what is now Xīnjiāng.

In order to prevent challenges to his leadership, Beg only delegated minor matters to his subordinates, deciding all important issues himself.[20] He personally appointed and promoted the army officers, many of whom were from his native Kokand.[21] While the mixed ethnicity of his soldiers made communication difficult, it acted as an insurance against insurrection.[22] In order to guarantee his country's independence, Beg played the superpowers Russia and Britain off against one another.[23] He first signed a commercial agreement with Russia in 1872, and then a treaty with Britain in 1873.[24] In order to establish the legitimacy of his government in the eyes of his countrymen and the world, he offered his allegiance to the Ottoman Empire in 1873, which was then regarded as the leading light in the Islamic world.[25] In return the Ottomans supplied Beg with weapons and military advisors.[26] Unfortunately, the cost of maintaining his army of 30,000 to 40,000 soldiers to ward off a Qīng attack alienated his subjects, many of whom looked forward to a return to Chinese rule.[27] Some Huí even joined the Qīng army.[28] By 1875, having crushed the Muslim rebellion in Shǎnxī and Gānsù (1862 – 1873), the vanguard of the Qīng forces was already installed in Qítái ("Seventh military stop"), just 150 km from Wūlǔmùqí.[29] Wūlǔmùqí itself fell with almost no resistance on 18/19 August 1876.[30] Until the end Beg believed that it would be possible to come to some sort of agreement with the Qīng leaders, whereby he would remain as ruler of the Muslim state of Kashgaria in return for his allegiance to China.[31] He therefore ordered his men not to fire on the Qīng forces.[32] However, as the Qīng military successes multiplied, the pacifistic fraction in the Qīng Court lost ground.[33] After Beg died suddenly of a stroke, internal disputes and a succession struggle broke out among Beg's followers, which developed into a civil war.[34] With the Muslim resistance severely weakened, the Qīng retook the towns and cities in quick succession.[35] The reconquest was complete with the fall of Hétián on 2 January 1878.[36] Beg's body was exhumed and publically burned, and the surviving members of his family were either killed, castrated or died in prison.[37] Many Huí and Uyghur fled to Kyrgyzistan, Kazakhstan and Uzbekistan.[38] After the suppression of the rebellion, China tightened its grip on Chinese Turkestan, which was given provincial status in 1884, and became known as Xīnjiāng ("New Dominion").[39] By encouraging massive immigation of Hàn Chinese, the stage was set for tension and outbreaks of violence between the ethnic groups.[40]

3. Being a much larger country than Japan, it tended to be more inward-looking, and therefore less interested in the technical advances being made elsewhere. Only a handful of influential Chinese, such as Lǐ Hóngzhāng (1823 – 1901; **Fig. 10.46**) realized that China had much to learn from the west.[166]

4. Although China was faced by the same external threats as Japan, it tended to rely on its capacity to absorb the foreign influences. After all, in the past it had always managed to sinicize its invaders. The ruling Qīng Dynasty with its roots in Manchuria was living evidence of this ability.

5. Being a much larger country inhabited by diverse ethnic groups, meant that the central government had less control over its territory. The further away from the capital, the more autonomous it was and is. It therefore proved harder to push through changes. In order not to jeopardize their careers or even their lives, provincial officials tended to supply Běijīng with information that they thought it wanted to hear. Even when a major disaster struck that could not be denied, the "facts" were often couched in harmless terms. This made it very difficult for the central government to assess the situation correctly and act accordingly. As a result, change only came slowly to the Middle Kingdom.

Under the Treaty of Tiānjīn (**Box 9.1**) Christian missionaries were free to travel anywhere in China.[167] As Catholic missionaries had already penetrated far inland, this clause did not initiate any major changes on the ground. It only legalized an already existing situation. Not that the treaty gave the missionaries more than a veil of official protection. Considering the political unrest, preaching Christianity could be a risky undertaking. While the Protestant missionaries were generally content to live and work in the "pacified" coastal provinces of China, some of the unmarried Catholic missionaries often risked their lives in the wilds of western China as Fortune was at pains to point out.[168] The torture and murder of missionaries and their Chinese catechists continued after Fortune's return to Britain. There were a number of examples in Guìzhōu in 1862, 1866 and 1869 respectively.[169] The case involving Jean-Pierre Néel (1832 – 1862) was particularly gruesome. The missionary was first dragged by his hair by a galloping horse before being beheaded along with his host and three of his Chinese catechists.[170] Anti-Christian sentiments ran high at Yǒuyáng (E Sichuan).[171] François Mabileau (1829 – 1865) was sent to investigate.[172] He had been there less than a month when a mob descended on his house, dragged him to the river, tried to drown him, but when this proved unsuccessful, showered blows on his head.[173] After his murder he was replaced by Jean-François Rigaud (1834 – 1869), who was stabbed to death and beheaded in the mission less than 4 years later.[174] His assistant Jean Hue (1837 – 1873) escaped, but to the surprise of many locals returned a few months later.[175] Learning that the inhabitants of Qiánjiāng ("Black River"), 80 km north of Yǒuyáng, were also interested in Christianity, Hue and a Chinese priest Michel Tay set off on what proved to be their final mission.[176] Just over a year later Jean-Joseph-Marie Baptifaud (1845 – 1874) and some of his Chinese converts were slaughtered in Yúnnán.[177] Some of these missionaries who risked their lives to spread the gospel were also fine naturalists, who opened western eyes to China's rich biodiversity.[178]

Although Fortune dreamed of the utopia which would follow once the Chinese had been converted to Christianity[179], the tactlessness of many Protestant missionaries, who claimed that Christianity was the one true religion and Buddhism and Daoism merely superstitions, did not endear them to the majority of the population.[180] Moreover, the behaviour of the foreigners, most of whom were nominally Christians, was hardly a good advertisement for Christianity.[181] As a result most Protestant missionaries only managed to convert a handful of Chinese.[182]

As the gateway to Běijīng, Tīanjīn had suffered unduly from foreign aggression. After the Anglo-French Expedition in 1860 it had been occupied by the European powers, who commandeered large tracts of land for their concessions (**Fig. 10.39**).[183] They did this with total disregard for the local farmers and their ancestors' graves.[184]

Fig. 10. 39 European-style house in the Foreign Concession in Tiānjīn.

The French were particularly disliked for using a former imperial villa as their consulate and erecting a grand cathedral on the site of a Buddhist temple and then provocatively naming it "Notre Dame des Victoires" ("Our Lady of Victories"; **Fig. 10.40**).[185]

Fig. 10. 40 Catholic Cathedral in Tiānjīn (Wànghǎi Lóu). Built in 1869, destroyed during the Tiānjīn Massacre (1870), and again during the Boxer Rebellion (1900). It was rebuilt in 1904, and renovated after the 1976 earthquake, but is now badly in need of restoration.

No wonder that Tīanjīn became a hotbed for anti-Christian sentiments. When the nuns at the St. Vincent de Paul Convent started offering cash rewards for orphans and destitute children, rumours started to circulate that the children's eyes and hearts were being gouged out to make magic potions.[186] According to Chinese beliefs, this gave the children no chance of going to heaven.[187] On 19 June 1870 the French Consul, Henri-Victor Fontanier (1830 – 1870) was able to convince ZHŌU Jiāxūn, the Tīanjīn Dàotái, that the accusations were groundless.[188] He had less success with the district magistrate, who demanded an immediate investigation.[189] The next day CHÓNG Hòu (1826 – 1893), the Superintendant of Trade, tried to calm the situation, and eventually persuaded the consul to allow an inspection in order to disprove the charges.[190] Father Claude-Marie Chévrier (1821 – 1870) agreed to inform the local authorities about any future deaths, so that the cause could be jointly verified.[191] However, when the dàotái, the prefect and the district magistrate came to carry out their investigation, the French Consul objected that he had not been given advanced warning.[192] Infuriated he rushed into the office of CHÓNG Hòu to demand an explanation, attempted to shoot the superintendant, and when some of CHÓNG's attendants tried to restrain him, broke loose and rushed out into the street.[193]

When the district magistrate, whom he accused of fomenting the trouble, approached him, the irate consul attempted to shoot him, and inadvertently killed one of his attendants.[194]

The crowd which had gathered went berserk, beating the consul and his assistant to death, stripping them naked and mutilating their corpses before throwing them into a nearby canal.[195] After burning down the French Consulate, and killing all its inmates, the mob proceeded to the orphanage (**Fig. 10.41**), where the ten nuns were speared, stabbed, cut down by swords, or felled by axes.[196]

Marie-Anne Pavillion (1823 – 1870) and Marie-Aimée Tillet (1836 – 1870) were roasted over a fire, while Marie-Séraphie Clavelin (1822 – 1870), who was in charge of the orphanage's pharmacy, had her eyes gouged out and her heart torn from her body while still alive.[197] Two of the nuns were then impaled on lances and put on display at the entrance to the orphanage.[198] During the frenzy, 30 – 40 Chinese employees and even some of the orphans were killed.[199] Father Chévrier and his Chinese assistant Père HÚ Vincent (1821 – 1870) were killed and the cathedral burned down (**Fig. 10.42**).[200]

Fig. 10.41 The gutted chapel of the St. Vincent de Paul Convent, Tiānjīn, where 10 nuns were brutally murdered in 1870.

Fig. 10.42 The shell of the Catholic Cathedral in Tiānjīn after the fire in 1870.

Although the mob were after the French, the Roman Catholics and their Chinese converts, other foreigners including a newly-married Russian couple and some Protestant churches suffered in the collateral damage.[201]

William Hyde Lay (1836 – 1876), the second son of George Tradescant Lay (**Box 3.9**), who was Acting British Consul at the time, was made temporary French Consul, and had the unenviable job of examining the mutilated putrefying bodies, and negotiating for the execution of the perpetrators with the Chinese authorities.[202] Since the Foreign Office classed Lay as one of its interpreters who could not interpret, one wonders how he managed.[203] France demanded an indemnity and an official apology.[204] If it had not been for the outbreak of the Franco-Prussian War on 15 July 1870, it seems likely that France would have used the "Tīanjīn Massacre" as a *casus belli*.[205] Although it could hardly be blamed for the incident, the Qīng government was once again forced to capitulate to the foreign demands.[206] CHÓNG Hòu was sent to France to apologize in person.[207]

While the French were mainly concerned with guarding Roman Catholic interests, the more commercially-minded British took advantage of the clause in the Treaty of Tīanjīn (**Box 9.1**) allowing foreigners with a passport countersigned by the Chinese to travel in the Middle Kingdom. In this way they could look for an overland trade route between India and China and thus save the time and expense of the roundabout sea voyage via the pirate-infested Strait of Malacca.[208]

10.4 Prizing the Chinese Interior Open

In 1867 the Chief Commissioner of British Burma, General Albert Fytche (1820 – 1892; **Fig. 10.43**) suggested the possibility of reopening the Bhamo trade route between Upper Burma and western China.[209] Captain Edward Bosc Sladen (1827 – 1890), Political Resident at Mandalay, was chosen to lead the expedition.[210] Dr. John Anderson (1833 – 1900), Professor of Comparative Anatomy and Curator of the Imperial Museum in Kolkata (Calcutta), accompanied the expedition as its medical and scientific officer.[211] After just over a month spent at Bhamo ("Potters' Village"; drom 22 January to 26 February 1868) to hire 120 mules and their drivers, Sladen's expedition finally set off.[212] They eventually made it through to Mángyǔnjiē ("Beard Permit Market", Manwyne) on the Chinese side of the border, but because of the hostility of the inhabitants only stayed for two days before pressing on to Téngchōng ("Ascending thoroughfare", Momien).[213] There the party was informed that the road to Dàlǐ ("Great truth"), the capital of the Panthay Sultanate (**Box 10.7**), was too infested with bandits to allow them to proceed any further.[214] Although they wondered whether this was an excuse, the Governor of Téngchōng seemed to be genuinely worried for their safety.[215] In the six weeks that they were there, they witnessed the execution of more than 16 robbers.[216] During this time, Anderson accompanied by an armed guard went shooting and collected a number of plants for his elder brother Thomas (1832 – 1870), who was Superintendent of the Royal Botanical Garden in Kolkata.[217] Their return trip to Bhamo was delayed until 13 July 1868 by a Chinese complaining that some of the boxes they had to carry were filled with useless "weeds" and skins.[218] These "weeds" represented the first herbarium specimens to be collected in Yúnnán![219] By the time they departed the summer monsoon had set in. The slippery tracks made the going difficult.[220] They were held up at Hotha for another 17 days but finally got back to Bhamo on 5 September 1868.[221] Although they had failed to penetrate far

Fig. 10.43 General Albert Fytche (1820 – 1892), an army officer who served in Burma before becoming Chief Commissioner of British Burma (1867 – 1871). In this function, he suggested the possibility of reopening the Bhamo trade route between Upper Burma and western China.

into Yúnnán, Sladen's mission had seemingly convinced the Kachins, Shans and Panthays of the advantages of cross-border trade.[222]

The adventurer Thomas Thornville Cooper (1839 – 1878), with the backing of some of the Shànghǎi merchants, undertook the journey in the opposite direction in 1868.[223] Considering the mountainous country through which he was to pass, the hostile authorities, the very real danger of being robbed and murdered by bandits, and the fact that he knew no Chinese to begin with, made this a highly risky undertaking.[224] He was advised by Fortune's Běijīng acquaintance Thomas F. Wade (**Box 10.10; Fig. 10.44**) to take no scientific instruments as this would only raise suspicions as to the object of his journey.[225]

Fig. 10.44 Sir Thomas Francis Wade (1818 – 1895). Trained as a soldier, he fought in the First Opium War. He acted as an interpreter in the negotiations leading up to the Treaty of Tiānjīn (1858), and during the Anglo-French invasion (1860). He rose from Chinese Secretary to British Ambassador (1871 – 1882). He became Professor of Chinese Language at the University of Cambridge, England, and is largely known today for the Wade-Giles System of romanization of the Chinese language.

Box 10.10

Sir Thomas Francis Wade (1818 – 1895) Elder son of Major Thomas Wade (Black Watch) and his Irish wife Anne Smythe.[1] In 1823 the family moved to Mauritius, but in 1827 Thomas returned with his mother and sisters to England, where he attended a private school in Richmond for two years.[2] From 1829 until 1832 he went to school at the Cape, where his father was then stationed.[3] After a few years at Harrow, he continued his studies at Trinity College, Cambridge.[4] His father then bought him a commission in the army in 1838, and after service in Ireland and the Ionian Islands, he was sent to China to fight in the First Opium War.[5] He was just in time to take part in the attack on Zhènjiāng ("Guard River") and advance towards Nánjīng.[6] After the Treaty of Nánjīng, he stayed on in China. The talents of this fun-loving but quick-tempered young man were soon recognized by successive governors.[7] In rapid succession he became interpreter to the Hong Kong garrison in 1843, Cantonese Interpreter to the Supreme Court of Hong Kong in 1846, Assistant Chinese Secretary to Sir John Francis Davis in 1846, Vice-Consul at Shànghǎi in 1852, and Chinese Secretary to Sir John Bowring in 1856.[8] When the Second Opium War broke out in 1857 (**Box 6.1**), he became Chinese Secretary to Lord Elgin and with Horatio Nelson Lay (1832 – 1898), the eldest son of George Tradescant Lay (**Box 3.9**), conducted the negotiations which culminated in the Treaty of Tiānjīn (**Box 9.1**).[9] In 1859 he published his "Peking Syllabary", a system of transliteration of Chinese pronunciation based on the Běijīng dialect.[10] In 1860 he helped to smooth the advance of the British and French troops to Tiānjīn and Běijīng, and interpreted once the Chinese capital was reached.[11] The war over, he worked as Chinese Secretary for the British Legation in Běijīng.[12] It is here that he and Fortune would have met in September 1861. When Sir Frederick Bruce (**Box 9.4**) returned to England in 1864, Wade remained as Chargé d'Affaires until Sir Rutherford Alcock (**Box 3.10**) arrived from Japan at the end of 1865.[13] He doubted whether China would be able to deal with its internal and external problems and make progress.[14] In 1868, while on leave in England, Wade married Amelia Herschel (1841 – 1926), daughter of the famous astronomer Sir John Herschel (1792 – 1871).[15] They had four sons.[16] When Alcock left China at the beginning of 1870, Wade succeeded him, first as Chargé d'Affaires, and then in 1871 as Plenipotentiary.[17] Believing that the prime duty of newcomers was to learn Chinese, he gave them a thorough grounding in the Chinese language.[18] Those that failed to apply themselves had to resign, those that failed the examination were barred promotion, while those who passed were given an extra allowance.[19] His period in charge of the legation was far from easy. During the Massacre of Tiānjīn in 1870 a British subject was said to have been killed, so he was involved in seeking justice.[20] He then mediated in the dispute between Japan and China following the Japanese invasion of Táiwān in 1874.[21] In 1875 and 1876, following the murder of Augustus Margary, he was involved in the protracted negotiations which led up to the Chefoo Convention.[22] Furthermore, with the increased number of Treaty Ports, there was always something happening which required his attention. Some of the ports like Shàntóu (Swatow) were initially so unruly, that he had to approve placing a howitzer in front of the consulate to curb possible violence![23] He had on occasion to reprimand the consular personnel for taking the law into their own hands, but showed his disapproval when they freed a murderer or gave him a light sentence.[24] Some of the Treaty Ports could be so quiet that to pass the time the personnel took to drink.[25] Alcoholism or mismanagement of personal affairs sometimes led a consul to embezzle consular funds, which resulted in imprisonment.[26] Dealing with all these cases wore Wade out, and he had increasing difficulty in putting pen to paper or even reading a book.[27] In 1882 he was recalled to England, leaving a backlog of unanswered correspondence.[28] The next year he retired from the diplomatic service, and moved to Cambridge, where he donated more than 4300 volumes of Chinese literature to his Alma Mater.[29] He was appointed the university's first Professor of Chinese Language in 1888, a position he held until his death at the age of 76.[30] He was succeeded by the diplomat Herbert Allen Giles (1845 – 1935) whose modification of Wade's system in 1892 is known as the Wade-Giles System.[31] It was not until 1979 that this system of romanization was officially replaced on mainland China by Hànyǔ Pīnyīn in 1958.[32]

Even a diary was suspect, as his notes about people and places could suggest that he was spying on behalf of the British government. On one occasion he had to chant from it as though it were his prayer book![226] For the journey through Húběi and Sìchuān he dressed as a Chinaman with large green spectacles to hide his blue eyes.[227] He departed from Hànkǒu ("Han River Mouth") by boat on 4 January 1868.[228] He travelled up the Cháng Jiāng ("Long River", Yangtze) through the Three Gorges as far as Chóngqìng with only his terrier Zeila, a Chinese guide, and his faithful interpreter and Chinese teacher George Phillips.[229]

Once he got to Chóngqìng, he decided to head north for Chéngdū and attempt to travel via Tibet to Sadiya, Darjeeling or Nepal.[230] Cooper left Chéngdū on 7 March 1868 armed with a passport from the Viceroy of Sìchuān addressed to the Chinese Minister in Lhasa ordering all Tibetan and Chinese officers to help him.[231] He headed west across the Sìchuān Basin until he reached the snow-clad mountains on its western perimeter. In places the snow was still so deep that it was impossible to proceed for more than a fortnight.[232] While he was waiting for the snow to melt, he looked round Kāngdìng ("Well-being established") and visited the lamasery outside the town.[233] He also dined with Bishop Joseph-Marie Chauveau (1816 – 1877), who loaned him a horse and lent him 200 taels.[234] On leaving Kāngdìng and entering Tibetan Sìchuān he donned European clothes and mounted the horse rather than remain cooped up in a chair for any longer.[235] He felt elated at the sense of freedom this gave him.[236] Had he known of all the difficulties which lay ahead, he might have felt otherwise. As the party climbed, the air became so rarefied that they experienced difficulty in breathing.[237] The glare of the sunlight reflected off the snow was so intense that their faces became severely sunburnt, while the ponies and mules had to have their eyes bandaged to prevent snow-blindness.[238] By the time they reached Lǐtáng, their sunburnt faces were bleeding.[239] Fearing that Cooper and Phillips represented the vanguard of a force sent to annex the country, the Tibetan lamas forbade the merchants to supply Cooper and his baggage animals with food.[240] After five days the starving men and animals finally reached the market town of Bātáng, where the population was inquisitive, but not hostile.[241] He was befriended by an officer of the Chinese Commissariat, who suggested that Cooper join him on his forthcoming journey to Lhasa.[242] Although the Viceroy of Sìchuān had given him permission to go to Lhasa, the lamas did not want him to enter Tibet as they feared that British trade would introduce Christianity into their country and thus undermine their grip on the population.[243]

Since he was forbidden to travel into Tibet, he decided to follow the Láncāng Jiāng ("Turbulent River") southwards as far as he could go.[244] Just outside Bātáng Cooper was tricked into marrying a 16-year-old Tibetan called Lotzung when he stopped for lunch![245] When Cooper made no attempt to consummate the marriage, and living in fear of attacks by bandits, Lotzung finally left him.[246] It was probably just as well, as his guard debunked with the provisions when Cooper's party was threatened by Tibetan cavalry.[247] At a number of villages they were refused food.[248] They were given a little food at a musk hunter's cabin, and tried to buy a lamb, but lost it in an ensuing fight.[249] At Xiálǐ (Tsali) they had their first square meal in three days, but had to barricade the door against a possible attack.[250] The next day they caught up with a tea caravan led by two of the chief Tibetan Mandarin's officers.[251] Seeing the poor state of Cooper and his men, the officers gave them a meal.[252] Completely worn out by the time he arrived in Déqīn (Deqen, Atenze), Cooper slept for 16 hours.[253]

In order to prevent him from getting through to Dàlǐ the Chinese mandarins at Wéixī ("Safeguard the West") spread a rumour that Cooper was an important mandarin who had just arrived from Běijīng ahead of a great imperial army sent to fight the Panthay rebels.[254] Although he got as far as Tōngdiàn ("Open Pasture"), the dangerous situation forced Cooper to return to Wéixī.[255] One wonders what Cooper would have done, had he known that Sladen's party had arrived in Téngchōng, just 250 km away.[256] Assuming that he was loaded with money, the civil mandarin inveigled him into his mansion, where he was held for a month on the pretence of protecting him from outside attack.[257] Little knowing that Cooper had only 80 taels left, the mandarin tried to force Cooper to loan him 2500 taels.[258] After brandishing his revolver to prevent the mandarin from searching his possessions, he dared not eat for two days in case the food was poisoned.[259] Phillips overheard the Déqīn Mandarin suggest that Cooper be murdered.[260] This prompted a decision to attempt an escape on 30 July.[261] However, they only got as far as Cākǎ before being recaptured and brought back to Wéixī.[262] Cooper's party was finally allowed to leave on 6 August.[263]

Fearing that he would be killed if he attempted to continue on to Dàlǐ, Cooper unwillingly retraced his steps to Bātáng.[264] As they approached Déqīn they saw how a flash-flood had destroyed part of the town and inundated the fields.[265] They themselves narrowly avoided being caught in a cloudburst and accompanying flash-flood, which washed away part of the mountainside over which they had just travelled.[266] Five days later, at the foot of the appropriately named Robbers Hill, they were accosted by a party of bandits, who only let Cooper through when he cocked his rifle at them.[267] Their reception in Bātáng was less friendly than that on the outward journey, because the lamas had been spreading a rumour that a great famine would follow if foreigners were allowed to enter Tibet.[268] After leaving Bātáng in the direction of Lǐtáng Cooper met Lotzung's mother, who suggested that she act as Lotzung's replacement.[269] Just before Lǐtáng Cooper's party got lost in the dark and were set upon by three savage Tibetan sheepdogs.[270] When they finally arrived at Lǐtáng, the inhabitants expressed surprise that Cooper was still alive.[271] On 18 September 1868 between Hékǒu ("River Mouth") and Kāngdìng Cooper's party was caught in a snowstorm, which obliterated the track and caused them to lose their way.[272] After a fortnight at Kāngdìng, during which time Cooper shed his European clothes and was once more transformed into a Chinaman with shaved forehead and pigtail, the party set off for Lúdìng in chairs.[273] Several of the villages which they had passed on the outward journey had been washed away by the monsoonal rains.[274] Cooper finally arrived back at Yǎ'ān ("Calm and safe") after crossing the snow-capped Dàxiàng ("Great Elephant") Mountains.[275] After travelling for a day and a half through plantations of Chinese tree privet (*Ligustrum lucidum*) grown for the candle wax that the male scale insects (*Ericerus pela*) secrete as protection, they hired a raft and travelled down the Qīngyī Jiāng ("Green-coated River") to Lèshān ("Happy Hill").[276] Rain prevented him from visiting the sacred mountain of Éméi Shān ("Softly-chant Eyebrow Mountain"), which was soon to become renowned for its rich plant diversity.[277] From Lèshān they took a small junk down the Mǐn Jiāng to Yíbīn ("Suitable Guest").[278] On the Cháng Jiāng between Yíbīn and Chóngqìng the junk got caught in a whirlpool, shipped a great deal of water, and almost sank before reaching the river bank.[279] After a few days in Chóngqìng Cooper boarded another junk bound for Shāshì ("Sand City").[280] However, Cooper's troubles were not yet over. On the way down the Cháng Jiāng they were stopped by a customs' gunboat, collided with another junk and nearly capsized, while Cooper was so weakened by fever that he had to be carried in a chair through Shāshì to the boat bound for Hànkǒu.[281] He arrived back in Hànkǒu on 11 November 1868, where he was given an English bed and was able to recuperate for a few days before travelling on to Shànghǎi.[282]

George Phillips joined him there in December 1868 in preparation for Cooper's attempt to reach China from Kolkata in 1869/1870.[283] They were lucky to reach India, as only a short distance out of Hong Kong an opium smoker hiding in the hold of the paddle steamer "Clan Alpine" caused the ship to catch fire.[284] Once in India the difficulty of obtaining mules to transport the baggage caused them to make an unnecessary trip of over 1500 km to Ambala at the foot of the Himalayas, which delayed their departure up the Brahmaputra River.[285] To make matters worse, the monsoon set in early in 1869, which prevented them from crossing into eastern Tibet before the snows fell.[286] This meant they had to wait first at fever-ridden Dibrugarh (Debrughur) and later at Sadiya (Sudiya) until the rains eased off in October.[287] Sadiya, as frontier station of the British Raj, was surrounded by dense leech-infested jungle which was home to many tigers, leopards, bears, rhinoceroses, wild elephants and treacherous hill tribes such as the Mishmi, who had already killed two French missionaries, Nicolas Michel Krick and Augustin Etienne Bourry, who had ventured into their domain in 1854.[288] Cooper managed to convince a local chief of the more peaceful Khamtis, recent Buddhist immigrants from northern Burma, to lead him through the Mishmi Hills.[289] Above Brahmakund ("Sacred Spring") the occurrence of numerous narrows, rapids and rocks in the Lohit River ("Red River") ruled out navigation.[290] The party were forced to follow the trails through the forest. The Mishmi chiefs, who not only bore a grudge against the British for giving protection to their runaway slaves, but acting on orders from Lhasa, did all they could to stop Cooper's advance.[291] By the time he got to Prun (New Year 1870; near present-day Walung or Walong) it had become clear that he stood no chance of reaching Bātáng.[292] A large party of soldiers had been stationed at Rìmǎ on the Tibetan side of the border to prevent his entry.[293] Moreover, his ankle was so swollen by an abscess that he could only move with the help of crutches.[294] There was no option but to turn back. Although disappointed at not having reached Bātáng, he was glad to return to Sadiya, where he had his first warm bath in months.[295] At Kolkata Phillips and Cooper parted company.[296] Phillips returned to China, while Cooper continued on to Britain to consider his next move.[297]

This second failure only hardened Cooper's resolve to complete the crossing. When DÙ Wénxiù's son Hassan was returning from his unsuccessful mission to Britain to gain support for the Panthay Rebellion (**Box 10.7**), the India Office allowed Cooper to accompany him back to China.[298] However, Cooper's dream was once more thwarted, when they learned in Yangon (Rangoon) that the rebellion had been crushed.[299] He ended his life as political agent in Bhamo.[300]

Fig. 10.45 Augustus Raymond Margary (1846 – 1875), the interpreter Wade selected to deliver the imperial passports to Colonel H. A. Browne in Burma. He crossed China from Yāntái to Bhamo in just five months. He was murdered in Mángyǔnjiē (Yúnnán) under mysterious circumstances. The "Margary Affair" led to the Chefoo Convention (1876).

Once the Panthay Rebellion was over, there was renewed hope of establishing a trade route between British Burma and China.[301] Colonel Horace Albert Browne (1831 – 1914) was asked to assess the possibilities, although this was not explicitly stated in the application for passports.[302] Imperial passports were obtained in Běijīng, but as Colonel Browne's orders were to enter China from Bhamo, the problem arose of how to deliver the passports.[303] The British Ambassador to China, Sir Thomas Wade (**Box 10.10; Fig. 10.44**) selected the optimistic interpreter Augustus Raymond Margary (1846 – 1875; **Fig. 10.45**) from Yāntái (Chefoo) for the challenging task, as he spoke Chinese fluently and was well-versed in Chinese customs.[304]

On 23 August 1874 Margary left Shànghǎi with the passports and three Chinese servants.[305] They boarded an American steamer bound for Hànkǒu.[306] There Margary hired ZHŌU Yúdǐng as his cook.[307] Leaving Hànkǒu on 4 September 1874, the group navigated the Cháng Jiāng as far as Yuèyáng in Húnán.[308] After crossing Dòngtíng Hú ("Profound Lake"), they sailed up the Yuán Jiāng.[309] Although the scenery along the meandering river was often magnificent, being cooped up in the boat became tedious.[310] Unfortunately, when he went on land, Margary invariably attracted the unwanted attention of bystanders, who had probably never seen a "foreign devil" before.[311] He frequently had fever and diarrhoea, followed by an attack of amoebic dysentery, and became so weak that he could not rise without assistance.[312] Just as he began to feel a little better, he was inflicted by rheumatism.[313] After 20 days the boat finally docked at the flourishing city of Hóngjiāng ("Big River"), situated at the confluence of the Yuán Jiāng and the Wǔ Shuǐ.[314] After buying provisions and hiring new hands, they pushed on up the Wǔ River.[315] During the 16 day journey from Hóngjiāng to Zhènyuǎn ("Garrison remote") Margary was plagued by a succession of illnesses: pleurisy, rheumatism, indigestion, bad neuralgia, toothache etc.[316] However, at Yùpíng ("Jade Screen"), Margary was pleasantly surprised to discover that the local magistrate was the former Head Writer at the British Legation in Běijīng.[317]

Margary's party disembarked at Zhènyuǎn on 27 October 1874, as they were heading overland for Guìyáng ("South of Guì Mountain"), the capital of Guìzhōu.[318] However, a mob tried to prevent the unloading of the baggage, so Margary had to ask the local magistrate for an armed guard.[319] Margary was impressed by the scenery in Guìzhōu, but the roads were rough going and everywhere there were signs of the destruction perpetrated during the Xián-Tóng Rebellion (**Box 10.6**).[320] In Guìyáng the population proved very civil, but Margary was besieged until midnight by relays of literati, who wanted to meet the foreigner and examine his curious possessions.[321] On 8 November 1874 the party headed WSW for Kūnmíng ("Offspring Míng"), the capital of Yúnnán. In order to escape the confines of his chair, Margary rode some of the way, but this mode of transport soon lost its charm, as he was unable to read while in the saddle.[322] It took the party a total of twenty days to reach Kūnmíng. The local magistrate was most helpful, providing Margary with an escort of two pleasant mandarins, the inveterate opium-smoker ZHŌU and the 65-year old YÁNG with his rough hands, deep voice and kindly eye, who went ahead to give the Chinese authorities advanced notice of Excellency MĀ's (Margary's) arrival.[323]

On the way to Dàlǐ he was well looked after by the Chinese officials, but saw ample evidence of the destruction caused during the Panthay Rebellion (**Box 10.7**).[324] In Dàlǐ Margary met General CÉN Yùyīng, who complemented him on his knowledge of Chinese, and quizzed him about England and Burma.[325] Although Margary had been warned about bad feelings towards the British, who had secretly backed the Huí rebels against the Qīng Government, he was treated with great respect by the inhabitants of Dàlǐ.[326]

Margary left Dàlǐ on 18 December 1874 to cross the snow-covered Qīngshuǐláng Shān ("Clear Water Láng Mountains").[327] He was held up for some time, while troops scoured the hills for robbers.[328] On 27 December 1874 he reached Yǒngpíng ("Perpetually peaceful"), where Margary described the mandarins as "brutes".[329] He therefore pressed on to Téngchōng, which was one of the last cities to fall to the Imperial troops at the end of the Panthay Rebellion (**Box 10.7**). At this city, famed for its hot springs, the mandarins were "delightfully civil".[330] Five more days along the lovely Dàyíng ("Large Surplus") Valley brought him to Mángyǔnjiē, where he was treated with the greatest respect.[331] Even the dreaded LǏ Zhēnguó kowtowed to him.[332] Nonetheless, he felt that there were intrigues brewing in this district directed against Browne's mission.[333] With a Burmese guard of forty men Margary's party was escorted over the Kachin (Kakhyen) Hills and arrived in Bhamo on 17 January 1875 after an epic journey of almost five months across China.[334]

Browne's mission got off to a false start. It only made it as far as the fortified village of Sawadi, a couple of hours from Bhamo, when it was held up for a week by bickering over the payment of the bullock-drivers, and the size and allocation of some of the boxes.[335] Browne decided to return to Bhamo and take the more direct Ponline Route via Mángyǔnjiē.[336] This route was not without its own dangers in the form of man-eating tigers and the Kachin hill tribes.[337] Since Browne's expedition represented a rich prize for robbers, there was a very real possibility of an attack. Moreover, it was rumoured that the real object of the undertaking was not travel, but to survey the route for a railway link between Bhamo and Téngchōng, an outrage for the hardliners ("ironhats") in the Qīng Government, and a source of competition for the cross-border traders.[338] Moreover, the Zǒnglǐ Yámén ("Premier's Office") in Běijīng, which had issued the passports, at no time sanctioned any trade.[339] If these parties were aware of the true purpose of Browne's mission, they would have had a vested interest in its failure.[340] If this was not bad enough, a monk forged a letter in the name of the King of Burma directing the Kachins to oppose Browne's party.[341] In order to ascertain the truth of the rumour about a possible attack, Margary offered to go ahead to Mángyǔnjiē, where he had been well-received on his outward journey.[342] He reached the village of Seray unharmed, but he and his five-man team were murdered once they got to Mángyǔnjiē on 21 February 1875.[343] Browne's group was attacked on the following day by Kachins and some Chinese.[344] The gun battle lasted from late morning until about 5 p.m., when the attackers were finally driven off by setting fire to the forest in which they were concealed.[345] Browne's party retreated along the Tàipíng River, but news that Chinese were massing at the northern and southern ends of the valley sent them scuttling over the hills to Burma.[346] After an absence of little over three weeks, they were glad to get back to the comparative safety of Bhamo on the 26 February 1875.[347]

As a result of the public outcry, an Anglo-Chinese commission consisting of Thomas G. Grosvenor (1842 – 1886), Second Secretary at the British Legation, the much-travelled Edward Colborne Baber (1843 – 1890), Arthur Davenport (1836 – 1916) and the Qīng officials LǏ Hànzhāng (LI Hanchang, 1821 – 1899, the elder brother of LǏ Hóngzhāng) and XUĒ Huàn (1815 –1880) was sent to investigate the "Margary Affair".[348] The results were inconclusive, because no witness of the murders was produced.[349] General CÉN's claim that savages were responsible was simply reiterated.[350] Although it was clear to Rutherford Alcock (**Box 3.10**) and Daniel Brooke Robertson (**Box 5.27**) that Britain was asking for trouble by sending Margary into an area that was just recovering from the Panthay Rebellion (**Box 10.7**)[351], Thomas Francis Wade (**Box 10.10**) seized on the "Margary Affair" to demand an indemnity of 200,000 taels or £ 10,

000, a mission of apology to Queen Victoria, four more treaty ports (Běihǎi [Guǎngxī], Wēnzhōu [Zhèjiāng], Wúhú [Ānhuī], Yíchāng [Húběi]) and six ports of call on the Cháng Jiāng.[352] The conciliatory Chinese soldier-diplomat LǏ Hóngzhāng (1823 – 1901; **Fig. 10.46**) eventually agreed to Wade's demands at the Treaty Port of Yāntái (Chefoo), although he managed to delay the opening of Chóngqìng to foreign trade until 1891.[353] When Fortune read about the "Chefoo Convention" (13 September 1876), he must have remembered the fortnight he had spent at the "Brighton of China" in August 1861.[354]

Fig. 10.46 LǏ Hóngzhāng (1823 – 1901), Chinese soldier-diplomat, who was often called upon to solve tricky problems, such as the Margary Affair, the Sino-Japanese War, and the Boxer Rebellion.

Fig. 10.47 Captain William John Gill (1843 – 1882), who was the first foreigner to complete the crossing from China into Burma. His murder put an end to his adventurous life.

Fortune was no doubt thrilled and just a little jealous of the possibilities for travel in inland China by the 1870s. Captain William John Gill (1843 – 1882; **Fig. 10.47**) was a case in point. He had served in India for several years, but having inherited a considerable fortune, decided to spend his time travelling to exotic places.[355] He consulted Thomas T. Cooper and travelled to Berlin to meet Ferdinand von Richthofen (1833 – 1905), who was considered to be the greatest authority on the exploration of China.[356] From Berlin Gill went straight to Marseille, where he caught a ship bound for China.[357] When he arrived in China in 1876, he knew little or no Chinese.[358] He therefore needed companions who spoke the language. Through his contacts at the British Legation, he met Edward Colborne Baber, who had just been appointed consul in Chóngqìng.[359] They travelled together on board the SS "Hankow" which left Shànghǎi on 23 January 1877.[360] Having transferred to a smaller vessel, they finally arrived in Hànkǒu on 30 January 1877.[361] During the week they were there it snowed continuously, so the travellers were happy to stay at the British Consulate with its blazing fires.[362] Rather than risk being overcharged, Baber and Gill let their Chinese servants hire a boat for the two-month journey from Hànkǒu to Chóngqìng.[363]

In the 17 days he spent in Chóngqìng Gill had a chair made for his overland trip to Chéngdū.[364] In fact, he only used it when entering or leaving a large town to emphasize his importance.[365] Like Fortune he preferred to walk or ride a pony.[366] As there were no mules available, Gill needed twenty coolies to carry his luggage, and another eight for his chair and those of his servants.[367] When he set off on 26 April, his procession must have been quite impressive. It took Gill's party a fortnight to reach the capital of Sìchuān.

In Chéngdū Gill was joined by the self-important William Mesny (1842 – 1919), who had been in China since 1860.[368] He had fought with the Sìchuān Army during the Xián-Tóng Rebellion (**Box 10.6**) and until March 1877 had been General Superintendent of Foreign Ordnance for Guìzhōu Province.[369] Having ruled out the other possibilities, the companions agreed to travel to Bātáng and Déqīn in the footsteps of T. T. Cooper.[370] On the way they both collected plants, Mesny for Henry Fletcher Hance (1827 – 1886), the British Vice-consul at Huángpǔ (Whampoa) (**Fig. 10.48**) and Gill for the British Museum in London.[371]

Fig. 10.48 ***Jasminum mesnyi*, named after Gill's companion, William Mesny (1842 – 1919).**

On arrival in Kāngdìng on 25 July 1877 they were warmly welcomed by Cooper's old acquaintance Bishop Joseph-Marie Chauveau.[372] Chauveau found a man (Peh-ma) who understood Tibetan to help on the next stage of the journey.[373]

After a hard trek over the mountains Gill and Mesny finally arrived in Bātáng on 25 August 1877.[374] The town was completely new, as earthquakes in 1871 had destroyed all the houses and the vineyards that Cooper had seen.[375] Mulberry trees grew well but no silk was manufactured, as killing the silk-worms was considered a mortal sin.[376] Gill was surprised at the number and size of the lamaseries they met with in this part of Sìchuān.[377] In Bātáng with only 300 families, the lamasery boasted 1300 lamas, who with slaves to work for them, lived a life of comparative ease.[378] Over the years the lamasery had accrued the greater part of the arable land in the Bātáng Plain, which was rented to peasants at a good price.[379] This income represented pure profit, as the lamaseries' lands were tax-free.[380] As a result, the lay population had to bear the brunt of national and local taxation.[381] By the 1870s this oppressive state of affairs was leading to widespread emigration, mostly to Yúnnán where, in the aftermath of the Panthay Rebellion (**Box 10.7**), there were ample opportunities for farming.[382]

Gill's first impression of Yúnnán was that its inhabitants "seem to divide their time between opium-smoking and eating sunflower seeds"![383] After almost a month the explorers finally reached Dàlǐ, where it rained incessantly.[384] In vain the north gate had been barred to chase away the Spirit of the Waters, and a gun was fired at the sky to frighten the rain god.[385] After a week they pushed on to Mángyǔnjiē where they were visited by LǏ Zhēnguó, who had been implicated in Margary's murder.[386] While at Mángyǔnjiē they received a warm letter of welcome from Cooper.[387] So, accompanied by a guard of 20 soldiers, they set off for Burma.[388] Gill paid his respects to Margary's memory near the spot where Margary had been murdered, little realizing that he himself would suffer the same fate less than 5 years later.[389] Near Ponsee they met a band of robbers, who let them pass, once they realized they were prepared to fight.[390] Cooper had sent an advanced party with newspapers, food, drink, tobacco and cigars to Ma-mou to meet the travellers.[391] From there they took Cooper's comfortable boat to Bhamo, where Cooper was waiting to congratulate them on their successful mission.[392] If Cooper felt jealous that Gill and Mesny had accomplished what he had failed to do, he did not show it. Gill was delighted to be once again in a clean house, with a damask tablecloth and numerous other long-forgotten luxuries.[393] On 6 November 1877 Gill and Mesny shook hands with Cooper, and with a sincere "see you again soon" boarded a boat bound for Yangon (Rangoon).[394] Little did they realize that they would never see Cooper again. Less than six months later on 24 April 1878 Cooper was shot dead by one of his guard, whom he had punished.[395] By the time Fortune died in 1880, this backdoor into China had become well-established. It was used by 20th Century plant hunters such as George Forrest (1873 – 1932) and Frank Kingdon-Ward (1885 – 1958).[396]

It was clearly impossible for a country the size of China to prevent infiltration and defend its borders adequately. On the pretext of revenging the deaths of the crew of the "Rover", who had been murdered by aborigines, the USA made an unsuccessful attempt to invade S Táiwān in 1867.[397] As we have seen, after Japan annexed the Ryūkyū Islands, which had long paid tribute to China (the Ryūkyūs continued to pay tribute until 1875[398]), they invaded Táiwān in 1874 because some Ryūkyū fishermen had been murdered there in 1871.[399] China had to pay for the cost of the Japanese expedition and compensate the relatives of the murdered men before the occupying forces would withdraw.[400] Likewise, Russia occupied Ili (Kulja) in NW China in June 1871 on the pretext of upholding the Sino-Russian trade agreement (Treaty of Kulja, signed in 1851) which was being threatened by muslim dissidents (**Box 10.9**).[401] However, the Russians held on to Ili after the muslim uprising was suppressed in 1878, and only returned the greater part of it to China in 1881 after the Chinese had agreed to pay 9 million roubles for occupation costs, and compensate those Russians who had lost property or relatives during the rebellion (Treaty of St. Petersburg, signed on 24 February 1881).[402] Realizing that they could carve up China with impunity, Japan and the western powers (Britain, France, Germany, Russia) started grabbing what land they could.[403]

Writing in 1852 Fortune had already prophesied "that a few years will see a vast change in China; it may be that another war and all its horrors is inevitable, and whenever that takes place this vast country will be opened up to foreigners of every nation."[404] Only a few years after Fortune died, China lost the remainder of its tributary states (Vietnam in 1885, Burma in 1886 and Korea in 1895), which had acted as a buffer against external aggression.[405]

10.5 Decline and Death

Fortune was no doubt saddened by the events taking place in China. Moreover, the loss of his mentors and other friends must have also depressed him. One by one his circle of scientific friends and acquaintances shrank. In 1865 Falconer (**Box 5.7**), Paxton (**Box 5.14**), Cuming (**Box 3.8**) and Lindley (**Box 2.1**) died. In 1866 followed Siebold (**Box 8.4**) and Perceval (**Box 5.23**). In 1867 it was the turn for Bruce (**Box 9.4**) and Girard (**Box 8.9**). On 4 June 1868 the inventor of the Wardian Cases (**Box 2.6**), on which Fortune relied so much, passed on. His rival in Japan (Veitch, **Box 8.1**) and his friend in China (Morrison, **Box 3.7**) followed in 1870. The death of John Standish (**Box 5.15**) on 24 July 1875 must have come as a shock. He no doubt read about the death of Townsend Harris (**Box 8.11**) in New York City in 1878, but was probably unaware that his Yokohama acquaintance William Aspinall (**Box 8.13**) had died in poverty on 3 October 1879. As these old friends and colleagues passed on, he must have realized that his time was also running out. In his last scientific article, published 7 weeks before his death[406], he mentioned that he sometimes took a walk in Brompton Cemetery, no doubt to pause at the grave of his two young daughters (**Figs. 10.49, 10.50**), and pay his respects to departed colleagues, such as Murchison (**Fig. 10.51**), James Veitch and his son John Gould (**Fig. 10.52**).

Fig. 10.49 Fortune family grave in Brompton Cemetery prior to its restoration in 2010. Only a few words were legible on the slab.

Fig. 10.50 The Fortune grave after restoration. The text reads: "In loving memory of Agnes Fortune 1847–1848/Mary Fortune 1852 – 1857/Robert Fortune, Renowned Botanist 1812 – 1880/Thomas Fortune 1857 – 1881/Jane Fortune 1816 – 1901/Helen Jane Fortune 1840 – 1910/This memorial was provided by the ancestors [sic] and supporters of Robert Fortune in 2010."

Fig. 10.51 Monumental grave of the geologist, Sir Roderick Murchison (1792–1871) in Brompton Cemetery.

Fig. 10.52 Grave of the horticulturalist James Veitch and his family in front of an obelisk in Brompton Cemetery.

Britain suffered a number of wet summers in 1875, 1878 and 1879.[407] 1879 "proved to be the wettest, most sunless year of the nineteenth century"[408], and not only caused widespread crop failures[409] and liver-rot in sheep[410], but dampened the spirits of many like Alfred Tennyson (1809–1892), who at midnight on 30 June 1879 was driven to write[411]:

> Midnight—and joyless June gone by,
> And from the deluged park
> The cuckoo of a worse July
> Is calling thro' the dark:
> ...
> And, now to these unsummer'd skies
> The summer bird is still,
> Far off a phantom cuckoo cries
> From out a phantom hill;

It seems likely that the weather would have affected Fortune's health, as he suffered from rheumatics.[412]

After all that Fortune had done to enrich British gardens, it is perhaps surprising that no honours were bestowed on him in the UK. In a class-conscious society this may have had something to do with his humble background. Although he was an Honorary Member of the Agri-Horticultural Society of India, and was awarded a medal by the Parisian Société d'Acclimatation in 1859 for his introductions, he never gained any official recognition in his native land.[413] "We can but chronicle the fact as a source of humiliation to the nation, or rather to its rulers, who have so much to give to the successful, or even unsuccessful, Briton militant, so little to the patriot whose services to the Empire are often so much more important, and so much more enduring, although, it may be, not so attractive to the general public."[414] In fact, he became largely forgotten in his own lifetime. Perhaps realizing this, Fortune made a conscious effort shortly before his death to emphasize the leading role he had played in the introduction of a large number of East Asiatic plants into cultivation. In "The Gardeners' Chronicle" of 3 January 1880 he listed those plants he had discovered in China while in the service of the Horticultural Society, those he had introduced at a later date, and those trees, shrubs and herbaceous plants he had discovered in Japan.[415] In "The Gardeners' Chronicle" of 21 February 1880 he supplied some notes on a number of these introductions. He concluded with his famous valedatory statement: "But enough: we will leave the plants to speak for themselves."[416] The editor of "The Gardeners' Chronicle", Maxwell Tylden Masters (1833 – 1907), little realizing that these were really his final words, appended a footnote: "If Mr. Fortune himself is not tired, we are quite sure our readers are not; and even the plants will be all the better if Mr. Fortune will interpret for them, and not leave them to speak for themselves."[417] It was not to be. Fortune passed away at his home on Tuesday, 13 April 1880, and was interred in the family grave in Brompton Cemetery four days later.[418] If he could have chosen, he would probably have decided to spend his new existence sitting under some bushes of *Osmanthus fragrans*. "One tree is enough to scent a whole garden. I have often sat down under the shade of these very bushes, in the midst of this perfumed atmosphere, and almost fancied myself in the garden of Eden."[419]

After his death a number of obituaries appeared.[420] Even by Victorian standards, they are remarkably colourless, confining themselves largely to Fortune's expeditions and listing his introductions. One gets the impression that most necrologists were talking about some figure from the distant past. This lack of information has given later commentators plenty of leeway to formulate their own opinions about Robert Fortune's character. He has been described as both discourteous[421] and suave/tactful[422], modest/self-deprecating/self-effacing[423] and self-important.[424] Of course, most people are full of contradictions, and as such cannot be easily categorized using a single adjective. So what sort of a man was he really?

10.6 Fortune's Personality

Born as a son of a hedger, he only received a basic education in Edrom Parish School.[425] His teacher would have drummed the fear of God into his pupils. Fortune referred to himself as a "consistent Protestant".[426] By applying himself, he learned to write good English, and thus improve his lot, a typically Scottish trait. Those letters which have survived, although clearly penned in haste, contain remarkably few grammatical errors (**Fig. 10.53**).[427]

Moreover, the handwriting betrays an energetic personality.[428] It was this quick-thinking, hard-working young man with his seemingly boundless energy who caught the attention of John Lindley (**Box 2.1**).

When Fortune was offered the job of collecting plants in China, he would only have had a vague idea of what this entailed. Nevertheless, realizing that this represented a unique opportunity to further his career, he took up the challenge. So, yes he could be said to have been ambitious.[429] He probably realized that this was one of the few ways to support his growing family. Without any knowledge of Chinese, and far from his family, he must have felt extremely lonely. However, only once in his four books does he admit this.[430] While other plant collectors in similar circumstances turned to drink, or returned from their expeditions prematurely, he grit his teeth and got on with his mission. This clearly called for determination[431], tenacity[432], and perseverance[433], something he may have learned in his childhood. He could not afford to let his family or his staunch supporter Lindley down.

Not only did he not know any Chinese to start with, but he had not the faintest idea of the Chinese psyche. Driven by a desire to fulfil as many of the Horticultural Society's wishes as possible, he was conned on a number of occasions. The "yellow" *Camellia* turned out to be anything but[434], and the expensive paeonies, that were said to have come all the way from Sùzhōu, were actually grown in the Moutan Gardens, about 9 km west of Shànghǎi.[435] However, "once bitten, twice shy", he soon learned to distrust those who told him what he wanted to hear, and were willing to supply whatever he wanted. Since he could not believe that his servants would actually go all the way to Huángshān in search of tea plants and seeds, when these could be obtained nearby, he decided to visit the tea-growing area himself.[436] As this area was out of bounds to foreigners, and he was trying to smuggle one of China's assets out of the country, he could have got into serious trouble.

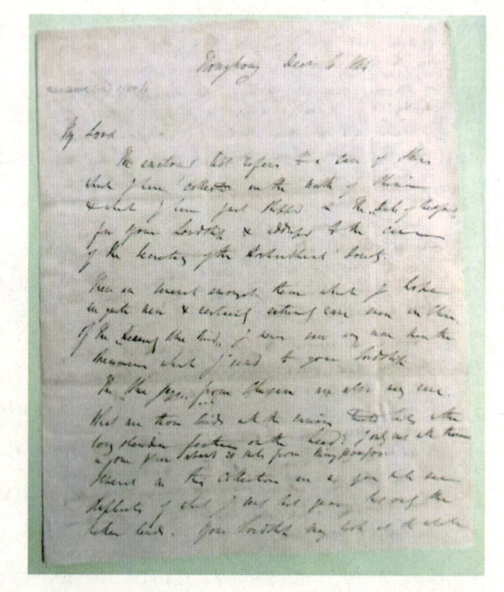

Fig. 10. 53 A sample of Fortune's handwriting, betraying his energetic personality.

When he ventured beyond the limits of the Treaty Ports in search of novelties, he demonstrated considerable courage.[437] It was not just the danger of being caught red-handed, but being out of reach of a doctor, should he be taken ill. For this reason he had "... a great horror of being touched by a Chinese beggar"[438], and disliked the custom of presenting the guest with tit-bits using the chopsticks that had already been in the host's mouth.[439] When he had to order food in an inn, he stuck to simple fare such as hard-boiled eggs and rice, which could not be contaminated.[440] However, his faith in the Almighty helped him to overcome his fear of travelling far from the civilized world ("committing myself to the care of Him who can preserve us alike in all places, I resolved to encounter the difficulties and dangers of the road with a good heart").[441]

With regard to food he became very partial to bamboo[442], but never ordered lotus roots, as he did not like them, and refused to drink sake or its Chinese equivalent ("little else than rank poison"), preferring to quaff French wine or brandy.[443] He also liked to take his tea without milk or sugar.[444] He ate both fish and meat, but balked at the idea of eating monkey.[445] He did not like sweet, insipid apples, and preferred melting pears to the crunchy Asiatic pears he invariably encountered in Japan.[446] Although he himself smoked cigars[447] and a pipe[448], he was averse to people smoking during meals.[449]

Fortune tells us of the excitement and pleasure he felt when he thought he was about to discover a new and beautiful plant.[450] The introduction of novel plants became such an obsession, that he sometimes overstepped the mark as, for instance, when he helped himself to cuttings in Siebold's nursery in Nagasaki.[451] When he was after particular plants, he did not let anything or anybody get in his way. He jumped a nunnery fence to examine "... a number of trees and bushes which seemed worth looking at..."[452] Neither did he have any respect for bureaucratic niceties. When he was ordered to leave Edo prematurely, because he had no right to be there without the British Chargé d' Affaires' permission, he got his own back on Dr. Francis Myburgh by elaborating the incident in detail in "Yedo and Peking".[453] One wonders why he was not taken to court for slander!

He had no such problems with the commoners he encountered on his journeys.[454] Although some were frightened by his strange appearance[455], he tried to allay their fears by talking and joking with them, and letting them examine his clothes and collections.[456] A fellow Scot, John Scarth, described him as "always good-humoured, and had made a lot of friends among the little brats that ran about the cottages".[457] When he was followed by a rabble, he attempted to stay calm, as he knew from experience that the hooligans could resort to stone-throwing if remonstrated with.[458] Wherever he went he made friends among the expat community, who welcomed him into their homes, sometimes for weeks on end.[459] It seems unlikely that they would have done this had he been taciturn[460], gloomy[461] or humourless.[462] Like Alistair Watt, I personally detect a dry sense of humour in his writings.[463] For instance, when he visited Shàoxīng he "saw many ornamental gates in the town, erected to the memory of virtuous women, who, judging from the number of these structures, must have been unusually numerous in the place..."[464] Similarly, when he left the Kawasaki tea-house with its attractive attendants, he wrote with tongue in cheek that "... the best of friends must part at last, so I was obliged to bid adieu to mine host and his fair waiting-maids of the "Ten Thousand Centuries," and pursue my way to Kanagawa".[465] Again, referring to the Japanese plants he had sent back to England by ship, he jokingly remarks "While they pursue their long and lonely voyage let us hope they may be favoured with fair winds and smooth seas, and with as little salt water as possible, which sadly disagrees with their constitution."[466] His contemporary Charles Gordon (**Box 9.3**) described those excursions he took in Fortune's company in 1861 as "some of my most pleasant walks".[467] He refers to the running commentary Fortune gave him as they passed from plant to plant in the nurseries near Tiānjīn[468], hardly what one would expect of a dour Scot.

Did Fortune have any interest in the arts? As a "consistent Protestant", he probably frowned upon theatricals, although he seems to have tolerated a little Chinese opera.[469] In his books he makes no allusions to Western composers, so he may not have been very musical. However, he did enjoy listening to the "pretty music" played and sung by the innkeeper's daughter in Xiákǒu[470], and found that the priest at the temple of Wú Shān produced "not unmusical sounds" on his recorder.[471]

> ... I could hear the wild and not unpleasing strains of my friend's flageolet as he wended his way homewards through the woods.[472]

He noted that the piano, that the Franciscan Padre Carlos Tena played in Dolores needed tuning[473], so he cannot have been completely tone-deaf!

Basically he loved all manner of beautiful things. To start with, he had a good eye for attractive plants.[474] He could also wax lyrical about the scenery he encountered on his travels.[475]

He must often have been reminded of Psalm 121:

> I to the hills will lift mine eyes,
> from whence doth come mine aid.
> My safety cometh from the Lord,
> who heav'n and Earth hath made.

He described his first view of the high Himalayas in glowing terms:

> The snowy mountains lay before me in all their grandeur, and the sun was shining on them...Their snowy peaks seemed to reach to heaven itself, and to pierce the deep-blue sky.
>
> Never in all my wanderings had such a view been presented to my eyes. It was indeed grand and sublime in the fullest sense of the words. How little the most gigantic works of man seemed when compared with these! The pyramids of Egypt themselves, which I had looked upon in wonder some years before, now sank into utter insignificance! I could have looked for hours upon such glorious objects, but the clouds soon closed in around me, and I saw the snowy range no more.[476]

He also loved to listen to birds singing, liked the bright green of young foliage[477], the freshness of the early morning with its dew-covered plants[478], and the light cast by a full moon.[479] Such experiences made him forget the hardships he experienced on the way.[480] Sometimes he was so enthralled by the scene that his servants had to remind him that it was time to move on.[481]

While he was in China, he learned to appreciate Míng pottery and other *objets d'art*. He clearly had a good eye for such works of art, judging by the prices they fetched at the auctions in London. "Having a kind of mania for collecting ancient works of Chinese art, such as porcelain vases, bronzes, enamels, and such things..."[482] also helped to line his pocket. So, while he could be a romantic soul, there was also a business side to his character.[483] He apparently had little artistic leaning, as he used drawings and latterly photographs by Dr. Barton, Peter Cracroft, Dr. Walter Dickson, Dr. Jones Lamprey, John Scarth, George Wyndham (**Box 9.6**) and native artists to illustrate his books.[484] He was most probably an avid reader. Besides "Punch" and "The Illustrated London News", he read scientific articles, as well as poetry and novels by his countrymen Robert Burns and Sir Walter Scott.[485]

Like most men, Fortune had a glad eye. He clearly had a foible for attractive Asiatic women. Fortune shared the widely held opinion that the women in Sùzhōu are particularly beautiful: "The ladies here are considered by the Chinese to be the most beautiful in the country, and, judging from the specimens which I had an opportunity of seeing, they certainly deserve their high character."[486] In Cíchéng (Zhèjiāng) pretty women seem to have been particularly numerous in Fortune's day. "It is a curious and striking fact that in this old city and its vicinity one rarely sees an unpleasing countenance. And this holds good with the lower classes as well as it does with the higher."[487] While he was visiting a mandarin in this city, some of the mandarin's wives and daughters were looking on. Fortune admitted that he "was willing to look upon their pretty faces as long as possible."[488] On one occasion he caught a glimpse of an attractive widow as she passed by in her chair. He "was so much struck with her beauty that I instantly stood still and looked after the chair."[489] One can almost see his jaw dropping! In order to see more of her, he even followed the widow to her husband's grave![490] There were only two detractions to Chinese women as far as Fortune was concerned: the absence of an intellectual expression caused by a lack of education[491], and their deformed feet.[492] "It is certainly a most barbarous custom that of deforming the feet of Chinese ladies, and detracts greatly from their beauty."[493] In Japan the feet were not bound, and it was much easier for a foreign man to come into contact with the opposite sex in daily life. The owners of the tea-houses used pretty young girls as decoys to lure customers to tarry.[494] Interestingly enough, Fortune considered the nuns, who begged for a living "... much the handsomest girls we saw in Japan"[495], so it would seem that he preferred women without make-up. One wonders what Jane Fortune made of her husband's musings on female physiognomy, or did she never bother to read his books?[496]

We know next to nothing about Robert Fortune's family. He made not even oblique references to his siblings, wife and children in his books. In Japan he mentions being asked by a gaggle of women if he was single or married, but withholds the answer![497] Was this a sign of the times, when one referred to people by their surnames, or was he trying to hide something? Was he attempting to cover up his humble origin, and the fact that he was conceived out of wedlock? The by now well-to-do Robert probably felt that this blot in the family history ought to be erased. Consequently, he let it be known that he was actually born in 1813, a date found in some sources.[498]

What sort of a father was he? Was he the archetypal paterfamilias, expecting complete subordination?[499] Judging from some of the statements made in his books, he was a firm believer in discipline. He advocated the use of force to teach the Chinese a lesson.[500] "The people of Tianjin and in the country around it are quiet and inoffensive, and particularly civil and polite to foreigners ... Having received a good flogging [in 1860], these children had now become very good boys; and if they did not love us, which we could scarcely expect, we were certainly feared and respected."[501] Unfortunately, we shall probably never know whether his children loved or feared him.

Chapter 11 Epilogue

When Robert Fortune died, he left a wife and four unmarried children behind. According to the British Census of 1881 his widow and their eldest daughter (Helen Jane, "Ellen") and youngest offspring (Alice) were living at Gilston Road.[1] At that time their elder son (John Lindley) was still farming at Elphinstone (**Fig. 11.1**).[2]

Fig. 11.1 View of fields near Elphinstone Tower Farm, once farmed by John Lindley Fortune (1842–1913).

Their younger son (Thomas) died, aged 23, in Tranent, Haddingtonshire on 4 March 1881.[3] It seems likely that he had been visiting his brother at Elphinstone when death overtook him. Had he contracted tuberculosis and was able to gain a certain amount of relief in Scotland away from the pea-soup smogs for which London was so renowned?

Tranent may not have been much better, because it is and was a coal-mining area. Elphinstone Colliery was owned by the Duries, so it is plausible that Alice met her future husband John Durie while she was visiting her brother John Lindley. John Durie (1851 – 1894) was described as an affable, kind and genial person, with an interest in a variety of scientific subjects and an admirer of Thomas Carlyle and Robert Burns.[4] The couple were married at St. James' Church, Westgate-on-Sea, a newly developed seaside resort in NE Kent, on 10 September 1884.[5] Their honeymoon was cut short by the death of John's father, Robert Hogg Durie, on 13 October 1884, at the age of 67.[6] John had now to take charge of "Barney Mains" (**Fig. 11.2**) with a total acreage of 3500 acres[7] and the coal-mining business employing 150 men and 10 boys.[8] The young couple rushed back to Scotland, leaving Alice's sister Helen Jane to look after their mother.

Fig. 11.2 Barney Mains, near Haddington, East Lothian, where Alice Fortune lived with her husband, John Durie, from 1884 to 1890.

The Duries had four children: Robert (1885 – 1949[9]), named in traditional Scottish style after his paternal grandfather, Robert Hogg Durie, John junior[10] (1886 – 1967), Jane Fortune after her maternal grandmother, and Norman Percival. In 1890 they moved into the newly built mansion "Stair Park" (**Fig. 11.3**) on the outskirts of Tranent.[11] Although John was a busy man with his farm and colliery to run, he still found time to be Chairman of the United East Lothian Agricultural Society, Chairman of the Church Defence Association (established in July 1890) to retain the Church of Scotland's status as the "established church"[12], Chairman of the Tranent School Board, Vice-convenor of the County Council, a director of the local gas company and an active member of the Conservative Party.[13] With all these commitments, Alice did not see much of her husband. They had been married for less than ten years when he died of an acute attack of rheumatic fever at "Stair Park" on 7 June 1894 at the age of 43.[14] Helen Jane and John Lindley Fortune came up from England to be at Alice's side. John Lindley Fortune was a pall-bearer at the funeral, which was attended by many prominent figures and ministers of all denominations. The streets in Tranent were lined by scores of inhabitants, as the shops and schools were closed as a sign of respect.[15] John Durie's younger brother, James Lauder Durie, could have taken over the business, had he not died three years previously, while returning from Australia (buried at sea on 2 June 1891).[16] Alice's mother-in-law, Catherine ("Kate") Taylor (1830 - 1895) followed soon afterwards on 2 March 1895[17], so with her only remaining brother in Surrey and just some of her sister-in-laws around, Alice moved to Edinburgh, where the four young children could be expected to get a better education. With an ample inheritance, Alice was able to afford a Georgian townhouse on the north side of Albany Street, diagonally opposite the house where her mother had worked (**Fig. 11.4**).

Fig. 11.3 Stair Park, Tranent, where the Duries lived from 1890 until John's death in 1894. It was empty for some time, before being set on fire by suffragettes. It was finally demolished to make way for a housing estate.

This address was to become the focal point of many a Fortune family reunion. It was here that Alice's mother and her elder sister died (on 27 November 1901 and 19 June 1910 respectively).[18] Alice herself passed away at her Albany Street home on 11 October 1927 and was buried with her husband in Tranent Parish Churchyard (**Fig. 11.5**).

Fig. 11.4 Alice Durie's Georgian House in Albany Street, Edinburgh. It was here that she passed away in 1927.

Alice's son Robert, who studied mine management, does not appear to have married, as he was still living with his mother in 1915[19], and is buried with his parents next the South Gate of Tranent Parish Churchyard (**Fig. 11.6**). Two of the other children wed in quick succession during the turmoil of World War I. Jane Fortune ("Daisy") Durie married the war hero Lieutenant-Colonel William Gemmill (ca. 1877 - 1918) DSO (**Fig. 11.7**), a son of the sheep farmer William Cunninghame Gemmill of "Greendykes" (Macmerry) and his wife Elizabeth on 21 October 1915.[20]

Fig. 11.5 Tranent Parish Churchyard, where the Duries were buried.

Fig. 11.6 Durie headstones in Tranent Parish Churchyard. The obelisk on the right marks the last resting place of John Durie's parents. John Durie, his wife Alice, and their son Robert lie buried next the Celtic cross.

Fig. 11.7 Lieutenant-Colonel William Gemmill (ca. 1878 – 1918), who married Jane Fortune ("Daisy") Durie in 1915.

Fig. 11.8 Roxburghe Hotel (now Crowne Plaza), Edinburgh, where two of Alice Durie's children held their wedding receptions.

It must have been a grand occasion with a reception in the Roxburghe Hotel (now Crowne Plaza), Edinburgh (**Fig. 11.8**).[21] Daisy gave birth to a baby daughter, Alice Fortune Gemmill on 22 August 1917.[22] Unfortunately, their wedded bliss was short-lived, for on 25 March 1918 the forty-year-old William Gemmill was killed at Loupart Wood, 20 km S of Arras (France) by a German shell, while encouraging his troops during a rearguard action.[23] The whole of Haddingtonshire was devastated by the news, with messages of condolance for his aged mother and young wife being read in Haddington Parish Church, Wishart Church (Tranent) and Gladsmuir Parish Church.[24] After the war Daisy placed a brass plaque to her husband's memory on the east wall of Gladsmuir Parish Church (**Fig. 11.9**).[25]

On 27 April 1921 she married Alexander Mackenzie, the younger son[26] of John Alexander Mackenzie (1857 – 1890) and Elizabeth MacDonald Sinclair of "Ardlair", Spylaw Road, Edinburgh.[27] They do not appear to have had any children.[28]

John had already married Eliza Bryson Mitchell (1885 – 1970), the second daughter of Thomas Bryson Mitchell (1852 – 1935) on 2 June 1915.[29] Their marriage was also celebrated in the Roxburghe

Hotel.[30] Initially the couple (**Fig. 11.10**) made their home at Dowlaw (Dulaw) Farm near Eyemouth, Berwickshire, where John had been the tenant since 1911 (**Figs. 11.11**, **11.12**).[31]

However, the rocky substrate and the exposed position next the North Sea, cannot have made it very profitable.[32] In 1918 John requested Sir John Richard Hall (1865 – 1928) of Dunglass Estate to relieve him of the tenancy.[33] His sister Jane had already encouraged John to lease "Greendykes" (**Fig. 11.13**), which was on the market after her husband William Gemmill had been killed.[34] John later regretted his decision as prices collapsed in the 1920s.[35] They weathered the Depression, but finally gave up farming in December 1942, when they turned over the farm to James Edwin Rennie of "East Fenton", Drem.[36] The couple had at least three sons, one that died in infancy in 1918[37], another who was to become an engineer, and Dr. Thomas Bryson Mitchell Durie (1919-2007), who was interviewed by Dawn Macleod in 1992.[38] Although Tommy would have gladly followed in his father's footsteps (**Fig. 11.14**), his disillusioned parents discouraged this, and suggested he become a doctor.[39]

Fig. 11.9 Brass plaque in memory of William Gemmill in Gladsmuir Parish Church.

Fig. 11.10 John Durie Jr. (1886–1967) and his wife Eliza Bryson Mitchell (1885–1970) in later life.

Fig. 11.11 Dowlaw Farm near Eyemouth, Berwickshire, where John Durie Jr. was tenant between 1911 and 1918.

Fig. 11.12 Fields at Dowlaw Farm exposed to the North Sea.

Fig. 11.13 Greendykes, Macmerry, run by William Gemmill, and after his death, leased by John Durie Jr.

So after attending the Royal High School in Edinburgh, he went on to study medicine in the Scottish capital (1938 – 1943).[40] During World War II he served with the Scottish Horse

Regiment.[41] When he was demobilized in 1947, he worked at Leith Hospital for a year before joining the Bacteriology Laboratory of the Edinburgh Royal Infirmary.[42] There he focussed on patient care and training nurses, but also taught medical students at the University of Edinburgh (1963 – 1984).[43] He is described as being "a nice man".[44] In 1948 Tommy married Ann Robertson Dunlop (1915 – 2007).[45] They had three children, Alison, John and Ann.[46]

Fig. 11.14 One of John Durie Jr.'s sons, Thomas Bryson Mitchell Durie (1919 – 2007), with "Clyde" in 1932. He became a physician.

Shortly after the end of the First World War on 7 January 1919 Norman Percival married Eliza Bryson Mitchell's younger sister Margaret.[47] They had two sons, Norman Percival Jr. and Thomas Mitchell Durie.[48] Neither seem to have married.[49]

When his father died, John Lindley Fortune left his "paradise"[50] at Elphinstone Tower and went to live at "Newhouse" (**Fig. 11.15**), south of Cranleigh in Surrey[51], just over 40 km from his mother at Brompton.

As John Lindley's elder sister Helen Jane never married[52] and was therefore free to take care of their mother, this could not have been the reason for moving south. It might have been the depression of the 1880s, caused by a combination of bad seasons in the 1870s and cheap imports that forced him and many Scottish farmers to move south.[53] However, based on subsequent events, it seems more likely that he was asked by Baron William Buller Fullerton-Elphinstone (**Box 10.3**) to leave.[54] On the 10 August 1882 he married his Scottish housekeeper, Alison Jane Lumsden (1845 – 1923) in London.[55] Having inherited enough money after his father died, John Lindley took up farming again.[56] John and Alison ("Alice") had a son, Robert Fortune (1884 – 1961), and two daughters, Margaret Alice (1885 – 1951) and Constance Edith Fortune (1887 – 1951).[57] Having married late, John Lindley never lived to see any of his children wed. He died on 23 May 1913 at the age of 70 and was buried at Brompton Cemetery on 28 May 1913 (**Fig. 11.16**).[58]

Fig. 11.15 Newhouse, south of Cranleigh in Surrey, where John Lindley Fortune farmed after he left the Elphinstone Estate.

The following year World War I broke out. His son Robert, who was a barrister, may have joined the Border Regiment, which was billeted in Carlisle.[59] While he was there he could have met Wilhelmina Purves (1883 – 1963), the youngest daughter of the local schoolmaster William Purves. The couple were married by Rev. H. Ernest Scott, Vicar of St. Mary's Carlisle, at Stanwix Parish Church on 3 January 1917.[60] The wedding was attended by the bridegroom's mother and his aunt Alice Durie.[61]

Alice was probably responsible for inserting the short announcement in "The Scotsman".[62] After their marriage the couple lived at "Little Withybush" on Knowle Lane in Cranleigh.[63] The couple had four daughters: Alison E. ("Betty") Fortune (1918 – 1926), Alice Purves Fortune (1920 – 1942), Zella Charlton Fortune (1922 – 1990) and Winifred M. Fortune.[64] When their second daughter was on her way, Alison magnanimously turned over "Newhouse" to the young couple, and went to live with her unmarried daughters at "Laurel Bank", 16 Midhope Road in Woking, Surrey.[65] On 27 November 1923 Alison, was "called home". After their mother died, Margaret Alice and Constance Edith Fortune continued to live at "Laurel Bank" for a number of years.[66] However, in 1930 they moved to "Loganlee" in Heathside Crescent, closer to the centre of Woking.[67] They lived there until 1951 when first Constance and then Margaret passed on.[68] Although there is no indication on the granite slab (**Fig. 11.16**), Constance was laid to rest along with her parents on 16 May 1951.[69] Margaret's last resting place remains unknown. Feeling that his end was near, her elder brother Robert sold "Newhouse" on 24 November 1960 to Frederick Reginald and Terence Patrick Courtney, before being taken into Warlingham Park Hospital.[70] He died there at the beginning of April 1961, and was buried in Cranleigh Cemetery (**Fig. 11.17**) on 8 April 1961.[71] His wife, who went to live at "Wayside", St. Marys Road, Leatherhead, followed 26 months later.[72] While Robert Fortune and his direct descendents have all passed away, the memory of the plant hunter lives on through the plants he introduced to England's green and pleasant land (**Fig. 11.18**).

Fig. 11.16 Grave of John Lindley Fortune (1842–1913), his wife Alison Jane Lumsden (1845–1923), and their younger daughter, Constance Edith Fortune (1887 – 1951) in Brompton Cemetery.

Fig. 11.17 Grave of Robert Fortune Jr. (1884 – 1961) and family in Cranleigh Cemetery, Surrey.

Fig. 11.18 A *Trachycarpus fortunei* near the pond at Newhouse Cottage as a living memento to the plant hunter, Robert Fortune.

Index to Place Names

Where possible the Pinyin transliteration of the Chinese names has been employed. Fortune's spellings are given in *italics*.

Aban Court, Cheltenham: Box 9.2 (Staveley)
Aberdeen: Box 9.3 (Gordon)
Aberdeen University: Box 5.7 (Falconer)
Acton: Box 4.1 (Royle)
Acton Green, London: Box 2.1 (Lindley)
Addiscombe: Box 3.4 (Balfour)
Aden: Ch. 5.1, Ch. 5.12, Box 5.33 (Bentinck)
Agra : Ch. 3.5 note 230, Ch. 5.6, Box 5.12 (Batten), Box 6.4, Box 9.9 (Grant)
Ah- sax- saw Temple , see Asakusa Kan' non Temple
Ah- yuh- wang Temple , see Ayù Wáng Sì
Alaska: Box 9.2 (Staveley)
Albany: Box 3.1 (Abeel)
Albany Street, Edinburgh: Ch. 1.3, Ch. 11, Figs. 1.13, 11.4
Albemarle Street, London: Acknowledgements, Ch. 10.1, Figs. 10.13, 10.14
Albert Bridge: Ch. 10.1 note 101
Aldermaston: Box 6.2 (Glendinning)
Alexandra Palace: Ch. 10.1
Alexandria: Ch. 5.1, Ch. 5.12, Ch. 5.12 note 714, Ch. 9.2, Box 9.10 (Ceylon), Fig. 5.2, 5.4, 9.54
All Saints Church, West Ham: Box 2.3 (Reeves)
Allahabad: Ch. 5.6, Box 5.8 (Jameson), Ch. 6.2, Box 9.3 (Gordon), Fig. 5.71
Almora: Ch. 5.6, Box 5.13 (Ramsay), Fig. 5.87
Almorah , see Almora
Ambala: Box 9.9 (Grant), Ch. 10.4
Amba Mariam, see Magdala
America: Ch. 1.1
Amherst College: Box 8.7 (Brown)
Amoy , see Xiàmén
Amsterdam: Box 8.2 (Kaempfer), Box 8.3 (Thunberg), Box 8.12 (Heusken)
Amur, see Hēilóngjiāng
Ancient Observatory, Běijīng: Ch. 9.2, Figs. 9.21, 9.24
Āndìngmén, Běijīng: Ch. 9.2, Box 9.9 (Grant), Fig. 9.37, 9.38
Ānhuī: Ch. 5.2, Ch. 5.5, Ch. 5.8, Ch. 5.4 note 127, Ch. 5.8 note 459, Box 5.19 (Tàipíng), Box 5.20 (Hóng), Ch. 10. 4, Box 10.5 (Niǎn), Fig. 5.17
Anoo Plantation : Ch. 5.6, Fig. 5.85, 5.86
Ansei Earthquake: Box 8.7
An- ting Gate , see Āndìngmén
Antwerp: Box 8.4 (Siebold)
Àomén, see Macao

Aózhù Pagoda, Zhènhǎi: Fig. 5.149
Apothecaries' Hall, London: Acknowledgements
Ardlair, Spylaw Road, Edinburgh: Ch. 11
Argentina: Ch. 1.1, Box 3.8 (Cuming), Box 5.14 (Paxton)
Aroostook: Box 9.4 (Bruce)
Arras, France: Ch. 11
Asakusa Kan'non Temple: Ch. 8.2, Box 8.7 (Brown), Box 8.9 (Girard), Box 8.15 (Introductions) note 17, Figs. 8.35, 8.36, 8.37
Asakusa Shrine: Ch. 8.2
Ascension Island: Box 2.1 (Lindley)
Ascot: Ch. 5.7, Box 5.15 (Standish), Box 8.15 (Introductions) note 60, Ch. 10.1, Ch. 10.1 note 2, Fig. 10.1
Asia: Box 8.2 (Kaempfer), Box 9.2 (Staveley), Ch. 10.1, Figs. 3.25, 3.40D, 3.40F, 3.40G
Assam: Box 5.31 (Beadon)
Astoria: Ch. 7.1, Box 7.1
Atago Jinja: Ch. 8.2
Atagoyama: Ch. 8.2, Figs. 8.30, 8.31
Atango-yama, see Atagoyama
Atami: Box 3.10 note 24 (Alcock)
Atenze, see Déqīn
Atfeh, Egypt: Ch. 5.1
Athens, Greece: Box 9.6 (Wyndham)
Athens, New York State: Box 3.1 (Abeel)
Atlantic Ocean: Ch. 1.1, Box 3.8 (Cuming), Ch. 5.1, Ch. 8.2 note 148
Auchterarder: Box 5.24 (Burdon)
Australasia: Fig. 3.40G
Australia: Box 2.6 (Ward), Box 5.18 (Bonham), Box 5.29 (Chusan), Box 6.2 (Glendinning), Box 8.1 (Veitch), Box 10.3 (Fullerton-Elphinstone), Ch. 11, Fig. 3.40E
Austria: Box 3.12 note 85 (Introductions), Box 9.7 note 67 (Cí Xǐ)
Awadh: Box 5.13 (Ramsay), Box 6.4 (Indian Mutiny)
Awashima Island: Ch. 8.2 note 139
Ayrshire: Box 1.2 (McNab)
Ayton: Ch. 1.2
Ayuka's Temple, see Ayù Wáng Sì
Ayù Wáng Sì: Ch. 5.8, Figs. 5.119, 5.120, 5.121
Azamgarh: Box 5.12 (Batten), Box 9.3 (Gordon)

Bādàchù: Ch. 9.2, Figs. 9.43, 9.45, 9.47, 9.49, 9.55
Bad Kissingen: Box 8.4 (Siebold), Box 8.4 note 51 (Siebold)
Bageshwar: Box 5.13 (Ramsay)
Bago Pegu: Box 9.3 (Gordon)
Bagshot: Box 5.15 (Standish), Box 8.15 (Introductions) notes 12, 30, 60
Bagshot Bridge: Box 5.15 (Standish)
Bagshot Park: Box 5.15 (Standish)
Bǎiguān, see Shàngyú
Báiquǎn Islands: Ch. 3.4

Baijnath: Ch. 5.6, Fig. 5.84
Báiyún Sì, Běijīng: Ch. 9.2, Figs. 9.27, 9.28
Baku: Box 8.2 (Kaempfer)
Balaclava: Box 8.8 note 4 (Hall)
Bālǐqiáo: Ch. 9.2, Figs. 9.51, 9.52
Bālǐzhuāng: Ch. 9.2, Fig. 9.43
Ballymote: Box 5.23 (Perceval)
Balmoral: Ch. 8.1 note 98
Baltic Sea: Box 8.2 (Kaempfer)
Bandar Abbas: Box 8.2 (Kaempfer)
Bangkok: Box 3.1 (Abeel)
Barbados: Box 9.3 (Gordon)
Barbuda: Box 3.1 (Abeel)
Bardstown: Box 7.3 (Holt)
Bareilly Jail: Box 5.13 note 32 (Ramsay)
Barhamville Young Ladies' Seminary: Box 8.7 (Brown)
Barnet, London: Box 2.1 (Lindley)
Barney Mains, Haddington: Ch. 11, Fig. 11.2
Basel: Box 2.1 (Lindley)
Basil Street: Ch. 10.1
Bātáng: Ch. 10.4
Bātáng Plain: Ch. 10.4
Batavia: Box 3.1 (Abeel), Box 3.11 (Lockhart), Box 5.10 (Lady Mary Wood), Box 8.2 (Kaempfer), Box 8.3 (Thunberg), Box 8.4 (Siebold), Box 8.6 (Hepburn)
Battlesden House: Box 5.14 (Paxton)
Battersea Bridge: Ch. 10.1
Battersea Fields: Ch. 10.1
Battersea Park: Acknowledgements, Ch. 10.1, Figs. 10.25, 10.27, 10.28, 10.29, 10.30
Bay, Luzon: Ch. 10.6 note 471
Bay of Bengal: Ch. 5.1
Bayonne: Box 5.28 (Harvey)
Baxter Park, Dundee: Box 5.14 (Paxton)
Beaufort Street: Ch. 10.1
Bedford: Box 5.24 (Burdon)
Bedfordshire: Box 5.14 (Paxton)
Běihǎi: Box 5.24 (Burdon), Ch. 10.4
Běihéyàn Dàjiē, Běijīng: Ch. 9.2
Běijīng: Acknowledgements, Ch. 2.2 note 45, Box 2.2 (Opium), Box 3.11 (Lockhart), Ch. 5.4, Box 5.19 (Tàipíng), Box 5.19 note 32 (Tàipíng), Box 5.20 (Hóng), Box 5.24 (Burdon), Box 8.7 (Brown), Ch. 9.1, Ch. 9.2, Ch. 9.1 note 42, Ch. 9.2 note 120, Box 9.1 (Tiānjīn), Box 9.2 (Staveley), Box 9.3 (Gordon), Box 9.4 (Bruce), Box 9.5 (Ricci), Box 9.6 (Wyndham), Box 9.7 (Cí Xǐ), Box 9.8 (Bowlby), Box 9.8 note 16 (Bowlby), Box 9.9 (Grant), Ch. 10.1, Ch. 10.3, Ch. 10.4, Box 10.5 (Niǎn), Box 10.7 (Panthay), Box 10.10 (Wade), Figs. 3.49, 9.13, 9.18, 9.21, 9.23, 9.34, 9.35, 9.36
Běijīng Administrative College: Box 9.5 note 25 (Ricci)
Běijīng Botanic Garden: Ch. 2.2 note 50

Běitáng: Box 9.4 (Bruce), Box 9.9 (Grant)
Běi Yùnhé: Ch. 9.1, Fig. 9.15
Belfast: Box 5.1 (P&O)
Belgium: Box 2.1 (Lindley), Box 5.22 (Meadows), Box 8.15 (Introductions) note 40, Box 9.8 (Bowlby)
Belgrade: Box 9.6 (Wyndham)
Belle Vue , Edo: Ch. 8
Bellevue Crescent, Edinburgh: Ch. 1.3, Fig. 1.15
Benares: Box 9.3 (Gordon)
Bengal: Box 2.2 note 7 (Opium), Box 4.1 (Royle), Box 5.31 (Beadon)
Benten dori: Ch. 10.2 note 138
Benten-kutsu, Kanagawa: Ch. 8.2 note 199
Beringia: Fig. 3.40G
Berkshire: Box 5.15 (Standish), Box 5.30 (Vansittart), Box 6.2 (Glendinning)
Berlin: Box 9.6 (Wyndham), Box 9.8 (Bowlby), Ch. 10.4
Berne: Box 9.9 (Grant)
Berwick-upon-Tweed: Ch. 1.2
Berwickshire: Ch. 11, Fig. 11.11
Bhamo, Burma: Ch. 10.4, Figs. 10.43, 10.45
Bheem Tal , see Bhimtal
Bhimtal: Ch. 5.6, Fig. 5.61
Bhurtpoor Plantation , see Bhurtpur Plantation
Bhurtpore Plantation, see Bhurtpur Plantation
Bhurtpur Plantation: Ch. 5.6 note 349, Box 5.7 (Falconer), Fig. 5.87
Bicton House: Box 6.2 (Glendinning)
Bihar: Box 2.2 note 7 (Opium)
Bing- bong , see Píngwàng
Birkenhead Park: Box 5.14 (Paxton)
Birmingham: Box 5.28 (Harvey)
Bisham: Box 5.30 (Vansittart)
Blackadder: Ch. 1.2, Ch. 1.2 note 131, Fig. 1.10
Blackadder Estate: Ch. 1.2, Ch. 10
Blackadder House: Ch. 1.2, Fig. 1.5
Blackadder Toun: Ch. 1.2
Blackheath: Box 3.11 (Lockhart), Ch. 10.2 note 138
Black Rock Hill Temple, see Wū Shí Shān Sì
Black Sea: Box 10.3 (Fullerton-Elphinstone)
Blackwater, see Blackadder
Blair Adam, Kinross-shire: Ch. 1.2 note 134, Box 10.3 (Fullerton-Elphinstone)
Bogue, see Hǔmén
Bóhǎi Region: Ch. 3.4 note 197
Bóhǎi Sea: Box 2.2 (Opium), Box 8.5 (Loureiro)
Bohea Mountains : Ch. 3.4, Ch. 5.4, Ch. 5.9
Bokengee , see Buken-ji
Bolivia: Ch. 1.1, Box 9.4 (Bruce)
Bolzano, see Bozen

Bombay, see Mumbai
Bo'ness: Box 9.10 (Ceylon)
Bonin Islands: Box 3.9 (Lay)
Bonn: Box 8.4 note 53
Bonnyrigg: Box 5.32 (Dalhousie)
Boppard am Rhein: Box 8.4 note 53
Bordeaux: Box 3.5 (Thom)
Borneo: Box 3.1 (Abeel), Box 3.9 (Lay), Box 8.8 note 4 (Hall)
Boshi Sen Field Research Laboratory: Ch. 5.6, Ch. 5.6 note 339
Boston: Box 8.6 (Hepburn), Box 8.8 (Hall), Box 9.4 (Bruce)
Boston Church, Duns: Box 1.1 note 13 (Buchan)
Boston Harbour: Ch. 7.1, Fig. 7.1
Boulogne: Box 9.2 (Staveley)
Bourges: Box 8.9 (Girard)
Bournemouth: Box 3.7 (Morrison)
Bowood House: Box 5.15 (Standish)
Bozen: Box 8.4 note 53
Brahmakund: Ch. 10.4
Brahmaputra River: Ch. 10.4
Bramley Station: Ch. 11 note 64
Brazil: Box 9.6 (Wyndham)
Breckenridge County: Box 7.3 (Holt)
Breda: Box 8.12 (Heusken)
Brest: Ch. 1.1
Bridge of Earn: Box 9.9 (Grant)
Bristol, Rhode Island: Box 8.8 (Hall), Box 8.8 note 7 (Hall)
Britain: Ch. 1.1, Ch. 1.3, Ch. 1 notes 53, 64, 85, Ch. 2.1, Ch. 2.2, Ch. 2.2 note 54, Box 2.2 (Opium), Box 2.2 note 3 (Opium), Box 2.3 (Reeves), Box 2.7 note 6 (Campbell-Johnston), Ch. 3.1, Box 3.4 (Balfour), Box 3.5 (Thom), Box 3.9 (Lay), Box 3.12 notes 20, 61, 65, 68, 72, 74, 79, Ch. 5.6, Ch. 5.8, Ch. 5.11, Ch. 5.4 note 170, Box 5.6 (Champion), Box 5.7 (Falconer), Box 5.8 note 18 (Jameson), Box 5.11 (Saharanpur), Box 5.13 (Ramsay), Box 5.15 (Standish), Box 5.16 notes 5, 9, 14 (Introductions), Box 5.28 (Harvey), Box 5.29 (Chusan), Box 5.32 (Dalhousie), Box 5.33 (Bentinck), Box 6.1 (Arrow), Box 7.3 note 24 (Holt), Ch. 8.2, Ch. 8.3, Box 8.8 (Hall), Box 8.13 note 6 (Aspinall), Box 8.15 (Introductions) notes 9, 13, 25, 26, 30, 38, 46, 62, Ch. 9.2, Box 9.1 (Tiānjīn), Box 9.2 (Staveley), Box 9.3 (Gordon), Box 9.9 (Grant), Ch. 10.1, Ch. 10.2, Ch. 10.3, Ch. 10.4, Ch. 10.5, Ch. 10.3 notes 202, 204, Ch. 10.5 note 423, Box 10.9 (Muslim), Figs. 1.4, 5.15, 6.3A, 6.3F, 8.26, 9.55
British Embassy (Constantinople): Box 9.6 (Wyndham)
British Empire: Ch. 1.1, Box 5.6 (Champion)
British Burma, see Burma
British India, see India
British Isles: Ch. 1.2, Ch. 1.1 note 45, Ch. 7.3
British Legation (Athens): Box 9.6 note 13 (Wyndham)
British Legation (Běijīng): Ch. 9.1 note 47, Ch. 9.2 notes 49, 76, Box 9.4 note 26 (Bruce), Box 9.6 note 4 (Wyndham), Box 10.10 note 12 (Wade), Fig. 9.19
British Legation (Berlin): Box 9.6 note 12 (Wyndham)
British Legation (Edo): Box 3.10 note 21 (Alcock), Ch. 8.2 notes 72, 82, 83, Ch. 8.3 notes 153, 168, 203, Box 8.4

note 73 (Siebold), Box 9.9 note 48 (Grant)
British Legation (Madrid): Box 9.6 note 13 (Wyndham)
British Museum, London: Ch. 8.2 note 182, Box 8.2 (Kaempfer)
British North America, see Canada
British Virgin Islands: Box 3.1 (Abeel)
Brittany: Ch. 5.1
Brixham Cave: Box 5.7 (Falconer)
Bromhall, Fife: Box 9.4 (Bruce)
Brompton: Ch. 11
Brompton Cemetery: Acknowledgements, Ch. 4.2, Ch. 6.2, Ch. 7.1 note 20, Box 9.2 (Staveley), Ch. 10.5, Ch. 11, Ch. 11 notes 3, 18, 69, Figs. 2.8, 4.6, 10.49, 10.50, 10.51, 10.52, 11.16
Brompton House: Ch. 10.1
Brompton Road: Ch. 10.1
Brompton Square: Ch. 10.1, Fig. 10.5
Bromptons, see Consumption Hospital
Brookwood Hospital: Ch. 11 note 68
Broughton Street, Edinburgh: Ch. 1.3, Fig. 1.14
Brussels: Box 3.6 (Sinclair), Box 8.4 (Siebold), Box 8.4 note 30 (Siebold)
Bryant University: Acknowledgements
Bucharest: Box 9.6 (Wyndham)
Buckinghamshire: Box 8.14 (Howard-Vyse)
Buenos Aires: Box 3.8 (Cuming), Box 3.8 note 4 (Cuming)
Buitenzorg Botanic Gardens: Box 8.15 (Introductions) notes 56, 61
Buken-ji, Yokohama: Ch. 8.2, Box 8.7 (Brown), Figs. 8.18, 8.19
Bungo Channel : Ch. 8.2
Burlington: Box 7.2 (Mason)
Burma: Box 5.7 (Falconer), Box 5.32 (Dalhousie), Box 6.4 (Indian Mutiny); Box 9.3 (Gordon), Ch. 10.4, Box 10.7 (Panthay), Figs. 10.43, 10.45, 10.47
Butterwick: Box 5.15 (Standish)
Byznath , see Baijnath

Cading , see Jiādìng
Cádiz: Ch. 1.1, Ch. 5.1 notes 3, 5, 12, Ch. 5.8 note 393
Cairo: Ch. 5.1, Ch. 5.12, Ch. 5.12 note 714
Cākǎ: Ch. 10.4
Calcutta , see Kolkata
California: Box 2.7 (Campbell-Johnston), Box 3.9 (Lay), Box 5.18 (Bonham), Ch. 7.1, Ch. 8.2 note 168, Box 8.11 (Harris), Box 9.2 (Staveley)
Cambridge: Box 2.5 (Smith-Stanley), Box 10.10 (Wade), Figs. 2.3, 10.44
Cambridge University: Box 10.10 (Wade)
Cam-poo , see Gǎnpǔ
Canada: Ch. 1.1, Box 5.30 (Vansittart), Box 5.32 (Dalhousie), Box 9.2 (Staveley), Box 9.4 (Bruce), Box 10.3 (Fullerton-Elphinstone)
Cancer Hospital (Royal Marsden Hospital): Ch. 5.7, Fig. 5.108
Canton , see Guǎngzhōu

Cáo'é: Ch. 5.3
Cape Chichakoff, see Sata-Misaki
Cape Coast, Ghana: Box 9.3 (Gordon)
Cape Colony: Ch. 2.2 note 70, Box 10.10 (Wade)
Cape of Good Hope: Ch. 4.1, Ch. 5.9 note 502, Box 5.1 (P&O), Box 8.3 (Thunberg), Box 8.7 (Brown), Ch. 9.2, Box 9.9 (Grant), Ch. 10.1 note 2
Cape Horn: Box 8.11 (Harris)
Cape Tourinan: Box 5.10 (Lady Mary Wood)
Cape Town: Box 5.17 (Ganges), Box 8.3 (Thunberg)
Cape York: Box 8.1 (Veitch)
Caracas: Box 3.5 (Thom)
Carberry Tower, East Lothian: Acknowledgements, Box 10.3 (Fullerton-Elphinstone), Figs. 10.20, 10.21
Cardiff: Box 3.6 (Sinclair)
Carlisle: Ch. 11
Carlisle Castle: Ch. 11 note 59
Caspian Sea: Box 8.2 (Kaempfer)
Castle Hill, Devonshire: Box 8.15 (Introductions) note 61
Catton: Box 2.1 (Lindley)
Cawnpore, see Kanpur
Celestial Empire, see China
Central America: Box 3.8 (Cuming)
Central Asia: Box 10.9 (Muslim)
Centre College, Danville: Box 7.3 (Holt)
Ceylon: Box 2.6 (Ward), Ch. 5.1, Ch. 5.12, Box 5.4 (Braganza), Box 5.6 (Champion), Box 5.10 (Lady Mary Wood), Box 5.33 (Bentinck), Box 8.1 (Veitch), Box 8.2 (Kaempfer), Box 8.3 (Thunberg), Box 8.11 (Harris), Figs. 5.7, 5.8
Chan-chow-wan, see Zhāngjiāwān
Chandannager: Acknowledgements
Cháng Jiāng: Box 2.2 (Opium), Ch. 3.1, Ch. 3.2, Box 3.2 (Schoedde), Box 3.4 (Balfour), Ch. 5.8, Ch. 5.9, Ch. 5.9 note 590, Box 5.19 (Tàipíng), Box 8.5 (Loureiro), Ch. 10.4, Ch. 10.4 notes 225, 244
Chāngpíng: Box 9.4 (Bruce), Box 9.6 (Wyndham), Box 9.8 (Bowlby)
Chángshā: Box 5.19 (Tàipíng)
Chang-shan, see Jiāngshān
Chapel Island, see Dōngdìng Dǎo
Chapoo, see Zhàpǔ
Chapu, see Zhàpǔ
Chargos Archipelago: Box 5.3 (Moresby)
Charleston: Ch. 7.1, Box 7.3 (Holt)
Charleville Hotel, Mussoorie: Ch. 5.6 note 319
Charterhouse School: Box 5.12 (Batten)
Château de Ferrières: Box 5.14 (Paxton)
Châteaumeillant: Box 8.9 (Girard)
Chatsworth: Box 5.14 (Paxton)
Chayu, see Rìmǎ
Chá yuán: Ch. 5.2

Chefoo , see Yāntái
Chefoo Convention: Ch. 10, Box 10.10 (Wade)
Chekiang , see Zhèjiāng
Chelsea: Ch. 4.1, Ch. 10.1, Box 10.1 (Moore), Fig. 4.6
Chelsea Embankment: Ch. 10.1, Box 10.1 (Moore)
Chelsea Physic Garden: Acknowledgements, Box 2.1 (Lindley), Box 2.6 (Ward), Ch. 4.1, Ch. 10.1, Ch. 10.1 note 74, Box 10.1 (Moore), Box 10.1 note 16 (Moore), Figs. 4.2, 4.3, 10.8, 10.9, 10.10
Chelsea Vestry: Ch. 10.1
Chelsham: Ch. 11, Ch. 11 note 70
Cheltenham: Box 9.2 (Staveley)
Chen-chiang, see Zhènjiāng
Chéngdé: Ch. 9.2, Box 9.7 (Cí Xǐ), Box 9.7 note 16 (Cí Xǐ), Box 9.8 (Bowlby), Fig. 9.36
Chéngdū: Ch. 10.4
Chenhai , see Zhènhǎi
Chennai: Box 1.1 (Buchan), Box 3.4 (Balfour), Ch. 5.12, Box 5.10 (Lady Mary Wood), Box 5.33 (Bentinck), Box 9.3 (Gordon)
Ch' en-yuan Fu, see Zhènyuǎn
Cheops Pyramid: Box 8.14 (Howard-Vyse)
Cher, France: Box 8.9 (Girard)
Cheshire: Box 5.14 (Paxton)
Chestnut Hill, Boston: Box 8.8 (Hall)
Chichester: Box 9.3 (Gordon)
Chíhé: Box 10.5 (Niǎn)
Chih-p' ing, see Chípíng
Chile: Box 3.8 (Cuming)
Chimoo Bay, see Shēnhù Gǎng
China: Frontispiece, Preface, Acknowledgements, Ch. 1.2, Ch. 2.1, Ch. 2.2, Ch. 2.2 notes 33, 40, 57, 65, 68, Box 2.1 (Lindley), Box 2.2 (Opium), Box 2.3 (Reeves), Box 2.4 (Dent), Box 2.5 (Smith-Stanley), Box 2.6 (Ward), Box 2.7 (Campbell-Johnston), Ch. 3.1, Ch. 3.2, Ch. 3.4, Ch. 3.5, Ch. 3.1 note 53, Ch. 3.2 note 140, Ch. 3.5 notes 223, 228, Box 3.1 (Abeel), Box 3.3 (Macgowan), Box 3.4 (Balfour), Box 3.4 note 8 (Balfour), Box 3.5 (Thom), Box 3.8 (Cuming), Box 3.9 (Lay), Box 3.10 (Alcock), Box 3.10 note 26 (Alcock), Box 3.11 (Lockhart), Box 3.12 (Introductions) notes 3, 9, 21, 85, 92, 97, 108, Ch. 4.2, Ch. 5.1, Ch. 5.2, Ch. 5.3, Ch. 5.4, Ch. 5.5, Ch. 5.6, Ch. 5.7, Ch. 5.8, Ch. 5.11, Ch. 5.12, Ch. 5.1 note 13, Ch. 5.4 notes 148, 206, 243, Ch. 5.5 note 295, Box 5.10 (Lady Mary Wood), Box 5.15 (Standish), Box 5.15 note 11 (Standish), Box 5.16 (Introductions) notes 1, 8, 9, 18, 21, 27, Box 5.18 (Bonham), Box 5. 19 (Tàipíng), Box 5.20 (Hóng), Box 5.22 (Meadows), Box 5.23 (Perceval), Box 5.24 (Burdon), Box 5.27 (Robertson), Box 5.28 (Harvey), Box 5.29 (Chusan), Ch. 6.1, Ch. 6.2 note 31, Box 6.1 (Arrow), Box 6.2 (Glendinning), Box 6.3 (Introductions) notes 1, 2, Ch. 7.1, Ch. 7.2, Ch. 7.3, Ch. 8.1, Ch. 8.2, Ch. 8.3, Ch. 8.2 note 168, Box 8.7 (Brown), Box 8.8 (Hall), Box 8.8 note 4 (Hall), Box 8.11 (Harris), Box 8.12 (Heusken), Box 8.15 (Introductions) note 46, Ch. 9.1, Ch. 9.2 note 106, Box 9.1 (Tiānjīn), Box 9.2 (Staveley), Box 9.3 (Gordon), Box 9.4 (Bruce), Box 9.5 (Ricci), Box 9.6 (Wyndham), Box 9.7 (Cí Xǐ), Box 9.7 note 28 (Cí Xǐ), Box 9.8 (Bowlby), Box 9.9 (Grant), Ch. 10.1, Ch. 10.2, Ch. 10.3, Ch. 10.4, Ch. 10.5, Ch. 10.6, Ch. 10.1 note 16, Ch. 10.3 notes 166, 182, Ch. 10.4 note 362, Ch. 10.6 note 491, Box 10.1 note 6 (Moore), Box 10.3 (Fullerton-Elphinstone), Box 10.7 (Panthay), Box 10.9 (Muslim), Box 10.10 (Wade), Figs. 2.2, 3.3, 3.24, 3.40E, 3.40H, 3.46, 3.57, 4.5, 5.9, 5.16, 5.20, 5.24, 5.81, 5.124, 5.126, 5.154, 5.165, 6.3, 7.5, 7.7B, 7.7D, 8.46, 9.7, 9.20, 9.54, 10.43, 10.45, 10.47
China Sea: Ch. 3, Box 3.3 (Macgowan), Box 5.18 (Bonham), Box 10.3 (Fullerton-Elphinstone)

Chinatown, Yokohama: Box 8.9 (Girard)
Chinchew Bay, see Quánzhōu Gǎng
Chinese Central Asia, see Chinese Turkestan
Chinese City, Beijing: Ch. 9.1, Ch. 9.2
Chinese Turkestan: Box 10.8 (Huí), Box 10.9 (Muslim)
Ching- hoo, see Qīnghú
Ching- wang- mun, see Qiánmén
Chinhae, see Zhènhǎi
Chinhai, see Zhènhǎi
Chinkiang, see Zhènjiāng
Chin-kiang-fu, see Zhènjiāng
Chípíng: Box 10.5 (Niǎn)
Chippenham: Box 3.8 note 2 (Cuming)
Chirnside: Acknowledgements, Ch. 1.2, Box 1.1 (Buchan)
Chiswick: Box 1.2 (McNab), Ch. 2.1, Ch. 3.1, Box 3.12 (Introductions) notes 10, 38, Box 5.14 (Paxton), Ch. 6.1, Box 6.2 (Glendinning), Ch. 10.1
Chóng'ān: Ch. 5.4
Chóng Míng River: Ch. 5.8 note 398
Chóngqìng: Ch. 10.4
Christ's Hospital School, London: Box 2.3 (Reeves)
Chuānbí: Box 2.2 (Opium), Box 2.2 note 25 (Opium), Box 2.7 (Campbell-Johnston), Box 2.7 note 6 (Campbell-Johnston)
Chuānbí Fort: Box 2.2 note 34 (Opium)
Chu- chu, see Zǐxī
Chu- chu- foo, see Qúzhōu
Chuenpi, see Chuānbí
Chún-ān: Ch. 5.2
Church of the Sacred Heart, see Tenshu-do
Chusan, see Zhōushān
Cíchéng: Box 2.2 (Opium), Ch. 5.8, Ch. 5.9, Ch. 5.6 note 348, Ch. 10.6, Figs. 5.122, 5.124
Císhòu Pagoda: Ch. 9.2, Fig. 9.43
Clapham: Box 2.3 (Reeves)
Coal Hill, see Jǐngshān
Cockburnspath: Ch. 11
Coimbra: Box 9.5 (Ricci)
Cold Water Temple, see Léngshuǐ'Ān
Colombo: Box 2.7 (Campbell-Johnston)
Columbia, South Carolina: Box 8.7 (Brown)
Connecticut: Box 8.7 (Brown), Box 8.8 (Hall)
Constantinople: Box 9.6 (Wyndham)
Consumption Hospital: Ch 5.7, Fig. 5.107
Coombe Wood: Box 8.1 (Veitch)
Copenhagen: Box 9.10 (Ceylon)
Corbett National Park: Ch. 5.6 note 342
Corfu: Box 5.6 (Champion)

Cork: Box 7.1
Cornwall: Box 5.12 (Batten)
Covent Garden, London: Ch. 7.1 note 19, Box 8.10 (Stevens)
Coventry: Box 5.14 (Paxton)
Cranleigh, Surrey: Acknowledgements, Ch. 11, Fig. 11.15
Cranleigh Cemetery: Ch. 11, Ch. 11 note 64, Fig. 11.17
Cranley Place: Ch. 5.7
Cranshaws: Ch. 1.3
Crathie: Box 9.3 (Gordon)
Cremorne Pleasure-gardens: Acknowledgements, Ch. 10.1, Fig. 10.31
Crimea: Box 5.14 (Paxton)
Cromwell Road: Ch. 10.1
Crowne Plaza, Edinburgh, see Roxburghe Hotel
Croydon: Box 8.10 (Stevens)
Crystal Palace: Ch. 5.7, Box 5.14 (Paxton)
Cuba: Ch. 1.1, Box 7.3 (Holt)
Cumbria Archive Centre, Carlisle: Ch. 11 notes 51, 55, 56, 57, 60, 61, 63, 64

Dàbēi Sì: Ch. 9.2, Fig. 9.46
Dàbùtóu: Ch. 5.9, Fig. 5.128
Dàgū: Ch. 9.1, Ch. 9.2, Box 9.4 (Bruce)
Dàgū Forts: Ch. 7.2, Ch. 9.1, Box 9.4 (Bruce), Box 9.7 (Cí Xǐ), Box 9.9 (Grant)
Daigyo-ji, Kamakura: Ch. 8.3, Fig. 8.63
Dailly: Box 1.2 (McNab)
Dalhousie Castle: Box 5.32 (Dalhousie)
Dàlǐ: Ch. 10.4, Box 10.7 (Panthay)
Dangosaka: Ch. 8.2, Ch. 8.3
Dang- o- zaka , see Dangosaka
Dànshuǐ: Ch. 5.9, Ch. 5.1 note 2
Danville: Box 7.3 (Holt)
Dàochǎng Bāng: Ch. 5.10
Dàochǎngshān Tǎ: Ch. 5.10, Fig. 5.158
Dàoxiàn, see Dàozhōu
Dàozhōu: Box 5.19 (Tàipíng)
Darjeeling: Box 5.31 (Beadon), Box 9.3 (Gordon), Ch. 10.4
Dartford: Box 9.2 note 22 (Staveley)
Dàxiàng Mountains: Ch. 10.4
Dàyáng: Ch. 5.4, Ch. 5.4 note 130
Dàyíng Valley: Ch. 10.4
Dàyú: Box 5.26
Dà Yùnhé: Box 2.2 (Opium), Ch. 5.2, Ch. 5.10, Box 5.19 (Tàipíng), Ch. 9.1, Box 9.5 (Ricci), Box 10.5 (Niǎn), Fig. 9.15
Dàzhàlan: Ch. 9.2, Figs. 9.31, 9.32
Debrughur, see Dibrugarh
Decima , see Dejima

Dehradun: Ch. 5.6, Box 5.8 (Jameson), Fig. 5.74
Déhuì Bridge, Yúyáo: Ch. 5.9 note 575
Dejima: Ch. 8.1, Ch. 8.1 notes 21, 22, Box 8.2 (Kaempfer), Box 8.3 (Thunberg), Box 8.4 (Siebold), Fig. 8.9
Delhi : Ch. 5.6, Box 5.7 (Falconer), Ch. 6.2, Box 6.4 (Indian Mutiny)
Dēngzhōu, see Pénglái
Denmark: Box 5.22 (Meadows)
Deqen, see Déqīn
Déqīn: Ch. 10.4
Derbyshire: Box 5.14 (Paxton)
Deshima, see Dejima
Desima , see Dejima
Devonshire: Box 3.8 (Cuming), Box 6.2 (Glendinning), Box 8.15 (Introductions)
Deyra Doon , see Dehradun
Dhanaulti: Fig. 5.78
Diànshān Hú: Ch. 5.10, Fig. 5.152
Dibrugarh: Ch. 10.4
Dickinson's Nursery: Box 10.1 (Moore)
Dolores : Ch. 10.6, Ch. 10.6 note 472
Dōngchāngfǔ: Box 10.5 (Niǎn)
Dōngdìng Dǎo: Ch. 3.1
Dōngqián Hú: Ch. 5.3, Fig. 5.37
Dōngshāncūn: Ch. 5.9 [between notes 485 and 486]
Dōngtiáo Xī: Ch. 5.10
Dòngtíng Hú: Ch. 10.4
Dowlaw Farm: Ch. 11, Ch. 11 notes 19, 33, Figs. 11.11, 11.12
Drem: Ch. 11
Drum Tower , Běijīng: Ch. 9.2, Fig. 9.38, 9.39
Drum Tower, Tiānjīn: Fig. 9.10
Drum Wave Islet, see Gǔlàngyǔ
Dublin: Box 3.11 (Lockhart), Ch. 4.1
Dulaw Farm, see Dowlaw Farm
Dum-Dum: Box 4.1 (Royle)
Dunbar: Ch. 1.2
Dundee: Box 5.14 (Paxton)
Dunfermline Abbey: Box 9.4 (Bruce)
Dunglass Estate: Ch. 11, Ch. 11 note 19
Dunnikier House: Box 9.4 note 1 (Bruce)
Duns: Ch. 1.2, Ch. 1.3, Ch. 1.2 note 124, Box 1.1 (Buchan), Ch. 5.7, Fig. 5.113
Duns South Church: Ch. 1.2, Fig. 1.6
Dunse, see Duns
Dunse History Society: Acknowledgements
Dunure: Box 1.2
Durham: Box 9.8 (Bowlby)
Düsseldorf, Germany: Fig. 5.71
Dutch East Indies: Ch. 1.2, Box 5.10 (Lady Mary Wood), Box 8.4 (Siebold), Box 8.6 (Hepburn), Box 8.11 (Harris)

Ealing: Box 3.10 (Alcock)
East China Sea: Ch. 3.2, Ch. 5.9 note 590
East Fenton, Drem: Ch. 11
East Indies: Box 10.3 (Fullerton-Elphinstone)
East Lothian, see Haddingtonshire
East Orange, New Jersey: Box 8.6 (Hepburn)
East Windsor, Connecticut: Box 8.7 (Brown)
Easter Island: Box 3.8 (Cuming), Box 8.7 (Brown)
Eastern Chinese Trumpet Creeper Rivulet: Ch. 5.10
Eco- ying , see Eko-in
Ecuador: Ch. 1.1
Edinburgh: Acknowledgements, Ch. 1.3, Box 1.1 (Buchan), Box 1.2 (McNab), Box 4.1 (Royle), Box 5.6 (Champion), Box 5.32 (Dalhousie), Box 8.8 note 4 (Hall), Box 9.2 (Staveley), Box 9.3 (Gordon), Box 9.9 (Grant), Ch. 10.1, Ch. 11, Ch. 11 notes 33, 37, Figs. 1.13, 1.14, 1.15, 1.17, 11.4, 11.8
Edinburgh Academy: Box 5.13 (Ramsay)
Edinburgh High School: Box 5.8 (Jameson), Box 9.9 (Grant), Ch. 11 (Dr. Durie)
Edinburgh Royal Infirmary: Ch. 11
Edinburgh University: Box 5.7 (Falconer), Box 5.8 (Jameson), Box 8.8 note 4 (Hall), Box 9.3 (Gordon), Ch. 11
Edith Grove: Ch. 10.1
Edo: Box 3.10 (Alcock), Box 3.10 notes 21, 23 (Alcock), Ch. 8.1, Ch. 8.2, Ch. 8.3, Ch. 8.1 notes 69, 78, 103, 148, 164, Box 8.2 (Kaempfer), Box 8.3 (Thunberg), Box 8.4 (Siebold), Box 8.7 (Brown), Box 8.9 (Girard), Box 8.11 (Harris), Box 8.12 (Heusken), Box 8.14 (Howard-Vyse), Box 8.15 (Introductions) notes 7, 52, Box 9.9 (Grant), Ch. 10.2, Ch. 10.6, Box 10.2 (Namamugi), Figs. 3.45, 8.27, 8.47, 8.48
Edo Bay: Ch. 8.2, Ch. 8.3
Edo Castle: Ch. 8.2, Box 8.7 (Brown)
Edrom: Ch. 1.2, Ch. 1.3, Ch. 1.2 notes 124, 126, 131, Ch. 2.2, Ch. 2.2 note 71, Ch. 4.1 note 7, Ch. 11
Edrom Churchyard: Ch. 1.2, Ch. 1.2 note 126, Figs. 1.7, 1.8
Edrom Parish Church: Ch. 1.2, Ch. 1.2 note 110, Box 1.1 note 13 (Buchan)
Edrom Parish School: Ch. 1.2, Ch. 10.6, Fig. 1.9
Egypt: Ch. 5.1, Ch. 5.8, Box 7.3 note 24 (Holt), Ch. 9.2, Box 9.4 (Bruce), Ch. 10.6
Eko-in: Ch. 8.2, Figs. 8.35, 8.38, 8.41
Elba: Ch. 1.1
Elbe: Ch. 1.1
Elburz Mountains: Box 8.2 (Kaempfer)
Elizabethtown: Box 7.3 (Holt)
Ellingham, Woking: Ch. 11 note 65
Elm Park, London: Ch. 10.1, Fig. 10.3
Elmira College: Box 8.7 (Brown)
Elphinstone: Acknowledgements, Ch. 11
Elphinstone Colliery: Ch. 11
Elphinstone Estate: Acknowledgements, Ch. 10.1, Fig. 11.15
Elphinstone Tower: Ch. 10.1 note 57, Ch. 11
Elphinstone Tower Farm: Ch. 10.1, Box 10.3 (Fullerton-Elphinstone), Ch. 11, Figs. 10.22, 11.1
Éméi Shān: Ch. 10.4

England: Ch. 2.1, Ch. 2.2, Box 2.2 note 3 (Opium), Box 2.3 (Reeves), Box 2.4 (Dent), Ch. 3.1, Ch. 3.2, Ch. 3.4, Ch. 3.5, Ch. 3.1 note 8, Ch. 3.2 note 130, Box 3.1 (Abeel), Box 3.4 (Balfour), Box 3.5 (Thom), Box 3.7 (Morrison), Box 3.8 (Cuming), Box 3.9 (Lay), Box 3.10 note 26 (Alcock), Box 3.11 (Lockhart), Box 3.11 note 22 (Lockhart), Box 3.12 (Introductions) notes 23, 31, 34, Ch. 4.1, Box 4.1 (Royle), Ch. 5.6, Ch. 5.7, Box 5.6 (Champion), Box 5.7 (Falconer), Box 5.14 (Paxton), Box 5.16 (Introductions) notes 4, 15, 25, Ch. 5.18 (Bonham), Box 5.24 (Burdon), Box 5.27 (Robertson), Box 5.31 (Beadon), Ch. 6.1, Ch. 6.2 Box 6.2 (Glendinning), Box 6.3 (Introductions) note 12, 14, 20, Ch. 7.2, Box 7.1 (Smith), Ch. 8.1, Box 8.1 (Veitch), Box 8.8 note 4 (Hall), Box 8.13 (Aspinall), Box 8.13 notes 6 (Aspinall), Ch. 9.2, Box 9.3 (Gordon), Box 9.9 (Grant), Ch. 10.1, Ch. 10.4, Ch. 10.6, Box 10.10 (Wade), Ch. 11, Ch. 11 note 3, Figs. 3.15, 10.44
English Channel: Ch. 5.1, Box 8.3 (Thunberg)
Enniskillen: Box 9.3 (Gordon)
Enoshima: Ch. 8.3
Epunga , see Mount Inasa
Er- she- pa- tu , see Èrshíbādū
Èrshíbādū: Ch. 5.4, Fig. 5.56
Essex: Ch. 10.1
Esslingen am Neckar: Box 8.4 note 53 (Siebold)
Estonia: Box 8.2 (Kaempfer)
Ethiopia: Box 9.2 (Staveley)
Eton College: Box 2.5 (Smith-Stanley), Box 5.31 (Beadon), Box 9.4 (Bruce)
Europe: Ch. 1.1, Ch. 1.2, Box 2.2 (Opium), Box 2.3 (Reeves), Ch. 3.5, Box 3.1 (Abeel), Box 3.10 (Alcock), Box 3. 12 (Introductions) note 21, Ch. 5.9, Box 5.7 (Falconer), Box 5.16 (Introductions) note 28, Ch. 7.3, Box 7.3 (Holt), Ch. 8.1, Box 8.1 (Veitch), Box 8.3 (Thunberg), Box 8.4 (Siebold), Box 8.11 (Harris), Box 8.15 (Introductions) notes 5, 42, 56, Ch. 9.2, Box 9.9 (Grant), Figs. 3.15, 5.24, 5.35, 5.77, 8.11
Exeter: Box 5.12 (Batten), Box 6.2 (Glendinning), Box 8.1 (Veitch), Box 8.15 (Introductions)
Eyemouth, Berwickshire: Ch. 11, Fig. 11.11
Ezo: Box 8.14 (Howard-Vyse), Ch. 10.2

Falconer's Hall Estate: Box 5.15 (Standish)
Far East: Preface, Box 3.10 (Alcock), Box 5.17 (Ganges), Box 8.6 (Hepburn), Box 8.9 (Girard), Box 9.5 (Ricci); Box 9.6 (Wyndham), Ch. 10.1, Ch. 10.2, Ch. 10.1 note 72, Fig. 4.1
Fa- tee nurseries , see Huādì nurseries
Faversham: Box 5.18 (Bonham)
Fayetteville: Box 8.6 (Hepburn)
Fǎyǔ Sì: Ch. 3.2, Figs. 3.32, 3.33
Fēiyīng Tǎ, Húzhōu: Ch. 5.10, Figs. 5.154, 5.155
Fènghuà District: Ch. 5.9
Fènghuà Jiāng: Ch. 5.9, Fig. 5.128
Fengshan (near Kaohsiung, Taiwan): Ch. 5.9
Fife: Ch. 1.2, Ch. 1.3, Box 9.4 (Bruce), Box 9.4 note 1 (Bruce)
Fiji: Box 8.1 (Veitch)
Finland: Box 8.2 (Kaempfer)
Fitzwilliam Museum, Cambridge: Fig. 2.3
Five Tiger Gate, see Wǔhǔmén
Flanders: Ch. 1.1

Florida: Box 8.8 (Hall)
Flushing, New York: Box 8.8 (Hall)
Fódǐng Shān: Ch. 3.2, Ch. 3.2 note 145, Fig. 3.34
Fokien , see Fújiàn
Foo- chow- foo , see Fúzhōu
Forbidden City, Běijīng: Ch. 9.1, Ch. 9.2, Box 9.7 (Cí Xǐ), Figs. 9.25, 9.26
Foreigners' Cemetery, Yokohama: Box 8.13
Formosa , see Táiwān
Forres: Box 5.7 (Falconer)
Forres Academy: Box 5.7 (Falconer)
Fort George Island, Florida: Box 8.8 (Hall)
Forth Valley: Ch. 1.3
Fow- ching gate , see Fùxīngmén
Fow- ching- mun , see Fùchéngmén
Foxholes: Box 5.15 (Standish)
France: Ch. 1.1, Box 1.1 (Buchan), Box 3.5 (Thom), Box 3.12 (Introductions) note 28, 57, Box 5.7 (Falconer), Box 5.22 (Meadows), Box 5.28 (Harvey), Box 6.3 (Introductions) note 23, Ch. 7.1, Box 8.9 (Girard), Box 9.1 (Tiānjīn), Box 9.3 (Gordon), Box 9.7 note 28 (Cí Xǐ), Ch. 10.3, Ch. 10.4, Ch. 11
French Legation (Edo): Ch. 8.2 note 92
Fùchéng Lù, Běijīng: Ch. 9.2
Fùchéngmén, Běijīng: Ch. 9.2
Fùchūn Jiāng: Ch. 5.2, Ch. 5.4
Fújiàn: Ch. 2.2, Ch. 2.2 note 54, Ch. 3.1, Ch. 3.5, Ch. 3.4 note 191, Ch. 5.4, Ch. 5.5, Ch. 5.9, Ch. 5.10, Ch. 5.9 note 521, Box 5.7 (Falconer), Box 5.16 (Introductions) note 1, Box 5.21 (Small Sword), Box 8.6 (Hepburn)
Fukue-jima: Ch. 8.1
Fulham Road: Ch. 5.7, Ch. 10.1, Fig. 5.107
Fung- hwa , see Fènghuà
Fusi- yama , see Mount Fuji
Fùxīngmén: Ch. 9.2
Fùyáng: Ch. 5.2
Fúyuánshuǐ: Box 5.20 (Hóng)
Fúzhōu: Ch. 2.1, Ch. 2.2, Ch. 2.2 note 41, Box 2.2 (Opium), Ch. 3.4, Ch. 3.4 notes 180, 188, Box 3.6 (Sinclair), Box 3.7 (Morrison), Box 3.9 (Lay), Box 3.10 (Alcock), Box 3.11 (Lockhart), Ch. 5.3, Ch. 5.4, Ch. 5.9, Ch. 5.10, Ch. 5.1 note 2, Ch. 5.16 (Introductions) notes 5, 16, 26, Ch. 10.6 note 480, Figs. 3.41, 3.42, 3.43, 3.44, 3.45, 3.47, 5. 28

Gadoli Plantation: Ch. 5.6, Ch. 5.6 note 327, Box 5.8 (Jameson), Figs. 5.79, 5.80, 5.87
Galapagos Islands: Box 3.8 (Cuming)
Galicia: Ch. 5.1
Galle: Ch. 5.1, Ch. 5.12, Box 5.4 (Braganza), Box 5.10 (Lady Mary Wood), Box 5.33 (Bentinck)
Ganga: Ch. 5.6, Box 6.4 (Indian Mutiny), Box 9.3 (Gordon), Box 9.9 (Grant), Fig. 5.69
Ganges , see Ganga
Gǎngkǒu Zhèn: Ch. 5.2
Gǎngzīhòu Beach: Ch. 3.1, Fig. 3.6
Gǎnpǔ: Ch. 5.9, Ch. 5.9 note 588, Fig. 5.151

Gānsù: Box 10.8 (Huí), Box 10.9 (Muslim)
Gāoxióng, see Kaohsiung
Garhwal: Ch. 5.6, Box 5.12 (Batten)
Gdansk: Box 8.2 (Kaempfer)
Gé'ěrmù: Ch. 5.4 note 160
Geneva, New York State: Box 3.1 (Abeel)
Genua: Box 8.4 note 53 (Siebold)
Georgia: Box 3.1 (Abeel), Ch. 7.1, Box 8.8 (Hall)
Germany: Box 3.12 (Introductions) note 48, Box 5.22 (Meadows), Box 8.2 (Kaempfer), Box 8.4 (Siebold), Box 9.7 notes 28, 67 (Cí Xǐ), Ch. 10.4
Ghana: Box 9.3 (Gordon)
Ghent: Ch. 1.1, Box 8.4 (Siebold), Box 8.15 (Introductions) note 41
Ghent University: Box 8.4 (Siebold)
Gibraltar: Box 9.8 (Bowlby)
Gifford: Ch. 1.3
Gilston Road: Ch. 6.2, Ch. 10.1, Ch. 11, Figs. 6.7, 6.8
Giza: Ch. 5.1, Fig. 5.5
Gizeh, see Giza
Gladsmuir Parish Church: Ch. 11, Fig. 11.9
Glasgow: Box 3.5 (Thom), Box 5.14 (Paxton), Box 5.17 (Ganges), Box 7.1 (Smith)
Goa: Box 9.5 (Ricci)
Gohyaku Rakan-ji: Box 8.7 (Brown)
Golden Grove Tea Plantation, Greenville: Box 7.1 (Smith), Figs. 7.2, 7.4
Golmud, see Gé'ěrmù
Goodtrees Estate, see Moredun Estate
Goolongsoo, see Gǔlàngyǔ
Goryōkaku Fortress: Ch. 10.2
Goto islands , see Fukue-jima
Goto-Retto Islands, see Fukue-jima
Göttingen Botanic Garden: Ch. 9.2 note 128
Gotto Islands, see Fukue-jima
Grand Canal , see Dà Yùnhé
Grande Vue , Edo: Ch. 8, Figs. 8.30, 8.31
Grange Cemetery: Box 9.9 (Grant)
Grange School: Box 9.8 (Bowlby)
Gravesend: Box 9.3 (Gordon)
Great Britain, see Britain
Great Wall: Ch. 5.4, Ch. 5.4 note 195, Box 9.6 (Wyndham), Box 9.7 (Cí Xǐ)
Greece: Box 5.14 (Paxton)
Greendykes, Macmerry: Ch. 11, Ch. 11 note 34, Fig. 11.13
Greenlaw: Ch. 1.3
Green Park, London: Ch. 5.7, Ch. 5.7 note 382, Ch. 10, Fig. 5.109
Greenville, South Carolina: Ch. 7.1, Box 7.1 (Smith), Figs. 7.2, 7.3, 7.4
Greenwich: Fig. 5.60
Grove Station Baptist Church, Greenville: Fig. 7.3

Guam: Box 8.7 (Brown)
Guāndì Miào: Ch. 5.4
Guāndǐng Reservoir: Ch. 5.9 note 553, Figs. 5.146, 5.147
Guāndǐng Sì: Ch. 5.9, Ch. 5.9 note 553, Figs. 5.146, 5.147
Guāndǐng Temple, see Guāndǐng Sì
Guāndǐng Valley: Ch. 5.9
Guǎngdōng: Ch. 2.1, Ch. 2.2, Ch. 3.2, Ch. 5.6 note 327, Box 5.20 (Hóng), Box 5.21 (Small Swords), Box 6.1 (Arrow), Ch. 8.3, Box 9.5 (Ricci), Ch. 10.3 note 166, Fig. 6.2
Guǎngqúmén, Běijīng: Ch. 9.1
Guǎngxī: Box 5.19 (Tàipíng), Box 5.20 (Hóng), Box 5.24 (Burdon), Box 6.1 (Arrow), Ch. 10.4
Guǎngzhōu: Box 1.2 (McNab), Ch. 2.1, Ch. 2.2 notes 56, 70, Box 2.2 (Opium), Box 2.3 (Reeves), Ch. 3.1, Ch. 3.4, Ch. 3.1 notes 60, 188, Box 3.1 (Abeel), Box 3.4 (Balfour), Box 3.5 (Thom), Box 3.5 note 25 (Thom), Box 3.7 (Morrison), Box 3.9 (Lay), Box 3.10 (Alcock), Box 3.11 (Lockhart), Box 3.12 (Introductions) notes 31, 58, 71, 86, Ch. 5.3, Ch. 5.5, Ch. 5.6, Ch. 5.9, Ch. 5.11, Ch. 5.8 note 399, Box 5.5 (Beale), Box 5.14 (Paxton), Box 5.16 (Introductions) notes 15, 26, Box 5.20 (Hóng), Box 5.22 (Meadows), Box 5.27 (Robertson), Box 6.1 (Arrow), Box 6.1 note 7 (Arrow), Ch. 7.1, Ch. 7.2, Ch. 8.3, Box 8.8 note 4 (Hall), Box 8.9 (Girard), Box 9.9 (Grant), Fig. 7.7D
Guānlùbù: Box 5.20 (Hóng)
Gǔběikǒu: Box 9.6 (Wyndham)
Guddowli Plantation , see Gadoli Plantation
Guildford: Box 10.1 (Moore)
Guìlín: Box 5.19 (Tàipíng)
Guìyáng: Ch. 10.4
Guìzhōu: Box 5.26 (Wáng), Ch. 10.3, Ch. 10.4, Box 10.6 (Xián-Tóng)
Gǔlàngyǔ: Ch. 2.1, Ch. 2.2, Ch. 2.2 notes 43, 70, Box 2.2 (Opium), Ch. 3.1, Box 3.1 (Abeel), Figs. 3.4, 3.6, 3.7
Gulf of Aqaba: Box 5.3 (Moresby)
Gulf of Suez: Ch. 5.1
Gulf of Thailand: Box 3.1 (Abeel)
Gunter Grove: Ch. 10.1, Fig. 10.2
Gurhwal , see Garhwal
Gǔshān: Ch. 5.3, Figs. 5.26, 5.27, 5.28, 5.29
Gushing Spring Temple, see Yǒngquán Sì (Gǔshān)
Guy's Hospital, London: Box 3.11 (Lockhart)
Gwaldam: Fig. 5.84
Gyokusen-ji, Shimoda: Box 8.12 (Heusken)

Hachiman-gu, Kamakura: Ch. 8.3, Fig. 8.58
Hackney: Box 2.6 (Ward)
Haddington: Ch. 1.2, Ch. 11, Fig. 11.2, 11.3
Haddington Library: Acknowledgements
Haddington Parish Church: Ch. 11
Haddingtonshire (East Lothian after 1921): Box 1.2 (McNab), Box 5.32 (Dalhousie), Box 10.3 (Fullerton-Elphinstone), Ch. 11, Fig. 11.2
Hǎi Hé: Box 2.2 (Opium), Ch. 9.1, Box 9.9 (Grant)
Haileybury: Box 5.12 (Batten)
Hakodate: Box 8.14 (Howard-Vyse)

Halifax, Yorkshire: Box 5.14 (Paxton)
Hamburg: Box 8.2 (Kaempfer)
Hameln: Box 8.2 (Kaempfer)
Hampshire: Box 9.6 (Wyndham)
Hampstead: Box 3.9 (Lay)
Han-gang River, Korea: Ch. 10.2
Hǎn Nán Lì Kè, see Khan Ariq
Hangchow, see Hángzhōu
Hang- chow- foo, see Hángzhōu
Hángzhōu: Ch. 2.2 note 42, Box 2.2 (Opium), Ch. 3.2 note 131, Box 3.3 (Macgowan), Ch. 5.2, Ch. 5.8, Ch. 5.9, Box 5.24 (Burdon), Figs. 5.11, 5.12, 5.151
Hángzhōu Bay, see Hángzhōu Wān
Hángzhōu Wān: Ch. 5.2, Ch. 5.9, Ch. 5.9 notes 586, 590, Fig. 5.21
Hànkǒu: Ch. 8.3, Ch. 10.4, Ch. 10.4 note 223
Happy Valley, Hong Kong: Ch. 3.1 note 13
Haridwar: Box 9.3 (Gordon)
Harrow School: Box 5.32 (Dalhousie), Box 10.10 (Wade)
Hartford, Connecticut: Box 8.8 (Hall)
Harvard Medical School: Box 8.8 (Hall)
Hasedera-ji, Kamakura: Ch. 8.3, Figs. 8.60, 8.61
Haßfurt: Box 8.4 (Siebold)
Hawaii: Ch. 3.2 note 119
Hawalbagh Plantation: Ch. 5.6, Figs. 5.81, 5.87
Hawulbaugh Plantation, see Hawalbagh Plantation
Hazaribagh: Box 9.3 (Gordon)
Heathfield Lodge, Acton: Box 4.1 (Royle)
Heathside Crescent, Woking: Ch. 11
Heavitree, Exeter: Box 5.12 (Batten)
Héběi Province: Box 2.2 (Opium), Box 5.19 (Tàipíng), Box 9.7 (Cí Xǐ), Box 10.5 (Niǎn)
Héféi: Acknowledgements
Hēilóngjiāng: Box 2.2 (Opium)
Hékǒu, Jiāngxī, see Qiānshān
Hékǒu, Sìchuān: Ch. 10.4
Helsinki: Box 8.2 (Kaempfer)
Hénán: Box 10.5 (Niǎn)
Hénán Dǎo: Box 3.7 (Morrison), Box 6.1 (Arrow)
Hermitage, St. Petersburg: Fig. 1.1
Hertfordshire: Box 5.12 (Batten), Box 6.2 (Glendinning) note 15
Hétián: Ch. 5.4 note 160, Box 10.9 (Muslim)
Hexham: Box 3.10 (Alcock)
Héxī Corridor: Box 10.8 (Huí)
Hibiya Park: Ch. 8.1 note 79
Himalayas: Acknowledgements, Ch. 4.2, Box 4.1 (Royle), Ch. 5.5, Ch. 5.6, Ch. 5.8, Ch. 5.6 note 348, Box 5.7 (Falconer), Box 5.8 (Jameson), Box 6.3 (Introductions) note 7, Ch. 10.4, Ch. 10.6, Figs. 5.5, 5.88, 5.124, 5.162
Hofwil: Box 9.9 (Grant)

Hofwyl, see Hofwil
Hokkaidō: Box 8.14 (Howard-Vyse), Ch. 10.2
Hokow, see Qiānshān
Ho-kow, see Hékǒu
Honan Island, see Hénán Dǎo
Hongaku-ji, Kamakura: Ch. 8.3, Fig. 8.62
Hónghú: Ch. 10.4 note 244
Hóngjiāng: Ch. 10.4
Hong Kong: Ch. 2.2, Ch. 2.2 note 70, Box 2.2 (Opium), Box 2.2 note 38 (Opium), Box 2.4 (Dent), Box 2.7 (Campbell-Johnston), Box 2.7 notes 5, 6, Ch. 3.1, Ch. 3.2, Ch. 3.4, Ch. 3.1 note 5, Box 3.1 (Abeel), Box 3.5 (Thom), Box 3.9 (Lay), Box 3.10 (Alcock), Box 3.11 (Lockhart), Box 3.12 (Introductions) notes 4, 78, 80, 81, 83, 86, 90, 100, Ch. 5.1, Ch. 5.2, Ch. 5.5, Ch. 5.8, Ch. 5.9, Ch. 5.10, Ch. 5.11, Ch. 5.1 note 11, Box 5.1 (P&O), Box 5.4 (Braganza), Box 5.5 note 1 (Beale), Box 5.6 (Champion), Box 5.10 (Lady Mary Wood), Box 5.17 (Ganges), Box 5. 23 (Perceval), Box 5.24 (Burdon), Box 5.25 (Patridge), Box 5.27 (Robertson), Box 5.29 (Chusan), Box 5.30 note 13 (Vansittart), Box 6.1 (Arrow), Box 8.5 (Loureiro), Box 8.7 (Brown), Box 8.8 note 4 (Hall), Box 8.9 (Girard), Box 8. 12 (Heusken), Box 9.2 (Staveley), Box 9.3 (Gordon), Box 9.4 (Bruce), Box 9.4 note 18 (Bruce), Box 9.9 (Grant), Ch. 10.1, Ch. 10.4, Box 10.2 note 5 (Namamugi), Box 10.3 (Fullerton-Elphinstone), Box 10.10 (Wade), Figs. 2.8, 3. 1, 3.2, 5.8, 5.60, 5.116
Honjo: Box 8.7 (Brown)
Honren Temple: Box 8.4 (Siebold)
Honshu: Ch. 8.2, Box 8.7 (Brown), Box 8.15 (Introductions) note 12
Hoo- chow, see Húzhōu
Hoo- chow- foo, see Húzhōu
Hooghly River: Ch. 5.6, Fig. 5.69
Hoo- shan temple, see Wúshān Sì
Horsham: Box 2.3 (Reeves)
Hotan, see Hétián
Hotha: Ch. 10.4
Hòulōngcūn: Ch. 5.9
Houses of Parliament, London: Ch. 11
Hozo-mon, Asakusa Kan' non Temple: Ch. 8.2
Huādì Nurseries: Ch. 2.1, Ch. 3.1, Box 3.12 (Introductions) note 26, Ch. 5.9
Huādū: Box 5.20 (Hóng)
Huáng Hé: Box 10.5 (Niǎn), Box 10.8 (Huí)
Huángpǔ: Ch. 3.1, Box 3.1 (Abeel), Box 8.12 note 9 (Heusken), Ch. 10.4
Huángpǔ Jiāng: Ch. 5.2, Box 5.5 (Beale)
Huángshān: Ch. 5.2, Ch. 5.8, Ch. 5.2 notes 19, 68, Ch. 5.4 note 127, Box 5.16 (Introductions) note 18, Ch. 10.6 note 436
Huángshān Shì: Ch. 5.2
Huāyuán, see Yuánhuā
Húběi: Box 2.2 note 16 (Opium), Box 5.19 (Tàipíng), Box 5.20 (Hóng), Ch. 10.4, Box 10.5 (Niǎn)
Huddersfield: Box 7.1 (Smith)
Hudson Falls: Box 8.11 (Harris)
Hudson River: Box 3.1 (Abeel)
Huìjì Sì: Ch. 3.2, Fig. 3.34

Hŭmén: Box 2.2 (Opium), Box 2.2 note 34 (Opium), Ch. 3.1
Húnán: Box 2.2 note 16 (Opium), Box 5.19 (Tàipíng), Box 5.20 (Hóng), Ch. 8.3, Ch. 10.4, Box 10.6 (Xián-Tóng)
Hung Chiang Ssu, see Hóngjiāng
Hupeh, see Húbĕi
Húzhōu: Ch. 5.9, Ch. 5.10, Ch. 5.10 note 634, Figs. 5.154, 5.158
Hwuy- chow , see Huángshān
Hyde Park: Ch. 5.7, Box 5.14 (Paxton), Ch. 10.1
Hyde Park Corner: Ch. 5.7, Ch. 10.1

Iberian Peninsula: Ch. 1.1, Ch. 1.1 note 18, Box 3.2 (Schoedde), Box 3.10 (Alcock), Box 5.1 (P&O)
Ili District: Box 2.2 (Opium), Ch. 10.4
Imperial Museum Kolkata: Ch. 10.4
India: Preface, Box 1.1 (Buchan), Box 2.2 (Opium), Box 2.6 (Ward), Box 3.4 (Balfour), Box 3.6 (Sinclair), Ch. 4.2, Box 4.1 (Royle), Ch. 5.3, Ch. 5.4, Ch. 5.5, Ch. 5.6, Ch. 5.7, Ch. 5.8, Ch. 5.9, Ch. 5.10, Ch. 5.11, Ch. 5.12, Box 5.3 (Moresby), Box 5.7 (Falconer), Box 5.14 (Paxton), Box 5.27 (Robertson), Box 5.31 (Beadon), 5.32 (Dalhousie), Ch. 6.1, Ch. 6.2, Box 6.1 (Arrow), Box 6.3 (Introductions) notes 7, 8, 17, 23, Box 6.4 (Indian Mutiny), Ch. 7.1, Box 7.1 (Smith), Box 8.2 (Kaempfer), Box 8.11 (Harris), Box 9.2 (Staveley), Box 9.3 (Gordon), Box 9.5 (Ricci), Box 9.9 (Grant), Ch. 10.1, Ch. 10.3, Ch. 10.4, Ch. 10.5, Figs. 5.61, 5.72, 5.90, 5.124, 5.151, 5.162
Indian Ocean: Ch. 5.1, Ch. 5.12, Box 5.3 (Moresby)
Indochina: Fig. 3.40H
Indonesia, see Dutch East Indies
Indus: Box 5.7 (Falconer), Box 5.8 note 15 (Jameson)
Inkerman: Box 5.6 (Champion)
Inland Sea , Japan: Ch. 8.2
Inner Mongolia: Box 9.7 (Cí Xĭ)
Ino- sima , see Enoshima
Institution for the Deaf and Dumb, New York: Box 8.7 (Brown)
Inverleith House, Edinburgh: Ch. 1.3, Fig. 1.17
Ionian Islands: Box 5.6 (Champion), Box 10.10 (Wade)
Iowa: Box 7.2 (Mason)
Ireland: Ch. 1.2, Box 5.23 (Perceval), Box 9.3 (Gordon), Ch. 10, Ch. 10.1 note 16, Box 10.10 (Wade)
Irrawaddy: Box 5.7 (Falconer), Box 9.3 (Gordon)
Isfahan: Box 8.2 (Kaempfer)
Isle of Wight: Ch. 5.1
Islington College, London: Box 5.24 (Burdon)
Istanbul: Box 5.6 (Champion), Box 5.17 (Ganges), Box 9.6 (Wyndham)
Italy: Box 5.14 (Paxton), Box 5.32 (Dalhousie), Box 9.5 (Ricci), Box 9.9 (Grant)

Jacksonville, Florida: Box 8.8 (Hall)
Jakarta, see Batavia
Jamaica: Box 2.6 (Ward)
Japan: Preface, Acknowledgements, Ch. 2.2 note 42, Ch. 3.4, Ch. 3.2 note 105, Box 3.3 (Macgowan), Box 3.6 (Sinclair), Box 3.7 (Morrison), Box 3.8 (Cuming), Box 3.10 (Alcock), Box 3.10 note 16 (Alcock), Box 3.11 (Lockhart), Box 3.12 (Introductions) note 102, Ch. 5.3, Ch. 5.5 note 274, Ch. 5.6 note 348, Box 5.15 (Standish), Box 5.15 note 11 (Standish), Box 5.16 (Introductions) notes 14, 28, Box 5.29 (Chusan), Ch. 6.1, Box 6.3 notes 1, 2,

Box 6.3 (Introductions) notes 1, 2, 20, Ch. 8.1, Ch. 8.2, Ch. 8.3, Ch. 8.1 notes 64, 139, 176, Box 8.1 (Veitch), Box 8.2 (Kaempfer), Box 8.3 (Thunberg), Box 8.4 (Siebold), Box 8.4 note 83 (Siebold), Box 8.5 (Loureiro), Box 8.6 (Hepburn), Box 8.7 (Brown), Box 8.8 (Hall), Box 8.8 note 10 (Hall), Box 8.9 (Girard), Box 8.11 (Harris), Box 8.12 (Heusken), Box 8.13 (Aspinall), Box 8.15 (Introductions) notes 11, 24, 38, 43, 46, 53, 56, 60, 61, Ch. 9.1, Box 9.9 (Grant), Ch. 10.1, Ch. 10.2, Ch. 10.3, Ch. 10.4, Ch. 10.5, Ch. 10.6, Ch. 10.2 note 162, Box 10.1 note 6 (Moore), Box 10.4 (Meiji), Box 10.10 (Wade), Box 10.10 note 21 (Wade), Figs. 3.40H, 8.2, 8.10, 8.39, 8.47, 8.48, 8.68, 10.32

Java: Ch. 3.1, Box 3.1 (Abeel), Box 3.11 (Lockhart), Box 3.12 (Introductions) note 36, Box 5.10 (Lady Mary Wood), Box 8.2 (Kaempfer), Box 8.3 (Thunberg), Box 8.4 (Siebold), Box 8.4 note 53 (Siebold), Box 8.6 (Hepburn), Box 8.15 (Introductions) notes 42, 56, 61, Box 9.9 (Grant)

Jehol, see Chéngdé

Jempuku-ji, see Zenpuku-ji

Jhansi: Box 6.4 (Indian Mutiny)

Jiādìng: Ch. 3.2, Ch. 3.2 note 141, Ch. 5.10

Jiāngkǒu Zhèn: Ch. 5.9

Jiāngshān: Ch. 5.4

Jiāngshān Gǎng: Ch. 5.4

Jiāngsū: Box 2.2 (Opium), Ch. 3.2 note 120, Ch. 5.10, Box 5.20 (Hóng), Box 10.5 (Niǎn)

Jiāngxī: Ch. 5.4, Ch. 5.9, Ch. 5.10, Box 5.26 (Wáng), Box 9.5 (Ricci)

Jiànshuǐ: Box 10.7 (Panthay)

Jiāxīng: Ch. 5.2

Jiayi (near Kaohsiung, Taiwan): Ch. 5.9

Jífāng Pavilion, Jǐngshān: Fig. 9.25

Jìng Hǎi Temple (Nánjīng): Box 2.2 (Opium)

Jǐngshān: Ch. 9.2, Ch. 9.2 note 56, Figs. 9.25, 9.26

Jīnjī Fortress: Box 10.8 (Huí)

Jīnjīatáng: Ch. 5.10

Jīnmén Dǎo: Box 3.1 (Abeel)

Jīntáng Dǎo: Ch. 3.1 note 58, Ch. 5.3, Ch. 5.5, Ch. 5.9, Ch. 5.3 notes 72, 79, Fig. 5.25

Jiǔqū Xī, Wǔyíshān: Ch. 5.4, Ch. 5.4 note 181, Fig. 5.53

Jiǔquán: Box 10.8 (Huí)

Jiǔshí Wān: Ch. 5 (Nine stone meander)

John Gray Centre, Haddington: Fig. 11.3

Jönköping: Box 8.3 (Thunberg)

Joo- ne- shoo , see Juniso

Joryu-ji: Box 8.7 (Brown)

Juan Fernandez Islands: Box 3.8 (Cuming)

Juniso Shrine, Edo: Ch. 8.3, Fig. 8.52

Kabul: Ch. 5.6, Box 5.7 (Falconer)

Kachin Hills: Ch. 10.4

Kakhyen Hills, see Kachin Hills

Kagoshima: Ch. 10.1, Ch. 10.2, Fig. 10.16

Kalimantan: Box 3.1 (Abeel)

Kaliningrad: Box 8.2 (Kaempfer)

Kamakura : Acknowledgements, Ch. 8.3, Figs. 8.58, 8.59, 8.60, 8.62, 8.63, 8.64, 10.37

Kaminari-mon, Asakusa Kan' non Temple: Ch. 8.2, Fig. 8.36
Kanagawa : Ch. 8.2, Ch. 8.3, Ch. 8.2 note 161, Box 8.5 (Loureiro), Box 8.6 (Hepburn), Box 9.9 (Grant), Ch. 10.6, Box 10.2 (Namamugi), Fig. 8.27
Kanasawa : Ch. 8.3
Kandy: Box 5.6 (Champion)
Kan' ei-ji: Ch. 10.2
Kan-foo, see Gǎnpǔ
Kāngdìng: Ch. 10.4
Kang- koo , see Gǎngkǒu Zhèn
Kan- poo , see Gǎnpǔ
Kan-p' u, see Gǎnpǔ
Kanpur: Box 4.1 (Royle), Box 5.12 (Batten), Ch. 6.2, Box 9.9 (Grant), Fig. 6.6
Kaohsiung: Ch. 5.9
Kaolagir Plantation : Ch. 5.6, Fig. 5.74
Karachi: Box 5.30 (Vansittart)
Kashgar, see Kǎshí
Kashgaria: Box 10.8 (Huí), Box 10.9 (Muslim)
Kǎshí: Box 10.9 (Muslim)
Kashmir: Ch. 5.6
Kawasaki: Ch. 8.2, Ch. 10.6, Figs. 8.42, 8.43
Kawasaki Daishi: Ch. 8.2, Box 10.2 (Namamugi), Figs. 8.42, 8.43
Kawasaky, see Kawasaki
Kazakhstan: Box 10.9 (Muslim)
Kea- hing- foo , see Jiāxīng
Kelloe: Box 1.1 (Buchan)
Kelloe Bridge: Fig. 1.10
Kelloe Estate: Ch. 1.2, Ch. 1.2 note 131, Box 1.1 (Buchan)
Kelloe Farm: Fig. 1.11
Kelloe House: Ch. 1.2, Fig. 1.11
Kelso: Box 3.9 (Lay)
Kelvingrove Park, Glasgow: Box 5.14 (Paxton)
Kent: Box 5.18 (Bonham), Box 9.2 note 22 (Staveley), Box 9.3 (Gordon), Ch. 11
Kentucky: Box 7.3 (Holt)
Kew: Box 1.2 (McNab), Box 2.1 (Lindley), Box 2.3 (Reeves), Box 2.6 (Ward), Box 3.10 (Alcock), Box 3.10 note 26 (Alcock), Box 3.12 (Introductions) notes 19, 33, Ch. 4.1, Box 5.14 (Paxton), Box 5.16 (Introductions) notes 6, 15, Box 6.3 (Introductions) note 21, Ch. 8.2 note 146, Box 8.15 (Introductions) note 57, Fig. 8.56
Khali Estate: Box 5.13 (Ramsay)
Khan Ariq: Box 10.9 (Muslim)
Kiangse , see Jiāngxī
Kiating, see Lèshān
Kii-suido: Ch. 8.2
Kilgraston House: Box 9.9 (Grant)
Kincardineshire: Box 3.4 (Balfour)
King Street, London: Ch. 5.7, Ch. 10.1, Fig. 5.111
King's College, London: Box 4.1 (Royle)

King's Cross Station, London: Ch. 5.7, Ch. 10.1, Fig. 5.112
King's Road: Ch. 10.1, Ch. 10.1 note 106
Kingsbridge: Box 3.8 (Cuming)
King- shan , see Jǐngshān
Kingston, Massachusetts: Box 8.8 note 7 (Hall)
Kin- hwa , see Jīnjīatáng
Kin- keang , see Xìn Jiāng
Kinmen Island, seeJīnmén Dǎo
Kino Channel , see Kii-suido
Kinross-shire: Ch. 1.2 note 135, Box 10.3 (Fullerton-Elphinstone)
Kipling Road: Ch. 5.6
Kirkaldy: Box 9.4 note 1 (Bruce)
Kissena Park Historic Grove: Box 8.8
Kitang , see Jīntáng Dǎo
Kiu- siu , see Kyushu
Knightsbridge: Box 3.8 (Cuming), Box 3.8 note 2 (Cuming), Ch. 10.1
Knowle Cottage: Box 3.8 (Cuming)
Knowle Lane, Cranleigh: Ch. 11
Knowsley Hall: Box 2.5 (Smith-Stanley), Fig. 2.4
Knutsford: Box 5.14 (Paxton)
Kobe City Museum: Fig. 8.48
Koblenz: Box 3.6 (Sinclair)
Kokand: Box 10.9 (Muslim)
Kolkata: Acknowledgements, Box 2.2 (Opium), Box 2.3 (Reeves), Box 4.1 (Royle), Ch. 5.3, Ch. 5.5, Ch. 5.6, Ch. 5. 9, Ch. 5.10, Ch. 5.11, Ch. 5.12, Ch. 5.6 note 327, Box 5.1 (P&O), Box 5.3 (Moresby), Box 5.7 (Falconer), Box 5.10 (Lady Mary Wood), Box 5.29 (Chusan), Box 5.32 (Dalhousie), Box 5.33 (Bentinck), Box 6.1 (Arrow), Box 9.3 (Gordon), Box 9.9 (Grant), Ch. 10.4, Figs. 5.50, 5.60, 5.166
Kolkata Botanical Garden: Box 5.7 (Falconer), Box 5.11 (Saharanpur), Figs. 5.50, 5.62, 5.67, 5.68, 5.90
Kong- k'how- ta , see Shòu Fēng Tǎ
Kooasur Plantation : Ch. 5.6
Koo- lung- soo , see Gǔlàngyǔ
Koo- nu- hoo , see Kūnmíng Lake
Koo- shan , see Gǔshān
Korea: Box 8.15 (Introductions) note 46, Box 9.5 (Ricci), Ch. 10.2, Ch. 10.4, Fig. 3.40H
Korin-ji: Ch. 8.3, Box 8.12 (Heusken), Fig. 8.50
Koti-Ghur: Box 9.9 (Grant)
Kowloon: Box 2.2 note 25 (Opium), Box 5.27 (Robertson), Box 9.4 note 18 (Bruce), Box 9.9 (Grant)
Kraków: Box 8.2 (Kaempfer)
Kùchē: Box 10.9 (Muslim)
Kulangsu , see Gǔlàngyǔ
Kulja, see Ili
Kulla Grunden Shoal: Box 9.10 (Ceylon)
Kumamoto: Ch. 10.2
Kumaon : Ch. 5.6, Box 5.12 (Batten), Box 5.13 (Ramsay)
Kumaun, see Kumaon

Kūnlún Mountains: Ch. 5.4 note 160
Kūnmíng: Ch. 10.4, Box 10.7 (Panthay)
Kūnmíng Lake: Ch. 9.2, Box 9.7 (Cí Xǐ)
Kuppeana Plantation : Ch. 5.6
Kurnool: Box 3.4 (Balfour)
Kwangtung, see Guǎngdōng
Kyōto: Ch. 8.2, Ch. 8.1 note 78, Ch. 10.1, Ch. 10.2, Ch. 10.2 note 123, Box 10.4 (Meiji), Figs. 8.27, 10.33
Kyrgyzistan: Box 10.9 (Muslim)
Kyushu: Ch. 8.2, Box 8.4 (Siebold)

Lachmeshwar Plantation: Ch. 5.6, Ch. 5.6 note 340, Box 5.7 (Falconer)
Lackham House: Box 3.8 note 2 (Cuming)
La Flèche: Box 5.9 (Du Halde)
La Guaira: Box 3.5 (Thom)
Laguna de Bay, Luzon: Ch. 3.3, Box 3.12 (Introductions) note 5, Fig. 3.37
Laguna Province, Luzon: Ch. 10.6 note 472
Lal Bahadur Shastri National Academy of Administration, Mussoorie: Ch. 5.6 note 319
Lama Mosque , see Báiyún Sì
Lama Temple , see Yōnghé Gōng
Lammermuir Hills: Ch. 1.3, Fig. 1.12
Lanark: Box 6.2 (Glendinning)
Láncāng Jiāng: Ch. 10.4, Ch. 10.4 note 244
Lancashire: Ch. 1.1, Ch. 1.1 note 6, Box 2.5 (Smith-Stanley), Box 8.13 (Aspinall), Ch. 10.1
Lanchee , see Lánxī
Landour: Box 5.7 (Falconer)
Lang- shuy- ain , see Léngshuǐ' Ān
Lántián: Ch. 5.9, Fig. 5.134
Lánxī: Ch. 5.4, Ch. 5.4 note 142, Fig. 5.40
Lánzhōu: Box 10.8 (Huí)
Laos: Box 10.7 (Panthay)
Lasswade: Box 8.8
Latin America: Ch. 1.1
Latton: Box 5.31 (Beadon)
Lauder: Ch. 1.3
Laurel Bank, Woking: Ch. 11
Leang- kung- foo , see Liáng gōng fū
Lea Bridge nurseries: Box 10.1 (Moore)
Leatherhead: Ch. 11
Leatherhead Hospital: Ch. 11 note 72
Leiden: Box 8.4 (Siebold), Box 8.4 notes 36, 43 (Siebold)
Leiden University: Box 8.2 (Kaempfer)
Leiderdorp: Box 8.4 note 53 (Siebold)
Léifēng Tǎ: Ch. 5.2, Fig. 5.11
Leith: Box 5.8 (Jameson)
Leith Hospital: Ch. 11

Leith Walk, Edinburgh: Ch. 1.3, Box 1.2 (McNab)
Lemgo: Box 8.2 (Kaempfer)
Léngshuǐ' Ān: Ch. 5.9, Ch. 5.9 note 533, Figs. 5.144, 5.145
Lèshān: Ch. 10.4
Le- tsun , see Lǐcūn
Levant: Box 3.10 (Alcock)
Leyton, Essex: Box 10.1 (Moore)
Lhasa: Ch. 10.4
Liáng gōng fǔ, Běijīng: Ch. 9.1, Box 9.4 (Bruce), Box 9.4 note 26 (Bruce), Fig. 9.19
Liáng's Mansion, see Liáng gōng fǔ
Liánzhèn (Héběi): Box 5.19 (Tàipíng)
Liáodōng Peninsula: Box 9.7 (Cí Xǐ), Box 9.7 note 28 (Cí Xǐ)
Liberation Bridge: Figs. 3.41, 3.42
Liberton: Ch. 1.3
Lǐcūn: Ch. 5.9, Ch. 5.9 note 486, Figs. 5.134, 5.135, 5.136, 5.138, 5.140
Lieme: Box 8.2 (Kaempfer)
Limerick: Box 9.10 (Ceylon)
Lìngdìng: Ch. 3.1
Língfēng Shān: Ch. 5.8
Língguāng Sì: Ch. 9.2, Figs. 9.44, 9.45
Línglóng Gōngyuán: Ch. 9.2 note 87
Língshān Gǎng: Ch. 5.4
Língxù Shān, see Lóngquán Shān
Ling- yang- sze , see Língguāng Sì
Lintin, see Lìngdìng
Líntóng: Box 10.8 (Huí)
Línxià: Box 10.8 (Huí)
Lisbon: Ch. 1.1, Box 9.5 (Ricci)
Lǐtáng: Ch. 10.4
Litchfield Law School: Box 7.1 (Smith)
Little Withybush, Cranleigh: Ch. 11
Liùhé Tǎ, Hángzhōu: Ch. 5.2, Fig. 5.12
Liùlǐ Cūn: Ch. 5.9
Liúlichǎng, Běijīng: Ch. 9.2
Liverpool: Box 2.5 (Smith-Stanley), Box 3.5 (Thom), Box 3.5 note 44 (Thom), Box 3.11 (Lockhart), Box 5.10 (Lady Mary Wood), Box 5.14 (Paxton), Box 5.24 (Burdon), Box 5.33 (Bentinck), Box 7.1 (Smith), Box 8.13 (Aspinall)
Liverpool Central Library: Acknowledgements, Figs. 4.1, 4.5, 10.53
Liverpool World Museum: Acknowledgements, Figs. 2.5, 3.26, 3.40A-H
Lock Haven: Box 8.6
Loddiges' Nursery: Box 5.14 (Paxton)
Lo-fou-shan, see Luófú Mountains
Loga River , see Tama River
Loganlea, see Loganlee
Loganlee, Woking: Ch. 11

Lohba : Ch. 5.6
Lohit River: Ch. 10.4, Ch. 10.4 note 290
Lomonds: Ch. 1.3
London: Ch. 1.1 note 45, Box 1.2 (McNab), Ch. 2.2, Box 2.1 (Lindley), Box 2.2 (Opium), Box 2.3 (Reeves), Box 2.5 (Smith-Stanley), Box 2.6 (Ward), Ch. 3.4, Ch. 3.5, Box 3.3 (Macgowan), Box 3.4 (Balfour), Box 3.8 (Cuming), Box 3.10 (Alcock), Box 3.11 (Lockhart), Ch. 4.1, Ch. 4.2, Box 4.1 (Royle), Ch. 5.1, Ch. 5.7, Ch. 5.8, Ch. 5.8 note 428, Box 5.1 (P&O), Box 5.2 (Ripon), Box 5.7 (Falconer), Box 5.14 (Paxton), Box 5.18 (Bonham), Box 5.23 (Perceval), Box 5.24 (Burdon), Box 5.25 (Patridge), Box 5.27 (Robertson), Ch. 6.1, Ch. 6.2 note 34, Ch. 7.1, Ch. 7.2, Box 7.1 (Smith), Ch. 8.3, Ch. 8.2 note 182, Box 8.2 (Kaempfer), Box 8.3 (Thunberg), Box 8.5 (Loureiro), Box 8.8 note 4 (Hall), Box 8.10 (Stevens), Box 8.13 (Aspinall), Box 9.3 (Gordon), Box 9.8 (Bowlby), Box 9.9 (Grant), Box 9.10 (Ceylon), Ch. 10.1, Ch. 10.4, Ch. 10.6, Box 10.3 (Fullerton-Elphinstone), Box 10.10 (Wade), Ch. 11, Ch. 11 notes 3, 70, Figs. 1.16, 2.7, 2.8, 3.36, 3.39, 5.60, 5.102, 5.103, 5.106, 6.7, 8.45, 10.23
London Docks: Ch. 10.1
London Magdalen Hospital: Box 2.3 (Reeves)
London University: Box 2.1 (Lindley), Box 8.1 (Veitch)
Longformacus: Fig. 1.12
Long Island: Ch. 7.1, Box 7.1 (Smith)
Lóngquán Shān, Yúyáo: Ch. 5.4, Ch. 5.9, Box 5.26 (Wáng), Fig. 5.38
Lóng Tán Pagoda, Shàngráo: Fig. 5.43
Long- yeou , see Lóngyóu
Lóngyóu: Ch. 5.4
Loochoo, see Okinawa
Loo- co- jou , see Lúgōu Qiáo
Loo-din-chow, see Lúdìng
Loo- le- chang , see Liúlichǎng
Los Angeles: Box 2.7 (Campbell-Johnston), Fig. 10.46
Louisiana: Ch. 7.1, Ch. 7.3
Louisville: Box 7.3 (Holt)
Loupart Wood: Ch. 11
Louvre, Paris: Box 8.11 (Harris)
Lower Tank Pond, Chelsea Physic Garden: Ch. 4.1, Fig. 4.3
Lübeck: Box 8.2 (Kaempfer)
Luchmaiser Plantation, see Lachmeshwar Plantation
Lucknow: Acknowledgements, Box 5.13 note 32 (Ramsay), Box 6.4 (Indian Mutiny), Box 9.3 (Gordon)
Lúdìng: Ch. 10.4
Lúgōu Qiáo: Ch. 9.2, Box 10.5 (Niǎn)
Luh- le- heen , see Liùlǐ Cūn
Lüneburg: Box 8.2 (Kaempfer)
Lun- ke , see Xītiáo Xī
Luófú Mountains: Ch. 2.1, Ch. 2.2, Ch. 2.2 note 56
Luōwényùcūn: Box 9.6 (Wyndham)
Luòxīng Tǎ, Fúzhōu: Ch. 5.9, Fig. 5.141
Lutchmisser Plantation , see Lachmeshwar Plantation
Luzon: Ch. 3.3, Ch. 3.3 note 157, Box 3.12 (Introductions) notes 5, 79, Box 9.9 (Grant), Figs. 3.37, 3.38
Lycée St. Louis, Paris: Box 8.8 note 4 (Hall)

Lyndhurst: Box 3.2 (Schoedde)

Macao: Ch. 2.1, Ch. 2.2 note 70, Box 2.2 (Opium), Box 2.3 (Reeves), Box 2.4 (Dent), Ch. 3.1, Box 3.1 (Abeel), Box 3.7 (Morrison), Box 3.9 (Lay), Box 3.11 (Lockhart), Box 5.5 (Beale), Box 5.10 (Lady Mary Wood), Box 5.25 (Patridge), Box 5.27 (Robertson), Box 6.1 note 7 (Arrow), Box 8.5 (Loureiro), Box 8.6 (Hepburn), Box 8.7 (Brown), Box 8.7 note 18 (Brown), Box 8.12 (Heusken), Box 9.5 (Ricci)
Macau, see Macao
Macerata: Box 9.5 (Ricci)
Macmerry: Ch. 11, Fig. 11.13
Madagascar: Box 1.1 (Buchan)
Madeira: Box 5.1 (P&O)
Madras, see Chennai
Madrid: Box 9.6 (Wyndham)
Mae- yaski : Ch. 8
Magdala: Box 9.2 (Staveley)
Mahmoudieh Canal: Ch. 5.1
Maidenhead: Box 5.30 (Vansittart)
Maine: Box 9.4 (Bruce)
Mainpuri: Box 5.12 (Batten)
Malacca: Box 3.1 (Abeel), Box 3.4 (Balfour), Box 5.18 (Bonham)
Malay Peninsula: Box 3.1 (Abeel), Box 8.12 note 9 (Heusken)
Malaya: Box 2.6 (Ward), Box 8.11 (Harris)
Malaysia: Fig. 3.40H
Maldives: Box 5.3 (Moresby)
Malta: Ch. 5.1, Box 5.7 (Falconer), Box 5.14 (Paxton), Box 5.17 (Ganges), Fig. 5.4
Ma-mou: Ch. 10.4
Manchester: Box 3.5 (Thom), Box 8.13 (Aspinall), Ch. 10.1
Manchuria: Ch. 10.3
Mandalay: Ch. 10.4
Mángyǔnjiē: Ch. 10.4, Fig. 10.45
Manhattan: Box 8.6 (Hepburn), Box 8.7 (Brown)
Manila: Box 2.4 (Dent), Ch. 3.2, Ch. 3.3, Box 3.8 (Cuming), Ch. 4.1, Box 5.29 (Chusan)
Mannenya Inn: Ch. 8.2
Mansion of Plum- trees , see *Mae- yaski*
Manwyne, see Mángyǔnjiē
Máohuò Cūn, near Pǔxī: Ch. 5.11, Ch. 5.11 note 691, Figs. 5.163, 5.164
Marco Polo Bridge, see Lúgōu Qiáo
Mariaburghausen: Box 8.4 (Siebold)
Mariana Islands: Box 8.7 (Brown)
Marseille: Box 8.4 (Siebold), Ch. 10.4
Martyrs Monument, Meerut: Figs. 6.5, 6.6
Marylebone: Box 10.3 (Fullerton-Elphinstone)
Massachusetts: Box 8.7 (Brown), Box 8.8 (Hall), Box 8.8 note 7 (Hall)
Mauritius: Box 3.4 (Balfour), Box 5.14 (Paxton), Box 9.2 (Staveley), Box 10.10 (Wade)
Mǎwěi: Ch. 5.9, Fig. 5.141

Mayfield: Ch. 11
Meath Hospital: Box 3.11 (Lockhart)
Mecca: Box 6.4 (Indian Mutiny)
Mediterranean: Box 2.2 (Opium), Box 5.10 (Lady Mary Wood), Box 5.30 (Vansittart)
Meerut : Ch. 5.6, Box 5.7 (Falconer), Ch. 6.2, Box 6.4 (Indian Mutiny), Box 9.3 (Gordon), Figs. 5.89, 6.5, 6.6
Meguro District, Tokyo: Box 8.7 (Brown)
Mei- che , see Méi Xī
Méichéng: Ch. 5.2, Ch. 5.4, Figs. 5.14, 5.17
Meiji Gakuin: Box 8.6 (Hepburn)
Meireki: Ch. 8
Méi Xī: Ch. 5.10, Figs. 5.160, 5.161
Mekong River, see Láncāng Jiāng
Melbourne: Box 8.1 (Veitch)
Méngshān, see Yǒng'ān
Mentmore Towers: Box 5.14 (Paxton)
Mexico: Box 3.5 (Thom), Box 3.9 (Lay)
Miaco , see Kyōto
Mí Hé: Box 10.5 (Niǎn)
Miánfó Sì: Ch. 5.10 note 621
Miánfó Sì Jiē: Ch. 5.10 note 621
Middle East: Box 7.3 (Holt), Box 10.7 (Panthay)
Middle Kingdom, see China
Middlesex: Box 4.1 (Royle)
Middleton Barony: Ch. 7.1, Ch. 7.1 note 7
Middleton Place Gardens: Ch. 7.1 note 7
Midhope Road, Woking: Ch. 11
Midlands, England: Ch. 1.1
Midlothian: Box 5.32 (Dalhousie), Box 8.8
Millburn Bridge: Ch. 1.3
Milton, Massachusetts: Box 8.6 (Hepburn), Box 8.8 (Hall)
Milton Bryan: Box 5.14 (Paxton)
Mindanao: Box 3.9 (Lay)
Míng Tombs: Box 9.6 (Wyndham)
Mǐn Jiāng: Ch. 3.4, Ch. 3.4 note 183, Ch. 5.3, Ch. 5.4, Ch. 5.9, Ch. 5.9 note 508, Ch. 10.4, Fig. 5.28
Mishmi Hills: Ch. 10.4, Ch. 10.4 note 285
Mississippi: Box 7.3 (Holt)
Mito: Ch. 10
Miyozaki-cho, Yokohama: Fig. 8.16
Momien, see Téngchōng
Moncreiffe Hill: Ch. 1.3
Monghyr: Box 9.3 (Gordon)
Monson, Massachusetts: Box 8.7 (Brown)
Montacute House: Fig. 1.5
Montrose: Box 3.4 (Balfour)
Moredun Estate: Ch. 1.3

Morrison Hill: Box 8.7 (Brown)
Morrison School: Box 8.7 (Brown)
Moscow: Ch. 1.1, Box 8.2 (Kaempfer)
Moss Law: Ch. 1.3
Mount Banahaw, Luzon: Ch. 3.3, Ch. 3.3 note 161, Fig. 3.38
Mount Fuji: Box 3.10 (Alcock), Ch. 8.2, Ch. 8.3, Box 8.1 (Veitch), Box 10.4 (Meiji)
Mount Inasa: Ch. 8.1, Fig. 8.15
Mount Tambora: Ch. 1.2
Moutan Gardens: Ch. 10.6
Móxīn Tǎ, see Luóxīng Tǎ
Mozambique: Box 1.1 (Buchan), Box 9.5 (Ricci)
Mumbai: Box 5.3 (Moresby), Box 5.4 (Braganza), Box 9.2 (Staveley), Box 9.3 (Gordon), Box 9.9 (Grant)
Munich: Box 2.1 (Lindley), Box 4.1 (Royle), Box 8.4 (Siebold)
Mussooree , see Mussoorie
Mussoorie: Ch. 5.6, Box 5.7 (Falconer), Figs. 5.75, 5.76, 5.77, 5.78
Myanmar, see Burma
Mynpoory, see Mainpuri
Myogyo Temple: Ch. 8.1, Figs. 8.5, 8.6

Nagasaki : Acknowledgements, Box 3.3 (Macgowan), Box 3.10 (Alcock), Box 3.10 note 23 (Alcock), Box 6.3 (Introductions) notes 1, 2, Ch. 8.1, Ch. 8.2, Ch. 8.3, Ch. 8.1 notes 19, 160, 208, Box 8.2 (Kaempfer), Box 8.3 (Thunberg), Box 8.4 (Siebold), Box 8.5 (Loureiro), Box 8.14 (Howard-Vyse), Box 9.3 (Gordon), Box 9.9 (Grant), Ch. 10.6, Figs. 8.4, 8.5, 8.7, 8.8
Nagasaki Harbour: Figs. 8.3, 8.4
Nagasaki University: Acknowledgements
Nainee Tal , see Nainital
Nainital: Ch. 5.6, Box 5.12 (Batten), Figs. 5.61, 5.88
Nairnshire: Box 9.3 (Gordon)
Nakamise-dori, Asakusa Kan' non Temple: Ch. 8.2, Fig. 8.37
Namamugi: Ch. 10, Box 10.2 (Namamugi), Fig. 10.15
Namoa Island, see Nán' ào Dǎo
Nán' ān, see Dàyú
Nán' ào Dǎo: Ch. 3.1
Nán Cài Cūn: Ch. 9.1
Nánchāng: Box 9.5 (Ricci)
Nan- che , see Lánxī
Nánhǎi: Box 9.7 (Cí Xǐ)
Nánjīng: Box 2.2 (Opium), Box 2.2 note 52 (Opium), Box 2.3 (Reeves), Box 2.7 note 6 (Campbell-Johnston), Box 3.4 (Balfour), Box 3.5 (Thom), Box 3.9 (Lay), Box 3.11 (Lockhart), Ch. 5.8, Box 5.18 (Bonham), Box 5.19 (Tàipíng), Box 5.19 note 31 (Tàipíng), Box 5.20 (Hóng), Box 6.1 (Arrow), Box 8.8 (Hall), Box 9.5 (Ricci), Box 10.5 (Niǎn), Box 10.10 (Wade), Fig. 5.118
Nánjīng Institute of Geology and Palaeontology: Acknowledgements
Nanking , see Nánjīng
Nánpíng: Ch. 5.3
Nan- see- mun , see Yòu' ānmén

Nan- tsin , see Nánxún
Nánxiáng: Ch. 1.2 note 131, Ch. 5.10
Nánxún: Ch. 5.10, Ch. 5.10 note 610, Fig. 5.153
Nanziang , see Nánxiáng
Narutaki, Nagasaki: Ch. 8.1, Box 8.4 (Siebold), Figs. 8.11, 8.12
Narutaki-juku: Box 8.4 (Siebold)
Narva: Box 8.2 (Kaempfer)
National Gallery of Ireland: Ch. 10.1
National Library of Scotland: Acknowledgements
Natural History Museum, London: Ch. 5.8 note 428
Nechow , see Yìqiáo
Nepal: Box 6.4 (Indian Mutiny), Ch. 10.4
Netherlands: Ch. 5.4 note 202, Box 5.5 (Beale), Box 5.22 (Meadows), Box 6.3 (Introductions) notes 4, 23, Box 8.3 (Thunberg), Box 8.4 (Siebold), Box 8.4 note 61 (Siebold), Box 8.8 (Hall), Box 8.12 (Heusken), Box 8.15 (Introductions) notes 7, 9, 13, 26, 30, 37, 38, 45, 48, 51, 61, 62, 63, Figs. 8.10, 8.46
Newark: Box 8.7 (Brown)
New Brunswick: Box 3.1 (Abeel), Box 9.4 (Bruce)
New Brunswick Theological Seminary: Box 3.1 (Abeel)
Newcastle: Ch. 4.1
New England: Ch. 1.1, Ch. 1.1 note 48
Newfoundland: Box 9.4 (Bruce)
New Guinea: Box 8.7 (Brown)
New Haven: Box 7.1 (Smith), Box 8.7
Newhouse, Cranleigh: Ch. 11, Ch. 11 notes 63, 64, Fig. 11.15
Newhouse Arboretum: Ch. 11
Newhouse Cottage: Fig. 11.18
New Jersey: Box 3.1 (Abeel), Box 8.6 (Hepburn), Box 8.7 (Brown)
New Orleans: Box 7.3 (Holt)
New Summer Palace, see Yíhéyuán
New Town, Edinburgh: Ch. 1.3
New York: Box 3.1 (Abeel), Box 3.3 (Macgowan), Box 7.1 (Smith), Box 8.6 (Hepburn), Box 8.7 (Brown), Box 8.8 (Hall), Box 8.11 (Harris), Box 8.12 (Heusken), Ch. 10.5
New York State: Box 3.1 (Abeel), Box 8.7, Box 8.11 (Harris)
New Zealand: Box 8.11 (Harris), Box 10.3 (Fullerton-Elphinstone)
Niànbādū, see Èrshíbādū
Niigata: Box 8.7 (Brown)
Nile: Ch. 5.1
Níngbō: Acknowledgements, Ch. 2.2 note 48, Box 2.2 (Opium), Ch. 3.1, Ch. 3.2, Ch. 3.4, Ch. 3.4 note 197, Box 3.3 (Macgowan), Box 3.5 (Thom), Box 3.6 (Sinclair), Box 3.7 (Morrison), Box 3.12 (Introductions) notes 95, 96, 109, Ch. 5.3, Ch. 5.4, Ch. 5.5, Ch. 5.8, Ch. 5.9, Ch. 5.11, Box 5.16 (Introductions) notes 7, 11, Box 5.22 (Meadows), Box 5.24 (Burdon), Box 5.25 (Patridge), Box 5.27 (Robertson), Box 5.28 (Harvey), Box 6.3 (Introductions) note 16, Ch. 9.1 note 12, Figs. 3.20, 5.139, 5.149, 5.163
Níngbō Plain: Ch. 5.9, Figs. 5.125, 5.126, 5.135
Ningpo , see Níngbō
Ning- kang- jou , see Yínjiāng Zhèn

Níngxià, see Wúzhōng
Nipon , see Honshu
Nippon, see Honshu
Norristown Academy: Box 8.6 (Hepburn)
North America: Ch. 1.1, Ch. 1.2
North Carolina: Ch. 7.1
North Sea: Ch. 11, Fig. 11.12
Northumberland: Ch. 1.2, Box 3.10 (Alcock), Box 5.22 (Meadows)
Northwestern Provinces, India: Box 4.1 (Royle)
Norwich: Box 2.1 (Lindley)
Norwood: Box 5.13 (Ramsay)
Notre Dame des Victoires, Tiānjīn: Ch. 10.3, Figs. 10.40, 10.42
Novgorod: Box 8.2 (Kaempfer)
Nürnberg: Box 8.4 (Siebold)

Ogee, see Ogi
Ogi: Box 9.9 (Grant)
Oh-gee, see Ogi
Okinawa: Box 3.6 (Sinclair), Box 3.9 (Lay)
Old Battersea Bridge, London: Fig. 10.26
Old Summer Palace, see Yuánmíngyuán
Omora , see Omori
Omori: Ch. 8.2
Omorikaigan: Ch. 8.2
Orissa: Box 2.2 note 7 (Opium), Box 5.31 (Beadon)
Osaka: Ch. 8.3
Oslo: Box 9.10 (Ceylon)
Osumi Strait: Ch. 8.2, Ch. 8.3
Ottoman Empire: Box 10.9 (Muslim)
Oudh, see Awadh
Owasco Outlet: Box 8.7 (Brown)
Oxford: Ch. 5.7, Fig. 5.115
Oxford University: Box 3.10 (Alcock), Box 5.32 (Dalhousie)
Oyu Geyser: Box 3.10 note 24 (Alcock)

Pacific Ocean: Box 3.8 (Cuming), Ch. 8.2, Ch. 8.3, Box 10.4 (Meiji)
Pagoda Island, Fúzhōu: Ch. 5.9 note 508, Fig. 5.141
Pakhoi, see Běihǎi
Pak- wan , see Shàngyú
Palestine: Box 5.7 (Falconer)
Pale- twang , see Bālǐzhuāng
Pali- kao , see Bālǐqiáo
Paorie , see Paurie
Papenberg , see Takahoko Island
Paris: Ch. 2.2 note 65, Box 4.1 (Royle), Box 5.9 (Du Halde), Box 8.2 (Kaempfer), Box 8.3 (Thunberg), Box 8.8

note 4 (Hall), Box 8.9 (Girard), Box 8.11 (Harris), Box 8.14 (Howard-Vyse), Box 9.3 (Gordon)
Park Hill, Streatham: Box 10.1 (Moore)
Parthenon, Athens: Box 9.4 note 1 (Bruce)
Pasadena: Box 2.7 (Campbell-Johnston)
Pata- tshoo , see Bādàchǔ
Paterson, New Jersey: Box 8.6 (Hepburn)
Patna: Box 9.3 (Gordon)
Paurie: Ch. 5.6, Box 5.8 (Jameson), Fig. 5.78
Pearl River, see Zhū Jiāng
Pegli: Box 8.4 note 53 (Siebold)
Peiho River, see Hǎi Hé
Peitang, see Běitáng
Pekin, see Běijīng
Peking , see Běijīng
Penang: Box 5.18 (Bonham), Box 8.12 note 9 (Heusken)
Pénghú Islands: Box 9.7 (Cí Xǐ)
Pénglái: Box 3.7 (Morrison)
Pennsylvania: Box 8.6 (Hepburn)
Penzance: Box 5.12 (Batten)
People's Park, Halifax: Box 5.14 (Paxton)
Peradeniya Botanic Garden: Box 5.6 (Champion)
Persepolis: Box 8.2 (Kaempfer)
Persia: Box 8.2 (Kaempfer)
Persian Gulf: Box 5.3 (Moresby), Box 8.2 (Kaempfer)
Perth, Scotland: Ch. 1.3
Perthshire: Ch. 1.3, Box 9.9 (Grant)
Peru: Ch. 1.1
Philadelphia: Box 8.7 (Brown)
Philippines: Preface, Acknowledgements, Ch. 2.2 note 39, Ch. 3.2, Ch. 3.3, Ch. 3.3 note 158, Box 3.8 (Cuming), Box 3.9 (Lay), Box 8.1 (Veitch), Box 8.7 (Brown), Box 8.11 (Harris), Box 9.9 (Grant), Figs. 3.39, 3.40
Piccadilly: Ch. 3.2, Ch. 5.7
Pickadilly Hall: Ch. 5.7
Piedmont: Ch. 7.1
Ping-hoo, see Pínghú
Pínghú: Ch. 5.9, Ch. 5.10
Píngnán Guó: Box 10.7 note 11
Píngshuǐ: Ch. 5.8
Píngwàng: Ch. 5.10
Pitcairn Island: Box 3.8 (Cuming)
Plymouth, Connecticut: Box 7.1 (Smith)
Plymouth, Devonshire: Box 9.2 (Staveley)
Poland: Ch. 1.1
Poles Park: Box 6.2 note 15 (Glendinning)
Polwarth: Ch. 1.3
Polynesia: Ch. 8.2 note 176, Box 8.1 (Veitch)

Pompey, New York: Box 7.2 (Mason)
Ponline Route: Ch. 10.4
Ponsee: Ch. 10.4
Poo- in- chee , see Pǔxī
Poo- to , *Poo- too* , see Pútuó
Poplar, London: Box 5.4 (Braganza)
Port Suez: Ch. 9.2
Portland Bill: Box 9.10 (Ceylon)
Portsmouth: Box 5.27 (Robertson), Box 9.3 (Gordon)
Portugal: Ch. 1.1, Box 3.10 (Alcock), Box 5.1 (P&O), Box 9.5 (Ricci)
Portuguese Cemetery , see Zhàlán Mùdì
Pouching- hien , see Pǔchéng
Pou- shan Gardens , see Pǔshàn Gardens
Póyáng Hú: Ch. 5.10
Press Mains, Reston: Ch. 11 note 37
Prescot: Fig. 2.4
Preston: Box 2.5 (Smith-Stanley)
Prince's Park, Liverpool: Box 5.14 (Paxton)
Princeton University: Box 8.6 (Hepburn)
Prospect Hill , see Jǐngshān
Prun: Ch. 10.4
Prussia: Box 5.22 (Meadows), Box 8.12 (Heusken)
Pǔchéng: Ch. 5.4, Fig. 5.55
Puerto Rico: Ch. 1.1
Pǔjì Sì, Pútuó Shān: Ch. 5.5 note 290, Figs. 5.57, 5.58, 5.59
Punjab: Ch. 5.12, Box 5.8 note 15 (Jameson), Box 5.32 (Dalhousie), Box 9.9 (Grant)
Pǔshàn Gardens: Box 5.16 notes 23, 24
Pǔtuó Shān: Ch. 3.2, Ch. 3.2 note 143, Box 3.10 note 28 (Alcock), Ch. 5.5, Figs. 3.31, 3.32, 3.34, 3.35, 5.22, 5.57, 5.58
Pǔxī: Ch. 5.9, Ch. 5.11

Qazvin: Box 8.2 (Kaempfer)
Qiāndǎo Hú: Ch. 5.2
Qiánjiāng: Ch. 10.3, Ch. 10.3 note 176
Qiánmén, Běijīng: Ch. 9.2
Qiānshān: Ch. 5.4, Figs. 5.44, 5.45
Qiántáng Jiāng: Ch. 2.2 note 42, Box 2.2 (Opium), Ch. 5.2, Ch. 5.9 note 590
Qiānzhàng Yán Cataract, see Qiānzhàng Yán Pùbù
Qiānzhàng Yán Pùbù: Ch. 5.9, Fig. 5.133
Qīnghǎi: Ch. 5.4 note 160
Qīnghú: Ch. 5.4
Qīngniánhú Gōngyuán: Ch. 9.2 note 78
Qīngpǔ: Ch. 5.10
Qīng Shuǐ Ān: Ch. 5.9, Fig. 5.127
Qīngshuǐláng Shān: Ch. 10.4

Qīngyī Jiāng: Ch. 10.4
Qítái: Box 10.9 (Muslim)
Qom: Box 8.2 (Kaempfer)
Quan- sin- foo , see Shàngráo
Quan- ting Temple , see Guāndǐng Sì
Quánzhōu (Guǎngxī): Box 5.19 (Tàipíng)
Quánzhōu Gǎng: Ch. 3.1
Queen's Park, Glasgow: Box 5.14 (Paxton)
Queen's Road, Chelsea, see Royal Hospital Road
Queen's Road, Victoria: Ch. 3.1
Quito, see Ecuador
Qú Jiāng: Ch. 5.4
Qúzhōu: Ch. 5.4, Fig. 5.41

Ranger's House, Green Park: Ch. 5.7 note 382
Rangoon, see Yangon
Rawalpindi: Box 5.7 (Falconer)
Red Sea: Ch. 5.1, Ch. 5.12, Box 5.3 (Moresby)
Redheugh, Cockburnspath: Ch. 11
Regent's Park: Box 10.1 (Moore)
Rehdorf: Box 8.4 (Siebold)
Relief Church: Ch. 1.2
Reston: Ch. 11
Rhode Island: Box 3.1 (Abeel), Box 3.3 (Macgowan), Box 8.8 (Hall)
Richmond: Box 10.10 (Wade)
Rìmǎ: Ch. 10.4
Rio de Janeiro: Box 9.6 (Wyndham), Box 9.9 (Grant)
Ritha Tea Estate: Ch. 5, Fig. 5.84
Robbers Hill: Ch. 10.4
Roemah, see Rìmǎ
Rogate Lodge, Sussex: Box 9.6 (Wyndham)
Roman Catholic Cathedral , Běijīng: Ch. 9.2, Figs. 9.23, 9.30
Rome, Georgia: Box 8.8 (Hall)
Rome, Italy: Box 2.5 (Smith-Stanley), Box 9.5 (Ricci)
Rome Academy, New York State: Box 8.7 (Brown)
Roorkee: Box 5.8 note 17 (Jameson)
Rotterdam: Box 8.4 (Siebold)
Rouen: Box 9.3 (Gordon)
Roxburghe Hotel, Edinburgh: Ch. 11, Ch. 11 note 21, Fig. 11.8
Royal Albert Hall: Ch. 10.1
Royal Botanic Garden Edinburgh: Acknowledgements, Ch. 1.3, Ch. 1.3 note 150, Box 1.2 (McNab), Ch. 4.1, Fig. 1.17
Royal Botanic Gardens Kew, see Kew
Royal Exotic Nursery: Ch. 8.1, Ch. 10.1, Fig. 10.2
Royal High School, see Edinburgh High School

Royal Marsden Hospital, see Cancer Hospital
Royal Scottish Museum, Edinburgh: Ch. 10.1
Russia: Ch. 1.1, Box 5.7 (Falconer), Box 5.14 (Paxton), Box 8.15 (Introductions) notes 25, 31, Box 9.1 (Tiānjīn), Box 9.7 note 28 (Cí Xǐ), Ch. 10.4, Ch. 10.3 note 203, Box 10.9 (Muslim), Figs. 1.1, 8.46
Russia Plantation : Ch. 5.6
Russian border: Ch. 1.1
Russian Cemetery, Běijīng: Ch. 9.2, Box 9.6 (Wyndham), Box 9.8 (Bowlby)
Rutgers College: Box 8.7 (Brown)
Ryūkyū Islands: Box 8.9 (Girard), Ch. 10.2, Ch. 10.4
Ryūkyū Kingdom: Box 8.9 (Girard)

Sabah: Box 3.10 (Alcock)
Sadiya: Ch. 10.4
Sagami Bay: Ch. 8.3
Saharanpur: Box 4.1 (Royle), Ch. 5.6, Box 5.7 (Falconer), Box 5.8 (Jameson), Box 5.11 (Saharanpur), Box 5.12 (Batten), Figs. 5.50, 5.72, 5.73
Saharunpore, see Saharanpur
Sǎmén: Ch. 5.9
San Francisco: Ch. 8.2 note 168, Box 8.6 (Hepburn), Box 9.10 (Ceylon)
Sand Beach Church: Box 8.7 (Brown)
Sandhurst: Box 5.6 (Champion)
Sāngǎng: Ch. 5.4
Sān guān táng: Box 3.5 (Thom)
Sānlǐhé Dōnglù, Běijīng: Ch. 9.2
Sanmon, Shogun Cemetery: Ch. 8.2, Fig. 8.32
San Pablo: Ch. 3.3
San Remo: Box 2.5 (Smith-Stanley)
Saratov: Box 8.2 (Kaempfer)
Sata-Misaki: Ch. 8.3
Sawadi: Ch. 10.4
Sawtucket: Box 3.3 (Macgowan)
Saya de Malha Bank: Box 5.3 (Moresby)
Scarborough: Box 8.8 note 4 (Hall)
Schloss Freudenstein: Box 8.4 note 53 (Siebold)
Scotland: Acknowledgements, Ch. 1.2, Ch. 1.2 notes 88, 89, 135, Box 1.1 (Buchan), Box 1.2 (McNab), Box 3.4 (Balfour), Ch. 4.1, Ch. 4.2, Ch. 5.7, Ch. 5.5 note 283, Box 5.14 (Paxton), Ch. 6.2, Ch. 7.1, Ch. 9, Ch. 11, Figs. 3.12, 7.7A, 8.64
Scottish Borders: Ch. 1.2, Ch. 10.1
Scottish Highlands: Ch. 1.3, Ch. 3.1, Fig. 3.12
Scutari, see Üsküdar
Sea of Japan: Ch. 8.2 note 139
Seh-mun-yuen, see Shímén
Semarang: Box 5.10 (Lady Mary Wood), Box 8.4 note 53 (Siebold)
Seoul: Ch. 10.2
Seray: Ch. 10.4

Serbia: Box 9.6 (Wyndham)
Seue- tow- sze , see Xuědòu Sì
Seychelles: Box 5.3 (Moresby)
Shaanxi, see Shǎnxī
Shāchē: Box 10.9 (Muslim)
Sha- co , see Xiákǒu
Shājiǎo, see Chuānbí
Sha- k' he , see Shā Xī
Shāndōng: Ch. 3.4 note 197, Box 3.7 (Morrison), Ch. 9.1, Ch. 9.1 note 4, Ch. 9.2 note 130, Box 9.7 (Cí Xǐ), Box 10.5 (Niǎn)
Shanghae , see Shànghǎi
Shànghǎi: Ch. 2.2 notes 37, 45, 52, 59, Box 2.2 (Opium), Box 2.7 (Campbell-Johnston), Ch. 3.1, Ch. 3.2, Ch. 3.4, Ch. 3.2 note 120, Ch. 3.4 note 206, Ch. 3.5 note 223, Box 3.2 (Schoedde), Box 3.3 (Macgowan), Box 3.4 (Balfour), Box 3.5 (Thom), Box 3.5 note 44 (Thom), Box 3.6 (Sinclair), Box 3.7 (Morrison), Box 3.8 note 24 (Cuming), Box 3.10 (Alcock), Box 3.10 note 28 (Alcock), Box 3.11 (Lockhart), Box 3.12 (Introductions) notes 17, 26, 33, 35, 88, Ch. 5.2, Ch. 5.3, Ch. 5.4, Ch. 5.5, Ch. 5.8, Ch. 5.9, Ch. 5.10, Ch. 5.11, Ch. 5.1 note 2, Ch. 5.4 note 267, Ch. 5.8 note 399, Box 5.5 (Beale), Box 5.5 note 5 (Beale), Box 5.10 (Lady Mary Wood), Box 5.16 (Introductions) notes 2, 4, 13, 25, Box 5.17 (Ganges), Box 5.18 (Bonham), Box 5.19 (Tàipíng), Box 5.21 (Small Sword), Box 5.22 (Meadows), Box 5.23 (Perceval), Box 5.24 (Burdon), Box 5.25 (Patridge), Box 5.27 (Robertson), Box 5.28 (Harvey), Box 5.29 (Chusan), Box 6.3 (Introductions) notes 11, 18, 20, Ch. 7.2, Ch. 8.1, Ch. 8.2, Ch. 8.3, Box 8.1 (Veitch), Box 8.5 (Loureiro), Box 8.6 (Hepburn), Box 8.7 (Brown), Box 8.8 (Hall), Box 8.8 notes 4, 7 (Hall), Box 8.13 (Aspinall), Box 8.13 note 4 (Aspinall), Ch. 9.1, Ch. 9.2, Box 9.2 (Staveley), Box 9.9 (Grant), Ch. 10.1, Ch. 10. 4, Ch. 10.6, Ch. 10.3 note 166, Box 10.2 note 5 (Namamugi), Box 10.10 (Wade), Figs. 3.31, 3.49, 3.50, 5.22, 5.116, 5.123, 5.143, 5.160, 9.2, 10.15
Shang- i- yuen , see Chún-ān
Shang- o , see Cáo' é
Shàngráo: Ch. 5.4, Fig. 5.43
Shàngyú: Ch. 5.3, Ch. 5.4
Shānhòu: Ch. 5.9, Fig. 5.125
Shāntáng Canal, Sùzhōu: Fig. 3.27
Shan- te- Maou , see Guāndì Miào
Shàntóu: Box 10.10 (Wade)
Shānxī: Box 9.7 (Cí Xǐ)
Shǎnxī: Box 9.7 (Cí Xǐ), Box 10.5 (Niǎn), Box 10.8 (Huí), Box 10.9 (Muslim)
Sháoguān: Box 9.5 (Ricci)
Shaou- hing- foo , see Shàoxīng
Shàoxīng: Ch. 5.3, Ch. 5.4, Ch. 5.8, Box 5.24 (Burdon), Ch. 10.6, Fig. 5.23
Shāshì: Ch. 10.4
Shā Xī: Ch. 5.9
She- mun- yuen , see Shímén
Shēnhù Gǎng: Box 3.12 (Introductions) notes 1, 49, 82, Fig. 3.8
She- pa- ky , see Shípí
Shie- poo , see Shípǔ
Shikoku: Ch. 8.2
Shímén: Ch. 5.2

Shimoda: Box 8.11 (Harris), Box 8.12 (Heusken)
Shimonoseki: Box 8.9 (Girard), Box 9.7 (Cíxǐ)
Shimonoseki Straits: Ch. 10.1
Shinagawa, Tokyo: Ch. 8.2, Ch. 10.2, Fig. 8.28
Shinjuku Central Park, Tokyo: Ch. 8.3, Fig. 8.53
Shípí: Ch. 5.4
Shípǔ: Ch. 5.11, Box 5.30 (Vansittart)
Shiraz: Box 8.2 (Kaempfer)
Shizuoka: Ch. 10.2
Shòu Fēng Tǎ: Ch. 5.9, Figs. 5.125, 5.126
Shrewsbury School: Box 5.31 (Beadon)
Shùn-ān: Ch. 5.2
Shung-jay-sze, see Xiāngjiè Sì
Shùnjiāng Tower (Yúyáo): Ch. 5.9, Ch. 5.9 note 576
Siam: Box 3.1 (Abeel), Box 8.2 (Kaempfer), Box 8.12 (Heusken), Box 10.7 (Panthay)
Siberia: Box 3.12 (Introductions) note 20, Fig. 3.40H
Sìchuān: Ch. 10.3, Ch. 10.4, Box 10.6 (Xián-Tóng)
Sìchuān Basin: Ch. 10.4
Sicily: Box 5.7 (Falconer)
Sieu-wang-meou, see Xiāo Wáng Miào
Sikkim: Box 5.15 (Standish), Box 5.32 (Dalhousie)
Sikok, see Shikoku
Silver Island, see Jīntáng Dǎo
Sìmén, see Sǎmén
Sìmíng Shān: Ch. 5.9
Sìmíng Shān Cūn: Ch. 5.9
Simla: Box 9.9 (Grant)
Simonstown: Box 8.12 note 9 (Heusken)
Sinagawa, see Shinagawa
Singapore: Box 3.1 (Abeel), Box 3.9 (Lay), Ch. 5.1, Box 5.1 (P&O), Box 5.18 (Bonham), Box 5.29 (Chusan), Box 8.6 (Hepburn), Box 8.7 (Brown), Box 8.7 note 18 (Brown), Box 8.9 (Girard), Box 9.6 (Wyndham), Box 9.9 (Grant)
Sittoung River: Box 9.3 (Gordon)
Siwaliks: Box 5.7 (Falconer)
Sligo County: Box 5.23 (Perceval)
Somei, Tokyo: Ch. 8.2, Ch. 8.3, Fig. 8.34
Somerset: Box 5.31 (Beadon), Box 6.2 (Glendinning)
Sōngjiāng: Ch. 5.2, Fig. 5.10
Sōngluó: Ch. 5.2, Ch. 5.5, Fig. 5.20
Soochow, see Sùzhōu
Soo-chow-foo, see Sùzhōu
South Africa: Box 2.3 (Reeves), Ch. 3.1, Ch. 3.4, Box 5.1 (P&O), Box 5.3 (Moresby), Box 6.2 (Glendinning), Box 8.3 (Thunberg), Box 8.12 note 9 (Heusken), Fig. 8.10
South America: Box 3.8 (Cuming), Box 5.14 (Paxton), Ch. 8.2 note 176, Box 9.4 (Bruce), Fig. 1.3
Southampton: Box 3.2 (Schoedde), Ch. 4.2, Ch. 5.1, Ch. 5.7, Ch. 5.12, Ch. 5.1 note 4, Box 5.2 (Ripon), Box 5.3 (Moresby), Box 5.4 (Braganza), Box 5.10 (Lady Mary Wood), Box 5.17 (Ganges), Ch. 6.1, Ch. 9.2, Box 9.3

(Gordon), Box 9.10 (Ceylon), Ch. 10.1 note 2, Figs. 5.1, 9.54
South Carolina: Ch. 7.1, Ch. 7.3, Box 7.1 (Smith), Box 8.7 (Brown), Figs. 7.2, 7.3
South China Sea: Ch. 5.1
South Kensington Museum, see Victoria and Albert Museum
South Seas: Box 8.7 (Brown)
Spain: Ch. 1.1, Ch. 1.1 note 44, Box 3.10 (Alcock), Ch. 5.1, Box 5.1 (P&O), Box 5.10 (Lady Mary Wood), Box 5.14 (Paxton), Fig. 1.3
Spanish West Florida: Ch. 1.1 note 35
Spice Islands: Box 3.9 (Lay), Box 8.7 (Brown)
Springside Boarding School: Box 8.7 (Brown)
Spylaw Road, Edinburgh: Ch. 11
Sri Lanka, see Ceylon
St. Andrews: Ch. 10.1
St. Andrews University: Box 9.3 (Gordon)
St. Helena: Ch. 1.1
St. James' Church, Westgate-on-Sea: Ch. 11
St. James's Palace, London: Fig. 5.110
St. James's Street, London: Ch. 5.7, Ch. 10.1, Fig. 5.110
St. John's Church, Meerut : Fig. 5.89
St. Joseph's College, Bardstown: Box 7.3 (Holt)
St. Leonards: Box 2.6 (Ward)
St. Luke's Church, Chelsea: Ch. 4.2, Figs. 4.6, 4.7
St. Mary's Church, Carlisle: Ch. 11
St. Mary's Parish Church, Edinburgh: Ch. 1.3, Fig. 1.15
St. Marys Road, Leatherhead: Ch. 11
St. Petersburg: Box 8.15 (Introductions) note 29
St. Pierre-le-Guillard, Bourges: Box 8.9 (Girard)
St. Saviour's Church, London: Ch. 10.1, Fig. 10.6
St. Thomas: Box 3.1 (Abeel)
St. Thomas' Hospital: Box 5.14 note 24 (Paxton)
St. Vincent de Paul Convent, Tiānjīn: Ch. 10.3, Fig. 10.41
Stair Park, Tranent: Ch. 11, Fig. 11.3
Stanwix Parish Church, Carlisle: Ch. 11
Stockholm: Box 8.2 (Kaempfer), Box 9.10 (Ceylon)
Stoke: Box 10.1 (Moore)
Stoke Damerel, Plymouth: Box 9.2 note 28 (Staveley)
Stoke Poges: Box 8.14 (Howard-Vyse)
Strait of Malacca: Ch. 5.1, Ch. 10.3
Straits of Banka: Box 8.7 (Brown)
Straits Settlements: Box 5.18 (Bonham)
Streatham: Box 10.1 (Moore)
Sudiya, see Sadiya
Suez: Ch. 5.1, Ch. 5.8, Ch. 5.12, Box 5.1 (P&O), Box 5.3 (Moresby), Box 5.33 (Bentinck), Box 5.33 note 3 (Bentinck), Ch. 9.2, Figs. 5.6, 5.7, 5.166
Suez Canal: Box 5.1 (P&O)

Suffolk: Box 2.7 (Campbell-Johnston)
Suídé: Box 10.8 (Huí)
Suiy- kow , see Nánpíng
Sulawesi: Box 3.9 (Lay)
Su- mae- yah , see Somei
Sumatra: Ch. 5.1
Sumida River: Ch. 8.2, Box 8.7 (Brown), Box 8.9 (Girard)
Summerville: Ch. 7.1
Sumner Place: Ch. 5.7
Sundarbans: Ch. 5, Fig. 5.70
Sunderbunds , see Sundarbans
Sung- kiang- foo , see Sōngjiāng
Sung- lo , see Sōngluó
Sunlight Rock: Fig. 3.6
Sunningdale (Nursery): Ch. 5.7, Box 5.15 (Standish)
Suōyī Ford: Box 5.19 (Tàipíng)
Surrey: Box 3.10 (Alcock), Box 5.13 (Ramsay), Box 5.15 (Standish), Box 8.1 (Veitch), Box 8.10 (Stevens), Ch. 11, Figs. 3.10, 11.15, 11.17
Surrey Docks: Ch. 10.1 note 103
Surrey History Centre, Woking: Acknowledgements, Ch. 11
Surrey Infantry Museum, Woking: Fig. 3.10
Sussex: Box 2.6 (Ward), Box 9.6 (Wyndham)
Suwa Shrine (Nagasaki): Ch. 8.1, Fig. 8.8
Sùzhōu, Ānhuī: Ch. 2.2 note 48, Ch. 3.2, Ch. 3.5, Ch. 3.2 note 131, Box 3.11 (Lockhart), Box 3.12 (Introductions) notes 42, 95, 109, Box 5.19 note 32 (Tàipíng), Ch. 10.6, Figs. 3.27, 3.28, 3.29, 3.30
Sùzhōu Creek, see Sùzhōu Hé
Sùzhōu Hé: Ch. 3.2, Ch. 5.10, Box 5.5 (Beale)
Suzugamori execution ground: Ch. 8.2, Figs. 8.39, 8.40
Swan Walk Gate, Chelsea Physic Garden: Ch. 4.1, Ch. 4.1 note 15, Fig. 4.2
Swatow, see Shàntóu
Sweden: Box 8.2 (Kaempfer), Box 8.3 (Thunberg), Box 9.10 (Ceylon)
Swi-foo, see Yíbīn
Swinton: Ch. 1.3, Ch. 2.2 note 71
Swinton Parish School: Ch. 10.6 note 495
Switzerland: Box 5.14 (Paxton), Box 5.32 (Dalhousie), Box 9.9 (Grant)
Sydenham: Box 5.14 (Paxton)
Sydenham College: Box 3.10 (Alcock)
Sydenham Hill: Box 5.14 (Paxton)
Sydney: Box 2.6 (Ward), Box 5.29 (Chusan)
Sydney Street: Ch. 4.2

Ta-chien-lu, see Kāngdìng
Tahiti: Box 3.8 (Cuming), Box 10.3 (Fullerton-Elphinstone)
Tàicāng: Ch. 3.2
T'ai- hu , see Tàihú

Tàihú: Ch. 5.10, Ch. 5.10 note 634
Táinán: Box 5.25 (Patridge)
Tàipíng River: Ch. 10.4
Táiwān: Acknowledgements, Ch. 2.2 note 49, Ch. 3.5, Ch. 5.9, Ch. 5.9 notes 507, 521, 522, Box 5.25 (Patridge), Box 8.4 (Siebold), Box 9.7 (Cí Xǐ), Ch. 10.2, Ch. 10.4, Ch. 10.2 note 162, Box 10.10 (Wade)
Taiwan-Foo, see Táinán
Táiwānfǔ, see Táinán
Táiwān Strait: Ch. 5.9 note 521
Taj Mahal: Ch. 3.5 note 230, Ch. 5.6, Ch. 5.6 note 357, Box 9.3 (Gordon), Box 9.9 (Grant)
Takahoko Island, Nagasaki: Ch. 8.1, Fig. 8.3
Taklamakan Desert, see Tǎlèlāmǎgān Desert
Taku , see Dàgū
Tǎlèlāmǎgān Desert: Box 10.9 (Muslim)
Tǎlǐmù Péndì: Box 10.9 (Muslim)
Tama River: Ch. 8.2, Ch. 8.3
Tamkang University: Acknowledgements
Tamshui, see Dànshuǐ
Tam- shuy , see Dànshuǐ
Tangchow, see Pénglái
Tángqī: Ch. 5.2
Tantabin: Box 9.3 (Gordon)
Tan- see , see Tángqī
Taou- chang- shan- ta , see Dàochǎngshān Tǎ
Ta- pae- sze , see Dàbēi Sì
Tarim Basin, see Tǎlǐmù Péndì
Tartar City , Běijīng: Ch. 9.1, Ch. 9.2
Ta- sha- lar , see Dàzhàlan
Ta-tsian-loo, see Kāngdìng
Ta- tsong- tseu , see Tàicāng
Tatton Park: Box 5.14 (Paxton)
Ta- yang , see Dàyáng
Tcien- tang- kiang , see Qiántáng Jiāng
Tea Horse Route: Ch. 10.4 note 251
Tein- muh- shan , see Tiānmù Shān
Tein- tin , see Tiāntán
Tein- tung , see Tiāntóng
Temple House, Ballymote: Box 5.23 (Perceval)
Temple of Earth: Box 9.6 (Wyndham)
Temple of Heaven, see Tiāntán
Temple of Sun: Box 9.6 (Wyndham)
Téngchōng: Ch. 10.4
Tengyueh, see Téngchōng
Tenshu-do, Yokohama: Box 8.9 (Girard)
Ten Thousand Centuries Inn , see Mannenya
Terai: Box 5.13 (Ramsay)

Ternate: Box 3.9 (Lay)
Te- sye- mun , see Sămén
Texas: Ch. 7.1
Thailand, see Siam
Thames: Box 9.3 (Gordon), Ch. 10.1, Fig. 5.115
Thames Embankment: Ch. 10.1
Thayat Myo: Box 9.3 (Gordon)
Thistle Grove Lane, South Kensington: Ch. 5.7, Ch. 6.1, Ch. 6.2, Figs. 4.6, 5.103
Thistle Mountain, see Zĭjīn Shān
Thousand-Island Lake, see Qiāndăo Hú
Three Gorges: Ch. 10.4
Tiān' ānmén Square: Ch. 9.1
Tiān fèng Tă, Níngbō: Ch. 3.1, Fig. 3.20
Tiānjīn: Acknowledgements, Box 5.19 (Tàipíng), Box 5.22 (Meadows), Ch. 9.1, Ch. 9.2, Ch. 9.1 notes 16, 18, Box 9.2 (Staveley), Box 9.3 (Gordon), Box 9.4 (Bruce), Box 9.4 note 18 (Bruce), Ch. 10.3, Ch. 10.6, Box 10.5 (Niăn), Box 10.10 (Wade), Figs. 9.1, 9.7, 9.8, 9.9, 9.10, 9.11, 9.13, 9.15, 10.39, 10.40, 10.41, 10.42, 10.44
Tiānjīn Plain: Ch. 9.2
Tiānjīng: Box 5.19 (Tàipíng)
Tiānjĭng Sì: Ch. 5.9, Ch. 5.11
Tiānmù Shān: Ch. 5.10
Tiāntán, Běijīng: Ch. 9.2, Ch. 9.2 note 66, Fig. 9.33
Tiāntóng: Ch. 3.2, Box 3.12 (Introductions) notes 33, 43, Ch. 5.3, Ch. 5.5, Ch. 5.8, Figs. 3.24, 3.25, 5.36
Tiānxìn Yŏnglè Chán Sì: Ch. 5.4, Figs. 5.48, 5.49
Tibet: Ch. 10.4, Ch. 10.4 note 251
Tidore: Box 3.9 (Lay)
Tien- tsin , see Tīanjīn
Tiger Gate, see Hŭmén
Tilsit: Ch. 1.1
Tinghae , see Zhōushān
Toba-Fushimi: Fig. 10.33
Tokaido , see Tōkaidō
Tōkaidō: Ch. 8.2, Ch. 8.3, Box 10.2 (Namamugi), Figs. 8.27, 8.28, 8.33
Tokio, see Tokyo
Tokugawa Mausoleums: Ch. 8.2, Box 8.7 (Brown)
Tokyo: Acknowledgements, Ch. 8.2, Box 8.5 (Loureiro), Box 8.6 (Hepburn), Box 8.7 (Brown), Box 8.15 (Introductions) note 60, Ch. 10.2, Box 10.4 (Meiji), Figs. 3.45, 8.28, 8.33
Tokyo Railway Station: Ch. 8.1 note 79
Tong- chow , see Tōngzhōu
Tōngjì Qiáo, Yúyáo: Ch. 5.9, Figs. 5.38, 5.39
Tōngzhōu: Ch. 9.2
Too- poo- dow , see Dà bù tóu
Torinana, see Cape Tourinan
Torin-ji, Uraga: Ch. 8.2, Figs. 8.21, 8.23, 8.24, 8.25
To- rin- gee , see Torin-ji
Torrich: Box 9.3 (Gordon)

Torun: Box 8.2 (Kaempfer)
Tozen-ji: Ch. 8.2, Fig. 8.29, 8.65
Trafalgar: Ch. 1.1
Tranent: Ch. 11, Fig. 11.3
Tranent Parish Churchyard: Acknowledgements, Ch. 11, Figs. 11.5, 11.6
Treaty Ports: Ch. 2.1, Ch. 2.2 note 45, Box 2.2 (Opium), Box 2.3 (Reeves), Ch. 3.1, Ch. 3.4, Ch. 3.5, Box 3.1 (Abeel), Box 3.5 (Thom), Box 3.10 (Alcock), Ch. 5.1, Ch. 5.3, Box 6.1 (Arrow), Box 6.1 note 20 (Arrow), Box 8.6 (Hepburn), Box 8.8 (Hall), Box 9.2 (Staveley), Box 9.4 note 18 (Bruce), Ch. 10.6, Box 10.5 (Niǎn), Box 10.10 (Wade)
Treaty of Tīanjīn: Box 9.1, Ch. 10.3, Box 10.10 (Wade)
Trelleborg: Box 9.10 (Ceylon)
Trincomalee: Box 5.10 (Lady Mary Wood)
Trinidad: Box 3.4 (Balfour), Box 5.2 (Ripon)
Trinity College, Cambridge: Box 2.5 (Smith-Stanley), Box 5.12 (Batten), Box 10.10 (Wade)
Trinity College, Hartford, see Washington College
Tsai- tsoun , see Nán Cài Cūn
Tsali, see Xiálǐ
Tsan- tsin , *Tsan- tsing Temple* , see Tiānjǐng Sì
Tsaou- o , see Cáo' é
Tsasa- poo , see Zǒu yá bù
Tsa- yuen , see Chá yuán
Tse- kee , see Cíchéng
Tsin- tsun , see Xīngcūn
Tsing- poo , see Qīngpǔ
Tsong- gan , see Chóng' ān
Tsong- gan- hien , see Sāngǎng
Tsong- so , see Zhōngxìn
Túhài River: Box 10.5 (Niǎn)
Tùlǔfān: Box 10.9 (Moslem)
Tunaborg: Box 8.3 (Thunberg)
Tung- che , see Túnxī
Tung- hoo , see Dōngqián Hú
Tunglan (? Tōngdiàn): Ch. 10.4
Tung- t' ing- shan , see Dōngshān
Túnxī: Ch. 5.2, Figs. 5.18, 5.19
Turkey: Box 5.6 (Champion), Box 5.14 (Paxton)
Turku: Box 8.2 (Kaempfer)
Turnham Green: Ch. 4.1, Box 6.2 (Glendinning)
Turpan, see Tùlǔfān
Tver: Box 8.2 (Kaempfer)
Twyford School: Box 9.6 (Wyndham)
Tyburn: Ch. 5.7 note 382
Tyninghame House: Box 1.2 (McNab)
Tzekee River, see Yúyáo Jiāng
Tzeki, see Cíchéng

Ueno: Ch. 10, Figs. 10.35, 10.36
Ulm: Box 8.4 note 53 (Siebold)
Umballa, see Ambala
Ume-yashiki, Omori: Ch. 8.2
United Kingdom (UK), see Britain
United States of America: Ch. 1.1, Ch. 1.1 notes 53, 85, Box 3.1 (Abeel), Box 3.1 note 49 (Abeel), Box 3.3 (Macgowan), Box 3.12 (Introductions) note 41, Ch. 5.6, Box 5.30 (Vansittart), Box 5.30 note 13 (Vansittart), Box 7.1 (Smith), Box 7.2 (Mason), Box 7.3 note 24 (Holt), Ch. 8.1, Ch. 8.2, Box 8.6 (Hepburn), Box 8.7 (Brown), Box 8.8 (Hall), Box 8.11 (Harris), Box 8.12 (Heusken), Box 8.15 (Introductions) note 61, Box 9.1 (Tiānjīn), Box 9.4 (Bruce), Ch. 10.1, Ch. 10.4, Ch. 10.3 note 166, Figs. 6.3B, 7.3, 7.6, 7.7, 8.26
United States Legation (Edo): Ch. 8.3 notes 155, 163, Box 8.12 notes 20, 23 (Heusken), Fig. 8.49
United States Virgin Islands: Box 3.1 (Abeel)
University College, London: Box 4.1 (Royle)
University of the City of New York: Box 8.7 (Brown)
University of Edinburgh, see Edinburgh University
University of Pennsylvania: Box 8.6 (Hepburn)
University of St. Andrews, see St. Andrews University
University of Vienna, see Vienna University
Upper Burma, see Burma
Upper Peru, see Bolivia
Upper St. James's Park, see Green Park
Uppsala: Box 8.2 (Kaempfer), Box 8.3 (Thunberg)
Uppsala University: Box 8.3 (Thunberg)
Uraga Harbour: Ch. 8.2, Figs. 8.20, 8.21
Uruguay: Box 9.4 (Bruce)
Urumqi, see Wūlǔmùqí
USA, see United States of America
Üsküdar, Turkey: Box 5.6 (Champion)
Utrecht: Box 8.4 note 43 (Siebold)
Uttar Pradesh, see Awadh
Uzbekistan: Box 10.9 (Moslem)

Valençay: Ch. 1.1
Valparaiso: Box 3.8 (Cuming)
Van Diemen's Strait, see Osumi Strait
Vauxhall Bridge Station, London: Ch. 10.1, Fig. 10.1
Vicksburg: Box 7.3 (Holt)
Victoria: Ch. 3.1
Victoria Harbour: Ch. 5.1
Victoria and Albert Museum: Ch. 10.1, Ch. 10.1 note 9, Fig. 10.4
Vienna University: Acknowledgements
Vietnam: Ch. 10.4
Virgin Islands: Box 3.1 (Abeel)
Virginia: Ch. 7.3

Vlissingen: Box 8.4 (Siebold)
Volga: Box 8.2 (Kaempfer)

Wae-ping, see Wēipíng Zhèn
Wakamiya Oji, Kamakura: Ch. 8.3
Wales: Ch. 5.7
Walker-on-Tyne: Box 5.29 (Chusan)
Walong, see Walung
Walton Street: Ch. 10.1
Walung: Ch. 10.4
Wan-sheu-si, see Wànshòu Sì
Wànshòu Qiáo: Ch. 3.4, Fig. 3.41
Wànshòu Sì: Ch. 5.10, Figs. 5.156, 5.157
Wan-show-jou, see Wànshòu Qiáo
Wànghǎi Lóu, see Notre Dame des Victoires
Warlingham Park Hospital: Ch. 11, Ch. 11 note 70
Washbrook: Box 3.8 (Cuming)
Washington DC: Ch. 1.1, Ch. 7.1, Ch. 7.2, Box 7.3 (Holt), Box 9.4 (Bruce)
Washington College, Hartford: Box 8.8 (Hall)
Waterloo, Belgium: Ch. 1.1, Ch. 1.2
Waterloo Bridge Station: Ch. 10.1
Watton Street: Ch. 10
Wayside, St. Marys Road, Leatherhead: Ch. 11, Ch. 11 note 72
Wazirabad: Box 9.9 (Grant)
Wèi Valley: Box 10.8 (Huí)
Wēipíng Zhèn: Ch. 5.2
Wéixī: Ch. 10.4
Wellclose Square, London: Box 2.6 (Ward)
Wellcome Library, London: Figs. 3.41, 3.47, 10.41, 10.42
Wells, Somerset: Box 5.31 (Beadon)
Wēnzhōu: Ch. 10.4
West Africa: Box 9.3 (Gordon)
West Bromich: Box 5.28 (Harvey)
West End Park, see Kelvingrove Park
West Ham: Box 2.3 (Reeves)
West India Docks: Ch. 10.1 note 97
West Indies: Box 3.1 (Abeel), Box 10.3 (Fullerton-Elphinstone)
West London Cemetery, see Brompton Cemetery
West Point: Box 3.1 (Abeel), Box 7.2 (Mason)
Western Hills, Běijīng: Ch. 9.2, Box 9.6 (Wyndham), Fig. 9.50
Westgate-on-Sea, Kent: Ch. 11
Westminster Abbey: Ch. 5.7 note 379, Ch. 10.1 note 10
Westphalia: Box 8.2 (Kaempfer)
Whampoa, see Huángpǔ
White House, Washington: Ch. 1.1, Fig. 1.4

Wiesbaden: Box 8.4
Wiltshire: Box 3.8 note 2 (Cuming), Box 5.15 (Standish), Box 5.31 (Beadon)
Wimbledon: Ch. 3.5, Box 5.14 (Paxton)
Wishart Church, Tranent: Ch. 11
Woburn: Box 5.14 (Paxton)
Woking: Acknowledgements, Ch. 5.7, Ch. 10.1, Ch. 10.1 note 2, Ch. 11, Fig. 3.10
Woodhall Park: Box 5.14 (Paxton)
Woo- e- san , see Wǔyíshān
Woo- hoo- mun , see Wǔhǔmén
Woosung, see Wúsōng
Worthing: Box 5.30 (Vansittart)
Wotton: Box 5.14 (Paxton)
Wǔhàn: Box 5.19 (Tàipíng), Ch. 8
Wúhú: Box 5.19 (Tàipíng), Ch. 10.4
Wǔhǔmén: Ch. 3.4, Ch. 3.4 note 183
Wūlǔmùqí: Box 10.9 (Muslim)
Würzburg: Box 8.4 (Siebold), Box 8.4 note 51 (Siebold)
Würzburg University: Box 8.4 (Siebold)
Wúshān Sì: Ch. 5.10, Ch. 5.10 notes 639, 640, 642, Ch. 10.6, Fig. 5.159
Wū Shí Shān: Box 3.9 (Lay)
Wū Shí Shān Sì: Ch. 3.4, Fig. 3.43
Wǔ Shuǐ: Ch. 10.4
Wúsōng: Box 2.2 (Opium), Box 3.2 (Schoedde), Box 3.9 (Lay)
Wǔyíshān: Ch. 3.4, Ch. 5.3, Ch. 5.4, Ch. 5.5, Ch. 5.9, Ch. 5.4 notes 181, 201, 202, Box 5.7 (Falconer), Box 5.8 (Jameson), Figs. 5.47, 5.48, 5.49, 5.53, 5.54, 5.101
Wuzeerabad, see Wazirabad
Wúzhōng: Box 10.8 (Huí)

Xiákǒu: Ch. 5.4, Ch. 10.6
Xiálǐ: Ch. 10.4
Xiàmén: Box 2.2 (Opium), Ch. 3.1, Ch. 3.4, Ch. 3.1 note 20, Ch. 3.4 note 197, Box 3.1 (Abeel), Box 3.5 (Thom), Box 3.6 (Sinclair), Box 3.7 (Morrison), Box 3.9 (Lay), Box 3.10 (Alcock), Box 3.12 (Introductions) note 43, Ch. 5. 5, Ch. 5.10, Box 5.25 note 6 (Patridge), Box 5. 27 (Robertson), Box 5.28 (Harvey), Box 8.6 (Hepburn), Box 8.7 (Brown), Ch. 9.1, Ch. 10.1 note 16
Xī' ān: Box 9.7 (Cí Xǐ), Box 10.8 (Huí)
Xiāng Jiāng: Box 5.19 (Tàipíng)
Xiāngjiè Sì, Bādàchù: Ch. 9.2, Figs. 9.47, 9.48
Xiàngzhōu: Box 5.20 (Hóng)
Xiānyáng Zhèn: Ch. 5.4
Xiǎotāngshān: Box 9.6 (Wyndham)
Xiāo Wáng Miào: Ch. 5.9, Fig. 5.129
Xiàzhèn: Ch. 5.4, Ch. 5.4 note 165
Xīcháng' ān Jiē, Běijīng: Ch. 9.2
Xīkǒu Zhèn: Ch. 5.9
Xīn' ān Jiāng: Acknowledgements, Ch. 5.2, Fig. 5.17

Xīn'ān Jiāng Shuǐkù, see Qiāndǎo Hú
Xìng Cí Sì, Lǐcūn: Ch. 5.9, Fig. 5.138
Xīngcūn: Ch. 5.4
Xīníng: Box 10.8 (Huí)
Xìn Jiāng (River): Ch. 5.4, Fig. 5.44
Xīnjiāng Province: Box 2.2 (Opium), Ch. 5.4 note 160, Box 10.9 (Muslim), Fig. 10.38
Xīshuāngbǎnnà Tropical Botanical Garden: Acknowledgements
Xītiáo Xī: Ch. 5.10, Fig. 5.160
Xuānwǔmén, Běijīng: Ch. 9.2
Xuědòu Sì: Ch. 5.9, Figs. 5.131, 5.132, 5.133
Xǔ Shì Mínjū: Ch. 5.5 note 279

Yǎ'ān: Ch. 10.4
Yale University: Box 7.1 (Smith), Box 8.7 (Brown)
Yangon: Box 6.4 (Indian Mutiny), Box 9.3 (Gordon), Ch. 10.4
Yang- tse- kiang , see Cháng Jiāng
Yangzi, see Cháng Jiāng
Yánshān, see Qiānshān
Yāntái: Box 3.7 (Morrison), Ch. 9.1, Ch. 9.2 note 130, Ch. 10.4, Ch. 10.3 note 202, Box 10.5 (Niǎn), Figs. 9.2, 9.3, 9.4, 9.5, 9.6, 9.7, 10.45
Yāntáishān: Fig. 9.4
Yan- ting Gate , see Āndìngmén
Yao: Ch. 8.3
Yarkant, see Shāchē
Yedashe: Box 9.3 (Gordon)
Yeddo, see Edo
Yedo , see Edo
Yellow River, see Huáng Hé
Yellow Sea: Box 3.6 (Sinclair), Ch. 5.8 note 398
Yen- chow- foo , see Méichéng
Yentae , see Yāntái
Yêw-keuén, see Yóuquán
Yíbīn: Ch. 10.4
Yíchāng: Box 3.12 (Introductions) note 3, Ch. 10.4
Yíhéyuán: Box 9.7 (Cí Xǐ)
Yíng Tái: Box 9.7 (Cí Xǐ)
Yínjiāng Zhèn: Ch. 5.9, Fig. 5.139
Yìqiáo: Ch. 5.2, Ch. 5.4
Yīshān, see Huángshān
Yokohama: Acknowledgements, Box 3.10 note 23 (Alcock), Ch. 8.2, Ch. 8.3, Ch. 8.2 notes 148, 160, 208, Box 8.1 (Veitch), Box 8.4 (Siebold), Box 8.6 (Hepburn), Box 8.7 (Brown), Box 8.8 (Hall), Box 8.9 (Girard), Box 8.11 (Harris), Box 8.13 (Aspinall), Box 8.14 (Howard-Vyse), Box 8.15 (Introductions) note 17, Ch. 10.2, Ch. 10.5, Ch. 10.2 note 138, Box 10.2 (Namamugi), Fig. 8.16
Yokuhama , see Yokohama
Yǒng'ān: Box 5.19 (Tàipíng) (since 1915 Méngshān)

Yǒngdìngmén: Ch. 9.2
Yǒngdìngménnèi Dàjiē: Ch. 9.2
Yōnghé Gōng: Ch. 9.2, Box 9.6 (Wyndham), Box 9.9 (Grant), Fig. 9.41
Yǒng Jiāng: Ch. 3.1, Box 3.5 (Thom), Ch. 5.9
Yǒngpíng: Ch. 5.4, Ch. 10.4, Fig. 5.45
Yǒngquán Temple (Gǔshān): Ch. 5.3, Fig. 5.27, 5.29, 5.30, 5.31, 5.32
York: Ch. 5.7, Ch. 10.1, Fig. 5.112
Yorkshire: Ch. 1.1 note 6, Box 5.14 (Paxton), Box 5.15 (Standish)
Yòu'ānmén: Ch. 9.2
Yóuquán: Ch. 5.8 note 399
Yǒuyáng: Ch. 10.3
Yoxford: Box 2.7 (Campbell-Johnston)
Yuánhuā: Ch. 5.9
Yuán Jiāng: Ch. 10.4
Yuánmíngyuán: Ch. 9.2, Box 9.4 (Bruce), Box 9.7 (Cí Xǐ), Box 9.8 (Bowlby), Box 9.9 (Grant)
Yuen-hwa, see Yuánhuā
Yuen-ming-yuen, see Yuánmíngyuán
Yuen-shan, see Yǒngpíng
Yuèyáng: Ch. 10.4
Yu-eou, see Yúyáo
Yuk-shan, see Yùshān
Yung Chang Fu, see Yǒngpíng
Yúnnán: Box 9.3 (Gordon), Ch. 10.3, Ch. 10.4, Ch. 10.4 note 251, Box 10.7 (Panthay), Fig. 10.45
Yúnnánfǔ, see Kūnmíng
Yùpíng: Ch. 10.4
Yü-ping Hsien, see Yùpíng
Yùshān: Ch. 5.4, Ch. 5.4 note 160, Fig. 5.42
Yúyáo: Box 2.2 (Opium), Ch. 5.3, Ch. 5.4, Ch. 5.9, Box 5.24 (Burdon), Box 5.26 (Wáng), Figs. 5.38, 5.39, 5.150
Yúyáo Jiāng: Box 3.5 (Thom), Ch. 5.8
Yu-yeou, see Yúyáo
Yúzé Jiāng: Ch. 5.9, Fig. 5.130

Zayu, see Rìmǎ
Zenpuku-ji, Tokyo: Ch. 8.3, Box 8.11 (Harris), Fig. 8.49
Zhàlán Mùdì: Ch. 9.2, Box 9.5 (Ricci), Fig. 9.29
Zhāngjiāwān: Ch. 9.1, Figs. 9.16, 9.17
Zhāng Xī: Ch. 5.9, Figs. 5.139, 5.140
Zhàoqìng: Box 9.5 (Ricci)
Zhāo Xiān Tǎ, Bādàchù: Fig. 9.44
Zhàpǔ: Ch. 2.1, Ch. 2.2, Ch. 2.2 note 42, Box 2.2 (Opium), Ch. 3.4, Ch. 3.5, Ch. 3.4 note 197, 206, Box 3.2 (Schoedde), Ch. 5.3
Zhèjiāng: Box 2.2 (Opium), Ch. 3.4, Box 3.12 (Introductions) notes 2, 6, 43, 56, Ch. 5.2, Ch. 5.4, Ch. 5.5, Ch. 5.8, Ch. 5.9, Ch. 5.10, Ch. 5.11, Ch. 5.6 note 348, Ch. 5.8 note 459, Box 5.16 (Introductions) note 19, Box 5.21 (Small Sword), Box 5.29 (Chusan), Box 6.3 (Introductions) note 22, Ch. 10.4, Ch. 10.6, Fig. 5.37

Zhènhǎi: Box 2.2 (Opium), Ch. 3.1, Box 3.5 (Thom), Ch. 5.9, Box 5.25 (Patridge), Fig. 5.149
Zhènjiāng: Box 2.2 (Opium), Box 3.2 (Schoedde), Box 5.28 (Harvey), Ch. 9.1, Box 9.9 (Grant), Box 9.9 note 11 (Grant), Box 10.10 (Wade)
Zhènyuǎn: Ch. 10.4
Zhǐ Zhǐ Sì, Wǔyíshān: Ch. 5.4, Fig. 5.52
Zhōngtiān Pavilion: Box 5.26 (Wáng)
Zhōngxìn: Ch. 5.4
Zhōushān: Acknowledgements, Ch. 2.1, Ch. 3.1, Ch. 3.2, Ch. 3.4, Box 5.25 (Patridge), Fig. 3.10
Zhōushān Archipelago: Ch. 3.2, Ch. 5.4, Ch. 5.9, Box 5.29 (Chusan)
Zhōushān Dǎo: Ch. 2.2 note 70, Box 2.2 (Opium), Ch. 3.1, Ch. 3.4, Ch. 3.1 notes 58, 149, Box 3.2 (Schoedde), Box 3.5 (Thom), Box 3.11 (Lockhart), Box 3.12 (Introductions) notes 2, 6, 28, 91, 101, Ch. 5.5, Figs. 3.10, 3.11, 3.12, 3.17, 3.21, 3.22, 3.23
Zhū Jiāng: Ch. 3.1, Box 3.7 (Morrison), Box 6.1 (Arrow), Box 6.1 note 17
Zǐjīn Shān: Box 5.19 (Tàipíng), Box 5.20 (Hóng)
Zǐxī: Ch. 5.4
Zojo-ji, Tokyo: Fig. 8.32
Zorapore, see Zorapur
Zorapur: Box 3.4 (Balfour)
Zǒu yá bù: Ch. 5.2

Index to Persons, Firms and Vessels

Abeel, David Sr.: Box 3.1 (Abeel)

Abeel, David Jr.: Ch. 3.1, Ch. 3.1 note 31, Box 3.1 (Abeel), Box 5.25 (Patridge), Box 8.6 (Hepburn), Box 8.7 (Brown), Fig. 3.7

Abeel, Gustavus: Box 3.1 (Abeel)

A'chang: Ch. 5.9, Ch. 5.9 note 471, Figs. 5.129, 5.133

A-ching: Ch. 3.1, Ch. 3.1 note 88

Acka, Koko: Box 9.3 (Gordon)

Acka, Quako, see Acka, Koko

À Court, Annabella: Box 5.31 (Beadon)

Adam, Charles: Box 10.3 (Fullerton-Elphinstone)

Adam, Robert: Ch. 5.7 note 382

Adams, John: Box 3.8 (Cuming)

Aditya, Sandip: Acknowledgements

Adrian & Co.: Box 5.29 (Chusan)

Agassiz, Louis: Ch. 1.2

Ā' gŭ Băi, see Ya' qūb Beg

Aiton, William Townsend: Box 1.2 (McNab)

Àixīnjuéluó, Qíyīng: Ch. 2: Box 2.2 note 52 (Opium)

Àixīnjuéluó, Yīlǐbù: Ch. 2: Box 2.2 note 52 (Opium), Box 3.5 (Thom)

Alcock, John Rutherford: Ch. 3.4, Ch. 3.4 note 188, Box 3.9 (Lay), Box 3.10 (Alcock), Box 3.10 notes 16, 21, 26, Box 6.1 note 20 (Arrow), Ch. 8.2, Ch. 8.3, Ch. 8.3 notes 146, 164, 168, Box 8.1 (Veitch), Box 8.12 note 24 (Heusken), Box 9.9 (Hope Grant), Ch. 10.1, Ch. 10.4, Box 10.10 (Wade), Figs. 3.45, 8.31

Alcock, Thomas: Box 3.10 (Alcock)

Alexander, William: Fig. 5.24

Alford, Charles Richard: Box 5.24 (Burdon)

Ali, Muhamed: Ch. 5.1 note 7

Allen, Sarah: Box 7.1 (Smith)

Ān, Déhăi: Box 9.7 note 6 (Cí Xǐ)

Anderson, Arthur: Box 5.1 (P&O)

Anderson, John: Ch. 10.4

Anderson, Robert B.: Box 9.8 (Bowlby)

Anderson, Thomas: Ch. 10.4

Anderson, William: Box 2.1 (Lindley), Ch. 4.1

Andrew, Isabella: Box 2.3 (Reeves)

Andrew, John: Fig. 7.1

Angelelli, see Angiolelli

Angiolelli, Giovanna: Box 9.5 (Ricci)

"Ann": Box 5.10 note 11 (Lady Mary Wood), Box 5.25 (Patridge)

Anson, George: Box 9.9 (Hope Grant)

Arden, William (Baron Alvanley): Ch. 5.7

"Ariadne": Box 5.30 (Vansittart)

Arnold Arboretum Archives: Fig. 8.26
Arnott, George: Box 3.9 (Lay)
"Arrow": Box 6.1 (Arrow), Box 6.1 note 7 (Arrow)
Asako, Nyogo: Box 10.4 (Meiji)
Ashoka the Great: Ch. 5.8 note 407
Aspinall, Richard Jr.: Box 8.13 (Aspinall)
Aspinall, Richard Sr.: Box 8.13 (Aspinall)
Aspinall, William Gregson: Ch. 8.3, Box 8.13 (Aspinall), Ch. 10.5, Fig. 8.66
Aspinall, Cornes and Co.: Box 8.13 (Aspinall)
"Attalante": Ch. 9.1, Fig. 9.2
Auber, Daniel François: Box 8.4 note 30 (Siebold)
Aulick, John H.: Ch. 7.1

Baber, Edward Colborne: Ch. 10.4
Babington, Charles: Box 4.1 note 9 (Royle)
Bacon, Henrietta Mary: Box 3.10 (Alcock)
Badger, Albert: Ch. 4.2
Bái: Box 10.7 (Panthay)
Bai, Lakshmi: Box 6.4 (Indian Mutiny), Box 6.4 note 26
Bái, Yànhǔ: Box 10.8 (Huí)
Báihào: Box 10.6 (Xián-Tóng)
Bailes, Christopher: Acknowledgements
Baker, Robert: Ch. 5.7
Balfour, Edward Green: Box 3.4 note 20 (Balfour)
Balfour, George Jr.: Ch. 2.2 note 37, Ch. 3.1, Ch. 3.4, Box 3.4 (Balfour), Box 3.10 (Alcock)
Balfour, George Sr.: Box 3.4 (Balfour)
Balfour, Williamina Martha Arnold: Box 9.7 (Bowlby)
Ballou, Maturin Murray: Fig. 7.1
Banks, Dorothea: Fig. 8.56
Banks, Joseph: Box 1.2 (McNab), Box 2.1 (Lindley), Box 2.3 (Reeves), Box 8.3 (Thunberg)
Bannatyne, Kirkwood, France & Co.: Ch. 11 note 34
Baptifaud, Jean-Joseph-Marie: Ch. 10.3
Barclay, Maxwell V. L.: Box 3.8 note 24, Ch. 5.8 note 428, Ch. 8.3 note 182, Fig. 5.37A
Baring, Alexander: Box 9.4 (Bruce)
Barnard, Edward: Ch. 2.1, Ch. 2.2 note 70
Barnard, Ellen Emilia: Box 5.18 (Bonham)
Bartlett, Elizabeth Goodwin: Box 3.1 (Abeel), Box 8.7 (Brown)
Barton, Clara: Box 7.2 note 6 (Mason)
Barton, Dr.: Ch. 10.6
Bates, Henry Walter: Box 8.10 (Stevens)
Batten, John Hallet: Ch. 5.6, Box 5.12 (Batten), Box 5.13 (Ramsay)
Batten, Joseph Hallet: Box 5.12 (Batten)
Bazalgette, Joseph: Ch. 10.1, Fig. 10.23
Beach, William R.: Ch. 8.3
Beadon, Cecil: Ch. 5.12, Box 5.31 (Beadon)

Beadon, Richard: Box 5.31 (Beadon)
Beal, Helen: Box 8.8 note 7 (Hall)
Beale, Thomas Sr.: Box 5.5 (Beale), Box 5.5 note 1 (Beale)
Beale, Thomas Chaye: Ch. 5.2, Ch. 5.5, Ch. 5.8, Ch. 5.9, Box 5.5 (Beale), Box 5.18 (Bonham), Box 6.3 (Introductions) note 20, Ch. 8.2, Figs. 5.19, 5.99, 5.116
Bean, Edward: Ch. 4.1
Beato, Felice: Figs. 6.1, 8.27, 8.49, 10.15
"Beaver": Ch. 7.1
Beg, Ya' qūb: Box 10.9 (Muslim), Fig. 10.38
Bell, Captain: Ch. 8.3
"Belleisle": Box 9.9 (Grant)
Bellingham, John: Ch. 1.1 note 55
"Benares": Box 5.3 (Moresby)
"Bengal": Ch. 5.3 note 112
Bentham, George: Box 4.1 note 9 (Royle), Box 5.6 (Champion)
"Bentinck": Ch. 5.12, Box 5.33 (Bentinck), Box 9.3 (Gordon), Fig. 5.166
Bentinck, William: Box 5.33 (Bentinck)
Benzaiten: Ch. 8, Fig. 8.61
Berkeley, Miles Joseph: Box 2.1 note 19 (Lindley)
Bethune, Mr.: Ch. 5.6
"Bittern": Ch. 5.11, Box 5.30 (Vansittart)
"Blossom": Box 3.9 (Lay)
Boggs, Eli: Box 5.30 (Vansittart), Box 5.30 note 13 (Vansittart)
"Bombay": Ch. 4.1
Bonaparte, Joseph: Ch. 1.1
Bonaparte, Napoleon: Ch. 1.1, Ch. 1.1 notes 18, 53, Fig. 1.1
Bonham, George: Box 5.18 (Bonham)
Bonham, George Francis: Box 5.18 (Bonham)
Bonham, Samuel George: Box 3.6 (Sinclair), Box 3.10 (Alcock), Ch. 5.8, Box 5.18 (Bonham), Box 6.1 (Arrow), Fig. 5.116
Bonney, Samuel William: Ch. 8.3
Boone, Sarah Amelia: Box 3.1 (Abeel)
Boone, William Jones Sr.: Box 3.1 (Abeel)
Booth, John Wilkes: Box 7.3 (Holt), Box 7.3 notes 24, 25 (Holt)
Booth, Marijke: Acknowledgements
Borradaile, Margaret Watson: Box 10.2 (Namamugi), Box 10.2 note 5 (Namamugi)
Boswall, Alexander: Ch. 1.2 note 105
Boswall, Elizabeth: Ch. 1.2, Fig. 1.5
Boswall, Thomas: Ch. 1.2 note 105
"Bounty": Box 3.8 (Cuming)
Bourne, Richard: Box 5.1 (P&O)
Bourry, Augustin Etienne: Ch. 10.4
Bowlby, Russell: Box 9.8 (Bowlby)
Bowlby, Thomas: Box 9.8 (Bowlby)
Bowlby, Thomas William: Ch. 9.2, Box 9.4 (Bruce), Box 9.6 (Wyndham), Box 9.8 (Bowlby), Box 9.8 note 16

(Bowlby), Fig. 9.40
Bowman, Mr.: Ch. 5.3
Bown, Sarah: Box 5.14 (Paxton)
Bowring, John: Box 3.10 (Alcock), Box 5.28 (Harvey), Ch. 6, Box 6.1 (Arrow), Box 6.1 notes 7, 19 (Arrow), Box 10.10 (Wade)
Bowring, John Charles: Box 5.6 (Champion)
Bowring, Maria: Box 6.1 (Arrow)
Boxers: Box 9.7 (Cí Xǐ)
Boxer Rebellion: Box 9.7 (Cí Xǐ)
Bradbury, William: Box 5.14 (Paxton), Box 5.14 note 15 (Paxton)
Bradbury & Evans: Ch. 5.7 note 367
Brady, Mathew: Fig. 8.47
"Braganza": Ch. 5.1, Box 5.4 (Braganza), Fig. 5.8
Braine, George Thomas: Ch. 2.2 note 70
Bridgman, Elijah C.: Box 3.1 (Abeel)
Bright, John: Ch. 6.1, Ch. 6.1 note 7
Bristow, Naomi: Ch. 2.1 notes 14-22, 24-31, Ch. 2.2 note 32
British and American Steam Navigation Company: Ch. 7.1, Box 7.1 (Smith)
"British Queen": Box 7.1 (Smith)
Broderip, William John: Box 3.8 note 14 (Cuming)
Brooks, Charles William Shirley: Ch. 10.1
Brougham, Henry: Ch. 1.1
Broun, Christian: Box 5.32 (Dalhousie)
Brown, Aaron Venable: Box 7.3 (Holt)
Brown, Elizabeth, see Bartlett, Elizabeth Goodwin
Brown, Hattie: Box 8.7 (Brown)
Brown, Lancelot: Box 5.15 (Standish)
Brown, Robert (UK botanist): Box 2.1 (Lindley), Box 2.1 notes 7, 12 (Lindley), Fig. 3.9
Brown, Robert (USA): Box 8.7 (Brown)
Brown, Samuel Robbins: Box 3.1 (Abeel), Ch. 8.2, Box 8.7 (Brown), Fig. 8.18
Brown, Timothy H.: Box 8.7 (Brown)
Browne, Horace Albert: Ch. 10.4, Fig. 10.45
Bruce, Frederick William Adolphus: Box 3.7 (Morrison), Box 3.10 (Alcock), Box 8.14 (Howard-Vyse), Ch. 9.1, Ch. 9.2 note 66, Box 9.2 (Staveley), Box 9.4 (Bruce), Box 9.4 note 21 (Bruce), Box 9.6 (Wyndham), Box 9.9 (Hope Grant), Ch. 10.5, Box 10.10 (Wade), Fig. 9.20
Bruce, James (Lord Elgin): Box 3.10 (Alcock), Ch. 6.1, Box 6.1 (Arrow), Box 6.1 note 19 (Arrow), Ch. 8.1, Box 8.1 note 5 (Veitch), Box 8.8 note 4 (Hall), Box 8.12 (Heusken), Box 9.4 (Bruce), Box 9.8 (Bowlby), Box 9.8 note 16 (Bowlby), Box 9.9 (Hope Grant), Box 10.10 (Wade), Fig. 6.1
Bruce, Robert: Fig. 8.64
Bruce, Thomas: Box 9.4 (Bruce), Box 9.4 note 1 (Bruce)
Buchan, George Jr.: Ch. 1.2, 1.3, Box 1.1
Buchan, George Sr.: Box 1.1
Buchanan, James: Ch. 7.1, Box 7.3 (Holt)
Buckland, William: Ch. 1.2
Budden, John Henry: Box 5.13 (Ramsay)

Buddha: Ch. 3.2 note 143, Ch. 5.10, Ch. 8.3, Ch. 8.3 notes 191 – 199, Box 8.7 note 36 (Brown), Ch. 9.2 note 103, Figs. 5.34, 5.155, 8.59, 9.44
Buller, Anna Maria: Box 10.3 (Fullerton-Elphinstone)
Burdon, Edward Russell: Box 5.24 (Burdon)
Burdon, Isabella: Box 5.24 (Burdon)
Burdon, James: Box 5.24 (Burdon)
Burdon, John Shaw: Ch. 5.9, Box 5.24 (Burdon), Box 5.24 note 8 (Burdon), Fig. 5.148
Burgevine, Henry: Box 5.19 note 27 (Tàipíng), Box 9.2 (Staveley)
Burman, Johannes: Box 8.3 (Thunberg)
Burman, Nicolaas: Box 8.3 (Thunberg)
Burn, Robert: Box 3.1 (Abeel)
Burns, Alexander: Box 5.7 (Falconer)
Burns, Robert: Ch. 10.6, Ch. 11
Bùyī: Box 10.6 (Xián-Tóng)
Byron, George Gordon: Ch. 5.7

Caines, Sidney Richard Parry: Box 9.10 note 10 (Ceylon)
Caldbeck, Captain: Ch. 5.12
"Caldera": Box 5.10 (Lady Mary Wood)
"Cambridge": Box 3.1 (Abeel)
Campbell, Colin (Lord Clyde): Ch. 2.2 note 70, Ch. 3.2 note 149, Box 3.2 (Schoedde), Box 6.4 (Indian Mutiny), Box 9.9 (Grant), Box 9.9 note 35 (Grant)
Campbell, J. & G.: Box 3.5 (Thom)
Campbell, Louisa: Box 2.7 (Campbell-Johnston)
Campbell-Johnston, Alexander Robert: Ch. 2.2 note 70, Box 2.7 (Campbell-Johnston), Box 2.7 note 6 (Campbell-Johnston), Box 3.4 (Balfour), Fig. 2.8
Canning, Charles John: Ch. 2.1, Box 5.32 (Dalhousie), Box 6.4 (Indian Mutiny)
Capellen, Godert Alexander Gerard Philip van der: Box 8.4 (Siebold)
Carlyle, Thomas: Box 5.14 note 22 (Paxton), Ch. 11
Carmichael-Smyth, George Munro: Box 6.4 (Indian Mutiny)
Carnegie, Frances Mary: Box 5.6 (Champion)
"Caroline": Box 9.4 (Bruce)
"Castle Huntley": Box 3.1 (Abeel)
Catherine of Braganza: Box 2.2 note 3 (Opium)
Cattley, William: Box 2.1 (Lindley), Box 2.1 note 8 (Lindley)
Cautley, Proby T.: Box 5.7 (Falconer)
Cave, Kevin: Ch. 11 note 64
Cavendish, William George Spencer (Duke of Devonshire): Box 5.14 (Paxton)
Cecil, William (Lord Burghley): Ch. 10.1, Fig. 10.3
Cén, Yùyīng: Ch. 10.4, Ch. 10.4 note 342, Box 10.7 (Panthay)
"Ceylon": Ch. 9.2, Box 9.10 (Ceylon), Fig. 9.54
Champion, John Carey: Box 5.6 (Champion)
Champion, John George: Ch. 5.3, Box 5.6 (Champion)
Chang, Lo-hsing, see Zhāng, Luò-xíng
Chang, Tsung-yü, see Zhāng, Zōng-yǔ

Chapdelaine, Auguste: Box 6.1 (Arrow), Box 6.1 note 20 (Arrow)
Charlwood & Cummins: Ch. 6.2 note 34, Ch. 7.1
Chauveau, Joseph-Marie: Ch. 10.4
Chén, Ālín: Box 5.21 (Small Sword)
Ch' en, Alin, see Chén, Ālín
Chén, Lánbīn: Ch. 10.3 note 166
Cheng, Wan-chun, see Zhèng, Wàn-jūn
Chévrier, Claude-Marie: Ch. 10.3
Chiang, Chung- yüan, see Jiāng, Zhōng-yuán
Chiang, Kai-shek: Ch. 5.9 note 476
Chin, A-ling, see Chén, Ālín
Chinnery, George: Fig. 2.3
Ch' iu, Yüan-ts' ai, see Qīu, Yuān-cái
Ch' i-ying, see Àixīnjuéluó, Qíyīng
Chóng, Hòu: Ch. 10.3
Chosen-han: Box 5.29 (Chusan)
Chōshū Clan: Ch. 8.2 note 111, Ch. 10.1
Chou, Chia-hsün, see Zhōu, Jiā-xūn
Chow, Yu-ting, see Zhōu,Yú-dǐng
Chowfin, R. D.: Ch. 5.6 note 331
Christie & Manson: Ch. 5.7, Ch. 5.8, Ch. 5.8 note 437, Ch. 6.2, Ch. 8.1
Christie, Manson & Woods: Ch. 10.1, Figs. 5.111, 10.17
Chu, Hsi, see Zhū, Xī
Ch' ung-hou, see Chóng, Hòu
"Chusan": Ch. 4.1, Ch. 5.10, Box 5.29 (Chusan), Fig. 5.162
Cí Xǐ: Ch. 9.2, Box 9.7 (Cí Xǐ), Box 9.7 note 27 (Cí Xǐ), Fig. 9.36
"Clan Alpine": Ch. 10.4
Clark, David Oakes: Ch. 5.10
Clark, John: Box 9.10 (Ceylon)
Clarke, Amelia Helen: Box 5.27 (Robertson)
Clarke, Edward: Box 3.10 note 29 (Alcock), Ch. 8.3 note 208
Clarke, Woodthorpe Charles: Box 8.6 (Hepburn), Box 10.2 (Namamugi)
Clavelin, Marie-Séraphie: Ch. 10.3
Clavius, Christopher: Box 9.5 (Ricci)
Clay, Ann: Box 8.6 (Hepburn)
"Cleopatra": Ch. 10.1, Box 10.3 (Fullerton-Elphinstone)
Clipet: Fig. 8.16
Cobden, Richard: Ch. 6.1, Ch. 6.1 note 7
Cochrane, Alexander: Ch. 1.1
Cocks, Richard: Ch. 8.3
Collin, Lynda: Ch. 11 notes 51, 55, 56, 57, 60, 61, 63, 64, 72
Commercial Union: Box 8.13 (Aspinall)
Compton, Charles Spencer: Ch. 5.3
Confucius (person): Box 5.19 note 30, Box 9.5 (Ricci), Box 9.5 note 19 (Ricci), Fig. 5.123
"Confucius" (ship): Ch. 5.9, Fig. 5.142

Conover, Sanford, see Dunham, Charles
"Contest": Box 8.7 note 32 (Brown), Ch. 9.2 note 122
Cook, Dee: Acknowledgements
Cooke, George Wingrove: Ch. 10.1, Ch. 10.1 note 74
Cooke, Mr. (John Murray III): Ch. 5.7
Cooper, Thomas Thornville: Ch. 10.4, Ch. 10.3 note 168
Corbett, Boston: Box 7.3 note 24 (Holt)
Cornes, Frederick: Box 8.13 (Aspinall), Ch. 10.2 note 138
"Cornwall": Ch. 4.1
"Cornwallis": Box 2.2 (Opium), Box 5.30 (Vansittart)
Courtney, Frederick Reginald: Ch. 11, Ch. 11 note 70
Courtney, Terence Patrick: Ch. 11, Ch. 11 note 70
Cracroft, Peter: Ch. 10.6
Crawfurd, John: Box 5.7 (Falconer)
Crockett, Captain: Ch. 8.3
Cubitt, Thomas: Ch. 10.1
Cuming, Hugh: Ch. 3.3, Ch. 3.3 note 156, Box 3.8 (Cuming), Box 3.12 (Introductions) note 79, Ch. 4.2, Box 8.1 (Veitch), Ch. 10.5, Fig. 3.36
Cuming, Mary: Box 3.8 (Cuming)
Cuming, Richard: Box 3.8 (Cuming)
Cumming, William Henry: Box 3.1 (Abeel), Box 3.1 note 49 (Abeel), Box 8.6 (Hepburn)
Cunningham, Edward: Box 8.8 note 10 (Hall)
"Curacao": Box 8.1 (Veitch)
Cuthbertson, Alex: Ch. 1.2
Czar Alexander I: Ch. 1.1
Czar Nicholas I: Box 5.14 (Paxton)

Dǎi: Box 10.7 (Panthay)
D'Aiguebelle, Paul: Box 5.19 note 27 (Tàipíng)
Dana, James Dwight: Box 3.3 (Macgowan)
Dan-Kutch, see Denkichi
Darwin, Charles: Ch. 3.5 note 239, Box 5.7 (Falconer)
"Dartmouth": Ch. 7.1
"Dauntless": Box 10.3 (Fullerton-Elphinstone)
Davenport, Arthur: Ch. 10.4
David, Armand: Ch. 10.3 note 178
Davis, Jefferson: Box 7.3 (Holt)
Davis, John: Box 5.18 (Bonham), Box 9.4 (Bruce)
Davis, John Francis: Box 3.4 (Balfour), Box 3.9 (Lay), Box 3.9 note 21 (Lay), Box 10.10 (Wade)
Dawson, Thomas (Viscount Cremorne): Ch. 10.1
Dearborn, Captain: Ch. 5.9
De Bellecourt, Gustave Duchesne: Box 3.10 note 16 (Alcock), Box 8.9 (Girard)
De Bois, Annie: Box 5.23 (Perceval)
De Bourboulon, Alphonse: Box 9.4 (Bruce), Box 9.9 (Hope Grant)
De Candolle, Augustin Pyramus: Box 4.1 note 9 (Royle)

"De Drie Gezusters": Box 8.4 (Siebold)
De Haan, Willem: Box 8.4 (Siebold)
De Jussieu, Bernard: Box 8.3 (Thunberg)
Delamotte, William Alfred: Fig. 5.60
Delavay, Jean-Marie: Ch. 10.3 note 178
Delise, Mary: Box 5.13 (Ramsay)
Denham, Frank: Box 5.25 note 6 (Patridge)
"Denia": Ch. 5.3, Ch. 5.9 note 505
Denkichi, Kumano no: Ch. 8.3, Ch. 8.3 note 168
De Norman, William: Box 9.8 (Bowlby)
Dent, Lancelot: Ch. 2.2 note 70, Box 2.4 (Dent), Box 5.5 (Beale)
Dent, Wilkinson: Ch. 2.2 note 44, Box 2.4 (Dent)
Dent, William: Box 2.4 (Dent)
Dent & Co.: Ch. 2.1, Ch. 2.2, Ch. 2.2 note 70, Box 2.2 (Opium), Box 2.4 (Dent), Ch. 5.3, Box 5.28 (Harvey), Ch. 8. 2, Box 8.5 (Loureiro)
Dent, Beale & Co.: Box 5.5 (Beale)
Dent, Palmer & Co.: Box 3.10 note 29 (Alcock), Ch. 8.3 note 208
De Salis, John: Ch. 2.2 note 70
De Salis, Countess: Ch. 2.2 note 70
De Vries, W. Cores: Box 5.10 (Lady Mary Wood)
De Witt, Thomas: Box 8.12 (Heusken)
Dexter, Franklin Gordon: Box 8.8 (Hall)
Dickens, Charles: Ch. 4.2
Dickson, Walter (Edinburgh): Box 1.2 (McNab)
Dickson, Dr. Walter (Guǎngzhōu): Ch. 8.3, Ch. 10.6
Dilkie, Charles Wentworth: Box 5.14 (Paxton), Ch. 10.1
"Discoverer": Box 3.8 (Cuming)
Disraeli, Benjamin: Ch. 6.1, Box 10.3 (Fullerton-Elphinstone)
Dodge, Esther: Box 7.2 (Mason)
Doeve, Eppo: Fig. 5.166
Dōgen Zenji: Ch. 3.2 note 105, Fig. 3.25
Döllinger, Ignaz: Box 8.4 (Siebold)
Don, David: Box 4.1 note 9 (Royle)
Dòng: Box 10.6 (Xián-Tóng)
Dǒng, Fù-xiáng: Box 10.5 (Niǎn), Box 10.8 (Huí)
Dōngniǎn: Box 10.5 (Niǎn)
Donker Curtius, Jan Hendrik: Box 8.4 (Siebold)
D'Orléans, Louis: Box 5.9 (Du Halde)
Dorr, Eben M.: Box 8.7 (Brown), Box 8.12 note 24 (Heusken)
Dost Mohammad: Box 5.7 (Falconer)
Douglas, David: Ch. 3.2, Ch. 3.2 note 119
Dowager Kāngcí: Box 9.7 (Cí Xǐ), Box 9.7 note 6 (Cí Xǐ)
Downs, Henry: Box 5.29 (Chusan)
Drepper, Christine: Box 8.2 (Kaempfer)
Drummond, F. C.: Ch. 2.2 note 70

Drury-Lavin, Michael: Box 9.10 (Ceylon)
Dù, Hassan: Ch. 10.4
Dù, Wén-xiù: Ch. 10.4, Box 10.7 (Panthay)
Du Halde, Jean-Baptiste: Ch. 5.4, Box 5.9 (Du Halde), Box 5.9 note 6 (Du Halde)
Du Mans, Raphael: Box 8.2 (Kaempfer)
Duchess of Gloucester: Box 5.15 (Standish)
"Duke of Bedford": Ch. 3.2, Ch. 4.1
Duke of Brittany: Box 5.9 (Du Halde)
Duke of Cadore: Ch. 1.1 note 53
Duke of Devonshire, see Cavendish, William George Spencer
Duke of Northumberland, see Percy, Algernon
Duke of Orléans, see D'Orléans, Louis
Duke of Somerset: Box 5.14 (Paxton)
Duke of Wellington: Ch. 1.1
Dunant, Henry: Box 9.3 (Gordon)
Duncan, John: Box 5.16 note 6 (Introductions)
Duncan, Dr.: Box 3.12 note 58 (Introductions)
Dundas, Anne: Box 1.1 (Buchan)
Dundas, George: Box 1.1 (Buchan)
Dundas, Robert: Box 1.1 (Buchan)
Dundas, Captain: Ch. 8.2
Dunham, Charles: Box 7.3 note 25 (Holt)
Dunlop, Ann Robertson: Ch. 11
Duōlōngā: Box 10.8 (Huí)
Durie, Alice: Acknowledgements, Figs. 11.4, 11.6, 11.8
Durie, Alison: Ch. 11
Durie, Ann: Ch. 11
Durie, James Lauder: Ch. 11
Durie, Jane Fortune: Ch. 11, Fig. 11.7
Durie, John Jr.: Ch. 11, Ch. 11 notes 19, 34, 36, Figs. 11.10, 11.11, 11.13, 11.14
Durie, John Sr.: Ch. 11, Figs. 11.2, 11.3, 11.6
Durie, John III: Ch. 11
Durie, Norman Percival Jr.: Ch. 11
Durie, Norman Percival Sr.: Ch. 11
Durie, Robert: Ch. 11, Ch. 11 note 19, Fig. 11.6
Durie, Robert Hogg: Ch. 11
Durie, Thomas Bryson Mitchell: Ch. 11, Ch. 11 notes 35, 39, Fig. 11.14
Durie, Thomas Mitchell: Ch. 11
Dutch East India Company, see Verenigde Oost-Indische Compagnie
Duthie, John Firminger: Box 5.11 (Saharanpur)
Dutton, Thomas: Fig. 9.54
Dyer, Burella Hunter: Box 5.24 (Burdon)
Dyer, Samuel (businessman): Ch. 3.1 note 13
Dyer, Samuel (missionary): Box 5.24 (Burdon)
Dykes, Andrew: Ch. 11 note 37

Dykes, Thomas A.: Acknowledgements, Ch. 11 notes 32, 37

Earl, Charles: Box 5.29 note 7 (Chusan)
East India Company (EIC): Box 2.2 (Opium), Box 2.3 (Reeves), Ch. 4.2, Box 4.1 (Royle), Ch. 5.7 notes 383, 386, Ch. 5.10 note 664, Box 5.7 (Falconer), Box 5.11 (Saharanpur), Box 5.12 (Batten), Box 5.17 (Ganges), Box 5.18 (Bonham), Box 5.27 (Robertson), Box 5.31 (Beadon), Ch. 6.1, Ch. 6.2, Box 6.2 (Glendinning), Box 6.4 (Indian Mutiny), Ch. 7.1, Ch. 7.2, Figs. 4.4, 4.5, 5.9, 5.22
Eddis, Eden Upton: Fig. 2.1
Edgar, Thomas: Ch. 2.1
Edwards, Vivien: Ch. 11 notes 64, 70, 71
"Eleanor": Ch. 7.1
Elepoo, see Àixīnjuéluó, Yīlǐbù
Elliot, Charles: Box 2.2 (Opium), Box 2.7 (Campbell-Johnston), Box 2.7 note 6 (Campbell-Johnston), Ch. 3.1
Ellis, John: Box 5.14 (Paxton) note 21
Elsworthy, Thomas: Ch. 3.1 note 13
Elwon, Thomas: Box 5.3 (Moresby)
"Emily": Box 9.3 (Gordon)
Emperor Dàoguāng: Box 2.2 note 16 (Opium)
Emperor Gāozōng: Ch. 3.1 note 64
Emperor Gūangxù: Box 9.7 (Cí Xǐ), Box 9.7 note 38 (Cí Xǐ)
Emperor Huángdì: Ch. 5.2 note 19
Emperor Kāngxī: Ch. 9.2
Emperor Kein-lung, see Emperor Qiánlóng
Emperor Kōmei: Ch. 10.1, Ch. 10.2, Box 10.4 (Meiji)
Emperor Meiji: Ch. 10.2, Box 10.4 (Meiji), Box 8.6 (Hepburn), Box 8.11 (Harris), Fig. 10.32
Emperor Qiánlóng: Ch. 5.5, Ch. 9.2, Figs. 5.58, 9.48
Emperor Qíxiáng, see Emperor Tóngzhì
Emperor Shènmǔ Shénhuáng, see Empress Zé-tiān Wǔ
Emperor Tewodros II: Box 9.2 (Staveley)
Emperor Theodore II, see Emperor Tewodros II
Emperor Tóngzhì: Ch. 9.2 note 78, Box 9.7 (Cí Xǐ), Box 10.6 (Xián-Tóng)
Emperor Wànlì: Ch. 9.2, Box 9.5 (Ricci), Fig. 9.30
Emperor Xiánfēng: Ch. 9.2 note 77, Box 9.7 (Cí Xǐ), Box 9.7 note 6 (Cí Xǐ), Box 9.8 note 16 (Bowlby), Box 10.6 (Xián-Tóng)
Emperor Zhèngdé: Box 5.26 (Wáng)
Empress Alute: Box 9.7 (Cí Xǐ)
Empress Cí'ān: Box 9.7 (Cí Xǐ)
Empress Quán: Ch. 9.2 note 77
Empress Shōken, see Ichijō, Haruko
Empress Zétiān Wǔ: Ch. 3.1 note 64
"Emu": Ch. 2.1, Ch. 2.2, Ch. 3.1 note 13
"England": Ch. 8.2, Ch. 8.2 note 111, Box 8.1 (Veitch)
En Hai: Box 9.7 note 55 (Cí Xǐ)
"Erin": Ch. 5.9, Box 5.24 (Burdon), Box 5.25 (Patridge), Fig. 5.148
Espinasse, François: Ch. 1.2 note 88

Eulenburg, Friedrich Albrecht zu: Box 8.4 (Siebold), Box 8.12 (Heusken)
Evans, Captain: Ch. 9.2
Ewbank, Thomas: Ch. 7.1
Ewing, Henry: Box 6.2 (Glendinning)

Fairchild, David: Ch. 7.3
Falconer, David: Box 5.7 (Falconer)
Falconer, Hugh: Box 4.1 (Royle), Ch. 5.4, Ch. 5.6, Ch. 5.12, Box 5.7 (Falconer), Box 5.8 (Jameson), Box 5.11 (Saharanpur), Ch. 10.5, Fig. 5.50, 5.87
Fan, Ju-tseng, see Fàn, Rǔ-zēng
Fàn, Rǔ-zēng: Ch. 9.1 note 12, Box 10.5 (Niǎn)
Faraday, Michael: Ch. 10.1
Farnham, C.: Box 5.29 (Chusan)
Farrer, William: Box 3.12 (Introductions) note 87
"Fee-loong", see "Feilung"
"Feilung": Ch. 9.1
Féng, Yúnshān: Box 5.20 (Hóng)
Ferguson, David Kay: Figs. 1.6, 1.7, 1.8, 1.9, 1.10, 1.11, 1.12, 1.13, 1.14, 1.15, 1.17, 2.5, 2.8, 3.1, 3.2, 3.4, 3.8, 3.9, 3.11, 3.13, 3.14, 3.15, 3.16, 3.18, 3.22, 3.23, 3.25, 3.26, 3.27, 3.28, 3.30, 3.32, 3.34, 3.37, 3.38, 3.39, 3.40, 3.42, 3.46, 3.48, 3.50, 3.51, 3.52, 3.53, 3.54, 3.55, 3.56, 3.59, 4.1, 4.2, 4.3, 4.5, 4.6, 4.7, 5.3, 5.4, 5.15, 5.16, 5.17, 5.18, 5.19, 5.20, 5.35, 5.36, 5.37, 5.40, 5.41, 5.44, 5.49, 5.55, 5.58, 5.59, 5.62, 5.63, 5.68, 5.69, 5.73, 5.74, 5.75, 5.76, 5.77, 5.78, 5.79, 5.80, 5.81, 5.84, 5.85, 5.86, 5.87, 5.88, 5.89, 5.90, 5.93, 5.94, 5.103, 5.107, 5.108, 5.109, 5.110, 5.111, 5.112, 5.118, 5.124, 5.125, 5.126, 5.127, 5.128, 5.129, 5.130, 5.134, 5.135, 5.136, 5.137, 5.138, 5.139, 5.140, 5.142, 5.143, 5.144, 5.145, 5.146, 5.147, 5.149, 5.151, 5.152, 5.153, 5.156, 5.157, 5.158, 5.159, 5.160, 5.161, 5.163, 5.164, 6.3, 6.5, 6.6, 6.7, 6.8, 7.3, 7.4, 7.7A, 7.7B, 7.7D, 7.7E, 7.7F, 8.3, 8.4, 8.5, 8.6, 8.7, 8.8, 8.9, 8.11, 8.12, 8.14, 8.15, 8.19, 8.20, 8.21, 8.22, 8.23, 8.24, 8.25, 8.28, 8.29, 8.30, 8.31, 8.32, 8.33, 8.34, 8.36, 8.37, 8.38, 8.39, 8.40, 8.41, 8.42, 8.43, 8.46, 8.50, 8.52, 8.53, 8.56, 8.57, 8.58, 8.59, 8.60, 8.61, 8.62, 8.63, 8.64, 8.65, 8.67, 8.68, 9.2, 9.3, 9.4, 9.5, 9.6, 9.9, 9.10, 9.11, 9.12, 9.14, 9.15, 9.16, 9.17, 9.22, 9.24, 9.25, 9.26, 9.27, 9.28, 9.29, 9.31, 9.33, 9.34, 9.38, 9.44, 9.45, 9.46, 9.47, 9.48, 9.49, 9.50, 9.51, 9.55, 10.1, 10.3, 10.4, 10.5, 10.6, 10.9, 10.10, 10.13, 10.14, 10.17, 10.20, 10.21, 10.22, 10.25, 10.28, 10.29, 10.30, 10.31, 10.34, 10.36, 10.37, 10.39, 10.40, 10.49, 10.50, 10.51, 10.52, 10.53, 11.1, 11.2, 11.4, 11.5, 11.6, 11.8, 11.9, 11.10, 11.11, 11.12, 11.13, 11.15, 11.16, 11.17, 11.18
Field, Emily: Box 5.8 (Jameson)
"Fiery Cross": Ch. 8.3, Ch. 8.3 note 211, Box 9.9 (Grant)
Fillmore, Millard: Ch. 8.2 note 78
Finnis, John: Fig. 6.5
Fishbourne, Edmund: Box 5.20 note 25 (Hóng)
FitzClarence, Adolphus: Box 5.30 (Vansittart), Box 5.30 note 7 (Vansittart)
Fletcher & Fearnall: Box 5.4 (Braganza)
Fontanier, Henri-Victor: Ch. 10.3
Forbes, Paul Seimen: Ch. 5.8 note 399, Ch. 7.1
Ford, George Henry: Fig. 4.4
Forrest, George: Ch. 10.4, Ch. 10.4 note 330
Forster, Johann Reinhold: Box 8.4 (Siebold)
Fortescue, Captain: Box 8.15 (Introductions) note 61
Fortune, Agnes: Ch. 4.1, Ch. 4.2, Figs. 4.6, 10.50

Fortune, Alice: Ch. 10.1, Ch. 11, Fig. 11.2, see also Durie, Alice
Fortune, Alice Purves: Ch. 11, Ch. 11 note 64
Fortune, Alison E.: Ch. 11
Fortune, Alison Jane: Ch. 11, Ch. 11 note 63, see also Lumsden, Alison Jane
Fortune, Christian: Ch. 1.2 note 134
Fortune, Constance Edith: Ch. 11, Fig. 11.16
Fortune, Elizabeth: Ch. 1.2 note 134
Fortune, Helen Jane: Ch. 1.3, Ch. 6.1, Ch. 11, Fig. 10.50
Fortune, Isabella: Ch. 1.2, Ch. 1.2 note 134
Fortune, James: Ch. 1.2 note 134, Ch. 5.7
Fortune, Jane (sister): Ch. 1.2 note 134
Fortune, Jane (wife): Ch. 4.1, Ch. 4.2, Ch. 4.1 note 7, Ch. 5.7, Ch. 10.1, Ch. 10.6 note 495, Ch. 11, Fig. 10.50
Fortune, John: Ch. 1.2 note 134
Fortune, John Lindley: Ch. 2.1 note 23, Ch. 2.2, Box 2.1 (Lindley), Ch. 6.1, Ch. 10.1, Ch. 11, Figs. 10.22, 11.1, 11.15, 11.16
Fortune, Margaret: Ch. 1.2 note 134
Fortune, Margaret Alice: Ch. 11
Fortune, Mary: Ch. 5.7, Ch. 5.7 note 385, Ch. 6.1, Ch. 6.2, Ch. 7.1, Ch. 7.1 note 20, Fig. 10.50
Fortune, Robert Jr.: Ch. 11, Ch. 11 notes 63, 70, Fig. 11.17
FORTUNE, ROBERT: Frontispiece, Ch. 1.2, Ch. 1.3, Box 1.2 (McNab), Box 2.1 (Lindley), Box 2.2 (Opium), Box 2.3 (Reeves), Box 2.4 (Dent), Box 2.5 (Smith-Stanley), Box 2.6 (Ward), Box 3.1 (Abeel), Box 3.3 (Macgowan), Box 3.4 (Balfour), Box 3.5 (Thom), Box 3.8 (Cuming), Box 3.8 note 24 (Cuming), Box 3.9 (Lay), Box 3.11 (Lockhart), Box 3.12 (Introductions), Ch. 5.7, Box 5.2 (Ripon), Box 5.3 (Moresby), Box 5.4 (Braganza), Box 5.5 (Beale), Box 5.6 (Champion), Box 5.7 (Falconer), Box 5.8 (Jameson), Box 5.13 (Ramsay), Box 5.14 (Paxton), Box 5.15 (Standish), Box 5.16 (Introductions), Box 5.18 (Bonham), Box 5.21 (Small Sword), Box 5.24 (Burdon), Box 5.27 (Robertson), Box 5.31 (Beadon), Ch. 6.1, Box 6.2 (Glendinning), Box 6.3, Box 8.1 (Veitch), Box 8.4 (Siebold), Box 8.5, Box 8.6, Box 8.7, Box 8.8 (Hall), Box 8.9 (Girard), Box 8.11(Harris), Box 8.13 (Aspinall), Box 8.14 (Howard-Vyse), Box 9.4 (Bruce), Box 9.6 (Wyndham), Ch. 10, Box 10.1 (Moore), Box 10.3 (Fullerton-Elphinstone), Box 10.10 (Wade), Ch. 11, Figs. 1.1, 1.4, 1.6, 1.7, 1.9, 1.10, 1.14, 1.15, 1.16, 1.17, 2.1, 2.3, 2.6, 2.7, 2.8, 2.9, 3.2, 3.3, 3.4, 3.5, 3.7, 3.8, 3.10, 3.12, 3.13, 3.15, 3.17, 3.19, 3.20, 3.21, 3.22, 3.23, 3.24, 3.26, 3.27, 3.28, 3.29, 3.32, 3.34, 3.35, 3.36, 3.37, 3.38, 3.39, 3.40, 3.45, 3.46, 3.49, 3.50, 3.51, 3.52, 3.53, 3.54, 3.55, 3.56, 3.57, 3.58, 3.59, 4.1, 4.2, 4.3, 4.4, 4.5, 4.6, 5.1, 5.2, 5.4, 5.5, 5.8, 5.9, 5.10, 5.11, 5.13, 5.14, 5.15, 5.16, 5.17, 5.19, 5.20, 5.21, 5.22, 5.23, 5.24, 5.25, 5.27, 5.28, 5.30, 5.31, 5.33, 5.34, 5.35, 5.36, 5.37, 5.38, 5.40, 5.41, 5.42, 5.43, 5.44, 5.45, 5.46, 5.47, 5.48, 5.49, 5.50, 5.52, 5.53, 5.54, 5.55, 5.56, 5.57, 5.59, 5.60, 5.61, 5.67, 5.70, 5.71, 5.72, 5.74, 5.77, 5.78, 5.79, 5.81, 5.85, 5.87, 5.88, 5.89, 5.90, 5.92, 5.95, 5.99, 5.101, 5.102, 5.103, 5.104, 5.105, 5.110, 5.112, 5.113, 5.114, 5.116, 5.119, 5.120, 5.121, 5.122, 5.123, 5.124, 5.125, 5.126, 5.127, 5.128, 5.129, 5.132, 5.133, 5.135, 5.136, 5.137, 5.138, 5.139, 5.140, 5.142, 5.143, 5.144, 5.145, 5.146, 5.151, 5.152, 5.154, 5.157, 5.158, 5.159, 5.160, 5.162, 5.163, 5.164, 5.165, 5.166, 6.3, 6.4, 6.6, 6.7, 7.6, 7.7, 8.2, 8.3, 8.4, 8.5, 8.8, 8.9, 8.11, 8.13, 8.15, 8.17, 8.18, 8.19, 8.21, 8.22, 8.24, 8.25, 8.26, 8.30, 8.31, 8.33, 8.34, 8.35, 8.37, 8.39, 8.42, 8.44, 8.45, 8.46, 8.47, 8.48, 8.50, 8.52, 8.53, 8.58, 8.61, 8.63, 8.64, 8.66, 8.67, 8.68, 9.2, 9.5, 9.6, 9.7, 9.8, 9.9, 9.10, 9.11, 9.12, 9.13, 9.14, 9.15, 9.16, 9.18, 9.19, 9.20, 9.21, 9.27, 9.28, 9.29, 9.31, 9.32, 9.33, 9.34, 9.35, 9.36, 9.37, 9.38, 9.43, 9.44, 9.45, 9.47, 9.49, 9.50, 9.51, 9.53, 9.54, 9.55, 10.1, 10.2, 10.7, 10.8, 10.17, 10.18, 10.21, 10.22, 10.24, 10.49, 10.50, 10.53, 11.18
Fortune, Thomas: Ch. 11, Fig. 10.50
Fortune, Thomas Lawful: Ch. 1.2, Figs. 1.5, 1.8
Fortune, William Redpath: Ch. 1.2 note 134

Fortune, Winifred M.: Ch. 11
Fortune, Zella Charlton: Ch. 11
Fowke, Francis: Ch. 10.1, Fig. 10.11
Fowler, Sally: Box 8.6 note 5 (Hepburn)
Fraser, Alexander George: Ch. 2.2 note 70, Box 9.9 (Grant), Box 9.9 note 8 (Grant)
Fraser, John: Box 10.1 (Moore)
Freeman, Captain: Ch. 3.4
Freeman-Mitford, Algernon: Ch. 9.1 note 35
Freestone, Sarah: Box 2.1 (Lindley)
Frembly, John: Box 3.8 (Cuming)
Frenzeny, Paul: Fig. 9.13
Fullerton-Elphinstone, James Drummond: Box 10.3 (Fullerton-Elphinstone)
Fullerton-Elphinstone, William Buller: Ch. 10.1, Ch. 10.1 note 56, Box 10.3 (Fullerton-Elphinstone), Ch. 11, Figs. 10.19, 10.20, 10.21
Fytche, Albert: Ch. 10.4, Fig. 10.43

"Gaelic": Box 8.6 (Hepburn)
Gandhi, Indira: Box 5.13 note 32 (Ramsay)
Gang of Eight: Box 9.7 (Cí Xǐ)
"Ganges": Ch. 5.8, Box 5.17 (Ganges)
Garaway, Thomas: Box 2.2 note 3 (Opium)
Gardner, George: Box 5.6 (Champion)
Gaselee, Alfred: Box 9.7 (Cí Xǐ)
Gear, Angelica: Box 7.2 (Mason)
Gemmill, Alice Fortune: Ch. 11
Gemmill, Elizabeth: Ch. 11
Gemmill, William: Ch. 11, Figs. 11.7, 11.9, 11.13
Gemmill, William Cunninghame: Ch. 11
Genkei, Shoun: Box 8.7 note 36 (Brown)
Gerard, see Girard
Getty Research Institute, Los Angeles: Fig. 10.46
Ghosh, Ruby: Acknowledgements
Gibbon, Edward: Ch. 5.7
Gibbs, Livingstone & Co.: Ch. 2.2 note 70
Gibson, John: Ch. 10.1, Fig. 10.27
Gilbert, Elizabeth: Ch. 1.3 note 159
Giles, Herbert Allen: Box 10.10 (Wade)
Gill, William John: Ch. 10.4, Figs. 10.47, 10.48
Gilman, R. J.: Ch. 2.2 note 70
Gilpin, William Sawrey: Box 6.2 note 7 (Glendinning)
Giquel, Prosper: Box 5.19 note 27 (Tàipíng)
Girard, Prudence Séraphin Barthélémy: Ch. 8.2, Ch. 8.3 note 161, Box 8.9 (Girard), Box 8.9 note 9 (Girard), Box 8.12 (Heusken), Ch. 10.5, Fig. 8.35
Girdwood, Claude: Box 7.1 (Smith)
"Gladiator": Box 5.30 (Vansittart)

Gladstone, William: Ch. 6.1
Glendinning, Robert: Ch. 5.7 note 365, Box 5.15 note 12 (Standish), Ch. 6, Box 6.2 (Glendinning), Box 6.2 notes 7, 15, Box 6.3 (Introductions) notes 9 – 12, 14 – 16, 21 – 22, Box 8.10 note 6 (Stevens)
Glendinning, Robert Pince: Box 6.2 (Glendinning)
Gold, Rudolf: Acknowledgements, Figs. 3.5, 5.10, 5.11, 5.12, 5.14, 5.42, 5.43, 5.45, 5.47, 5.48, 5.51, 5.52, 5.53, 5.54, 5.56, 5.64, 5.65, 5.66, 5.82
Gordon, Charles Alexander: Ch. 9.1, Ch. 9.1 notes 18, 24, Box 9.3 (Gordon), Ch. 10.6, Fig. 9.8
Gordon, Charles George: Box 5.19 (Tàipíng), Box 9.2 (Staveley)
Gordon, George: Ch. 5.9 note 542
Gordon, Henry King: Box 9.3 (Gordon), Box 9.3 note 19 (Gordon)
Gordon, Marianne: Box 5.13 (Ramsay)
Gordon, William Alexander: Box 9.3 (Gordon)
Gorer, Richard: Box 3.12 (Introductions) notes 26, 66, 69, 77
Gough, Hugh: Box 2.2 (Opium), Box 3.5 (Thom)
Goulburn, Martha: Box 8.13 (Aspinall)
Gould, Harriot Reynolds: Box 8.1 (Veitch)
Govan, George: Box 4.1 (Royle), Box 5.11 (Saharanpur)
Gowan, G. E.: Box 5.12 (Batten)
Gowan, James Robert: Ch. 2.1
Goya, Francisco: Ch. 1.1 note 14, Fig. 1.2
Graeffer, Johann Andreas: Box 8.15 (Introductions) note 5
Graham, John: Box 6.2 (Glendinning)
Graham, Robert: Ch. 1.3
"Grampus": Box 10.3 (Fullerton-Elphinstone)
Grant, Francis Jr.: Box 9.9, Box 9.9 note 15 (Hope Grant), Fig. 9.42
Grant, Francis Sr.: Box 9.9 (Hope Grant)
Grant, James Hope: Box 6.4 (Indian Mutiny), Ch. 9.2, Box 9.2 (Staveley), Box 9.4 (Bruce), Box 9.8 note 16 (Bowlby), Box 9.9 (Grant), Box 9.9 notes 8, 11, 35 (Grant), Figs. 9.41, 9.42
Grant, Ulysses S.: Fig. 8.32
"Great Western": Box 7.1 (Smith)
Great Western Steamship Company: Box 7.1 (Smith)
Greathed, Edward Harris: Box 6.4 (Indian Mutiny)
Greaves, Walter: Ch. 10.1, Fig. 10.26
Greville, Charles: Box 3.12 (Introductions) note 19
Griffin, William: Box 5.14 (Paxton), Box 5.14 note 5 (Paxton)
Griffith, William: Ch. 5.6 note 306, Fig. 5.67
Grimshaw, John Atkinson: Ch. 10.1
Grimwood, Captain: Box 3.8 (Cuming)
Gros, Antoine-Jean: Fig. 1.1
Gros, Jean-Baptiste Louis: Box 6.1 (Arrow), Box 8.1 note 5 (Veitch)
Grosvenor, Thomas G.: Ch. 10.4
Guì Lán, see Cí Xǐ
Gully, Robert: Box 5.25 notes 7, 9, 11 (Patridge)
Gupta, Deepa: Acknowledgements
Gupta, Sudha: Acknowledgements

Gützlaff, Karl Friedrich: Box 3.11 note 13 (Lockhart)

Habu, Genseki: Box 8.4 note 18 (Siebold)
Haines, Frederick Paul: Box 9.3 (Gordon)
Hall, Captain: Ch. 3.1
Hall, Benjamin: Box 8.8 (Hall)
Hall, Chandler Prince: Box 8.8 note 7 (Hall)
Hall, Edward Cunningham: Box 8.8 note 7 (Hall)
Hall, Elizabeth: Box 8.8 note 7 (Hall)
Hall, Francis: Ch. 8.3, Ch. 8.3 note 161, Box 8.7 (Brown), Box 8.12 note 24 (Heusken)
Hall, George Rogers Jr.: Box 8.8 note 7 (Hall)
Hall, George Rogers Sr.: Box 5.21 note 7 (Small Sword), Ch. 8.2, Box 8.8 (Hall), Box 8.15 (Introductions) notes 5, 61, Fig. 8.26
Hall, Helen Beal: Box 8.8 note 7 (Hall)
Hall, John Richard: Ch. 11, Ch. 11 note 19
Hamilton, Charles: Box 1.2 (McNab)
Hamilton, Elizabeth: Box 2.5 (Smith-Stanley)
Hamilton, James: Box 2.5 (Smith-Stanley)
Hamilton-Gordon, George: Ch. 2.1, Ch. 2.2 note 70
Hàn: Ch. 10.1 note 16, Box 10.6 (Xián-Tóng), Box 10.7 (Panthay), Box 10.9 (Muslim)
Hán, Dòngxū: Ch. 5.4
Hance, Henry Fletcher: Ch. 10.4
Hānī: Box 10.7 (Panthay)
"Hankow": Ch. 10.4
Harada, Hiroji: Acknowledgements, Ch. 8.1 note 9
Harbauer, Franz Joseph: Box 8.4 (Siebold)
Hardin, Benjamin: Box 7.3 (Holt)
Hardy, Graham: Acknowledgements
Harris, Jonathan: Box 8.11 (Harris)
Harris, Townsend: Ch. 8.3, Ch. 8.2 notes 64, 69, 78, 103, Ch. 8.3 notes 148, 161, Box 8.4 note 25 (Siebold), Box 8.6 (Hepburn), Box 8.7 (Brown), Box 8.11 (Harris), Box 8.11 notes 7, 14, 22 (Harris), Box 8.12 (Heusken), Ch. 10.5, Figs. 8.47, 8.48, 8.50, 8.53
Harrison, Mary Louisa: Box 7.3 (Holt)
Hart, Henry: Ch. 2.2 note 70
Hart, Robert: Box 2.2 note 4 (Opium), Ch. 3.1 note 61, Box 5.22 note 16 (Meadows), Box 5.25 (Patridge)
Hartless, Amos C.: Box 5.11 (Saharanpur)
Harvard College: Fig. 8.26
Harvey, Frederick E. B.: Ch. 5.10, Box 5.28 (Harvey)
Harvey, James Vigers: Box 5.28 (Harvey)
Harvey, John: Box 9.4 (Bruce)
Hassert, Jane: Box 3.1 (Abeel)
Hastie, John: Ch. 1.2, Ch. 1.2 note 126, Box 1.1 (Buchan), Fig. 1.7
Havelock, Henry: Ch. 6.2, Box 6.4 (Indian Mutiny)
Hay, Susan Georgina: Box 5.32 (Dalhousie)
Heard & Co.: Box 8.6 (Hepburn)

Heco, Joseph: Ch. 8.3 note 161
Hedge, Ian: Acknowledgements
Hely, F. G.: Ch. 5.3, Ch. 5.3 note 96
Hemard, Mr.: Ch. 4.2
Henderson, Alexander: Ch. 2.1, Ch. 2.2 note 36
Hepburn, Clara, see Leete, Clarissa Mary
Hepburn, James Curtis: Ch. 8.2, Box 8.6 (Hepburn), Box 8.7 (Brown), Box 10.2 (Namamugi), Fig. 8.17
Hepburn, Samuel: Box 8.6 (Hepburn)
Hepburn, Samuel David: Box 8.6 (Hepburn), Box 8.6 note 16 (Hepburn)
Herbert, Thomas: Ch. 2.2 note 70
"Hermes": Box 5.18 (Bonham)
"Hero": Box 5.30 (Vansittart)
Herschel, Amelia: Box 10.10 (Wade)
Herschel, John: Box 10.10 (Wade)
Hesketh, Frances: Box 8.14 (Howard-Vyse)
Heusken, Henricus Conradus Joannes: Ch. 8.3, Ch. 8.2 notes 69, 82, 103, Ch. 8.3 note 148, Box 8.7 (Brown), Box 8.9 (Girard), Box 8.11 (Harris), Box 8.12 (Heusken), Figs. 8.50, 8.51
Heusken, Johannes Franciscus: Box 8.12 (Heusken)
Heusken-Smit, T. F.: Box 8.12 (Heusken)
Hibata, Ōsuke: Fig. 8.48
"Hibernia": Box 5.30 (Vansittart)
Hideyoshi, Toyotomi: Box 9.5 (Ricci)
Hilton, Brian: Ch. 3.2 note 145
"Himalayah": Box 9.3 (Gordon)
"Himmaleh": Box 3.9 (Lay)
"Hindostan": Box 5.1 (P&O), Box 5.3 (Moresby)
Hinsdale, Phoebe Allen: Box 8.7 (Brown)
Hobson, John: Ch. 5.9
Hodge, Jane: Box 8.1 (Veitch)
Hoffmann, Johann Joseph: Box 8.4 (Siebold)
Hogarth, Catherine: Ch. 4.2
Hogg, Tom: Acknowledgements
Holliday Wise and Company: Box 8.13 (Aspinall)
Holt, John W.: Box 7.3 (Holt)
Holt, Joseph: Ch. 7.1, Ch. 7.2, Box 7.3 (Holt), Fig. 7.6
Hóng, Huǒquán, see Hóng, Xiù-quán
Hóng, Huǒxiù, see Hóng, Jìng-yáng
Hóng, Jìngyáng: Box 5.20 (Hóng)
Hóng, Réndá: Box 5.20 (Hóng)
Hóng, Rénfā: Box 5.20 (Hóng)
Hóng, Réngān: Box 5.20 (Hóng), Box 5.20 note 18 (Hóng)
Hóng, Rénkūn, see Hóng, Xiùquán
Hóng, Xiùquán: Box 5.19 (Tàipíng), Box 5.19 note 31 (Tàipíng), Box 5.20 (Hóng), Box 5.20 note 19 (Hóng), Fig. 5.118
Hooker, Joseph Dalton: Box 5.7 (Falconer), Box 5.15 (Standish), Box 5.32 (Dalhousie)

Hooker, William J.: Box 2.1 (Lindley), Box 2.6 (Ward), Box 3.8 (Cuming), Box 3.9 (Lay), Box 3.10 notes 26, 29 (Alcock), Box 4.1 note 9 (Royle), Box 5.14 (Paxton), Box 5.16 note 28, Ch. 8.3 notes 146, 164, 208
Hope, Charles Southern: Ch. 8.3 note 186
Hope, R. C.: Ch. 8.3 note 186
Hope, Mr.: Ch. 8.3
Hornby, Charlotte Margaret: Box 2.5 (Smith-Stanley)
Hornby, Geoffrey: Box 2.5 (Smith-Stanley)
Hornby, Robert: Box 2.5 (Smith-Stanley)
"Hornet": Box 10.3 (Fullerton-Elphinstone)
Horsburgh, James: Box 5.3 (Moresby)
Howard-Vyse, Francis: Ch. 8.3, Box 8.14 (Howard-Vyse), Box 10.2 (Namamugi)
Howard-Vyse, Richard William: Box 8.14 (Howard-Vyse)
Hsüeh, Huan, see Xuē, Huàn
Hú, Vincent: Ch. 10.3
Hú, Xīng, see Sing-Hoo
Hú, Yǎqín: Ch. 5.9 note 532, Figs. 3.6, 3.43, 3.58, 5.23, 5.25, 5.26, 5.27, 5.28, 5.29, 5.30, 5.31, 5.32, 5.33, 5.38, 5.39, 5.122, 5.141, 5.154, 5.155, 8.54, 8.55
Huá, Héng-fāng: Box 3.3 (Macgowan)
Huáng, Lián-kāi: Box 6.1 (Arrow)
Huc, Évariste-Régis: Ch. 3.1 note 27
Hudson, Caroline: Box 8.13 (Aspinall)
Hue, Jean: Ch. 10.3, Ch. 10.3 note 176
Huí: Boxes 10.6 (Xián-Tóng), 10.7 (Panthay), 10.8 (Huí), 10.9 (Muslim)
Hume, Abraham: Box 3.12 (Introductions) note 31
Hume, Everard: Box 3.12 (Introductions) note 33
Hume, Charlotte Isabella: Box 3.4 (Balfour)
Hume, Susan: Box 3.4 (Balfour)
Hung, Hsiu-ch'uan, see Hóng, Xiùquán
Hunt, Arthur Ackland: Fig. 2.7
Hutcheon, Jane: Ch. 1.2 note 108
Hyde, Lenox & Co.: Ch. 2.2 note 70

Ichijō, Haruko: Box 10.4 (Meiji)
Ichimura, Joanne: Box 3.11 note 1 (Lockhart)
Ikeda, Inshu: Fig. 8.33
Ilipu, see Àixīnjuéluó, Yīlǐbù
"Illustrious": Box 10.3 (Fullerton-Elphinstone)
Imamura, Gen'emon Eisei: Box 8.2 (Kaempfer)
Imery, Eric: Acknowledgements
"Island Queen": Ch. 5.5
Itō, Hirobumi: Box 9.7 (Cí Xǐ)

Jackman, George Jr.: Box 10.1 (Moore)
Jameson, Janet Jane Helen: Box 5.8 (Jameson)
Jameson, Laurence: Box 5.8 (Jameson)

Jameson, William: Ch. 5.4, Ch. 5.6, Ch. 5.6 note 327, Box 5.8 (Jameson), Box 5.8 notes 14, 15 (Jameson), Box 5. 11 (Saharanpur), Figs. 5.50, 5.79
Jamieson, Captain: Ch. 5.11
Jardine, Robert: Box 5.23 (Perceval)
Jardine, William: Box 2.2 (Opium), Box 2.4 (Dent), Box 2.4 note 5 (Dent)
Jardine, Matheson & Co.: Ch. 2.2 note 70, Box 2.2 (Opium), Box 2.4 (Dent), Box 3.5 (Thom), Box 3.10 note 29 (Alcock), Ch. 5.9, Box 5.5 (Beale), Box 5.23 (Perceval), Box 5.25 (Patridge), Ch. 8.3 note 142, Fig. 5.148
"Jaseur": Box 5.30 (Vansittart)
Jaurès, Constant Louis Jean Benjamin: Box 8.9 (Girard)
"Java": Box 8.4 (Siebold)
Jefferson, Thomas: Ch. 1.1, Ch. 1.1 note 48
Jerrold, Douglas William: Box 5.14 (Paxton)
Jiǎ, Huì: Acknowledgements, Ch. 5.10 note 628, Figs. 5.61, 8.2
Jiāng, Yúyīng: Acknowledgements
Jiāng, Zhōngyuán: Box 5.19 (Tàipíng)
Jìn (Dynasty): Ch. 5.4 note 219
Jǐngpō: Box 10.7 (Panthay)
"John Cooper": Ch. 3.4
Johnson, Andrew: Box 7.3 (Holt), Box 7.3 note 25 (Holt)
Johnson, Stephen: Fig. 3.10
Johnston, Alexander: Ch. 2.2, Ch. 2.2 note 70, Box 2.7 (Campbell-Johnston)
"Jonge Adriana": Box 8.4 (Siebold)
Jordan, Dorothy: Box 5.30 note 7 (Vansittart)
Junglu, see Rónglù
Junot, Jean-Androche: Ch. 1.1

Kachins: Ch. 10.4
Kaempfer, Engelbert: Ch. 8.1, Box 8.2 (Kaempfer), Box 8.2 note 18 (Kaempfer), Box 8.3 (Thunberg)
Kaempfer, Johann Hermann: Box 8.2 (Kaempfer)
Kāng, Yǒuwèi: Box 9.7 (Cí Xǐ)
"Kayo", see "Chusan"
Keasberry, Benjamin: Box 3.1 (Abeel)
Keasberry, Charlotte: Box 3.1 (Abeel)
Kemper, Johannes: Box 8.2 (Kaempfer)
Kennedy, Arthur Edward: Box 5.27 (Robertson)
Kennedy, Thomas (Danure): Box 1.2 (McNab)
Kennedy, Thomas (Captain of "Arrow"): Box 6.1 (Arrow), Box 6.1 note 7
Kennedy, T.: Ch. 11 note 70
Kerr, Crawford: Ch. 2.2 note 70
Kerr, William: Box 1.2 (McNab), Box 5.16 (Introductions) note 15, Fig. 8.56
Keswick, William: Box 3.10 note 29 (Alcock), Ch. 8.3 note 142, Box 8.15 (Introductions) note 50
Keying, see Àixīnjuéluó, Qíyīng
Khamtis: Ch. 10.4
Khoja, Rāshidīn: Box 10.9 (Muslim)
Khwaja, Rāshidīn, see Khoja, Rāshidīn

King Charles I: Ch. 5.7 note 379
King Charles II: Box 2.2 note 3 (Opium), Ch. 10.1 note 106
King of Delhi: Box 6.4 (Indian Mutiny)
King Edward VII: Box 5.30 (Vansittart)
King Ferdinand VII of Spain: Ch. 1.1, Ch. 1.1 note 44, Fig. 1.3
King Gàn: Box 5.20 (Hóng)
King George III: Box 5.15 (Standish)
King Henry VIII: Ch. 5.7 note 379
King Hóngchù Qián: Ch. 5.2 note 33
King João VI of Portugal: Ch. 1.1, Box 5.1 (P&O)
King Willem I: Box 8.4 (Siebold)
King Willem II: Box 8.4 note 47 (Siebold), Fig. 8.46
King William IV: Box 5.30 note 7 (Vansittart)
King, Henry: Box 9.3 note 19 (Gordon)
Kingdon-Ward, Frank: Ch. 10.4
Kipling, Rudyard: Ch. 5.6 note 319
Kirk, Thomas: Ch. 3.4, Ch. 5.3
Kishen, see Qíshàn
Kitamura, Keiichi: Acknowledgements
Knight, Thomas Andrew: Box 2.1 note 12
Knollys, Henry: Box 9.9 (Grant)
Knox, William: Ch. 1.2, Ch. 1.2 note 126
"Kofuyo", see "Chusan"
Kǒng zǐ, see Confucius
Krafft, Per Jr.: Fig. 8.10
Krick, Nicolas Michel: Ch. 10.4
Kuo, Cheng-chang: Box 8.4 (Siebold)
Kusumoto, Ine: Box 8.4 (Siebold), Box 8.4 note 54 (Siebold)
Kusumoto, Oine, see Kusumoto, Ine
Kusumoto, Taki: Box 8.4 (Siebold), Box 8.4 note 54 (Siebold)
Kuwabara, Setsuko: Acknowledgements

"Lady Mary Wood": Ch. 5.5, Box 5.10 (Lady Mary Wood), Fig. 5.60
Lài, Wénguāng: Box 10.5 (Niǎn)
Lai, Wenkuang, see Lài, Wénguāng
Laird, Macgregor: Box 7.1 (Smith)
Lambert, Aylmer Bourke: Box 3.12 (Introductions) note 20
Lamprey, Jones: Ch. 10.6
"Lancefield": Ch. 5.11
Larpent, George: Box 3.12 (Introductions) note 25
Lassel[l], Elizabeth: Box 3.11 (Lockhart)
Lauchert, Richard: Fig. 9.35
Lawrence, Crowdy and Bowlby: Box 9.8 (Bowlby)
Lay, George Tradescant: Ch. 3.4, Ch. 3.5, Box 3.9 (Lay), Ch. 10.3, Box 10.10 (Wade), Figs. 3.44, 3.45
Lay, Horatio Nelson: Box 3.9 (Lay), Box 10.10 (Wade)

Lay, William Hyde: Ch. 10.3, Ch. 10.3 note 202
Leadbeater, Benjamin Sr.: Ch. 3.2, Ch. 4.1
Leadbeater, John: Ch. 3.2, Ch. 4.1
Lear, Edward: Box 2.5 (Smith-Stanley), Ch. 5.6 note 357
Le Brethron de Caligny, A. E.: Box 5.19 note 27 (Tàipíng)
Lee, Francis Lowell: Box 8.8 (Hall)
Lees, Nassau: Box 5.8 (Jameson)
Leete, Clarissa Mary: Box 8.6 (Hepburn)
Leete, Harvey: Box 8.6 note 5 (Hepburn)
Leete, Sarah: Box 8.6 note 5 (Hepburn)
Le Gobien, Charles: Box 5.9 (Du Halde)
Lěng, Qín: Acknowledgements
L'Estrange, Jane Anne: Box 5.23 (Perceval)
"L'Etoile": Ch. 9.1
Leys, Elizabeth: Box 9.3 (Gordon)
Li, Chaoshou, see Lǐ, Zhāoshòu
Li, Chenkuo, see Lǐ, Zhēnguó
Lǐ, Chéngsēn: Acknowledgements
Li, Chingfang, see Lǐ, Jìngfāng
Li, Hanchang, see Lǐ, Hànzhāng
Lǐ, Hànzhāng: Ch. 10.4
Lǐ, Hóngzhāng: Box 5.19 note 35 (Tàipíng), Box 9.2 (Staveley), Ch. 10.3, Ch. 10.4, Box 10.5 (Niǎn), Fig. 10.46
Li, Hsiuch'eng, see Lǐ, Xiùchéng
Li, Hungchang, see Lǐ, Hóngzhāng
Lǐ, Jīnfēng: Acknowledgements, Ch. 5.5 note 292, Ch. 5.9 notes 475, 549, Figs. 3.3, 3.20, 3.24, 3.33, 3.35, 3.57, 5.9, 5.13, 5.22, 5.102, 5.119, 5.131, 9.18, 9.21, 9.32, 9.39, 9.41, 9.43, 10.48
Lǐ, Jìngfāng: Box 5.20 (Hóng)
Lǐ, Liányīng: Box 9.7 (Cí Xǐ), Box 9.7 note 27 (Cí Xǐ)
Lì, Mǎdòu, see Ricci, Matteo
Lǐ, Méi: Acknowledgements
Lǐ, Shìzhōng, see Lǐ, Zhāoshòu
Li, Shihchung, see Lǐ, Zhāoshòu
Lǐ, Xiùchéng: Box 5.19 note 32 (Tàipíng)
Lǐ, Yàméng: Acknowledgements, Ch. 5.3 note 119, Figs. 3.12, 3.17, 3.31, 5.120, 5.121, 5.132, 5.133, 9.23, 9.30
Lǐ, Zhāoshòu: Box 10.5 (Niǎn)
Lǐ, Zhēnguó: Ch. 10.4, Ch. 10.4 note 343
Lian, Suchiu: Acknowledgements
Liáng Dynasty: Ch. 5.9 note 594
Liáng, Fā: Box 5.20 (Hóng)
Liáng, Shítài: Fig. 10.46
Liáng, Tíngnán: Ch. 9.1
Liberty, N.: Ch. 6.2, Ch. 7.1 note 20
Liddle, Jenny: Fig. 1.5
Lín, Jǐngxīng: Ch. 9.2 note 56
Lin, Tsehsü, see Lín, Zéxú

Lín, Zéxú: Box 2.2 (Opium), Box 2.2 note 16 (Opium), Box 2.4 (Dent), Box 3.11 (Lockhart), Fig. 2.2
Lincoln, Abraham: Box 7.3 (Holt), Box 7.3 note 25 (Holt), Box 8.11 (Harris), Box 9.4 (Bruce)
Lindley, Barbara: Box 2.1 (Lindley)
Lindley, George Jr.: Box 2.1 (Lindley)
Lindley, George Sr.: Box 2.1 (Lindley), Box 2.1 note 1 (Lindley)
Lindley, John: Ch. 2.1, Ch. 2.2, Ch. 2 notes 36, 70, Box 2.1 (Lindley), Box 2.3 (Reeves), Ch. 3.5, Ch. 3.5 note 238, Ch. 4.1, Ch. 4.2, Box 4.1 note 9 (Royle), Ch. 5.6, Ch. 5.7, Ch. 5.9 note 542, Box 5.14 (Paxton), Box 5.14 notes 10, 22 (Paxton), Box 8.1 (Veitch), Box 8.15 (Introductions) note 59, Ch. 9.2, Ch. 10.1, Ch. 10.5, Ch. 10.6, Box 10.1 (Moore), Fig. 2.1
Lindley, Nathaniel: Box 2.1 (Lindley)
Lindley, Sarah: Box 2.1 (Lindley)
Lindsay, H.: Ch. 2.2 note 70
Linnaeus, Carolus Jr.: Box 8.3 (Thunberg)
Linnaeus, Carolus Sr.: Box 8.3 (Thunberg), Fig. 8.10
Lippe-Detmold, Friedrich Adolf zur: Box 8.2 (Kaempfer)
Lìsù: Box 10.7 (Panthay)
Liú, Bīng: Fig. 7.7C
Liú, Jǐntáng: Box 10.8 (Huí)
Liú, Lìchuān: Box 5.21 (Small Sword)
Liu, Lich'uan, see Liú, Lìchuān
Liú, Qīniáng: Fig. 5.141
Liú, Sōngshān: Box 10.8 (Huí)
Livingstone, David: Ch. 3.5 note 239
Lobb, Thomas: Box 8.15 (Introductions) notes 42, 43, 56, 61
Lock, John: Box 5.13 (Ramsay)
Lock & Whitfield: Fig. 3.45
Locker, Frederick: Ch. 5.7
Lockhart, Samuel Black: Box 3.11 (Lockhart)
Lockhart, William: Ch. 3.4, Ch. 3.5, Ch. 3.1 notes 5, 19, Ch. 3.5 note 223, Box 3.11 (Lockhart), Box 3.11 note 20 (Lockhart), Box 8.7 note 18 (Brown), Ch. 9.2, Figs. 3.49, 9.21
Loddiges, George: Ch. 2.1, Box 2.6 (Ward)
London & Oriental: Box 8.13 (Aspinall)
London Stereoscopic Company: Fig. 5.106
Lóng Yù, see Xiàodìng
Lonsdale, William: Box 5.7 (Falconer)
López Piquer, Luis: Fig. 1.3
Lord Aberdeen, see Hamilton-Gordon, George
Lord Ashburton, see Baring, Alexander
Lord Burghley, see Cecil, William
Lord Castlereagh, see Stewart, Robert
Lord Dalhousie, see Ramsay, James Andrew Broun
Lord Elgin, see Bruce, James
Lord Elphinstone, see Fullerton-Elphinstone, William Buller
Lord Lansdowne: Ch. 10.1
Lord Macartney, see Macartney, George

Lord Nelson: Box 3.9 note 22 (Lay)
Lord Palmerston, see Temple, Henry John
Lord Prudhoe, see Percy, Algernon
Lord Salisbury: Box 10.3 (Fullerton-Elphinstone)
Lord Saltoun, see Fraser, Alexander George
Lotz, Franz Joseph: Box 8.4 (Siebold)
Lotz, Maria Apollonia Josephine: Box 8.4 (Siebold)
Lo-tzung: Ch. 10.4
Loudon, John Claudius: Box 6.2 (Glendinning)
Loureira, see Loureiro
Loureiro, José da Silva: Ch. 8.2, Ch. 8.3, Box 8.5 (Loureiro)
Loviot, Fanny: Box 5.10 (Lady Mary Wood), Box 5.10 note 11 (Lady Mary Wood), Box 5.33 (Bentinck)
Lowder, John: Box 3.10 note 28 (Alcock), Fig. 3.31
Lowder, Lucy: Box 3.10 note 28 (Alcock)
Lowrie, Walter Macon: Ch. 3.4 note 206, Box 8.6 (Hepburn)
Lucius, Robert: Box 8.12 (Heusken)
Lucombe & Pince: Box 6.2 (Glendinning)
Lugard, Edward: Box 9.3 (Gordon)
Lumsden, Alison Jane: Ch. 11, Fig. 11.16
Lund, Matt: Ch. 11 note 59
Lushington, George Thomas: Box 5.12 (Batten), Box 5.13 (Ramsay)
Lushington, Laura: Box 5.13 (Ramsay)
Lyell, Charles: Box 3.3 (Macgowan), Box 5.7 (Falconer), Box 5.7 note 30 (Falconer)
Lloyds of London: Box 8.13 (Aspinall)

Mǎ, Ānliáng: Box 10.8 (Huí)
Mǎ, Déxīn: Box 10.7 (Panthay)
Mǎ, Fùtú: Box 10.7 (Panthay)
Ma, Fut'u, see Mǎ, Fùtú
Mǎ, Huālóng: Box 10.8 (Huí)
Ma, Ju-lung, see Mǎ, Rú-lóng
Ma, Jung, see Mǎ, Róng
Mǎ, Liánshēng: Box 10.7 (Panthay)
Ma, Lien-sheng, see Mǎ, Lián-shēng
Mǎ, Róng: Box 10.7 (Panthay)
Mǎ, Rúlóng: Box 10.7 (Panthay)
Ma, Te-hsin, see Mǎ, Déxīn
Mǎ, Zhàn'áo: Box 10.8 (Huí)
Mabileau, François: Ch. 10.3
Macartney, George: Box 5.16 (Introductions) note 9, Box 9.6 (Wyndham)
Macaulay, Thomas Babington: Box 10.10 note 7 (Wade)
MacDonald, John: Ch. 8.2, Ch. 8.2 note 98
M[a]cDonald, Mr.: Ch. 5.9
Macdonnell, Richard Graves: Box 5.27 (Robertson)
Macgowan, Daniel Jerome: Ch. 3.1, Ch. 3.1 note 61, Box 3.3 (Macgowan), Fig. 3.19

Macgregor, F. C.: Box 3.4 (Balfour)
Mackenzie, Alexander: Ch. 11
Mackenzie, Charles D.: Ch. 3.2, Ch. 5.2, Ch. 8.1
Mackenzie, John Alexander: Ch. 11
Mackenzie, Kenneth Ross: Ch. 8.1
Mackenzie Brothers and Co.: Ch. 3.2, Ch. 5.2, Ch. 8.1, Box 8.13 (Aspinall)
Mackenzie, Richardson & Co.: Ch. 10.1 note 41
Mackintosh, Annie: Box 9.3 (Gordon)
Macleod, Dawn: Ch. 11
Macvicar, John: Box 3.5 (Thom)
Maddock, Edward K.: Ch. 7.1, Ch. 7.1 note 11
Madison, Dolley: Ch. 1.1 note 77
Madison, James: Ch. 1.1, Ch. 1.1 notes 53, 85, Fig. 1.4
Maejima, Hisoka: Ch. 10.2
Magner, Teresa: Acknowledgements, Ch. 11 note 65
Magniac & Co.: Box 5.5 (Beale)
Magniac, Jardine & Co.: Ch. 2.2 note 70
Malcolm, George A.: Ch. 2.2 note 70
Mallard, Charles: Box 2.6 (Ward)
Manchu: Ch. 5.9 notes 507, 521, Box 5.19 (Tàipíng), Box 5.20 (Hóng), Ch. 9.1 note 46, Ch. 9.2 notes 60, 75, 76, 80, 81, 87, Box 9.7 notes 1, 3 (Cí Xǐ)
Mǎnzú, see Manchu
Marcy, William Learned: Box 8.11 (Harris)
Margary, Augustus Raymond: Ch. 10.4, Ch. 10.4 note 338, Box 10.10 (Wade), Fig. 10.45
"Marmora": Ch. 8.1
Marnock, Robert: Box 10.1 (Moore)
Marquis of Lansdowne, see Petty-Fitzmaurice, Henry
Marsden, William: Ch. 5.7, Fig. 5.108
Marsh, Henry: Box 5.29 (Chusan)
Marshall, William: Box 8.6 (Hepburn), Box 10.2 (Namamugi)
Martin, Henry Byam: Box 10.3 (Fullerton-Elphinstone)
Martin, William Alexander Parsons: Box 8.7 (Brown)
"Mary Louisa": Box 8.7 (Brown)
Mason, Charles: Ch. 7.1, Box 7.2 (Mason), Fig. 7.5
Mason, Chauncey: Box 7.2 (Mason)
Massie, Thomas Lecke: Ch. 10.1
Masson, Francis: Box 8.3 (Thunberg)
Masters, Maxwell Tylden: Ch. 10.5
Mather, Sarah: Box 9.2 (Staveley)
Matheson, Alexander: Ch. 2.2 note 70
Matheson, James: Ch. 2.2 note 70, Box 2.2 (Opium), Box 2.4 (Dent), Box 5.23 (Perceval)
Matotski, see Motoski
Matsuda, Masako: Acknowledgements
Matsuyama Clan: Box 5.29 (Chusan)
Maull & Polyblank: Fig. 3.36

Maximowicz, Carl: Box 8.15 note 29 (Introductions)
Maxwell, Catherine: Box 5.12 (Batten)
Maxwell, William G.: Ch. 3.1, Ch. 3.2
McAtear, Valerie: Fig. 8.45
McBryde, Mary Williamson: Box 3.1 (Abeel), Box 3.1 note 49 (Abeel)
McBryde, Thomas Livingston: Box 3.1 (Abeel), Box 3.1 note 49 (Abeel)
McCartee, Divie Bethune: Box 3.5 (Thom)
McCarty, James: Acknowledgements
McClay, David: Ch. 10.1 note 56, Figs. 5.104, 5.105, 5.113, 6.4
McDonald, Julie: Ch. 11 note 70
McFarlane, Captain: Ch. 5.5
McIntire, Rufus: Box 9.4 (Bruce)
McLean, Kenneth: Acknowledgements, Ch. 1.2 note 107, Ch. 1.3 note 143
McNab, James: Box 1.2
McNab, William: Ch. 1.3, Box 1.2 (McNab), Ch. 2.1, Fig. 1.16
McNeill, John: Ch. 1.2 note 120
Mcrae, Isabel: Box 5.7 (Falconer)
Meadows, John A. T.: Ch. 5.9, Box 5.22 (Meadows)
Meadows, Thomas Taylor: Box 5.22 (Meadows)
Medhurst, Walter Henry Sr.: Box 3.1 (Abeel), Box 3.11 (Lockhart), Ch. 5.9 note 472
"Medusa": Box 3.5 (Thom), Box 3.7 (Morrison)
Medway, Sue: Fig. 10.8
Mein, Frances Marion: Box 9.8 (Bowlby)
Mein, Margaret Matilda: Box 9.8 (Bowlby)
Mein, Pulteney: Box 9.8 (Bowlby)
"Melampus": Ch. 1.1
Melton Prior Institute, Düsseldorf: Fig. 5.71
Mesny, William: Ch. 10.4, Fig. 10.48
Miáo: Box 10.6 (Xián-Tóng)
Miáo, Pèilín: Box 10.5 (Niǎn)
Miao, P'ei-lin, see Miáo, Pèilín
Michaux, François André: Ch. 7.1
Michel, John: Box 3.7 (Morrison), Box 9.2 (Staveley)
Michie, Alexander: Box 10.10 note 7 (Wade)
Middleton, Henry: Ch. 7.1
Miller, Philip: Box 8.15 (Introductions) note 5
Miller, Ravenhill & Co.: Box 5.29 (Chusan)
Minamoto, Kugyo: Fig. 8.58
Minamoto, Sanemoto: Fig. 8.58
Minamoto, Yoritomo: Ch. 8.3, Fig. 8.64
Minet, Charles William: Box 9.2 note 22 (Staveley)
Minet, Susan Millicent: Box 9.2 (Staveley)
Míng: Ch. 5.2 note 36, Ch. 5.4 notes 202, 210, 220, Ch. 5.5 note 291, Ch. 5.8, Ch. 5.9 note 533, 576, Ch. 9.1 note 8, Ch. 9.2 notes 52, 56, Box 9.4 note 27 (Bruce), Box 9.6 note 10, Ch. 10.6 note 482, Figs. 5.51, 9.3, 9.7, 9.25, 9.39
Miquel, Friedrich Anton Wilhelm: Box 8.4 note 43 (Siebold)

Mishmis: Ch. 10.4
Misumi, Koyo: Acknowledgements
Mitchell, Eliza Bryson: Ch. 11, Fig. 11.10
Mitchell, Margaret W.: Ch. 11
Mitchell, Thomas Bryson: Ch. 11
"Mohawk": Box 7.1 (Smith)
"Monarch": Box 9.3 (Gordon)
Montcreiffe, David Steuart: Ch. 1.3
Montagu, George: Box 3.8 (Cuming), Box 3.8 note 2 (Cuming)
Montagu, John: Ch. 2.2 note 70
Moore, Mary: Box 2.1 (Lindley)
Moore, Thomas: Ch. 4.2, Ch. 4.2 note 23, Ch. 10.1, Ch. 10.1 note 72, Box 10.1 (Moore), Box 10.1 note 6 (Moore), Figs. 10.8, 10.10
More, Thomas: Ch. 10.1
Moresby, Fairfax: Box 5.3 (Moresby)
Moresby, Robert: Ch. 5.1, Box 5.1 (P&O), Box 5.3 (Moresby), Box 5.3 note 13 (Moresby)
Morgan, William: Ch. 2.2 note 70
Morillo, Pablo: Ch. 1.1
Morinaga, Haruno: Acknowledgements
"Morrison": Box 8.7 (Brown)
Morrison, Eliza: Box 3.7 (Morrison)
Morrison, Irvine: Box 10.3 (Fullerton-Elphinstone)
Morrison, John Robert: Box 3.5 (Thom)
Morrison, Martin Crofton: Ch. 3.2, Box 3.7 (Morrison), Ch. 5.3, Ch. 9.1, Ch. 10.5
Morrison, Robert: Box 3.1 (Abeel), Box 3.7 (Morrison), Box 8.7 (Brown)
Morrison, Ronald: Acknowledgements, Ch. 1.2 note 105
Morrow, James: Box 3.3 (Macgowan)
Moss, Michael: Box 8.14 (Howard-Vyse), Box 8.14 note 3 (Howard-Vyse)
Motoski, Mr.: Ch. 8.1
Murchison, Roderick: Ch. 10.1, Ch. 10.5, Fig. 10.51
Murray, Alexander Edward, 6th Earl of Dunmore: Ch. 10.1 note 56
Murray, Constance Euphemia Woronzow: Box 10.3 (Fullerton-Elphinstone)
Murray, Duncan: Acknowledgements
Murray, John III: Ch. 2.2 note 34, Ch. 3.5, Ch. 3.5 notes 238, 239, Ch. 5.7, Ch. 6.2, Ch. 10.1, Ch. 10.1 note 56, Figs. 5.104, 5.105, 5.113, 6.4, 10.12, 10.13
Murray, John VII: Acknowledgements
Murray, John Ivor: Box 8.8 (Hall), Box 8.8 note 4 (Hall)
Mùzōng, see Emperor Tóngzhì
Myburgh, Francis Gerhard: Ch. 8.3, Ch. 8.3 notes 160, 164, Box 8.12 note 24 (Heusken), Ch. 10.6

Nagel, Doris: Ch. 5.3 note 108
Nakamura, Ine, see Kusumoto, Ine
Nakamura, Jirokichi: Ch. 8.2, Fig. 8.41
Nakayama, Tadayasu: Box 10.4 (Meiji)
Nakayama, Yoshiko: Box 10.4 (Meiji)

Napier, Robert: Box 9.2 (Staveley)
Napier, William John: Box 2.7 (Campbell-Johnston)
Napoleon, see Bonaparte, Napoleon
National Library of Scotland, Edinburgh: Figs. 5.104, 5.105, 5.113, 6.4
National Portrait Gallery, London: Fig. 3.36, 3.45, 5.148, 6.1, 10.18, 10.19
National Trust: Fig. 1.5
Neale, Edward St. John: Box 8.14 (Howard-Vyse), Ch. 10.1, Box 10.2 (Namamugi)
Needham, Margaret: Ch. 11 note 46
Néel, Jean-Pierre: Ch. 10.3
Nees von Esenbeck, Christian: Box 4.1 note 9 (Royle)
Nehru, Jawahar Lal: Box 5.13 (Ramsay)
Nehru, Nan: Box 5.13 (Ramsay)
Neill, James: Ch. 6.2
Nelson, Mary: Box 3.9 (Lay)
"Nemesis": Box 2.2 (Opium), Box 2.7 (Campbell-Johnston), Fig. 5.39
Nevill, Dorothy: Ch. 10.1, Fig. 10.18
Neville-Cuming, Richard Henry: Fig. 5.162
Niǎn: Box 10.5 (Niǎn), Box 10.8 (Huí)
Nightingale, Florence: Box 5.6 (Champion)
Noble, Charles: Ch. 5.7 note 365, Box 5.15 (Standish), Box 8.18 (Introductions) note 31
Noel, Baptist: Box 3.1 (Abeel), Box 3.1 note 36 (Abeel)
Noisette, Philippe: Ch. 7.1
Notehelfer, Fred: Figs. 8.18, 8.27, 8.49
Nugent, Richard: Box 3.8 (Cuming)

Ocean Steam Yachting Company: Box 9.10 (Ceylon)
"Oenarang", see "Lady Mary Wood"
Ogawa, Okisato: Ch. 10.2
Ohsawa, Masahiko: Acknowledgements
Ohsawa, Mutsuko: Acknowledgements
Okuno, Masatsuna: Box 8.6 (Hepburn)
Oldham, Richard: Ch. 8.3, Ch. 8.3 note 208, Box 3.10 note 29 (Alcock)
Oliphant, Anne: Box 9.9 (Hope Grant)
Oliphant, Laurence: Ch. 8.1 note 15
Olyphant, David Washington Cincinnatus: Box 8.7 (Brown)
Oshichi, Yaoya: Ch. 8.2
Oswald, Elizabeth: Box 9.4 (Bruce)
Oswald, James Townsend: Box 9.4 note 1 (Bruce)
Otakusa, see Kusumoto, Taki
Otsuru: Box 8.12 (Heusken)
Ottoman Empire: Box 10.9 (Muslim)
Ou, Vincent, see Hú, Vincent
Outram, James: Box 6.4 (Indian Mutiny)

Page-Turner, Gregory Osborne: Box 5.14 (Paxton), Box 5.14 note 2 (Paxton)

Paiva, Francisco José de: Ch. 2.2 note 70
"Palinurus": Box 5.3 (Moresby)
Palliser, Frances Helen: Box 2.7 (Campbell-Johnston)
Palmer & Co.: Ch. 2.1, Box 2.4 (Dent)
"Panama": Box 8.6 (Hepburn)
Pandit, Ranjit Sitaram: Box 5.13 note 32 (Ramsay)
Pandit, Vijaya Lakshmi, see Nehru Nan
Pant, Nana Govind Dhondu: Box 6.4 (Indian Mutiny)
"Paou-shan": Ch. 5.11
Paris, John Ayrton: Box 4.1 (Royle), Box 4.1 note 10 (Royle)
Parke, Mr.: Box 3.11 (Lockhart)
Parker, Peter: Box 3.11 (Lockhart)
Parker, Tony: Acknowledgements
Parker, William: Box 2.2 (Opium), Box 5.30 (Vansittart), Box 5.30 note 6 (Vansittart)
Parkes, Catharine: Box 3.11 (Lockhart), Box 8.7 note 12 (Brown)
Parkes, Harry Smith: Box 3.11 note 13 (Lockhart), Ch. 6, Box 6.1 (Arrow), Box 6.1 note 7 (Arrow)
Parkman, Francis: Box 8.8 (Hall)
Parks, John Damper: Ch. 2.1, Box 2.3 (Reeves)
Parsons, Samuel Bowne: Box 8.8 (Hall)
Partridge, Daniel, see Patridge, Daniel
Patridge, Daniel: Ch. 5.9, Box 5.25 (Patridge), Box 5.25 notes 4, 6, 7, 9, 11 (Patridge)
Paudayal, Khum N.: Ch. 5.5 note 287
Paul, William: Ch. 10.1
Pavillion, Marie-Anne: Ch. 10.3
Pavlovna, Anna: Fig. 8.46
Paxton, Ann: Box 5.14 (Paxton)
Paxton, Joseph: Ch. 5.7, Box 5.14 (Paxton), Box 5.14 notes 2, 10, 21 (Paxton), Ch. 6.1, Ch. 10.1, Ch. 10.5, Figs. 5.91, 10.27
Paxton, William: Box 5.14 (Paxton)
Payne, Janet: Ch. 10.1 note 24
Peacock, George: Ch. 1.2, Ch. 1.2 note 127
Peh-ma: Ch. 10.4
Pengelly, William: Box 5.7 note 30 (Falconer)
Peninsular & Oriental Steam Navigation Co.: Ch. 5.1 notes 3, 4, 5, 12, Ch. 5.8 note 393, Ch. 5.10 note 684, Ch. 5.11 note 685, Ch. 5.12, Box 5.1 (P&O), Box 5.3 (Moresby), Box 5.4 (Braganza), Box 5.10 (Lady Mary Wood), Box 5.17 (Ganges), Box 5.29 (Chusan), Box 5.33 (Bentinck), Box 8.13 (Aspinall), Ch. 9.2 note 126, Box 9.10 (Ceylon), Figs. 5.2, 5.8, 5.162, 9.54
Pennefather, Mary Charity: Box 5.30 (Vansittart)
Penny, Henry: Ch. 1.3
Penny, Jane: Ch. 1.3, Ch. 1.3 note 151, Figs. 1.13, 1.14, 1.15, see also under Fortune, Jane
Perceval, Alexander Jr.: Ch. 5.9, Ch. 5.9 note 562, Box 5.23 (Perceval), Ch. 10.5
Perceval, Alexander Sr.: Box 5.23 (Perceval)
Perceval, Mary Jane: Box 5.23 (Perceval)
Perceval, Philip: Box 5.23 (Perceval)
Perceval, Spencer: Ch. 1.1 note 55

Percy, Algernon: Ch. 2.1, Ch. 2.2 note 70
Perelsztejn, Diane: Box 5.8 (Jameson)
Perrottet, Georges Samuel: Box 3.12 (Introductions) note 57
Perry, Matthew: Ch. 8.1, Ch. 8.2, Ch. 8.2 note 78, Ch. 8.3 notes 148, 168, Box 8.4 (Siebold), Box 8.11 (Harris), Ch. 10.2, Fig. 8.20
Peterson, Rebecca Ewing: Fig. 3.19
Petty-Fitzmaurice, Henry: Box 5.15 (Standish)
Phillips, George: Ch. 10.4
Phipps, John: Box 9.8 (Bowlby)
"Phlegethon": Box 2.2 (Opium), Fig. 5.39
Pierce, Franklin: Ch. 8.2 note 78, Box 8.11 (Harris)
Pilkington, Jeffrey: Ch. 5.7 note 384
Pinchbeck, Richard: Box 2.3 (Reeves)
Piquer, Luis López: Fig. 1.3
Pitcher, Philip Wilson: Ch. 3.1 note 20
Plimley, Kevin: Fig. 8.44
Polo, Marco: Ch. 5.9, Fig. 5.151
Pompe van Meerdervoort, Johannes Lijdius Catharinus: Box 3.3 (Macgowan)
Pope, Alexander: Box 2.2 note 3 (Opium)
Portman, Anton L. C.: Ch. 8.3, Ch. 8.3 note 148, Fig. 8.48
"Potomac": Box 8.6 (Hepburn)
Pottinger, Henry: Ch. 2.2 note 70, Box 2.2 (Opium), Box 2.2 notes 38, 52 (Opium), Box 2.7 (Campbell-Johnston), Ch. 3.1, Box 3.4 (Balfour), Box 3.5 (Thom), Box 3.9 (Lay), Box 5.25 (Patridge)
Potts, Captain: Ch. 5.1
Potts, John: Ch. 2.1, Box 2.3 (Reeves)
"President": Ch. 7.1
Prévost, George: Ch. 1.1
Price, Rulinda: Box 7.1 note 19
Prince Albert: Ch. 10.1
Prince Chún: Box 9.7 (Cí Xǐ)
Prince Duān: Box 9.7 (Cí Xǐ)
Prince Gōng: Ch. 9.2, Box 9.7 (Cí Xǐ)
Prince João, see King João VI of Portugal
Prince Mutsuhito, see Emperor Meiji
Prince of Satsuma: Ch. 8.2 note 111, Ch. 8.3 note 211, Box 10.2 (Namamugi)
Prince Yi Eun: Box 10.4 (Meiji)
Pǔ, Jùn: Box 9.7 (Cí Xǐ), Box 9.7 note 40 (Cí Xǐ)
Pǔ, Yí: Box 9.7 (Cí Xǐ)
Pulford, Russell Richard: Box 5.8 (Jameson), Box 5.8 note 17 (Jameson)
Purves, Sarah June: Ch. 11 note 61
Purves, Wilhelmina: Ch. 11, Ch. 11 notes 63, 72
Purves, William: Ch. 11

Qadir, Ghulam: Box 5.11 (Saharanpur)
Qián, Chù, see King Hóngchù Qián

Qián, Hóngchù, see King Hóngchù Qián
Qīng (Dynasty): Ch. 5.4 notes 211, 220, Ch. 5.9 notes 507, 576, Box 5.19 (Tàipíng), Box 5.20 (Hóng), Ch. 9.2, Box 9.7 note 83 (Cí Xǐ), Ch. 10.3, Ch. 10.4 note 338, Box 10.5 (Niǎn), Box 10.6 (Xián-Tóng), Box 10.7 (Panthay), Box 10.8 (Huí), Box 10.9 (Muslim), Figs. 9.36, 9.39
Qíshàn: Box 2.2 (Opium), Box 2.7 (Campbell-Johnston)
Qīu, Yuāncái: Box 10.5 (Niǎn)
Queen Anne: Box 2.2 note 3 (Opium)
Queen Elizabeth I: Ch. 10.1, Fig. 10.3
Queen Insurance: Box 8.13 (Aspinall)
Queen Isabella II of Spain: Box 5.1 (P&O)
Queen Maria II of Portugal: Box 5.1 (P&O)
Queen Victoria: Box 5.14 note 10 (Paxton), Box 5.25 (Patridge), Ch. 8.2, Ch. 8.2 note 98, Box 8.12 (Heusken), Box 9.2 (Staveley), Box 9.3 (Gordon), Box 9.6 (Wyndham), Ch. 10.1, Ch. 10.1 note 74, Fig. 10.25

Raeburn, Henry: Fig. 1.5
Ralston, John: Ch. 1.2
Ramsay, Edith Christian: Box 5.32 (Dalhousie)
Ramsay, George: Box 5.32 (Dalhousie)
Ramsay, Henry: Ch. 5.6, Box 5.12 (Batten), Box 5.13 (Ramsay), Box 5.32 (Dalhousie), Fig. 5.83
Ramsay, Henry Lushington: Box 5.13 (Ramsay)
Ramsay, James Andrew Broun: Ch. 5.12, Box 5.31 (Beadon), Box 5.32 (Dalhousie), Box 6.4 (Indian Mutiny), Fig. 5.165
Ramsay, John (father): Box 5.13 (Ramsay)
Ramsay, John (son): Box 5.13 (Ramsay)
Ramsay, Susan Georgiana: Box 5.32 (Dalhousie)
Rawes, Dr.: Ch. 3.1
Rè Xīdīng Hé Zhuó, see Khoja, Rāshidīn
Redpath, Agnes: Ch. 1.2, Ch. 5.7
Reeves, John Sr.: Ch. 2.1, Ch. 2.2, Ch. 2.2 note 70, Box 2.3 (Reeves), Box 2.3 notes 13, 16 (Reeves), Box 2.7 (Campbell-Johnston), Box 5.16 (Introductions) note 26, Fig. 2.3
Reeves, Jonathan: Box 2.3 (Reeves)
Reeves, Rachel (Rachel-Louisa): Box 2.3 (Reeves), Box 2.3 note 7 (Reeves)
Reid, George: Fig. 10.12
Reinwardt, Caspar Georg Carl: Box 8.4 (Siebold)
Rennie, James Edwin: Ch. 11, Ch. 11 note 36
Rennie & Co.: Box 5.17 (Ganges)
Renwick, James: Box 8.11 (Harris)
Ricci, Giovanni Battista: Box 9.5 (Ricci)
Ricci, Matteo: Ch. 9.2, Box 9.5 (Ricci), Figs. 9.23, 9.29, 9.30
Richardson, Charles Lenox: Box 8.6 (Hepburn), Box 8.14 (Howard-Vyse), Ch. 10.1, Ch. 10.1 note 41, Box 10.2 (Namamugi), Box 10.2 note 5 (Namamugi), Figs. 8.43, 10.15, 10.16
"Ridesdale": Box 9.10 (Ceylon)
Ridout, John: Ch. 4.1
Ridpath, Agnes, see Redpath, Agnes
Rigaud, Jean-François: Ch. 10.3, Ch. 10.3 note 174

"Ringdove": Box 8.14 (Howard-Vyse)
"Ripon": Ch. 5.1, Box 5.2 (Ripon), Figs. 5.1, 5.2, 5.3
Rivers, Thomas: Ch. 10.1
Robert, Frederick: Box 5.3 (Moresby)
Roberts, Issachar Jacox: Box 5.20 (Hóng), Box 5.20 note 19 (Hóng)
Robertson, Daniel: Box 5.27 (Robertson)
Robertson, Daniel Brooke: Ch. 5.10, Box 5.27 (Robertson), Ch. 10.4
Robinson, Frederick John: Box 5.2 (Ripon)
Rocheid Family: Ch. 1.3
Rogers, Ruth Miller: Box 8.8 (Hall)
Rolle, John: Box 6.2 (Glendinning)
Rolle, Louisa: Box 6.2 (Glendinning)
"Roman": Box 3.1 (Abeel)
Rónglù: Box 9.7 (Cí Xǐ)
Roob, Alexander: Fig. 5.71
Roos, Jay: Acknowledgements, Ch. 5.7 note 385, Ch. 6.1 note 32, Ch. 7.1 note 20, Ch. 10.5 note 418, Ch. 11 notes 3, 18, 69
Rose, Hugh Henry: Box 6.4 (Indian Mutiny)
Ross, J. B.: Ch. 8.3, Ch. 8.3 note 186
Rothschild: Box 5.14 (Paxton)
Rotton, Mary: Box 5.3 (Moresby)
Rouse, Simone: Ch. 11 notes 56, 60, 61, 63, 64, 72
"Rover": Ch. 10. 4
Roxburgh, William: Ch. 5.6
"Royal Albert": Box 10.3 (Fullerton-Elphinstone)
Royal Botanic Garden Edinburgh: Acknowledgements, Fig. 1.17
Royal Entomological Society: Fig. 8.45
Royal Horticultural Society: Figs. 1.16, 2.1, 2.6, 2.9, 3.3, 3.39, 3.52, 3.54, 3.59
Royle, John Forbes: Ch. 4.1, Ch. 4.2, Box 4.1 (Royle), Box 4.1 note 10 (Royle), Ch. 5.6, Ch. 5.6 note 327, Box 5.7 (Falconer), Box 5.8 note 14 (Jameson), Box 5.11 (Saharanpur), Ch. 7.1, Figs. 4.4, 5.72
Royle, William Henry: Box 4.1 (Royle)
Rudbeck, Olof Sr.: Box 8.2 (Kaempfer)
Ruskin, John: Box 5.14 note 22 (Paxton)
Russell, Henry Robin: Box 3.10 note 26 (Alcock)
Russell, Ian: Acknowledgements, Ch. 11 notes 19, 33
Russell, Lord John: Box 5.32 (Dalhousie), Ch. 6.1
Russell, Sarah: Box 2.3 (Reeves)
Russell, Sarah: Acknowledgements, Ch. 11
Russell & Co.: Box 2.2 (Opium), Ch. 5.10, Box 5.21 note 7 (Small Sword)
Rutherford, Daniel: Box 1.2 (McNab)

Sabine, Joseph: Ch. 2.1 note 8, Box 2.1 note 12 (Lindley)
Sacharissa, see Sidney, Dorothy
Sachinomiya, see Emperor Meiji
Sackville, John Frederick: Box 2.5 (Smith-Stanley)

Saigō, Takamori: Ch. 10.2, Fig. 10.34
Sam-qua, see Wú, Jiànzhāng
Samuda Brothers: Box 9.10 (Ceylon)
"San Jacinto": Box 8.12 (Heusken)
"Sancho Panza": Box 8.6 (Hepburn)
Santos, Maria de los: Box 3.8 (Cuming)
"São Luiz": Box 9.5 (Ricci)
Saris, John: Ch. 8.3
Sassoon, R. D.: Box 5.29 (Chusan)
Sato, Fumi: Box 8.13 note 18 (Aspinall)
Satow, Ernest: Ch. 8.2 note 69
Satsuma Clan: Ch. 10.1
Saunders, William Wilson: Ch. 10.1
Scarth, John: Ch. 3.1 note 20, Ch. 5.3 note 95, Ch. 5.5 note 286, Ch. 5.10 note 622, Ch. 10.6, Box 10.10 note 7 (Wade)
Schereschewsky, Samuel Isaac Joseph: Box 5.24 (Burdon)
Scheuchzer, Johann Gaspar: Box 8.2 (Kaempfer)
Schiller, Friedrich: Box 8.4 (Siebold)
Schlegel, Hermann: Box 8.4 (Siebold)
Schoedde, James Holmes: Box 2.2 (Opium), Ch. 3.1, Box 3.2 (Schoedde), Fig. 3.10
"Scotland": Ch. 8.3
Scott, Elizabeth: Box 9.6 (Wyndham)
Scott, H. Ernest: Ch. 11
Scott, Walter: Ch. 5.4, Ch. 10.6
Scott, Winfield: Box 9.4 (Bruce)
Sen, Boshi: Ch. 5.6
Sēnggélínqìn: Box 10.5 (Niǎn)
Seng-ko-lin-ch'in, see Sēnggélínqìn
"Serpent": Box 5.30 (Vansittart)
Seward, William Henry: Box 8.11 (Harris)
Seymour, Michael: Box 6.1 (Arrow)
Shah, Alam II: Box 5.11 (Saharanpur)
Shah, Firoz: Box 6.4 (Indian Mutiny)
Shah Suleiman I of Persia: Box 8.2 (Kaempfer)
Shanghai Steam Navigation Co.: Box 5.17 (Ganges), Box 8.5 (Loureiro)
Shans: Ch. 10.4
Shaw, Charles: Ch. 3.4
Shèngbǎo, General: Box 10.8 (Huí)
Shimazu, Hisamutsu: Box 10.2 (Namamugi)
Shimoto, Ine, see Kusumoto, Ine
Shimoto, Itoku, see Kusumoto, Ine
Shimoto Oine, see Kusumoto, Ine
Shortrede, Andrew: Ch. 5.8 note 399
Shugert, Samuel T.: Ch. 7.2
Shuǐ: Box 10.6 (Xián-Tóng)

Sidney, Dorothy: Ch. 5.7
Siebold, see Von Siebold
Sinclair, Charles: Box 3.6 (Sinclair)
Sinclair, Charles Anthony: Ch. 3.2, Box 3.6 (Sinclair), Ch. 5.10, Box 5.22 (Meadows)
Sinclair, Elizabeth MacDonald: Ch. 11
Sindhia, Mahadji: Box 5.11 (Saharanpur)
Sing-Hoo: Chs. 5.3, 5.4, Fig. 5.55
"Sirius": Box 7.1 (Smith)
"Sir Robert Peel": Box 9.4 (Bruce)
Skillet, Stephen Dadd: Fig. 5.8
Sladen, Edward Bosc: Ch. 10.4
Sladen, Percy: Ch. 10.4 note 210
Sloane, Hans: Box 8.2 (Kaempfer)
Smith, David Jr.: Box 7.1 (Smith)
Smith, David Sr.: Box 7.1 (Smith)
Smith, George: Box 5.32 note 12 (Dalhousie)
Smith, Henry: Ch. 7.1
Smith, Junius: Ch. 7.1, Box 7.1 (Smith), Fig. 7.2
Smith, J. Caldecott: Box 5.21 note 7 (Small Sword)
Smith, Lucinda: Ch. 7.1, Box 7.1 (Smith)
Smith, Ruth: Box 7.1 (Smith)
Smith, Samuel: Box 5.14 (Paxton), Box 5.14 note 5 (Paxton)
Smith, Mr.: Ch. 5.9
Smith-Stanley, Edward Jr. (13th Earl of Derby): Ch. 2.1, Ch. 2.2 note 70, Box 2.5 (Smith-Stanley), Ch. 3.1, Ch. 3.2, Ch. 3.3, Ch. 3.3 note 163, Box 3.4 (Balfour), Box 3.8 (Cuming), Box 3.8 note 24 (Cuming), Figs. 2.4, 2.5, 3.10, 3.26, 3.40, 4.1, 4.5
Smith-Stanley, Edward Sr. (12th Earl of Derby): Box 2.5 (Smith-Stanley)
Smith-Stanley, Edward George (14th Earl of Derby): Ch. 6.1
Smythe, Anne: Box 10.10 (Wade)
Sneyd, Harriet: Box 5.31 (Beadon)
Sneyd, Ralph Henry: Box 5.31 (Beadon)
Snow, Elizabeth: Acknowledgements, Ch. 1.2 notes 112, 139, Ch. 1.3 note 143, Ch. 4.1 note 7, Ch. 10.1 notes 60, 61, 97, Ch. 11 notes 1, 2
Snow, John: Ch. 10.1 note 74
Solly, Annette: Box 4.1 (Royle)
Solly, Edward Jr.: Ch. 2.2 note 62
Solly, Edward Sr.: Box 4.1 (Royle)
Somerville, Alexander: Ch. 1.2
Somerville, Margaret: Ch. 11 note 25
Somes, Joseph: Ch. 2.2 note 70
Sòng (Dynasty): Ch. 5.4 notes 219, 220, Ch. 5.5 note 291, Ch. 5.9 note 509
Sonogi, see Kusumoto, Taki
Soulié, Jean André: Ch. 10.3 note 178
Sowerby, George Brettingham Sr.: Box 3.8 note 14 (Cuming)
Standish, John: Ch. 5.7, Ch. 5.7 note 365, Box 5.15 (Standish), Box 5.15 note 11 (Standish), Box 5.16

(Introductions) notes 9, 13, Ch. 8.3, Box 8.15 (Introductions) notes 3, 10, 11, 12, 13, 23, 44, 48, 53, 60, 62, 64, 65, Ch. 9.2, Ch. 10.1, Ch. 10.5, Figs. 5.92, 10.1
Standish, Lucy: Box 5.15 (Standish)
Standish & Ashby: Box 5.15 (Standish)
Standish & Noble: Ch. 5.7, Box 5.15 (Standish), Box 5.15 note 12 (Standish), Box 5.16 (Introductions) notes 5, 8, 12, 15, 16, 17, 20, 21, 22, 27
Starkman, Margaretha: Box 8.3 (Thunberg)
Statham, Craig: Acknowledgements
Staveley, Charles William Dunbar: Ch. 9.1, Box 9.2 (Staveley), Box 9.2 note 15 (Staveley), Box 9.3 (Gordon), Fig. 9.7
Staveley, William: Box 9.2 (Staveley)
Stephens, Eleanor Jennings: Box 7.3 (Holt)
Steven, Martin: Acknowledgements, Ch. 11 notes 34, 35, 36, 39
Stevens, Edwin: Box 5.20 (Hóng)
Stevens, Augusta: Box 8.10 (Stevens)
Stevens, John Jr.: Box 8.10 (Stevens)
Stevens, John Sr.: Box 8.10 (Stevens)
Stevens, Samuel: Ch. 5.8 note 428, Ch. 8.3, Box 8.10 (Stevens), Fig. 8.45
Stewart, Robert: Ch. 1.1
Strauchon, John: Ch. 10.6 note 495
Stuessy, Tod: Ch. 8.3 note 173
Suí (Dynasty): Ch. 5.10 note 638, Ch. 10.1 note 16
Sullivan, George G.: Box 5.27 (Robertson)
"Sultan": Box 5.30 (Vansittart)
"Surprise": Box 8.7 (Brown)
Surratt, John: Box 7.3 note 24 (Holt)
Surratt, Mary: Box 7.3 (Holt)
Sùshùn: Box 9.7 (Cí Xǐ)
Swingle, Walter Tennyson: Ch. 2.2 note 53, Ch. 3.1
Swinhoe, Robert: Ch. 9.2 note 66

Taiken Japan: Fig. 8.44
Tàipíng: Boxes 5.19 (Tàipíng), 5.20 (Hóng), 5.21 (Small Sword), Box 8.8 (Hall), Box 8.13 (Aspinall), Ch. 9.1, Box 9.4 (Bruce), Box 10.5 (Niǎn), Box 10.6 (Xián-Tóng), Box 10.8 (Huí), Figs. 5.117, 5.118
Takahashi, Sakuzaemon Kageyasu Jr.: Box 8.4 (Siebold), Box 8.4 note 25 (Siebold)
Tamabayashi, Yoshio: Acknowledgements
Táng, Yènà: Ch. 5.9 note 489
Táng (Dynasty): Ch. 5.2, Ch. 5.4 note 219, Ch. 5.5 note 287
Tay, Michel: Ch. 10.3, Ch. 10.3 note 176
Tayler, Elizabeth Helen: Box 9.9 (Grant), Box 9.9 note 31 (Grant)
Taylor, Captain: Ch. 8.2
Taylor, Catherine: Ch. 11
Taylor, Liz: Ch. 2.1 notes 9, 12, 25
Teijsman, J. E.: Box 8.15 (Introductions) note 42
Temminck, Coenraad Jacob: Box 8.4 (Siebold)

Temple, Henry John (Lord Palmerston): Box 2.2 (Opium), Box 5.14 (Paxton), Ch. 6.1
Tena, Carlos: Ch. 10.6, Ch. 10.6 note 472
Tennent, Archibald Hay: Ch. 11 note 34
Tennyson, Alfred: Ch. 10.5
Thackeray, William Makepeace: Box 5.10 (Lady Mary Wood)
Thérond, Émile Théodore: Fig. 9.19
Thom, David: Ch. 3.5 note 44 (Thom)
Thom, John: Box 3.5 (Thom)
Thom, Robert: Ch. 3.2, Box 3.5 (Thom), Box 3.6 (Sinclair), Box 3.7 (Morrison)
Thomas, Rachel: Acknowledgements
Thompson, John: Ch. 4.1
Thomson, Amy M.: Ch. 1.2 note 108
Thomson, Antony Todd: Box 4.1 (Royle)
Thomson, John: Box 1.2 (McNab)
Thomson, John (photographer): Figs. 3.41, 3.47, 10.41, 10.42
Thorburn, Mr.: Ch. 8.3
Thornton, Thomas: Box 5.15 (Standish)
Thunberg, Carl Peter: Ch. 8.1, Box 8.2 (Kaempfer), Box 8.3 (Thunberg), Box 8.15 (Introductions) note 61, Fig. 8.10
Thunberg, Johan: Box 8.3 (Thunberg)
Tillet, Marie-Aimée: Ch. 10.3
Tod & McGregor: Box 5.17 (Ganges)
Tods, Murray & Jamieson: Ch. 11 notes 33, 37
Tokugawa, Ieharu: Box 8.3 (Thunberg)
Tokugawa, Iemochi: Ch. 10.1, Ch. 10.2
Tokugawa, Iesada: Box 8.11 (Harris)
Tokugawa, Ieyasu: Ch. 8.2
Tokugawa, Tsunayoshi: Box 8.2 (Kaempfer)
Tokugawa, Yoshinobu: Ch. 10.2, Ch. 10.2 note 126
Tomi: Ch. 8.2, Fig. 8.17
Tomlin, Jacob: Box 3.1 (Abeel)
Tóng, Zǐshī, see Zhāng, Zōngyǔ
Tope, Ramachandra Pandurang: Box 6.4 (Indian Mutiny)
Tope, Tantia, see Tope, Ramachandra Pandurang
Tope, Tatya, see Tope, Ramachandra Pandurang
Toward, Andrew: Box 5.15 (Standish)
Tradescant, John Jr.: Box 3.9 note 1 (Lay)
Tradescant, John Sr.: Box 3.9 note 1 (Lay)
Trefusis, Louisa, see Rolle, Louisa
Trivedi, Anjali: Fig. 5.61
"True Briton": Box 5.18 (Bonham)
Ts'en, Yü-ying, see Cén, Yùyīng
Tseng, Kuo-fan, see Zēng, Guófān
Tso, Tsung-t'ang, see Zuǒ, Zōngtáng
Tu, Wen-hsiu, see Dù, Wénxiù

Tunga: Ch. 8.3
"Tung-yu": Ch. 8.2
Turner, Guy: Acknowledgements
Turner, Joseph M. W.: Ch. 10.1
Turner, Sally: Acknowledgements

Ud Daula, Intizam: Box 5.11 (Saharanpur)
"United States": Box 8.6 (Hepburn)
Universal Marine: Box 8.13 (Aspinall)
Urquhart, Elizabeth Harries: Box 5.6 (Champion)
Uyghur: Box 10.9 (Muslim)

Van Houtte, Louis: Box 8.4 (Siebold)
Van Rensselaer, Stephen: Ch. 1.1
Vansittart, Edward Westby: Box 3.6 (Sinclair), Ch. 5.11, Box 5.30 (Vansittart), Box 5.30 note 6 (Vansittart)
Vansittart, Henry: Box 5.30 (Vansittart)
Varjoghe, Alina: Acknowledgements
Veitch, James Jr.: Box 8.1 (Veitch), Ch. 10.1, Ch. 10.5, Fig. 10.52
Veitch, James Herbert: Box 8.1 (Veitch)
Veitch, John Gould: Box 3.10 (Alcock), Ch. 8.1, Ch. 8.3, Box 8.1 (Veitch), Box 8.1 note 6 (Veitch), Box 8.15 (Introductions) notes 9, 11, 26, 36, 46, 50, 56, 61, 65, Ch. 10.1, Ch. 10.5, Figs. 8.1, 10.52
Veitch, John Gould Jr.: Box 8.1 (Veitch)
Verbiest, Ferdinand: Ch. 9.2
Verenigde Oost-Indische Compagnie (V.O.C.): Box 2.2 (Opium), Box 8.2 (Kaempfer), Box 8.3 (Thunberg), Fig. 8.10
Veselovsky, Nikolai Ivanovich: Fig. 10.38
"Victoria and Albert": Box 5.30 (Vansittart)
Vivekananda Institute of Hill Agriculture: Ch. 5.6, Fig. 5.81
Von Blume, Carl Ludwig: Box 8.4 (Siebold), Box 8.4 note 36 (Siebold), Ch. 9.2
Von Bunge, Alexander Georg: Ch. 9.2, Fig. 9.46
Von Fallenberg, Philipp Emanuel: Box 9.9 (Hope Grant)
Von Gagern, Hans: Box 8.4 note 51 (Siebold)
Von Gagern, Karoline Ida Helene: Box 8.4 (Siebold)
Von Humboldt, Alexander: Box 8.4 (Siebold)
Von Ketteler, Clemens: Box 9.7 (Cí Xǐ), Box 9.7 note 52 (Cí Xǐ)
Von Pufendorf, Esaias: Box 8.2 (Kaempfer)
Von Pufendorf, Samuel: Box 8.2 (Kaempfer)
Von Richthofen, Ferdinand: Ch. 10.4
Von Siebold, Alexander Georg: Box 8.4 notes 53, 73 (Siebold)
Von Siebold, Carl Caspar: Box 8.4 note 1 (Siebold)
Von Siebold, Clara Barbara: Box 8.4 note 1 (Siebold)
Von Siebold, Heinrich Philipp: Box 8.4 note 53 (Siebold)
Von Siebold, Helene Jr.: Box 8.4 note 53 (Siebold)
Von Siebold, Helene Sr, see Von Gagern, Karoline Ida Helene
Von Siebold, Johann Georg Christoph: Box 8.4 (Siebold)

Von Siebold, Mathilde Apollonia: Box 8.4 note 53 (Siebold)
Von Siebold, Maximilian August Constantin: Box 8.4 note 53 (Siebold)
Von Siebold, Philipp Franz Balthasar: Box 3.12 (Introductions) note 102, Box 5.16 (Introductions) notes 14, 28, Box 6.3 (Introductions) notes 1, 2, 4, Ch. 8.1, Ch. 8.3, Box 8.4 (Siebold), Box 8.4 notes 19, 36, 42, 43, 54, 61, Box 8.8 (Hall), Ch. 10.5, Ch. 10.6, Figs. 8.11, 8.12, 8.13, 8.14, 8.46
Von Siebold & Co.: Box 8.4 (Siebold)

Wade, Thomas: Box 10.10 (Wade)
Wade, Thomas Francis: Box 5.24 (Burdon), Box 5.28 (Harvey), Ch. 10.4, Ch. 10.4 notes 343, 353, Box 10.10 (Wade), Figs. 10.44, 10.45
Wadman, Mr.: Ch. 5.9
Wailes, William: Ch. 4.1
Walker, James Thomas: Ch. 3.4 note 188
Walkinshaw, William: Ch. 5.9, Ch. 5.9 note 572
Wallace, Alfred Russel: Box 8.10 (Stevens)
Wallace, William: Fig. 8.64
Waller, Edmund: Ch. 5.7
Wallich, Nathaniel: Box 4.1 (Royle), Ch. 5.6, Box 5.7 (Falconer)
Walsh & Co.: Box 8.8 (Hall)
Wáng, Ā'nào: Ch. 5.9
Wang-a-nok, see Wáng, Ā'nào
Wang, Chao-ping: Acknowledgements
Wáng, Lì: Acknowledgements
Wáng, Pàn: Box 9.5 (Ricci)
Wáng, Qìngqí: Box 9.7 note 19 (Cí Xǐ)
Wang, San-lang: Acknowledgements
Wáng, Shǒurén: Ch. 5.9, Box 5.26 (Wáng), Fig. 5.150
Wáng, "Sōngluó": Ch. 5.2, Ch. 5.3
Wáng, Shì: Box 5.20 (Hóng)
Wáng, Yángmíng, see Wáng, Shǒurén
Wáng, Yǔfēi: Acknowledgements
Wanstall, Mary: Box 3.11 note 13 (Lockhart)
Ward, Frederick Townsend: Box 5.19 note 27 (Tàipíng), Box 9.2 (Staveley)
Ward, Nathaniel Bagshaw: Ch. 2.2 note 35, Box 2.6 (Ward), Ch. 4.2, Ch. 10.1, Ch. 10.5, Box 10.1 (Moore), Fig. 2.7
Ward, Stephen Smith: Box 2.6 (Ward)
Wardle, Libby: Ch. 6.2 note 31
Warren, John Borlase: Ch. 1.1
Washington, George: Ch. 1.1 note 77
Watson, Eleanor: Box 8.11 (Harris)
Watson, Jane: Box 5.8 (Jameson)
Watt, Henry James: Ch. 11 note 61
Webb, Edward (Dent & Co.): Ch. 8.2, Ch. 8.3
Webb, Mr. (Antiquarian): Ch. 6.2, Ch. 8.1
Webster, Daniel: Box 9.4 (Bruce)
Webster, William: Ch. 10.1 note 85

Webster, Mr.: Ch. 10.1 note 60
Wéi, Chānghuī: Box 5.19 note 31 (Tàipíng)
"Wellesley": Box 5.30 (Vansittart)
Wendland, Hermann A.: Box 8.1 (Veitch)
Wermuth, van Heeckeren & Co.: Box 5.10 (Lady Mary Wood)
West of London and Pimlico Railway Company: Box 10.1 (Moore), Box 10.1 note 10 (Moore)
Wheeler, Thomas: Box 2.6 (Ward)
Whistler, James McNeill: Ch. 10.1
White, James: Ch. 2.2 note 70
Whiteman, Elizabeth: Box 1.2 (McNab)
Whyte, Thomas: Ch. 1.3
Wickliffe, Margaret: Box 7.3 (Holt)
Wickliffe, Robert: Box 7.3 (Holt)
Wigram, Money: Box 5.2 (Ripon)
Wilkinson, Jane: Box 2.4 (Dent)
Willcox, Brodie McGhie: Box 5.1 (P&O)
"William Fawcett": Box 5.1 (P&O)
Williams, Charles Wye: Box 5.1 (P&O), Box 5.33 (Bentinck), Box 5.33 note 4 (Bentinck)
Williams, Richard: Box 6.2 (Glendinning)
Williams, Samuel Wells: Box 8.6 (Hepburn)
Wills, Charles: Ch. 5.3, Ch. 5.3 note 112
Wilson, Archdale: Box 6.4 (Indian Mutiny)
Wilson, Arthur Ross: Box 5.13 (Ramsay)
Wilson, Bill: Acknowledgements, Fig. 11.3
Wilson, Ernest Henry: Box 3.12 (Introductions) note 3
Wilson, George Fergusson: Ch. 10.1
Wilson, Thomas: Box 5.10 (Lady Mary Wood), Box 5.33 (Bentinck)
Wilstach, Maria Sophia: Box 8.2 (Kaempfer)
"Winterton": Box 1.1 (Buchan)
Wishart, George: Ch. 10.1
Witte, Charlotte Elizabeth: Box 2.6 (Ward)
Wolmar, Christian: Ch. 10.1 notes 1, 2, 98
Wood, Charles: Box 5.10 (Lady Mary Wood)
Wood, Frances: Box 8.10 (Stevens)
Wood, Mary: Box 5.10 (Lady Mary Wood)
Wood, Nicholas: Fig. 2.7
Woodgate, Isabella Baines: Box 5.18 (Bonham)
Woodman, Richard: Fig. 3.44
Worshipful Society of Apothecaries, London: Fig. 2.7
Wrench, Mr.: Box 2.1 (Lindley)
Wright, H.: Ch. 2.2 note 70
Wú, Chéng'ēn: Ch. 5.10
Wu, Chia-li: Acknowledgements
Wu, Chienchang, see Wú, Jiànzhāng
Wú, Jiànzhāng: Box 5.18 (Bonham), Box 5.21 (Small Sword)

Wyndham, Charles: Box 9.6 (Wyndham)
Wyndham, George Hugh: Box 8.14 (Howard-Vyse), Ch. 9.2, Box 9.6 (Wyndham), Ch. 10.6, Fig. 9.35

Xī, Mèngméng: Ch. 5.5 note 289
Xiàodìng: Box 9.7 (Cí Xǐ)
Xiàozhé, see Empress Alute
Xiàozhēn, see Empress Cí'ān
Xīn, Hǔ, see Sing-Hoo
Xíng, Hǔ, see Sing-Hoo
Xìng, Huā, see Fortune, Robert
Xíng, Liè: Box 5.26 (Wáng)
Xīniǎn: Box 10.5 (Niǎn)
Xuē, Huàn: Ch. 10.4

Yale University Art Gallery: Fig. 7.1
Yamaguchi, Fumiko: Acknowledgements, Ch. 8.1 notes 16, 19
Yang, Hsiu-ch'ing, see Yáng, Xiùqīng
Yáng, Jiàn: Ch. 3.5 note 228
Yáng, Xiùqīng: Box 5.19 note 31 (Tàipíng), Box 5.20 (Hóng)
Yáng, Yùkē: Box 10.7 (Panthay)
Yang, Yü-k'o, see Yáng, Yùkē
Yáng, Mandarin: Ch. 10.4
Yao, Betty: Figs. 3.41, 3.47, 10.41, 10.42
Yè, Míngchēn: Box 6.1 (Arrow), Box 6.1 notes 1, 4 (Arrow), Fig. 6.2
Ye, Mingchoo, see Yè, Míngzhū
Yè, Míngzhū: Ch. 3.4
Yeh, Ming-ch'en, see Yè, Míngchēn
Yehe-Nara, see Yeho-Nala
Yeho-Nala, Huìzhēng: Box 9.7 (Cí Xǐ)
Yí: Box 10.7 (Panthay)
Yī-lǐ-bù, see Àixīnjuéluó, Yī-lǐ-bù
Yì, Concubine, see Cí Xǐ
Yìhétuán, see Boxers
Yǐn, Jiànlóng: Acknowledgements
Yīng, Jùn-shēng: Ch. 5.4 note 148
Ying, Tsun-shen, see Yīng, Jùnshēng
Yù, Xūnlíng: Fig. 9.36
Yùxiàn: Box 9.7 (Cí Xǐ)
Yüan, Tsu-te, see Yuán, Zūdé
Yuán, Yǒng: Ch. 5.3 notes 101, 105
Yuán, Zūdé: Box 5.21 (Small Sword)
Yuán (Dynasty): Ch. 5.5 note 291, Fig. 9.39

Zǎichún, see Emperor Tóngzhì

Zāitián, see Emperor Gūangxù
Zēng, Guófān: Box 5.19 (Tàipíng), Box 5.19 note 35 (Tàipíng), Box 10.5 (Niǎn)
Zēng, Guóquán: Box 5.20 (Hóng)
Zhāng, Liàngjī: Box 10.6 (Xián-Tóng)
Zhāng, Luòxíng: Box 10.5 (Niǎn)
Zhāng, Zōngyǔ: Box 10.5 (Niǎn)
Zhèng, Ā-fú, see Zhèng, Tóngchūn
Zhèng, Tóngchūn: Box 5.28 (Harvey)
Zhèng, Wànjūn: Ch. 3.2 note 147
Zhōu, Jiāxūn: Ch. 10.3
Zhōu,Yúdǐng: Ch. 10.4
Zhōu, Mandarin: Ch. 10.4
Zhū, Xī: Box 5.26 (Wáng)
Zhuàng: Box 10.7 (Panthay)
Zì, Yángmíng, see Wáng, Shǒurén
Zollmann, Philip Heinrich: Box 8.2 (Kaempfer)
Zuccarini, Joseph Gerhard: Ch. 8.1, Box 8.4 (Siebold), Box 8.4 note 36 (Siebold), Fig. 8.46
Zulueta & Co.: Ch. 5.1 notes 3, 12, Ch. 5.8 note 393
Zuǒ, Zōngtáng: Box 5.19 note 35 (Tàipíng), Box 10.8 (Huí)

Index to Plants and Animals

An attempt has been made to use the most recent nomenclature, although this was no easy matter with the constantly changing taxonomic concepts. Those species marked in bold are believed to have been introduced by Robert Fortune.

Aaron's Beard, see *Saxifraga stolonifera*

Abelia chinensis: Ch. 3.1, Box 3.12 (Introductions) , Fig. 3.9

Abelia rupestris , see *Abelia chinensis*

Abelia uniflora : Box 5.16 (Introductions)

Abies fortunei , see *Keteleeria fortunei*

Abies jezoensis sensu Fortune (1880a), see *Keteleeria fortunei*

Abies kaempferi , see *Pseudolarix amabilis*

Abies smithiana , see *Picea smithiana*

Abutilon avicennae , see *Abutilon theophrasti*

Abutilon theophrasti : Ch. 7.3

Acacia : Box 5.13 (Ramsay)

Acanthaceae: Ch. 5.9

Acer : Box 8.15 (Introductions)

Acer palmatum : Box 8.1 (Veitch), Box 8.15 note 1 (Introductions)

Aconitum autumnale , see *Aconitum fischeri*

Aconitum fischeri : Box 3.12 (Introductions)

Acorns, *see Quercus*

Actinidia chinensis : Box 3.12 (Introductions)

Actinidia deliciosa , see *Actinidia chinensis*

Adamia versicolor , see *Dichroa febrifuga*

Adenosma glutinosum : Box 3.12 (Introductions)

Adenosma grandiflorum , see *Adenosma glutinosum*

Aegle sepiaria , see *Citrus trifoliata*

Aerides jarckiana , see *Aerides leeana*

Aerides leeana : Box 3.12 note 5

Aerides quinquevulnera : Box 3.12 note 5

African marigold, *see Tagetes erecta*

African violet: Ch. 3, Box 5.6 (Champion)

Aglaia odorata : Ch. 5.9 note 500

Aix galericulata : Ch. 4.1

Akebia quinata: Box 3.12 (Introductions)

Aleurites fordii , see *Vernicia fordii*

Aleuritopteris argentea : Ch. 9.2

Allium : Ch. 8.3 note 176

Allspice, see *Chimonanthus praecox*

Almond, see *Amygdalus communis*

Amaranth, see *Amaranthus*

Amaranthus : Ch. 2.1, Ch. 2.2, Ch. 2.2 note 57, Ch. 3.5
Amaranthus blitum : Ch. 7.3
Amaranthus tricolor : Box 3.12 (Introductions)
Amelanchier racemosa , see *Exochorda racemosa*
Amherstia nobilis : Ch. 5.6
Ampelopsis tricuspidata , see *Parthenocissus tricuspidata*
Amygdalus communis : Ch. 5.6
Amygdalus pedunculata var. *multiplex* , see *Amygdalus triloba* var. *plena*
Amygdalus persica : Ch. 1.3, Ch. 2.2, Ch. 3.5, Ch. 3 notes 74, 191, 228, Box 3.4 note 8 (Balfour), Ch. 5.3, Ch. 8.3, Ch. 8.2 note 57, Ch. 9.2
***Amygdalus persica* 'Alboplena'** : Box 3.12 (Introductions)
Amygdalus persica 'Camelliaeflora' : Box 6.3 (Introductions)
Amygdalus persica 'Dianthiflora' : Box 6.3 (Introductions)
Amygdalus persica var. *nectarina* : Ch. 1.3
***Amygdalus persica* 'Sangineoplena'** : Box 3.12 (Introductions)
***Amygdalus persica* 'Shanghai Peach'** : Box 3.12 (Introductions)
Amygdalus triloba Ch. 9.1, Ch. 9.2
Amygdalus triloba* var. *plena: Box 6.3 (Introductions), Fig. 6.3A
Anacardiaceae: Ch. 3.1 note 84
Ananas comosus : Ch. 1.3, Box 6.2 (Glendinning)
Anas crecca : Ch. 3.1
Anemone hupehensis: Ch. 3.5, Box 3.12 (Introductions), Ch. 6.2, Figs. 3.50, 6.8
Anemone japonica var. *hupehensis* , see *Anemone hupehensis* [*A. japonica* not in China]
Aniseed, see *Illicium anisatum*
Annual Poa, see *Poa annua*
Ant: Box 5.25 (Patridge)
Apatura : Ch. 8.2 note 181
Aplonis panayensis : Ch. 3.3 note 163
Apple, see *Malus*
Aquebia , see *Akebia*
Aquilegia flabellata : Box 8.15 (Introductions)
Arabian jasmine, see *Jasminum sambac*
Arachis hypogaea : Ch. 3.1, Ch. 3.4, Ch. 5.4
Arachniodes standishii: Box 8.15 (Introductions), Box 10.1 (Moore)
Aralia : Ch. 8.2 note 90
Aralia elata 'Variegata' : Box 8.15 (Introductions)
Aralia papyrifera , see *Tetrapanax papyrifer(us)*
Aralia variegata , see *Aralia elata* 'Variegata'
Araliaceae: Ch. 5.9 note 517
Araucaria heterophylla : Box 8.1 (Veitch)
Arbor-vitae, see *Thuja*
Arctium lappa : Ch. 8.3 note 176
Ardisia lindleyana , see *Ardisia punctata*
Ardisia punctata : Ch. 2.1 note 9
Armeniaca mume : Box 8.4 note 46 (Siebold)

Arnica tussilaginea , see *Farfugium japonicum*
Artamus leucorhynchus (Artamidae): Ch. 3.3 note 163
Artemisia : Ch. 5.4 note 157
Arundina bambusifolia , see *Arundina graminifolia*
Arundina chinensis , see *Arundina graminifolia*
Arundina graminifolia : Box 3.12 (Introductions)
Arundina "sinensis", see *Arundina graminifolia*
Arundinaria auricoma , see *Pleioblastus viridistriatus*
Arundinaria fortunei , see *Pleioblastus fortunei*
Arundinaria fortunei var. *aurea* , see *Pleioblastus viridistriatus*
Ascaris lumbricoides : Ch. 7.3 note 77
Asclepias : Ch. 9.1 note 23
Asian glossy starling, *see Aplonis panayensis*
Aspidistra punctata : Ch. 2.1 note 8
Atlas moth, see *Attacus atlas*
Attacus atlas : Ch. 2.1, Ch. 2.2
Aucuba japonica : Ch. 8.1, Ch. 8.2, Ch. 8 notes 25, 90, Box 8.8 (Hall), Box 8.15 (Introductions), Ch. 10.1, Ch. 10.5 note 423, Figs. 8.26, 10.24
***Aucuba japonica* 'Limbata'** : Box 8.15 (Introductions)
*Aucuba japonica*forma *limbata* , see *Aucuba japonica* 'Limbata'
Autumn-flowering anemone, see *Anemone hupehensis*
Avena sativa : Ch. 1.2
Azalea , see *Rhododendron*
Azalea amoena , see *Rhododendron* x *obtusum* 'Amoenum'
Azalea bealei , see *Rhododendron simsii* 'Bealei'
Azalea "chinensis", see *Rhododendron molle*
Azalea crispiflora , see *Rhododendron indicum* 'Crispiflorum'
Azalea indica 'Bealei', see *Rhododendron simsii* 'Bealei'
Azalea indica vittata punctata , see *Rhododendron simsii* 'Vittatum Punctatum'
Azalea japonicum , see *Rhododendron japonicum*
Azalea mollis , see *Rhododendron molle*
Azalea narcissiflora , see *Rhododendron mucronatum* 'Narcissiflorum'
Azalea obtusa , see *Rhododendron* x *obtusum*
Azalea ovata , see *Rhododendron ovatum*
Azalea ramentacea , see *Rhododendron* x *obtusum* 'Album'
Azalea sinensis , see *Rhododendron molle*
Azalea squamata , see *Rhododendron farrerae*
Azalea vittata , see *Rhododendron simsii* 'Vittatum'
Azalea vittata var. *bealei* , see *Rhododendron simsii* 'Bealei'
Azalea vittato- punctata , see *Rhododendron simsii* 'Vittatum Punctatum'

Balicassiao, see *Dicrurus balicassius*
Balsam, see *Impatiens*
Bamboo: Ch. 2.2, Ch. 2.2 note 59, Ch. 3.2, Ch. 3.2 note 108, Ch. 5.3, Ch. 5.4, Ch. 5.5, Ch. 5.9, Ch. 5.10, Ch. 8.1, Ch. 8 notes 25, 26, Ch. 10.6, Figs. 5.36, 5.42, 5.125B, 5.126, 5.136, 5.137, 8.14

Bambusa edulis , see *Phyllostachys edulis*
Bambusa fortunei , see *Pleioblastus fortunei*
Bambusa fortunei var. *aurea* , see *Pleioblastus viridistriatus*
Bambusa variegata , see *Pleioblastus fortunei*
Bambusa viridistriata , see *Pleioblastus viridistriatus*
Bambusa vulgaris ' Vittata' : Box 3.12 (Introductions)
Bananas: Fig. 10.26
Bandicoot rat, see *Bandicota bengalensis*
Bandicota bengalensis : Box 5.32 (Dalhousie)
Banksian Rose, see *Rosa banksiae*
Banyan, see *Ficus*
Baphicacanthus cusia , see *Strobilanthes cusia*
Bar-bellied cuckooshrike, see *Coracina striata*
Barberry, see *Berberis*
Barbula sinensis , see *Caryopteris incana*
Barley, see *Hordeum*
Bean: Ch. 9.1, Ch. 9.2, Ch. 9.1 note 4
Bear, see *Ursus*
Beautyberry, see *Callicarpa*
Bee: Ch. 5.9, Box 5.6 (Champion)
Beech, see *Fagus sylvatica*
Beef, see *Bos primigenius*
Beetle: Ch. 5.3 note 428, Box 5.6 (Champion), Ch. 8.3, Ch. 8 notes 181, 182
Belamcanda chinensis : Ch. 9.1
Berberis : Ch. 5.4
Berberis anhweiensis : Ch. 5.4 note 148
Berberis bealei , see *Mahonia bealei*
Berberis consanguinea , see *Mahonia bealei*
Berberis fortunei , see *Mahonia fortunei*
Berberis japonica var. *bealei* , see *Mahonia bealei*
Berberis trifurca , see *Mahonia trifurca*
Berberis vulgaris : Ch. 5.4 note 148
Bignonia tomentosa , see *Paulownia tomentosa*
Bindweed, see *Calystegia*
Biophytum sensitivum : Ch. 2.1, Ch. 2.2, Ch. 3.5
Biota fortunei , see *Thuja orientalis* var. *falcata*
Bird: Ch. 1.2, Box 1.2, Box 2.3 (Reeves), Box 2.5 (Smith-Stanley), Ch. 3.1, Ch. 3.2, Ch. 3.3, Ch. 4.1, Ch. 4.2, Ch. 5.2, Ch. 5.4, Ch. 5.9, Box 8.6 (Hepburn), Ch. 10.6, Ch. 10.6 note 476, Figs. 2.4, 3.26, 3.40, 4.1, 8.57
Black-naped monarch, see *Hypothymis azurea styani*
Black-naped oriole, see *Oriolus chinensis*
Blackthorn, see *Prunus spinosa*
Bleeding heart, see *Lamprocapnos spectabilis*
Bluebeard, see *Caryopteris*
Blue jay: Ch. 3.2
Blue-tailed bee-eater, see *Merops philippinus*

Boehmeria nivea : Ch. 5.4, Ch. 5.4 note 175, Ch. 7.3, Figs. 5.43, 7.7D
Bolpopsittacus lunulatus : Fig. 3.40A
Bombyx mori : Ch. 5.10, Ch. 10.4
Bonsai: Ch. 2.2, Ch. 2.2 note 63, Ch. 3.1, Ch. 3.5, Ch. 8.2
Bos primigenius : Box 1.2, Ch. 3.2 note 119, Box 6.4 (Indian Mutiny), Ch. 9.2, Box 9.9 (Hope Grant), Ch. 10.1, Ch. 10.4
Boston Ivy, see *Parthenocissus tricuspidata*
Bougainvillea : Box 5.11 (Saharanpur), Fig. 5.73
Box, see *Buxus*
Bramble, see *Rubus*
Brassica chinensis , see *Brassica rapa* var. *chinensis*
Brassica rapa : Ch. 8.3 note 176
Brassica rapa subsp. *chinensis* , see *Brassica rapa* var. *chinensis*
Brassica rapa var. *chinensis* : Box 3.12 (Introductions), Ch. 7.3, Ch. 8.3
Brassica rapa* var. *glabra [Chinese cabbage, bai cai]: Ch. 3.5, Box 3.12 (Introductions), Ch. 9.2
Brassica "sinensis", see *Brassica rapa* var. *chinensis*
Broussonetia papyrifera : Ch. 5.10, Ch. 5.10 note 631, Fig. 5.161
Brown-breasted kingfisher, see *Halcyon smyrnensis fusca*
Buckthorn, see *Rhamnus*
Buckwheat, see *Fagopyrum*
Buddha's Hand, see *Citrus medica* 'Fingered'
Buddhist Pine, see *Podocarpus macrophyllus*
Buddlea, see *Buddleja*
Buddleja lindleyana: Ch. 3.2, Box 3.12 (Introductions), Ch. 5.2 note 42, Fig. 3.22
Bug: Box 5.25 (Patridge)
Bullock, see *Bos primigenius*
Butastur indicus : Ch. 3.3 note 163, Fig. 3.40H
Buttercup witch hazel, see *Corylopsis pauciflora*
Butterfly: Ch. 8.3, Ch. 8.3 note 181
Butterfly bush, see *Buddleja*
Buxus microphylla : Box 8.15 (Introductions)
Byrsopteris standishii , see *Arachniodes standishii*

Cabbage Oil-plant, see *Brassica rapa* var. *chinensis*
Calidris ruficollis : Fig. 3.40G
Callicarpa dichotoma : Box 6.3 (Introductions)
Callicarpa rubella : Ch. 2.1 note 8
Callista secunda , see *Dendrobium secundum*
Callistephus chinensis : Ch. 9.1 note 31
Callistephus sinensis , see *Callistephus chinensis*
Calystegia hederacea 'Flore Pleno', see *Calystegia pubescens* 'Flore Pleno'
***Calystegia pubescens* 'Flore Pleno'**: Box 3.12 (Introductions)
Camel, see *Camelus bactrianus*
Camellia : Ch. 2.2, Ch. 2.2 note 52, Box 2.3 (Reeves), Ch. 3.1, Ch. 3.4, Ch. 5.3, Ch. 5.4, Ch. 8.3, Ch. 8 notes 25, 90, Ch. 10.6, Fig. 8.50

Camellia euryoides : Ch. 2.1 note 8
Camellia japonica : Ch. 3.2
***Camellia japonica* 'Anemoniflora'** : Box 5.16 (Introductions)
***Camellia japonica* 'Cup of Beauty'** : Box 6.3 (Introductions)
***Camellia japonica* 'Fortune's Yellow'** : Ch. 3.5, Box 3.12 (Introductions)
Camellia japonica hexangularis , see *Camellia japonica* 'Myrtifolia'
Camellia japonica 'Myrtifolia' : Box 3.12 (Introductions)
***Camellia japonica* 'Princess Frederick William'** : Box 6.3 (Introductions)
Camellia reticulata : Ch. 2.1 note 9
Camellia sinensis : Ch. 1.2, Ch. 2.2, Ch. 2.2 note 46, Box 2.2 (Opium), Box 2.2 notes 2, 3, 4, Box 2.3 (Reeves), Box 2.6 (Ward), Ch. 3.1, Ch. 3.2, Ch. 3.4, Ch. 3.5, Ch. 3.4 note 195, Box 3.6 (Sinclair), Box 3.7 (Morrison), Box 4.1 (Royle), Ch. 5.1, Ch. 5.3, Ch. 5.4, Ch. 5.5, Ch. 5.6, Ch. 5.7, Ch. 5.8, Ch. 5.9, Ch. 5.10, Ch. 5 notes 68, 79, 127, 327, Box 5.7 (Falconer), Box 5.8 (Jameson), Box 5.10 (Lady Mary Wood), Box 5.13 (Ramsay), Box 5.31 (Beadon), Ch. 7.1, Ch. 7.2, Box 7.1 (Smith), Ch. 8.2, Ch. 8.3, Box 8.4 (Siebold), Box 8.13 (Aspinall), Box 9.9 (Hope Grant), Ch. 10.1, Ch. 10.4, Figs. 2.3, 5.20, 5.46, 5.51, 5.60, 5.63, 5.64, 5.65, 5.66, 5.74, 5.76, 5.82, 5.84, 5.85, 5.87, 5.137, 7.5, 7.6, 7.7, 8.42
***Camellia sinensis* 'Variegata'** : Ch. 8.15 (Introductions)
Camelus bactrianus : Ch. 9.2
Camelus dromedarius : Box 5.1
Campanula grandiflora , see *Platycodon grandiflorus*
Campanula grandiflora semi-double white, see *Platycodon grandiflorus* 'Albus Plenus'
Campanula nobilis , see *Campanula punctata*
Campanula punctata : Box 3.12 (Introductions), Ch. 9.2
Camphor tree, see *Cinnamomum camphora*
Campsis grandiflora : Ch. 5.10 note 628
Canarium album : Ch. 3.4 note 191
Canis lupus familiaris : Ch. 3.1, Ch. 3.1 note 27, Ch. 5.7, Ch. 5.9, Box 5.10 (Lady Mary Wood), Box 5.27 (Robertson), Box 8.2 (Kaempfer), Box 9.7 (Cí Xǐ), Box 9.8 (Bowlby), Box 9.9 (Hope Grant), Ch. 10.4, Ch. 10.6
Cannabis gigantea , see *Cannabis sativa*
Cannabis sativa : Ch. 3.5, Box 3.12 (Introductions)
Cape Jasmine, see *Gardenia jasminoides* var. *fortuneana*
Caper, see *Capparis spinosa*
Capparis spinosa : Ch. 5.9 note 502
Capra aegagrus hircus (goat): Ch. 3.1
Caprifolium fragrantissimum , see *Lonicera fragrantissima*
Carabidae: Ch. 5.10, Ch. 8.3 note 181
Carp, see *Cyprinus carpio*
Carpinus putoensis : Ch. 3.2, Fig. 3.35
Carrot, see *Daucus carota*
Carthamus tinctorius : Ch. 5.10
Caryopteris incana: Box 3.12 (Introductions)
Caryopteris mastacanthus , see *Caryopteris incana*
Castanea : Ch. 5.6, Ch. 5.8, Ch. 5.9, Ch. 5 notes 348, 548, Box 6.3 (Introductions), Ch. 7, Ch. 8.2 note 57, Figs. 5.85, 5.86
Castanea crenata : Ch. 5 note 348

Castanea japonica , see *Castanea crenata*
Castanea mollissima: Box 6.3 (Introductions), Fig. 5.124
Castanea seguinii: Box 6.3 (Introductions)
Castanopsis fargesii : Ch. 5.3
Castanopsis sclerophylla: Ch. 5.4, Box 5.16 (Introductions)
Cat, see *Felis catus*
Cattle, see *Bos primigenius*
Cattleya : Box 2.1 note 7 (Lindley)
Cedar, see *Cedrus*
Cedrus deodara : Ch. 5.6, Fig. 8.32
Celosia : Ch. 2.1, Ch. 2.2, Ch. 2 note 58
Celosia cristata : Ch. 9.1 note 31
Celtis orientalis , see *Trema orientalis*
Centipede: Box 5.25 (Patridge)
Cephalotaxus : Ch. 5.11 note 690, Fig. 8.23
Cephalotaxus fortunei: Box 3.12 (Introductions), Box 5.16 (Introductions)
Cerambycidae: Ch. 8.3 note 181
***Cerasus glandulosa* 'Alboplena'**: Box 3.12 (Introductions), Ch. 5.3
Cerasus serrulata : Box 2.3 (Reeves)
Cerasus x *yedoensis* : Ch. 8.2, Fig. 8.34
Ceratostigma plumbaginoides : Box 3.12 (Introductions)
Cercidiphyllum japonicum : Box 8.4 note 46
Cercis : Ch. 5.9
Cercis chinensis : Box 3.12 (Introductions), Ch. 9.2
Chaenomeles japonica : Ch. 8.3
Chain fern, see *Woodwardia*
Chamaecyparis : Ch. 8.1, Ch. 8.2 note 90
Chamaecyparis funebris , see *Cupressus funebris*
Chamaecyparis obtusa : Box 8.15 (Introductions)
***Chamaecyparis obtusa* 'Argentea'**: Box 8.15 (Introductions)
Chamaecyparis pisifera : Box 8.4 note 61 (Siebold), Box 8.15 (Introductions)
***Chamaecyparis pisifera* 'Argentea'**: Box 8.15 (Introductions)
Chamaecyparis pisifera 'Aurea': Box 8.15 (Introductions)
***Chamaecyparis pisifera* 'Filifera'**: Box 8.15 (Introductions)
***Chamaecyparis pisifera* 'Plumosa Argentea'**: Box 8.15 (Introductions)
***Chamaecyparis pisifera* 'Plumosa Aurea'**: Box 8.15 (Introductions)
Chamaerops fortunei , see *Trachycarpus fortunei*
Championella : Box 5.6 note 13 (Champion)
Championia : Box 5.6 (Champion)
Chenopodium : Ch. 9.1 note 23
Chestnut, see *Castanea*
Chicken, see *Gallus gallus domesticus*
Chimonanthus fragrans , see *Chimonanthus praecox*
Chimonanthus fragrans var. *grandiflorus* , see *Chimonanthus praecox*
Chimonanthus praecox : Ch. 3.1 note 74, Ch. 5.2 note 42

Chimonanthus praecox var. *grandiflorus* , see *Chimonanthus praecox*
China Aster, see *Callistephus chinensis*
Chinese Arbor-vitae, see *Thuja orientalis*
Chinese Ash, see *Fraxinus chinensis*
Chinese Bellflower, see *Platycodon grandiflorus*
Chinese Cabbage, see *Brassica rapa* var. *glabra*
Chinese Cedar, see *Cryptomeria japonica*
Chinese Date, see *Ziziphus jujuba*
Chinese Fringetree, see *Chionanthus retusus*
Chinese Gooseberry, see *Actinidia chinensis*
Chinese Green Indigo, see *Rhamnus utilis*
Chinese Indigo, see *Indigofera decora*
Chinese Jute, see *Abutilon theophrasti*
Chinese Lacquertree, see *Toxicodendron vernicifluum*
Chinese Nutmeg Yew, see *Torreya grandis*
Chinese Olive, see *Canarium album*
Chinese Plum yew, see *Cephalotaxus fortunei*
Chinese Rose, see *Hibiscus rosa- sinensis*
Chinese Sheep, see *Ovis aries*
Chinese Torreya, see *Torreya grandis*
Chinese Tree privet, see *Ligustrum lucidum*
Chinese Trumpet Creeper, see *Campsis grandiflora*
Chinese Varnish Tree, see *Toxicodendron vernicifluum*
Chinese Weeping Cypress, see *Cupressus funebris*
Chinese Wisteria, see *Wisteria sinensis*
Chionanthus chinensis , see *Chionanthus retusus*
Chionanthus retusus : Box 5.16 (Introductions)
Chirita sinensis , see *Primulina dryas*
Chrysanthemum : Ch. 2.2, Ch. 2.1 note 8, Box 2.3 (Reeves), Ch. 3.1, Ch. 8.2, Ch. 8.3, Ch. 9.1 note 31, Box 9.7 (Cí Xǐ)
Chrysanthemum indicum : Box 8.15 (Introductions)
***Chrysanthemum indicum* 'Chinese Minimum'** : Box 3.12 (Introductions)
***Chrysanthemum indicum* 'Chusan Daisy'** : Box 3.12 (Introductions)
Chrysanthemum morifolium 'Bronze Dragon' : Box 8.15 (Introductions)
Chrysanthemum morifolium 'Yellow Dragon' : Box 8.15 (Introductions)
Chrysocolaptes (lucidus) haematribon : Ch. 3.3 note 163, Fig. 3.40C
Chrysolarix amabilis , see *Pseudolarix amabilis*
Chusan Palm, see *Trachycarpus fortunei*
Cinchona : Box 4.1 (Royle), Box 5.7 (Falconer)
Cinnamomum camphora : Ch. 3.2 note 146, Ch. 5.4, Ch. 5.8, Ch. 5.10 note 631, Ch. 7.3, Figs. 5.121, 7.7C
Citrus x *aurantiifolia* : Ch. 9.1
Citrus x *aurantium* : Ch. 3 notes 84, 192, Ch. 5.4, Ch. 5.9 note 500, Box 8.8 (Hall), Ch. 8.2 note 57, Ch. 9.1
Citrus decumana , see *Citrus maxima*
Citrus japonica: Ch. 2.1, Ch. 2.2, Ch. 2 note 53, Ch. 3.5, Box 3.12 (Introductions), Ch. 10.1, Fig. 3.13
Citrus x *limon* : Ch. 3.4 note 191

Citrus margarita , see *Citrus japonica*
Citrus maxima : Ch. 3 notes 84, 191
***Citrus medica* ' Fingered'** : Ch. 2.1, Ch. 2.2, Ch. 3.4, Ch. 3.5, Ch. 3.4 note 190, Box 3.12 (Introductions), Fig. 3.46
Citrus medica var. *sarcodactylis* , see *Citrus medica* ' Fingered'
Citrus reticulata : Ch. 2.1, Ch. 2.2, Ch. 3.5, Box 3.12 (Introductions)
Citrus sarcodactylis , see *Citrus medica* ' Fingered'
Citrus x *sinensis* , see *Citrus* x *aurantium*
Citrus trifoliata: Box 5.16 (Introductions) , Fig. 5.93, 5.94
Clausena lansium : Ch. 3 notes 84, 191
Clematis : Ch. 3.1, Ch. 5.8, Ch. 8.3
Clematis coerulea , see *Clematis patens* ' John Gould Veitch'
Clematis florida var. *lanuginosa* , see *Clematis lanuginosa*
Clematis fortunei , see *Clematis patens* ' Fortunei'
Clematis fortunei coerulea , see *Clematis patens* ' John Gould Veitch'
Clematis x *jackmanii* : Box 10.1 (Moore)
Clematis lanuginosa: Box 5.16 (Introductions), Box 10.1 (Moore), Fig. 5.95
***Clematis patens* ' Fortunei'** : Box 8.15 (Introductions)
***Clematis patens* ' John Gould Veitch'** : Box 8.15 (Introductions)
***Clematis patens* ' Standishii'** : Box 8.15 (Introductions)
Clematis standishii , see *Clematis patens* ' Standishii'
Clematis viticella : Box 10.1 (Moore)
Clematis williamsii : Box 8.15 note 20 (Introductions)
Clerodendron , see *Clerodendrum*
Clerodendrum : Ch. 5.4
Clerodendrum bungei : Box 5.16 (Introductions)
Clerodendrum foetidum , see *Clerodendrum bungei*
Cleyera : Ch. 8 notes 25, 90
Cleyera fortunei , see *Cleyera japonica* ' Fortunei'
***Cleyera japonica* ' Fortunei'** : Box 8.15 (Introductions)
Cleyera japonica ' Tricolor' , see *Cleyera japonica* ' Fortunei'
Clover, see *Trifolium*
Clubmoss, see *Lycopodiella*
Coccus pela , see *Ericerus pela*
Cockroach: Box 5.25 (Patridge), Box 8.12 (Heusken)
Cockscomb, see *Celosia cristata*
Coelogyne chinensis , see *Pholidota chinensis*
Coelogyne fimbriata : Ch. 2.1 note 8
Coleto, see *Sarcops calvus*
Columba palumbus : Ch. 1.2
Columbine, see *Aquilegia*
Convallaria keiskei : Box 8.15 note 24 (Introductions)
Convallaria majalis ' Variegata' : Box 8.15 (Introductions), Box 8.15 note 24 (Introductions)
Convallaria variegata , see *Convallaria majalis* ' Variegata'
Cookia punctata , see *Clausena lansium*
Coracina striata : Ch. 3.3 note 163

Cormorant, see *Phalacrocorax carbo*
Corn, see *Zea mays*
Cornus officinalis : Box 8.4 note 46
Coronilla : Ch. 3.1
Corvus frugilegus : Ch. 1.2
Corylopsis pauciflora : Box 8.4 note 46, Box 8.15 (Introductions)
Corylopsis spicata : Box 8.4 note 46, Box 8.15 (Introductions)
Cotinus coggygria : Fig. 9.49
Cotton, see *Gossypium*
Cow, see *Bos primigenius*
Cowslip, see *Primula sieboldii*
Crape myrtle, see *Lagerstroemia*
Crataegus indica , see *Rhaphiolepis indica*
Crataegus monogyna : Ch. 1.2 note 124
Crepidiastrum denticulatum : Box 3.12 (Introductions)
Crown vetch, see *Coronilla*
Crustacean: Box 3.8 note 24 (Cuming)
Cryptomeria fortunei , see *Cryptomeria japonica*
Cryptomeria japonica: Ch. 3.1 note 69, Box 3.12 (Introductions), Ch. 5.3, Ch. 5.4, Ch. 5.8, Ch. 5.9, Ch. 5.9 note 548, Ch. 7.3, Ch. 8.2 note 67, Fig. 3.51
***Cryptomeria japonica* 'Nana'** : Box 3.12 (Introductions)
Cuckoo, see *Cuculus canorus*
Cuculus canorus : Ch. 10.5
Cucumber, see *Cucumis sativus*
Cucumis sativus : Ch. 8.3 note 176
Cumquat, see *Citrus japonica*
Cunninghamia lanceolata : Ch. 3.4, Ch. 3.2 note 146, Ch. 5.2, Ch. 5.4, Ch. 5.9, Ch. 5.9 note 548, Fig. 3.48
Cupressus : Ch. 3.2 note 146, Ch. 9.2 note 89
Cupressus funebris : Ch. 5.2, Ch. 5.4, Ch. 5.8, Ch. 5.9, Ch. 5.4 note 267, Box 5.5 (Beale), Box 5.16 (Introductions), Ch. 7.3, Fig. 5.15
Cupressus torulosa : Ch. 5.6
Cyclobalanopsis myrsinifolia: Box 6.3 (Introductions)
Cypress, see *Cupressus*
Cyprinus carpio : Ch. 8.3

Dahlia : Ch. 5.6
Damaster blaptoides : Ch. 8.3, Ch. 8.3 note 181
Damaster blaptoides fortunei : Ch. 8.3 note 139
Damaster blaptoides oxuroides : Ch. 8.3 note 139
Daphne : Ch. 5.2 note 42
Daphne fortunei , see *Daphne genkwa*
Daphne genkwa: Ch. 3.2, Box 3.12 (Introductions), Ch. 5.9, Fig. 3.21
Daphne genkwa var. *fortunei* , see *Daphne genkwa*
Daphne odora 'Aureomarginata' : Box 8.15 (Introductions)
Daphne variegata , see *Daphne odora* 'Aureomarginata'

Date palm, see *Phoenix dactylifera*
Daucus carota : Ch. 8.3 note 176, Ch. 9.2
David's Deer, see *Elaphurus davidianus*
Deer: Ch. 5.3, Ch. 5.7
Dendranthema , see *Chrysanthemum*
Dendranthema indica , see *Chrysanthemum indicum*
Dendranthema indica ' Chinese Minimum ', see *Chrysanthemum indicum* ' Chinese Minimum '
Dendranthema indica ' Chusan Daisy ', see *Chrysanthemum indicum* ' Chusan Daisy '
Dendranthema morifolia ' Bronze Dragon ', see *Chrysanthemum morifolium* ' Bronze Dragon '
Dendranthema morifolia ' Yellow Dragon ', see *Chrysanthemum morifolium* ' Yellow Dragon '
Dendrobium secundum : Box 3.12 (Introductions)
Dermorhytis fortunei , see *Abirus fortunei*
Deutzia crenata , see *Deutzia scabra*
Deutzia crenata var. *flore pleno* , see *Deutzia scabra* ' Plena '
Deutzia fortunei , see *Deutzia scabra* ' Plena '
Deutzia gracilis : Box 8.4 note 46
Deutzia scabra : Ch. 8.3, Box 8.4 note 46
***Deutzia scabra* ' Plena '** : Box 8.15 (Introductions)
Dianthus : Ch. 5.6
Dicentra spectabilis , see *Lamprocapnos spectabilis*
Dichroa febrifuga : Box 3.12 (Introductions)
Dichroa versicolor , see *Dichroa febrifuga*
Diclytra spectabilis , see *Lamprocapnos spectabilis*
Dicrurus balicassius : Ch. 3.3 note 163
Didymocarpus sinensis , see *Primulina dryas*
Dielytra spectabilis , see *Lamprocapnos spectabilis*
Diervilla florida , see *Weigela florida*
Digitalis : Box 2.1 note 7 (Lindley)
Digitalis sinensis , see *Adenosma glutinosum*
Dimocarpus lichi , see *Litchi chinensis*
Dimocarpus longan : Ch. 3 notes 84, 191
Dioscorea : Ch. 8.3 note 176
Dioscorea batatas : Box 3.12 (Introductions)
Diospyros kaki : Ch. 8.2 note 57
Diptera: Ch. 8.3 note 182
Dog, see *Canis lupus familiaris*
Dollarbird, see *Eurystomus orientalis*
Donkey, see *Equus africanus asinus*
Downy bamboo, see *Phyllostachys edulis*
Dromedary, see *Camelus dromedarius*
Dryandra cordata , see *Vernicia fordii*
Drynaria fortunei , see *Neolepisorus fortunei*
Dryocopus javensis : Fig. 3.40D
Dryopteris filix- mas : Box 2.6 (Ward)
Duck: Ch. 3.1, Ch. 9.2

Ducula aenea : Ch. 3.3 note 163
Ducula carola : Ch. 3.3 note 163, Fig. 3.36H
Dynastes dichotoma : Ch. 8.3 note 181
Dynastine beetle, see *Dynastes dichotoma*

Edgeworthia chrysantha: Ch. 3.1 note 74, Box 3.12 (Introductions), Ch. 5.2 note 42, Fig. 3.52
Edgeworthia papyrifera , see *Edgeworthia chrysantha*
Egg plant, see *Solanum melongena*
Elaeagnus : Ch. 8 notes 25, 90
***Elaeagnus pungens* ' Variegata'** : Box 8.15 (Introductions)
Elaeagnus variegata , see *Elaeagnus pungens* ' Variegata'
Elaeodendron fortunei , see *Euonymus fortunei*
Elaphurus davidianus : Box 3.10 note 26 (Alcock)
Elder, see *Sambucus nigra*
Elephant, see *Elephas maximus*
Elephas maximus : Box 9.3 (Gordon), Ch. 10.4
Elm, see *Ulmus*
Enkianthus : Ch. 2.2
Enkianthus quinqueflorus : Ch. 3.1, Fig. 3.4
Enkianthus reticulatus , see *Enkianthus quinqueflorus*
Equus africanus asinus : Ch. 9.2
Equus caballus : Ch. 5.1, Box 5.26 (Wáng), Ch. 8.2, Ch. 8.2 note 64, Box 8.12 (Heusken), Ch. 9.2, Box 9.3 (Gordon), Box 9.9 (Hope Grant), Ch. 10.1, Ch. 10.2, Ch. 10.3, Ch. 10.4, Box 10.2 (Namamugi), Box 10.5 (Niǎn)
Ericerus pela : Ch. 7.3, Ch. 10.4, Ch. 10.4 note 276
Eriobotrya japonica : Ch. 8.2 note 57
Eucalypt, see *Eucalyptus*
Eucalyptus : Box 5.13 (Ramsay)
Eugenia : Ch. 5.4
Euonymus : Ch. 8.2 note 90
Euonymus fortunei var. *radicans* : Box 8.15 (Introductions)
Euonymus radicans , see *Euonymus fortunei* var. *radicans*
Eupatorium fortunei : Box 5.16 (Introductions)
Eurya japonica forma *variegata* , see *Cleyera japonica* ' Fortunei'
Eurya latifolia var. *variegata* , see *Cleyera japonica* ' Fortunei'
Euryale ferox : Ch. 5.10, Fig. 5.152
Eurystomus orientalis : Fig. 3.40E
Evergreen chestnut, see *Castanopsis fargesii*
Evergreen oak: Box 3.10 note 25 (Alcock), Fig. 8.50
Exochorda grandiflora , see *Exochorda racemosa*
Exochorda racemosa: Ch. 5.3, Ch. 5.8, Box 5.16, Fig. 5.35
Exocoetidae: Box 8.6 (Hepburn)

Fagopyrum : Ch. 5.4, Ch. 5.9
Fagopyrum esculentum : Ch. 7.3
Fagus sylvatica : Ch. 1.2 note 124

False holly, see *Osmanthus heterophyllus*
Farfugium grande , see *Farfugium japonicum*
Farfugium japonicum : Ch. 5.11, Box 6.3 (Introductions)
Farfugium kaempferi , see *Farfugium japonicum*
Farfugium tussilagineum , see *Farfugium japonicum*
Felis catus : Box 5.27 (Robertson)
Ferns: Box 9.9 (Hope Grant), Box 10.1 (Moore), Fig. 10.10
Ficus : Ch. 3.1, Ch. 3.4, Box 3.4 note 8, Ch. 5.9 note 575
Ficus benghalensis : Ch. 5.6, Ch. 5.6 note 307, Fig. 5.68
Ficus benjamina : Ch. 3.4 note 192
Ficus carica : Box 3.4 note 7 (Balfour)
Ficus nitida , see *Ficus benjamina*
Fig, see *Ficus*
Filipendula multijuga : Box 8.15 (Introductions)
Fingered citron, see *Citrus medica* 'Fingered'
Firmiana simplex : Ch. 7.3
Fish: Ch. 1.2, Box 2.3 (Reeves), Ch. 5.10
Flax, see *Linum usitatissimum*
Fleas: Box 5.25 (Patridge)
Flowering Almond, see *Amygdalus triloba*
Flowering Cherry: Ch. 8.3
Flowering Peach, see *Amygdalus persica*
Fly: Ch. 5.1
Flying fish, see Exocoetidae
Forsythia fortunei , see *Forsythia suspensa*
Forsythia suspensa: Ch. 9.2, Ch. 9.1 note 128, Fig. 9.34
Forsythia suspensa var. *fortunei* , see *Forsythia suspensa*
Forsythia viridissima: Ch. 3.5, Ch. 3.1 note 74, Box 3.12 (Introductions), Ch. 5.8, Ch. 5.9, Ch. 5.2 note 42, Ch. 9.2, Ch. 9.1 note 128, Ch. 10.1
Fortunaea chinensis , see *Platycarya strobilacea*
Fortunella , see *Citrus japonica*
Fortunella japonica , see *Citrus japonica*
Fortunella margarita , see *Citrus japonica*
Fow-Show, see *Citrus medica* 'Fingered'
Foxglove, see *Digitalis*
Fragaria x *ananassa* : Ch. 8.3 note 186
Fraxinus chinensis : Box 3.12 (Introductions), Ch. 5.8, Ch. 7.3
French bean, see *Phaseolus vulgaris*
Frog, see *Rana*
Fuchsia : Ch. 5.6
Fuh-show, see *Citrus medica* 'Fingered'
Fuji, see *Wisteria floribunda*
Fumaria spectabilis , see *Lamprocapnos spectabilis*
Funkia fortunei , see *Hosta sieboldiana* var. *fortunei*

Galium aparine : Ch. 1.2 note 131
Gallus gallus domesticus : Ch. 9.2
Gardenia : Ch. 3.2, Ch. 5.2 note 42
Gardenia florida var. *fortuneana* , see *Gardenia jasminoides* var. *fortuneana*
Gardenia fortunei , see *Gardenia jasminoides* var. *fortuneana*
Gardenia jasminoides* var. *fortuneana: Ch. 3.1 note 74, Box 3.12 (Introductions), Ch. 5.9 note 500, Box 5.5 (Beale), Fig. 3.30
Gentian, see *Gentiana*
Gentiana : Ch. 5.11
Gentiana fortunei , see *Gentiana scabra*
Gentiana scabra: Box 5.16 (Introductions), Fig. 5.96
Gentiana scabra var. *fortunei* , see *Gentiana scabra*
Gentiana Section Pneumonanthe: Fig. 5.96
Gerardia glutinosa , see *Adenosma glutinosum*
Ghent Azaleas: Box 8.4 (Siebold)
Giant millet, see *Sorghum bicolor*
Giant redwood, see *Sequoiadendron giganteum*
Giant waterlily, see *Victoria*
Ginger, see *Zingiber officinale*
Ginkgo biloba : Ch. 5.10 note 632, Ch. 7.3, Ch. 8.2, Ch. 8.3, Ch. 8 notes 57, 90, Ch. 9.2, Figs. 7.7B, 8.58, 9.45
Gleditsia sinensis : Ch. 5.8, Ch. 7.3, Fig. 7.7E
Glycine max : Ch. 5.4, Ch. 8.3 note 176
Glycine sinensis , see *Wisteria sinensis*
Goat, see *Capra aegagrus hircus*
Gobo, see *Arctium lappa*
Golden Larch, see *Pseudolarix amabilis*
Golden-rayed lily, see *Lilium auratum*
Goldfussia cusia , see *Strobilanthes cusia*
Gordonia axillaris , see *Polyspora axillaris*
Gossypium : Box 2.2 note 4 (Opium), Ch. 3.2, Box 3.6 (Sinclair), Ch. 5.10, Ch. 8.3 note 176, Box 8.13 (Aspinall), Ch. 9.2, Ch. 10.1
Gourd: Box 9.7 (Cí Xǐ)
Grape-vine, see *Vitis*
Grass: Ch. 3.2, Ch. 5.9 note 517, Ch. 8.3, Ch. 8.2 note 64
Grass-cloth, see *Boehmeria nivea*
Greater flameback, see *Chrysocolaptes (lucidus) haematribon*
Green imperial-pigeon, see *Ducula aenea*
Grey-faced buzzard, see *Butastur indicus*
Grey-necked imperial-pigeon, see *Ducula carola*
Ground beetle, see Carabidae
Guava, see *Psidium guajava*

Habenaria susannae , see *Pecteilis susannae*
Halcyon smyrnensis fusca : Ch. 3.3 note 163, Fig. 3.40B
Hamamelis japonica : Box 8.4 note 46

Haong Yune, see *Citrus medica* 'Fingered'
Hawthorn, see *Crataegus monogyna*
Heavenly blue bird: Ch. 3.2
Helianthus annuus : Ch. 10.4
Helix japonica : Ch. 8.3 note 181
Helix quaesito : Ch. 8.3 note 181
Hemp, see *Cannabis sativa*
Hemp-palm, see *Trachycarpus fortunei*
Hevea brasiliensis : Box 2.4 (Ward)
Hibiscus rosa- sinensis : Ch. 3.1
Hibiscus syriacus : Ch. 3.1 note 74, Box 3.12 (Introductions)
Hibiscus syriacus var. *chinensis* , see *Hibiscus syriacus*
Hinoki cypress, see *Chamaecyparis obtusa*
Holly, see *Ilex*
Honeysuckle, see *Lonicera*
Hordeum : Ch. 1.2, Ch. 5.4, Ch. 5.9, Ch. 8.3
Hornbeam, see *Carpinus*
Horned holly, see *Ilex cornuta*
Horse, see *Equus caballus*
Hosta fortunei , see *Hosta sieboldiana* var. *fortunei*
Hosta sieboldiana var. *fortunei* : Box 8.15 (Introductions)
Hoya carnosa : Ch. 8.1, Ch. 8.1 note 17
Hoya motoskei , see *Hoya carnosa*
Hyalonema : Ch. 10.1
Hydrangea : Ch. 3.4, Box 3.12 (Introductions), Ch. 5.2 note 43, Fig. 8.12
Hydrangea macrophylla var. *macrophylla* : Box 8.4 (Siebold), Box 8.4 note 14
Hydrangea otaksa , see *Hydrangea macrophylla* var. *macrophylla*
Hydrangea petiolaris : Box 8.4 note 46 (Siebold)
Hyoscyamus niger : Box 5.8 (Jameson)
Hypothymis azurea styani : Ch. 3.3 note 163
Hystrix indica : Box 5.32 (Dalhousie)

Ilex : Box 10.1 (Moore)
Ilex cornuta: Box 5.16 (Introductions), Box 6.2 (Glendinning), Box 6.3 (Introductions), Box 6.3 note 7, Fig. 3.53
Ilex cornuta var. *fortunei* , see *Ilex cornuta*
Ilex fortunei , see *Ilex cornuta*
Ilex heterophylla , see *Osmanthus heterophyllus*
Ilex latifolia : Box 5.16 (Introductions)
Ilex reevesiana Fortune 1851, see *Ilex cornuta* Lindley & Paxton 1850 [e.g. Flora of China, 11: 389; www.theplantlist.org] or *Skimmia reevesiana* (Fortune 1851) Fortune 1852 [see Flora of China, 11: 78]
Illicium : Ch. 2.1, Ch. 2.2, Ch. 2.2 note 58
Illicium anisatum : Ch. 8.2, Ch. 8.2 note 90
Illicium verum : Ch. 3.5
Impatiens : Ch. 9.1 note 31
Indian Hawthorn, see *Rhaphiolepis indica*

Indigo: Box 2.4 note 5 (Dent)
Indigofera decora: Ch. 3.1 note 74, Box 3.12 (Introductions)
Insects: Box 1.2, Ch. 5.3, Ch. 5.8, Ch. 5.10, Ch. 5.8 note 428, Box 5.6 (Champion), Ch. 7.3 note 59, Ch. 8.3, Box 8.10 (Stevens), Figs. 5.37, 8.53
Ipomoea batatas : Ch. 3.1, Ch. 3.4, Ch. 5.2, Ch. 5.8, Ch. 5.9, Ch. 8.3 note 176
Iris : Ch. 8.3
Isatis indigotica , see *Isatis tinctoria*
Isatis tinctoria : Ch. 3.2, Ch. 3.5, Box 3.12 (Introductions), Ch. 5.10
Isatis tinctoria var. *indigotica* , see *Isatis tinctoria*
Ixora : Ch. 3.4
Ixora chinensis : Ch. 3.1, Fig. 3.2
Ixora coccinea sensu Fortune, see *Ixora chinensis*

January jasmine, see *Lonicera fragrantissima*
Japan Jay: Ch. 4.2
Japanese Angelica tree, see *Aralia elata*
Japanese Cedar, see *Cryptomeria japonica*
Japanese Honeysuckle, see *Lonicera japonica*
Japanese Laurel, see *Aucuba japonica*
Japanese Maple, see *Acer palmatum*
Japanese Mock-orange, see *Pittosporum tobira*
Japanese Privet, see *Ligustrum japonicum*
Japanese Toad lily, see *Tricyrtis hirta*
Jasminum lanceolaria : Ch. 5.9 note 500
Jasminum mesnyi : Ch. 10.4, Fig. 10.48
Jasminum nudiflorum: Ch. 3.5, Box 3.12 (Introductions), Ch. 9.1, Ch. 9.2, Fig. 3.54
Jasminum paniculatum , see *Jasminum lanceolaria*
Jasminum sambac : Ch. 3.1, Ch. 5.4, Ch. 5.9 note 500, Ch. 9.1
Javanese peafowl, see *Pavo muticus imperator*
Jonesia asoca , see *Saraca asoca*
Jujube, see *Ziziphus jujuba*
Juniper, see *Juniperus*
Juniperus : Ch. 5.8, Ch. 5.4 note 157, Ch. 8.1 notes 25, 90, Ch. 9.2, Figs. 8.50, 9.29
Juniperus chinensis : Ch. 3.1 note 74
Juniperus chinensis var. *chinensis* : Box 5.16 (Introductions)
Juniperus chinensis var. *fortunei* , see *Juniperus chinensis* var. *chinensis*
Juniperus fortunei , see *Juniperus chinensis* var. *chinensis*
Juniperus sphaerica , see *Juniperus chinensis* var. *chinensis*

Kerria japonica : Ch. 8.3
***Kerria japonica* ' Aureovariegata'** : Box 8.15 (Introductions)
Kerria japonica var. *aureovariegata* , see *Kerria japonica* ' Aureovariegata'
Kerria japonica forma *aureovariegata* , see *Kerria japonica* ' Aureovariegata'
Keteleeria fortunei : Box 5.16 (Introductions)
Kingfisher: Fig. 3.36B

Kiwi, see *Actinidia chinensis*
Kumara, see *Ipomoea batatas*
Kumquat, see *Citrus japonica*
Kwei, see *Osmanthus fragrans*

Lacebark pine, see *Pinus bungeana*
Lactuca denticulata , see *Crepidiastrum denticulatum*
Lagerstroemia : Ch. 3.1 note 74
Lagerstroemia fordii : Ch. 3.1
Lagerstroemia indica : Ch. 3.1, Fig. 3.1
Lagerstroemia speciosa : Ch. 3.1
Lamb, see *Ovis aries*
Lamprocapnos spectabilis : Ch. 3.1 note 74, Box 3.12 (Introductions), Ch. 10.1
***Lamprocapnos spectabilis* 'Alba'** : Box 3.12 (Introductions), Fig. 3.55
Lan, see *Aglaia odorata*
Laricopsis fortunei , see *Pseudolarix amabilis*
Laricopsis kaempferi , see *Pseudolarix amabilis*
Larix amabilis , see *Pseudolarix amabilis*
Larix kaempferi , see *Pseudolarix amabilis*
Lastrea standishii , see *Arachniodes standishii*
Layia : Box 3.9 (Lay)
Leadwort, see *Ceratostigma plumbaginoides*
Leech: Ch. 10.4
Lemon, see *Citrus* x *limon*
Leopard, see *Panthera pardus*
Leopard plant, see *Farfugium japonicum*
Lepidoptera: Ch. 8.3 note 182
Lepisorus fortunei , see *Neolepisorus fortunei*
Lice: Box 5.25 (Patridge)
Libocedrus dolabrata , see *Thujopsis dolabrata*
Ligularia kaempferi , see *Farfugium japonicum*
Ligularia tussilaginea , see *Farfugium japonicum*
Ligustrum coriaceum , see *Ligustrum japonicum* 'Rotundifolium'
Ligustrum fortunei , see *Ligustrum sinense* var. *sinense*
Ligustrum japonicum forma *variegatum* , see *Ligustrum japonicum* 'Variegatum'
***Ligustrum japonicum* 'Rotundifolium'** : Box 8.15 (Introductions)
***Ligustrum japonicum* 'Variegatum'** : Box 8.15 (Introductions)
Ligustrum lucidum : Ch. 7.3, Ch. 10.4
Ligustrum sinense* var. *sinense: Box 6.3 (Introductions), Fig. 6.3B
Ligustrum vulgare : Ch. 1.2 note 124
Lilac, see *Syringa*
Lilium auratum : Box 8.15 (Introductions)
Lilium brownii : Ch. 2.1, Ch. 2.2, Ch. 2.2 note 54, Ch. 3
Lilium brownii var. *brownii* : Ch. 2.2 note 54
Lilium brownii var. *viridulum* : Ch. 2.2 note 54

Lilium concolor var. *concolor* : Box 5.16 (Introductions)
Lilium concolor var. *sinicum* , see *Lilium concolor* var. *concolor*
Lilium formosanum : Ch. 5.9, Box 6.3 (Introductions), Fig. 5.142
Lilium fortunei , see *Lilium* x *maculatum*
Lilium japonicum auct., see *Lilium formosanum*
Lilium x *maculatum* : Box 8.15 (Introductions)
Lilium sinicum , see *Lilium concolor* var. *concolor*
Lily of Fokien, see *Lilium brownii*
Lily-of-the-valley, see *Convallaria*
Lime, see *Citrus aurantiifolia*
Limonia trifoliata , see Citrus trifoliata
Limonium sinense: Box 3.12 (Introductions)
Linden viburnum, see *Viburnum dilatatum*
Líng xiāo huā, see *Campsis grandiflora*
Liquidambar formosana : Ch. 5.10 note 631
Litchi chinensis : Ch. 3 notes 84, 191
Lithocarpus glaber : Ch. 5.4
London Plane, see *Platanus* x *hispanica*
Longan, see *Dimocarpus longan*
Longhorn, see Cerambycidae
Longicorn, see Cerambycidae
Lonicera : Ch. 3.1, Ch. 9.1, Ch. 9.2
Lonicera aureo- reticulata , see *Lonicera japonica* ' Aureoreticulata'
Lonicera fragrantissima* var. *fragrantissima: Box 3.12 (Introductions), Fig. 3.56
Lonicera fragrantissima subsp. *standishii* , see *Lonicera fragrantissima* var. *fragrantissima*
Lonicera japonica : Ch. 8.3, Ch. 8.3 note 173, Fig. 8.55
Lonicera japonica ' Aureoreticulata' : Box 8.15 (Introductions)
Lonicera standishii , see *Lonicera fragrantissima* var. *fragrantissima*
Loquat, see *Eriobotrya japonica*
Lotus, *see Nelumbo nucifera*
Lucanus : Ch. 8.3 note 181
Luzon flameback, see *Chrysocolaptes (lucidus) haematribon*
Luzon hornbill, see *Penelopides manillae*
Lychee, see *Litchi chinensis*
Lychnis senno: Box 8.4 note 46 (Siebold), Box 8.15 (Introductions), Fig. 8.67
Lychnis senno ' Variegata' : Box 8.15 (Introductions)
Lycopodiella cernua: Ch. 2.1, Ch. 2.2, Box 3.12 (Introductions)
Lycopodium caesium , see *Selaginella uncinata*
Lycopodium cernuum , see *Lycopodiella cernua*
Lycopodium dilatatum , see *Selaginella uncinata*
Lycopodium uncinatum , see *Selaginella uncinata*
Lycopodium willdenowii , see *Selaginella willdenowii*
Lycoris straminea: Box 3.12 (Introductions)
Lysimachia candida: Box 3.12 (Introductions)

Machilus thunbergii : Box 8.4 note 46 (Siebold)
Macropygia tenuirostris : Ch. 3.3 note 163, Fig. 3.36G
Maggot: Ch. 3.1, Box 9.8 (Bowlby)
Magnolia liliiflora , see *Yulania liliiflora*
Magnolia purpurea , see *Yulania liliiflora*
Magnolia stellata , see *Yulania stellata*
Mahonia bealei : Ch. 5.2, Box 5.5 (Beale), Box 5.16 (Introductions), Fig. 5.19
Mahonia bealei var. *planifolia* , see *Mahonia bealei*
Mahonia fortunei: Box 3.12 (Introductions), Fig. 3.57
Mahonia trifurca : Box 5.16 (Introductions)
Maidenhair Tree, see *Ginkgo biloba*
Maize, see *Zea mays*
Male fern, see *Dryopteris filix- mas*
Malus : Ch. 5.10, Ch. 9.1, Ch. 10.6
Malva : Ch. 9.1 note 23
Mammal: Box 2.3 (Reeves), Ch. 3.1, Box 5.7 (Falconer)
Mandarin duck, see *Aix galericulata*
Mandarin orange, see *Citrus reticulata*
Mangifera indica : Ch. 3.1 note 84
Mango, see *Mangifera indica*
Mangrove: Ch. 5.6
Man-neen-chang, see *Lycopodiella cernua*
Máozhú, see *Phyllostachys edulis*
Maou-chok, see *Phyllostachys edulis*
Maple, see *Acer*
Mastacanthus sinensis , see *Caryopteris incana*
Medicago minima: Box 3.12 (Introductions)
Melon: Ch. 5.10, Ch. 8.3 note 176
Merops philippinus : Ch. 3.3 note 163, Fig. 3.40F
Mespilus japonica , see *Eriobotrya japonica*
Microsorum fortunei , see *Neolepisorus fortunei*
Millet: Ch. 5.4, Ch. 9.1
Mílù, see *Elaphurus davidianus*
Mimulus : Ch. 5.6
Monkey: Ch. 10.6
Monkshood, see *Aconitum*
Morning star lily, see *Lilium concolor*
Morus : Ch. 5.4
Morus alba : Ch. 5.9, Ch. 5.10, Ch. 5.10 note 610, Ch. 10.4, Fig. 5.151
Morus alba var. *latifolia* , see *Morus alba* var. *multicaulis*
Morus alba var. *multicaulis* : Box 3.12 (Introductions)
Morus latifolia , see *Morus alba* var. *multicaulis*
Mosquito: Ch. 5.4, Box 5.13 (Ramsay), Box 5.25 (Patridge), Box 10.4 (Meiji)
Mosquito tobacco: Ch. 5.4
Moss: Ch. 5.6

Moth: Ch. 5.6
Moth orchid, see *Phalaenopsis amabilis*
Mow-chok, see *Phyllostachys edulis*
Mulberry, see *Morus*
Mule: Ch. 10.4, Ch. 10.4 note 285
Mulleripicus funebris : Ch. 3.3 note 163
Musa : Fig. 10.29
Mutton, see *Ovis aries*
Myrica nagi , see *Nageia nagi*
Myrica rubra : Ch. 5.5, Ch. 7.3, Box 8.4 note 46 (Siebold), Fig. 7.7A
Myrtaceae: Ch. 3.1 note 84

Nageia nagi **'Variegata'** : Box 8.15 (Introductions)
Nageia ovata , see *Nageia nagi*
Nectarine, see *Amygdalus persica* var. *nectarina*
Nelumbium , see *Nelumbo*
Nelumbo nucifera : Ch. 2.2, Ch. 2.2 note 50, Ch. 3.1, Ch. 5.4, Ch. 5.5, Ch. 5.8, Ch. 5 notes 295, 409, Ch. 10.6, Fig. 5.59
Neolepisorus fortunei : Box 10.1 note 7 (Moore)
Nepenthes : Ch. 2.2, Ch. 3.5
Nepeta incana , see *Caryopteris incana*
Nephelium litchi , see *Litchi chinensis*
Nephelium longana , see *Dimocarpus longan*
Nicotiana tabacum : Ch. 3.4, Ch. 5.3, Ch. 5.4, Ch. 5.4 note 157, Ch. 10.4
Norfolk Island Pine, see *Araucaria heterophylla*
Northern Sooty Woodpecker, see *Mulleripicus funebris*
Nymphaea tetragona : Ch. 5.10

Oak: Ch. 5.4, Ch. 5.8, Ch. 5.9, Ch. 5 note 548, Ch. 7.1
Oats, see *Avena sativa*
Olea aquifolia , see *Osmanthus* x *fortunei*
Olea europaea : Box 3.4 note 8 (Balfour)
Olea fragrans , see *Osmanthus fragrans*
Olea ilicifolia , see *Osmanthus heterophyllus*
Olive, see *Olea europaea*
Onion, see *Allium*
Opium poppy, see *Papaver somniferum*
Orange, see *Citrus* x *aurantium*
Orchid: Ch. 2 notes 39, 60, Box 2.1 (Lindley), Ch. 3.3, Box 3.8 (Cuming), Box 3.12 note 12 (Introductions), Ch. 8 note 90, Fig. 3.35
Orchidaceae: Ch. 2.2
Orchis susannae , see *Pecteilis susannae*
Oriolus chinensis : Ch. 3.3 note 163, Figs. 3.26
Orontium, see *Rohdea japonica*
Oryza sativa : Ch. 2.2 note 64, Box 2.2 note 4 (Opium), Ch. 3.1, Ch. 3.2, Ch. 3.4, Ch. 3.5, Ch. 5.3, Ch. 5.4, Ch. 5.9

note 517, Ch. 8.3, Ch. 8.3 note 176, Ch. 9.1, Ch. 10.6, Figs. 5.31, 8.25
Osmanthus* x *fortunei (*O. fragrans* x *heterophyllus*): Box 8.15 (Introductions)
Osmanthus fragrans : Ch. 3.1, Ch. 5.9 note 500, Box 5.5 (Beale), Ch. 9.1, Ch. 10.5, Fig. 3.5
Osmanthus heterophyllus : Box 8.15 (Introductions)
***Osmanthus heterophyllus* 'Rotundifolius'** : Box 8.15 (Introductions)
Osmanthus heterophyllus 'Variegatus' : Box 8.15 (Introductions), Fig. 8.68
Osmanthus ilicifolius , see *Osmanthus heterophyllus*
Osmanthus variegatus , see *Osmanthus heterophyllus* 'Variegatus'
Osmanthus variegatus nanus , see *Osmanthus heterophyllus* 'Rotundifolius'
Ovis aries : Box 1.2, Ch. 9.2, Ch. 10.4, Ch. 10.5, Ch. 11, Fig. 1.12
Oxalis sensitiva , see *Biophytum sensitivum*
Oxyrhynchus fortunei , see *Cryptoderma fortunei*

Pachysandra terminalis : Box 8.4 note 46 (Siebold)
Paederia foetida : Ch. 3.1, Fig. 3.4
Paederia scandens , see *Paederia foetida*
Paeonia : Ch. 2.2, Ch. 2.2 note 51, Box 2.3 (Reeves), Ch. 3.1, Ch. 3.2, Ch. 10.6
Paeonia atrosanguinea , see *Paeonia suffruticosa* 'Atrosanguinea'
Paeonia globosa , see *Paeonia suffruticosa* 'Globosa'
Paeonia japonica , see *Paeonia obovata* subsp. *obovata*
Paeonia moutan , see *Paeonia suffruticosa* subsp. *suffruticosa*
Paeonia obovata subsp. *japonica* , see *Paeonia obovata* subsp. *obovata*
Paeonia obovata var. *japonica* , see *Paeonia obovata* subsp. *obovata*
Paeonia obovata subsp. *obovata* : Ch. 8.3
Paeonia suffruticosa : Ch. 2.1, Ch. 5.3
Paeonia suffruticosa subsp. *suffruticosa* : Box 3.12 (Introductions)
***Paeonia suffruticosa* 'Atropurpurea'** : Box 3.12 (Introductions)
***Paeonia suffruticosa* 'Atrosanguinea'** : Box 3.12 (Introductions)
***Paeonia suffruticosa* 'Bijou de Chusan'** : Box 3.12 (Introductions)
***Paeonia suffruticosa* 'Caroline d'Italie'** : Box 3.12 (Introductions)
***Paeonia suffruticosa* 'Colonel Malcolm'** : Box 3.12 (Introductions)
***Paeonia suffruticosa* 'Globosa'** : Box 3.12 (Introductions)
***Paeonia suffruticosa* 'Glory of Shanghai'** : Box 3.12 (Introductions)
***Paeonia suffruticosa* 'Ida'** : Box 3.12 (Introductions)
***Paeonia suffruticosa* 'Lilacina'** : Box 3.12 (Introductions)
***Paeonia suffruticosa* 'Lord Macartney'** : Box 3.12 (Introductions)
***Paeonia suffruticosa* 'Osiris'** : Box 3.12 (Introductions)
***Paeonia suffruticosa* 'Parviflora'** : Box 3.12 (Introductions)
***Paeonia suffruticosa* 'Picta'** : Box 3.12 (Introductions)
***Paeonia suffruticosa* 'Pride of Hong Kong'** : Box 3.12 (Introductions)
***Paeonia suffruticosa* 'Reine des Violettes'** : Box 3.12 (Introductions)
***Paeonia suffruticosa* 'Robert Fortune'** : Box 3.12 (Introductions)
***Paeonia suffruticosa* 'Salmonea'** : Box 3.12 (Introductions)
***Paeonia suffruticosa* 'Versicolor'** : Box 3.12 (Introductions)
***Paeonia suffruticosa* 'Vivid'** : Box 3.12 (Introductions)

Paeony, see *Paeonia*
Palm: Box 9.3 (Gordon), Fig. 3.15
Panicum miliaceum : Ch. 5.9
Panthera pardus : Box 9.9 (Hope Grant), Ch. 10.4
Panthera tigris : Ch. 5.4, Ch. 10.4, Fig. 5.70
Papaver somniferum : Box 2.2 (Opium), Box 2.2 notes 7, 14, 16, 19, Box 2.4 (Dent), Box 3.9 (Lay), Box 3.10 (Alcock), Ch. 5, Box 5.19 (Tàipíng), Box 5.25 (Patridge), Box 5.27 (Robertson), Ch. 10, Box 10.7, Fig. 2.2
Paper mulberry, see *Broussonetia papyrifera*
Pardanthus chinensis , see *Belamcanda chinensis*
Parrot: Box 2.5 (Smith-Stanley), Fig. 3.36A
Parthenocissus tricuspidata : Box 8.1 (Veitch), Box 8.15 (Introductions)
Partridge: Ch. 3.1
Paulownia imperialis , see *Paulownia tomentosa*
Paulownia tomentosa : Ch. 8.3, Fig. 8.46
Pavo muticus imperator : Ch. 4.1
Pea, see *Pisum sativum*
Peach, see *Amygdalus persica*
Peacock, see *Pavo muticus imperator*
Peanut, see *Arachis hypogaea*
Pear, see *Pyrus*
Pecteilis susannae : Box 3.12 (Introductions)
Pedilonum secundum , see *Dendrobium secundum*
Pelargonium : Ch. 5.6, Box 10.1 (Moore)
Penelopides manillae : Ch. 3.3 note 163
Peony, see *Paeonia*
Persea thunbergii , see *Machilus thunbergii*
Persimmon, see *Diospyros kaki*
Phalaenopsis : Box 8.1 (Veitch)
Phalaenopsis amabilis : Ch. 3.3, Box 3.12 (Introductions), Fig. 3.39
Phaseolus vulgaris : Ch. 8.3 note 176
Pheasant: Ch. 3.1, Box 5.5 (Beale)
Philippine cuckoo-dove, see *Macropygia tenuirostris*
Phoenix canariensis : Ch. 11 note 65
Phoenix dactylifera : Box 8.2 (Kaempfer)
Phoenix sylvestris : Ch. 5.6 note 307
Phoenix Tree, see *Firmiana simplex*
Pholidota chinensis: Box 3.12 (Introductions)
Photinia glabra /serratifolia : Ch. 5.8
Photinia pustulata , see *Photinia serratifolia* var. *serratifolia*
Photinia serrulata , see *Photinia serratifolia* var. *serratifolia*
Phyllostachys edulis : Ch. 5.3, Ch. 5.9, Figs. 5.36, 5.136, 5.137
Phyllostachys fortunei , see *Pleioblastus fortunei*
Phytophthora infestans : Box 2.1 note 19 (Lindley)
Picea fortunei , see *Keteleeria fortunei*
Picea jezoensis sensu Fortune, see *Keteleeria fortunei*

Picea smithiana : Ch. 5.6
Pieris : Ch. 8.3 note 90
Pimela alba , see *Canarium album*
Pine, see *Pinus*
Pineapple, see *Ananas comosus*
Pink, see *Dianthus*
Pinus : Ch. 5.4, Ch. 5.6, Ch. 5.8, Ch. 5.10, Ch. 5.4 note 157, Ch. 8.2, Ch. 8.2 note 90, Ch. 9.1, Ch. 9.2, Ch. 9.1 note 130, Figs. 3.24, 9.29
Pinus bungeana: Box 6.3 (Introductions), Ch. 9.2, Box 9.6 (Wyndham), Figs. 6.3C, 6.3D
Pinus densiflora : Ch. 9.2 note 131
Pinus excorticata , see *Pinus bungeana*
Pinus fortunei , see *Keteleeria fortunei*
Pinus kaempferi , see *Pseudolarix amabilis*
Pinus massoniana : Ch. 3.4, Ch. 3.2 note 146, Ch. 5.2, Ch. 5.4, Ch. 5.8, Ch. 5.9, Ch. 5.10 note 631, Fig. 3.24
Pinus massoniana (Japan), see *Pinus thunbergii*
Pinus roxburghii : Ch. 5.6, Fig. 5.84
Pinus sinensis , see *Pinus massoniana*
Pinus tabuliformis : Ch. 9.2, Fig. 9.48
Pinus thunbergii : Ch. 8.3
Pisum sativum : Ch. 1.2, Ch. 8.3, Ch. 9.1, Ch. 9.2, Ch. 9.1 note 4
Pitcher plant, see *Nepenthes*
Pittosporum fortunei , see *Pittosporum glabratum* var. *glabratum*
Pittosporum glabratum* var. *glabratum: Box 3.12 (Introductions)
***Pittosporum tobira* 'Variegatum'** : Ch. 8.2 note 90, Box 8.15 (Introductions)
Pittosporum variegatum , see *Pittosporum tobira* 'Variegatum'
Planera acuminata , see *Zelkova serrata*
Platanthera susannae , see *Pecteilis susannae*
Platanus x *acerifolia* , see *Platanus* x *hispanica*
Platanus x *hispanica* : Ch. 5.10 note 631, Fig. 10.3
Platycarya strobilacea: Box 3.12 (Introductions), Box 8.4 note 46 (Siebold)
Platycladus dolabrata , see *Thujopsis dolabrata*
Platycladus orientalis , see *Thuja orientalis*
Platycladus orientalis 'Falcata', see *Thuja orientalis* var. *falcata*
Platycodon chinensis , see *Platycodon grandiflorus*
Platycodon grandiflorus : Ch. 3.1, Ch. 3.1 note 74, Box 3.12 (Introductions), Ch. 9.1, Fig. 3.8
Platycodon grandiflorus 'Albus Plenus' : Box 3.12 (Introductions)
Pleioblastus auricomus , see *Pleioblastus viridistriatus*
Pleioblastus fortunei : Box 8.15 (Introductions)
Pleioblastus variegatus , see *Pleioblastus fortunei*
Pleioblastus viridistriatus: Box 8.15 (Introductions)
Plum tree: Ch. 3.4 note 191, Ch. 5.6, Ch. 5.9 note 500, Ch. 8.2, Ch. 8.2 note 57, Ch. 8.2 note 71
Plum yew, see *Cephalotaxus*
Plumbago larpentae , see *Ceratostigma plumbaginoides*
Poa annua : Box 2.6 (Ward)
Poaceae: Ch. 5.9 note 517

Podocarpus : Ch. 8.1 note 25, Ch. 8.2 note 90
Podocarpus macrophyllus : Ch. 8.3, Fig. 8.57
Podocarpus nageia , see *Nageia nagi*
Polianthes tuberosa : Ch. 9.1 note 31
Polygonatum : Ch. 8.2 note 90
Polygonum tinctorium : Ch. 7.3
Polypodium fortunei , see *Neolepisorus fortunei*
Polyspora axillaris : Ch. 3.1, Fig. 3.4
Polystichopsis standishii , see *Arachniodes standishii*
Pomegranate, see *Punica granatum*
Pomelo, see *Citrus maxima*
Poncirus trifoliata , see *Citrus trifoliata*
Pongamia glabra , see *Pongamia pinnata*
Pongamia pinnata : Ch. 5.6, Fig. 5.63
Pony, see *Equus caballus*
Poplar, see *Populus*
Populus : Ch. 9.1
Porcupine, see *Hystrix indica*
Pork, see *Sus scrofa*
Porphyra dichotoma , see *Callicarpa dichotoma*
Porpoise: Box 8.6 (Hepburn)
Potato, see *Solanum tuberosum*
Potato blight, see *Phytophthora infestans*
Potentilla : Ch. 9.1
Primrose, see *Primula*
Primula cortusoides , see *Primula sieboldii*
Primula japonica: Ch. 8.3, Box 8.15 (Introductions)
Primula sieboldii : Ch. 8.3
Primula sinensis : Ch. 2.1 note 8, Box 2.3 (Reeves)
Primulina dryas: Ch. 3.1, Box 3.12 (Introductions), Fig. 3.3
Privet, see *Ligustrum*
Prunus : Ch. 5.9 note 500
Prunus amygdalus , see *Amygdalus communis*
Prunus glandulosa ' Alboplena' , see *Cerasus glandulosa* ' Alboplena'
Prunus mume , see *Armeniaca mume*
Prunus persica , see *Amygdalus persica*
Prunus persica ' Alboplena' , see *Amygdalus persica* ' Alboplena'
Prunus persica ' Camelliaeflora' , see *Amygdalus persica* ' Camelliaeflora'
Prunus persica ' Dianthiflora' , see *Amygdalus persica* ' Dianthiflora'
Prunus persica ' Sanguineoplena' , see *Amydalus persica* ' Sanguineoplena'
Prunus persica ' Shanghai Peach' , see *Amydalus persica* ' Shanghai Peach'
Prunus serrulata , see *Cerasus serrulata*
Prunus sinensis forma *albiplena* , see *Cerasus glandulosa* ' Alboplena'
Prunus spinosa : Ch. 1.2 note 124
Prunus triloba , see *Amygdalus triloba*

Prunus triloba forma *multiplex* , see *Amygdalus triloba* var. *plena*
Prunus triloba ' Multiplex' , see *Amygdalus triloba* var. *plena*
Prunus triloba var. *plena* , see *Amygdalus triloba* var. *plena*
Prunus x *yedoensis* , see *Cerasus* x *yedoensis*
Pseudolarix amabilis: Ch. 5.9, Ch. 5.11, Ch. 5 notes 542, 691, Ch. 6.1, Box 6.3 (Introductions), Ch. 7.3, Figs. 5.163, 6.3E
Pseudolarix fortunei , see *Pseudolarix amabilis*
Pseudolarix kaempferi , see *Pseudolarix amabilis*
Pseudotsuga fortunei , see *Keteleeria fortunei*
Psidium guajava : Ch. 3.3 note 84, Ch. 5.4
Pteris argentea , see *Aleuritopteris argentea*
Pterocarya rhoifolia : Box 8.4 note 46 (Siebold)
Pterostigma grandiflora , see *Adenosma glutinosum*
Pummelo, see *Citrus maxima*
Punica granatum : Ch. 9.1
Purple Emperor, see *Apatura*
Pyrus : Box 3.4 note 8 (Balfour), Ch. 5.6, Ch. 8.2 note 57, Ch. 9.1, Ch. 9.2, Ch. 10.6
Pyrus communis ' Stair Pear' , see *Pyrus communis* ' Williams Bon Cretien'
Pyrus communis ' Williams Bon Cretien' : Box 6.2 (Glendinning)

Quail: Ch. 3.1
Quercus : Ch. 5.4, Ch. 5.8, Ch. 5.9, Ch. 5.9 note 548, Ch. 7.1
Quercus bambusifolia , see *Cyclobalanopsis myrsinifolia*
Quercus inversa , see *Lithocarpus glaber*
Quercus myrsinifolia , see *Cyclobalanopsis myrsinifolia*
Quercus sclerophylla , see *Castanopsis sclerophylla*
Quercus sinensis , see *Quercus variabilis*
Quercus variabilis: Ch. 5.10 note 631, Ch. 9.2, Fig. 9.55

Radish, see *Raphanus sativus*
Ragged Robin, see *Lychnis*
Rajania quinata , see *Akebia quinata*
Ramie, see *Boehmeria nivea*
Rana : Box 2.3 (Reeves), Ch. 5.8
Rape-seed, see *Brassica rapa var. chinensis*
Raphanus sativus : Ch. 7.3, Ch. 8.3 note 176
Raphiolepis , see *Rhaphiolepis*
Raspberry, see *Rubus idaeus*
Rat: Box 5.25 (Patridge)
Red bayberry, see *Myrica rubra*
Redbud, see *Cercis*
Red collared dove, see *Streptopelia tranquebarica humilis*
Red-necked stint, see *Calidris ruficollis*
Red spinach, see *Amaranthus*
Red turtle dove, see *Streptopelia tranquebarica humilis*

Reevesia : Box 2.3 (Reeves)
Reevesia thyrsoidea : Ch. 2.1 note 8
Reeves' Spiraea, see *Spiraea cantoniensis*
Retinospora , see *Chamaecyparis*
Retinospora aurea , see *Chamaecyparis pisifera* 'Aurea'
Retinospora obtusa , see *Chamaecyparis obtusa*
Retinospora obtusa var. *variegata* , see *Chamaecyparis obtusa* 'Argentea'
Retinospora pisifera var. *aurea* , see *Chamaecyparis pisifera* 'Aurea'
Retinospora pisifera var. *variegata* , see *Chamaecyparis pisifera* 'Argentea'
Rhamnus chlorophora , see *Rhamnus globosa*
Rhamnus globosa : Ch. 5.8 note 459
Rhamnus utilis : Ch. 5.8, Ch. 5.8 note 459, Ch. 7.3
Rhaphiolepis indica : Box 3.12 (Introductions)
Rhaphiolepis ovata , see *Rhaphiolepis umbellata*
Rhaphiolepis umbellata : Box 8.15 (Introductions)
***Rhapis excelsa* 'Variegata'**: Box 8.15 (Introductions), Box 8.15 notes 52, 90
Rhapis flabelliformis var. *variegata* , see *Rhapis excelsa* 'Variegata'
Rhinoceros, see *Rhinoceros sumatrensis*
Rhinoceros sumatrensis : Ch. 10.4
Rohdea japonica : Ch. 8.1
Rhododendron : Ch. 2.2, Box 2.3 (Reeves), Ch. 3.1, Ch. 3.2, Ch. 5.4, Ch. 5.2 note 42, Box 5.15 (Standish), Ch. 8.1, Ch. 8.3, Fig. 5.35
Rhododendron degronianum* var. *heptamerum: Box 8.15 (Introductions)
Rhododendron farrerae: Box 3.12 (Introductions)
Rhododendron fortunei: Box 6.3 (Introductions)
Rhododendron heptamerum , see *Rhododendron degronianum* var. *heptamerum*
***Rhododendron indicum* '*Crispiflorum*'**: Box 5.16 (Introductions), Fig. 5.97
Rhododendron japonicum : Box 8.15 note 53
Rhododendron ledifolium var. *narcissiflorum* , see *Rhododendron mucronatum*
Rhododendron metternichii , see *Rhododendron degronianum* var. *heptamerum*
Rhododendron metternichii var. *heptamerum* , see *Rhododendron degronianum* var. *heptamerum*
Rhododendron metternichii var. *pentamerum* , see *Rhododendron degronianum* var. *heptamerum*
Rhododendron narcissiflorum , see *Rhododendron mucronatum* 'Narcissiflorum'
Rhododendron molle : Box 3.12 (Introductions), Ch. 5.3, Ch. 5.8
Rhododendron molle subsp. *japonicum* : Box 8.15 note 53
Rhododendron mucronatum 'Narcissiflorum': Box 5.16 (Introductions)
Rhododendron* x *obtusum: Box 3.12 (Introductions)
***Rhododendron* x *obtusum* 'Album'**: Box 3.12 (Introductions)
***Rhododendron* x *obtusum* 'Amoenum'**: Box 5.16 (Introductions), Fig. 5.98
Rhododendron ovatum: Box 3.12 (Introductions)
***Rhododendron simsii* 'Bealei'**: Box 5.16 (Introductions), Fig. 5.99
Rhododendron simsii var. *vittatum* forma *bealei* , see *Rhododendron simsii* 'Bealei'
Rhododendron simsii 'Vittatum': Box 3.12 (Introductions)
***Rhododendron simsii* 'Vittatum Punctatum'**: Box 5.16 (Introductions), Fig. 5.100
Rhododendron sinense , see *Rhododendron molle*

Rhododendron vittatum , see *Rhododendron simsii* ' Vittatum'
Rhododendron vittatum var. *bealei* , see *Rhododendron simsii* ' Bealei'
Rhododendron vittatum var. *punctatum* , see *Rhododendron simsii* ' Vittatum Punctatum'
Rhus vernicifera , see *Toxicodendron vernicifluum*
Rhus verniciflua , see *Toxicodendron vernicifluum*
Rhynchospermum jasminoides , see *Trachelospermum jasminoides*
Rice, see *Oryza sativa*
Rice-paper plant, see *Tetrapanax papyrifer*
Ring-cupped oak, see *Cyclobalanopsis*
Roettlera sinensis , see *Primulina dryas*
Rook, see *Corvus frugilegus*
Rosa : Ch. 2, Box 2.1 note 7 (Lindley), Box 2.3 (Reeves), Ch. 3.1, Ch. 3.1 note 74, Ch. 5.4, Ch. 5.9 note 500, Ch. 8. 3, Ch. 9.1, Ch. 9.2, Fig. 3.11
Rosa anemoniflora: Box 3.12 (Introductions)
Rosa banksiae : Ch. 2.2, Ch. 8.3, Fig. 8.56
Rosa banksiae var. *lutea* : Ch. 2.1 note 8
Rosa chinensis var. *pseudindica* , see *Rosa* x *odorata* ' Fortune's Double Yellow'
Rosa x fortuniana: Box 3.12 (Introductions)
Rosa indica var. *ochroleuca* , see *Rosa* x *odorata* forma *ochroleuca*
Rosa x *odorata* ' Beauty of Glazenwood', see *Rosa* x *odorata* ' Fortune's Double Yellow'
***Rosa* x *odorata* ' Fortune's Double Yellow'**: Ch. 2.2, Ch. 3.2 note 139, Box 3.12 (Introductions), Fig. 3.29
***Rosa* x *odorata* ' Fortune's Five Colour'**: Box 3.12 (Introductions)
Rosa x *odorata* ' Gold of Ophir', see *Rosa* x *odorata* ' Fortune's Double Yellow'
Rosa x *odorata* forma *ochroleuca* : Ch. 2.1 note 8
Rosa x *odorata* ' Pseudindica', see *Rosa* x *odorata* ' Fortune's Double Yellow'
Rosa x *odorata* var. *pseudindica* , see *Rosa* x *odorata* ' Fortune's Double Yellow'
Rosa pseudindica , see *Rosa* x *odorata* ' Fortune's Double Yellow'
Rosa rubiginosa : Ch. 1.2 note 124
Rosa sempervirens var. *anemoniflora* , see *Rosa anemoniflora*
Rosa triphylla , see *Rosa anemoniflora*
Rosaceae: Box 2.1 note 7 (Lindley)
Rose, see *Rosa*
Roundworm, see *Ascaris lumbricoides*
Roylea : Box 4.1 (Royle)
Rubber, see *Hevea brasiliensis*
Rubus : Ch. 5.5 note 283, Ch. 7.3, Fig. 7.7A
Rubus idaeus : Ch. 5.6
Ruellia indigotica , see *Strobilanthes cusia*
Rutaceae: Ch. 3.1 note 84

Sabina chinensis , see *Juniperus chinensis*
Saccharum officinarum : Ch. 3.1, Ch. 3.4, Ch. 5.3
Safflower, see *Carthamus tinctorius*
Sal, see *Shorea robusta*
Salisburia adiantifolia , see *Ginkgo biloba*

Salix babylonica : Ch. 5.8
Salmo : Ch. 1.2
Salsola : Ch. 9.1 note 23
Sambucus nigra : Ch. 1.2 note 124
Sapindaceae: Ch. 3.1 note 84
Sapium sebiferum , see *Triadica sebifera*
Saraca asoca : Ch. 5.6
Sarcops calvus : Ch. 3.3 note 163
Sasa fortunei , see *Pleioblastus fortunei*
Sasa variegata , see *Pleioblastus fortunei*
Sawara cypress, see *Chamaecyparis pisifera*
Saxifraga cortusifolia var. *fortunei* , see *Saxifraga fortunei*
Saxifraga fortunei: Box 8.15 (Introductions)
Saxifraga fortunei var. *tricolor* , see *Saxifraga stolonifera* ' Tricolor'
Saxifraga sarmentosa var. *tricolor* , see *Saxifraga stolonifera* ' Tricolor'
***Saxifraga stolonifera* ' Tricolor'** : Box 8.15 (Introductions), Ch. 9.2, Fig. 9.53
Saxifrage, see *Saxifraga*
Schizophragma hydrangeoides : Box 8.4 note 46 (Siebold)
Sciadopitys verticillata : Ch. 8.1, Ch. 8.2, Ch. 8.2 note 90, Box 8.15 (Introductions), Fig. 8.19
Scolopax rusticola : Ch. 3.1
Scutellaria : Box 3.12 (Introductions)
Scutigera : Box 3.8 note 24 (Cuming)
Sedum : Ch. 5.9 note 575
Selaginella uncinata : Box 3.12 (Introductions)
Selaginella willdenowii : Box 3.12 (Introductions), Ch. 5.2 note 42
Senecio kaempferi , see *Farfugium japonicum*
Sequoiadendron giganteum : Box 10.3
Sesame, see *Sesamum indicum*
Sesamum indicum : Ch. 8.3 note 176
Setaria italica : Ch. 5.9
Shaddock, see *Citrus maxima*
Shantung Cabbage, see *Brassica rapa* var. *glabra*
Sheep, see *Ovis aries*
Shell: Box 3.8 (Cuming), Box 3.8 note 13, Ch. 8.3
Shorea robusta : Ch. 5.6, Box 5.13 (Ramsay)
Sida tiliifolia , see *Abutilon theophrasti*
Siebold's Blue-leaved hosta, see *Hosta sieboldiana*
Silk: Box 2.2 note 4 (Opium), Ch. 3, Box 3.6 (Sinclair), Box 3.10 (Alcock), Ch. 5.10, Box 8.13 (Aspinall), Ch. 9.2, Ch. 10.1
Silkworm, see *Bombyx mori*
Silver apricot, see *Ginkgo biloba*
Silver spruce, see *Abies*
Skimmia fortunei , see *Skimmia reevesiana*
Skimmia fragrans , see *Skimmia japonica* ' Fragrans'
Skimmia japonica : Box 8.15 (Introductions)

Skimmia japonica subsp. *reevesiana* , see *Skimmia reevesiana*
Skimmia japonica var. *reevesiana* , see *Skimmia reevesiana*
Skimmia japonica ' Fragrans' : Box 8.15 note 43 (Introductions)
Skimmia oblata , see *Skimmia japonica*
Skimmia reevesiana: Box 5.16 (Introductions)
Small-flowered China jasmine, see *Jasminum lanceolaria*
Snail: Box 3.8 (Cuming), Ch. 5
Snipe: Ch. 3.1
Soap-bean tree, see *Gleditsia sinensis*
Solanum melongena : Ch. 8.3 note 176
Solanum tuberosum : Ch. 1.2
Song-pee-leen, see *Citrus reticulata*
Sooty woodpecker, see *Mulleripicus funebris*
Sorghum bicolor : Ch. 9.1, Fig. 9.14
Soybean, see *Glycine max*
Spathoglottis fortunei , see *Spathoglottis pubescens*
Spathoglottis pubescens : Box 3.12 (Introductions)
Spider lily, see *Lycoris straminea*
Spikemoss, see *Selaginella*
Spinach, see *Spinacia oleracea*
Spinacia oleracea : Ch. 7.3
Spiraea : Ch. 5.4, Ch. 5.11
Spiraea callosa , see *Spiraea japonica* ' Fortunei'
Spiraea cantoniensis : Box 2.3 (Reeves), Ch. 3.1 note 74, Ch. 5.4, Ch. 5.8, Ch. 5.2 note 42, Ch. 8.3
Spiraea cantoniensis ' Flore Pleno' , see *Spiraea cantoniensis* ' Lanceata'
Spiraea cantoniensis ' Lanceata' : Box 5.16 (Introductions)
Spiraea chinensis: Box 3.12 (Introductions)
Spiraea fortunei , see *Spiraea japonica* ' Fortunei'
Spiraea grandiflora , see *Exochorda racemosa*
Spiraea japonica : Ch. 5.4
***Spiraea japonica* ' Fortunei'** : Box 5.16 (Introductions), Fig. 5.101
Spiraea japonica var. *fortunei* , see *Spiraea japonica* ' Fortunei'
Spiraea palmata , see *Filipendula multijuga*
Spiraea prunifolia : Ch. 5.9, Ch. 5.2 note 42, Box 8.4 note 46 (Siebold)
Spiraea prunifolia var. *plena* , see *Spiraea prunifolia* ' Flore Pleno'
Spiraea prunifolia var. *prunifolia* , see *Spiraea prunifolia* ' Flore Pleno'
Spiraea prunifolia ' Flore Pleno' : Box 3.12 (Introductions)
Spiraea reevesiana , see *Spiraea cantoniensis*
Spiraea reevesiana var. *florepleno* , see *Spiraea cantoniensis* ' Lanceata'
Spotted bellflower, see *Campanula punctata*
Spotted imperial-pigeon, see *Ducula carola*
Stachyurus praecox : Box 8.4 note 46 (Siebold)
Stag-beetle, see *Lucanus*
Star anise, see *Illicium*
Star jasmine, see *Trachelospermun jasminoides*

Star magnolia, see *Yulania stellata*
Statice : Ch. 9.1 note 23
Statice fortunei , see *Limonium sinense*
Statice sinensis , see *Limonium sinense*
Stegodon : Ch. 5.3
Sterculia platanifolia , see *Firmiana simplex*
Sticky willie, see *Galium aparine*
Stillingia sebifera , see *Triadica sebifera*
Strawberry, see *Fragaria* x *ananassa*
Streptopelia tranquebarica humilis : Ch. 3.3 note 163
Strobilanthes cusia: Ch. 5.9, Box 6.3 (Introductions)
Styrax japonicus : Ch. 8.3, Box 8.4 note 46 (Siebold), Fig. 8.54
Styrax obassis : Box 8.4 note 46 (Siebold)
Sugar cane, see *Saccharum officinarum*
Sunflower, see *Helianthus annuus*
Sus scrofa : Ch. 3.2, Ch. 5.4, Box 6.4, Ch. 9.2, Box 9.8 (Bowlby), Fig. 5.36
Sweet briar, see *Rosa rubiginosa*
Sweet chestnut, see *Castanea*
Sweet potato, see *Ipomoea batatas*
Synaedrys sclerophylla , see *Castanopsis sclerophylla*
Syncerus caffer : Box 8.3 (Thunberg)
Syringa oblata: Box 6.3 (Introductions), Fig. 6.3F
***Syringa oblata* ' Alba'** : Box 6.3 (Introductions)
Syringa vulgaris var. *oblata* , see *Syringa oblata*

Tagetes erecta : Ch. 9.1 note 31
Tallow tree, see *Triadica sebifera*
Tamarix : Ch. 9.1 note 23
Taxus : Ch. 3.3 note 146, Ch. 8.2
Taxus baccata subsp. *cuspidata* , see *Taxus cuspidata*
Taxus baccata var. *cuspidata* , see *Taxus cuspidata*
Taxus cuspidata : Box 6.3 (Introductions), Box 8.4 note 46 (Siebold)
Taxus fortunei , see *Cephalotaxus fortunei*
Taxus verticillata , see *Sciadopitys verticillata*
Tea, see *Camellia sinensis*
Teak, see *Tectona grandis*
Teal, see *Anas crecca*
Tectona grandis : Box 5.7 (Falconer)
Teen-tsan, see *Attacus atlas*
Termite: Box 5.32 (Dalhousie)
Tetrapanax papyrifer : Ch. 2.2, Ch. 3.5, Ch. 5.9, Ch. 5.9 note 517, Box 6.3 (Introductions), Fig. 5.143
Thea bohea , see *Camellia sinensis*
Thea viridis , see *Camellia sinensis*
Thea viridis variegata , see *Camellia sinensis* ' Variegata'
Thorny olive, see *Elaeagnus pungens*

Thuja : Ch. 5.8, Ch. 8.2, Ch. 8.1 note 25, Ch. 9.1, Ch. 9.2 note 89, Ch. 9.1 note 130
Thuja dolabrata , see *Thujopsis dolabrata*
Thuja falcata , see *Thuja orientalis* var. *falcata*
Thuja orientalis : Ch. 5.10 note 631, Ch. 7.3, Box 8.15 note 59 (Introductions), Ch. 9.2, Ch. 9.2 note 131, Fig. 7.7F
Thuja orientalis* var. *falcata: Ch. 8.3, Box 8.15 (Introductions), Box 8.15 note 59
Thuja standishii: Box 8.15 (Introductions)
Thujopsis dolabrata : Ch. 1.2 note 131, Ch. 8.1, Ch. 8.2, Box 8.15 (Introductions), Figs. 8.21, 8.22, 8.23
Thujopsis dolabrata ' Variegata' : Box 8.4 note 61 (Siebold), Box 8.15 (Introductions)
Thujopsis laetevirens , see *Thujopsis dolabrata*
Thujopsis standishii , see *Thuja standishii*
Tiáo, see *Campsis grandiflora*
Tiger, see *Panthera tigris*
Tilia : Ch. 1.3
Tobacco, see *Nicotiana tabacum*
Tóng tree, see *Vernicia fordii*
Torenia concolor: Box 3.12 (Introductions)
Torreya grandis: Ch. 5.11, Ch. 5 notes 690, 691, Box 6.3 (Introductions), Ch. 7, Fig. 5.163
Tortoise: Box 2.3 (Reeves)
Toxicodendron vernicifluum : Ch. 5.8, Box 6.3 (Introductions)
Trachelospermum jasminoides: Box 3.12 (Introductions), Fig. 3.58
Trachycarpus fortunei : Ch. 3.1, Ch. 3.1 note 74, Ch. 5.2, Ch. 5.8, Box 5.16 (Introductions), Ch. 7.3, Ch. 10.1, Ch. 11, Ch. 11 note 65, Figs. 3.15, 3.16, 10.30, 11.18
Trapa bicornis , see *Trapa natans*
Trapa natans : Ch. 5.10
Trapa natans var. *bicornis* , see *Trapa natans*
Tree paeony, see *Paeonia suffruticosa*
Trefoil, see *Medicago*
Trema orientalis : Ch. 8.2 note 67
Triadica sebifera : Ch. 3.1, Ch. 3.2 note 146, Ch. 5.2, Ch. 5.3, Ch. 5.4, Ch. 5.4 note 136, Ch. 7.3, Figs. 3.14, 5.13
Tricyrtis hirta : Box 8.15 (Introductions)
Trifoliate orange, see *Citrus trifoliata*
Trifolium : Ch. 3.1
Triticum : Ch. 5.4, Ch. 5.6, Ch. 5.9, Ch. 8.3
Trout, see *Salmo*
Tuberose, see *Polianthes tuberosa*
Tung (-oil) tree, see *Vernicia fordii*
Turnip, see *Brassica rapa*
Tussilago japonica , see *Farfugium japonicum*

Ulmus : Ch. 1.2 note 124, Ch. 10.3
Ulmus keaki , see *Zelkova serrata*
Umbrella Pine, see *Sciadopitys verticillata*
Upside-down fern, see *Arachniodes standishii*
Ursus : Ch. 10.4
Urtica nivea , see *Boehmeria nivea*

Uvularia hirta , see *Tricyrtis hirta*

Veitchia joannis : Box 8.1 (Veitch)
Vernicia fordii : Ch. 5.2, Ch. 5.3, Ch. 5.4, Ch. 7.3, Fig. 5.16
Vernonia fortunei , see *Vernonia solanifolia*
Vernonia solanifolia : Box 5.16 (Introductions)
Veronica : Ch. 9.1
Viburnum dilatatum : Box 3.12 (Introductions)
Viburnum macrocephalum: Ch. 3.1 note 74, Box 3.12 (Introductions)
Viburnum plicatum: Ch. 3.1 note 74, Box 3.12 (Introductions), Fig. 3.59
Viburnum tomentosum var. *plicatum* , see *Viburnum plicatum*
Viburnum tomentosum var. *sterile* , see *Viburnum plicatum*
Viburnum tomentosum forma *sterile* , see *Viburnum plicatum*
Vicia faba : Ch. 1.2
Victoria : Ch. 2.2 note 60, Figs. 5.90, 5.91
Victoria amazonica : Ch. 5.6, Box 5.14 (Paxton), Box 5.14 note 11, Figs. 5.90, 5.91
Victoria regia , see *Victoria amazonica*
Viola : Ch. 5.6, Ch. 8.3
Violet, see *Viola*
Vitex negundo var. *heterophylla* : Ch. 9.2
Vitis : Ch. 8.2
Vitis vinifera : Ch. 1.3, Ch. 9.1, Ch. 9.2

Wampi, see *Clausena lansium*
Water-chestnut, see *Trapa natans*
Wax-insect tree, see *Fraxinus chinensis*
Weeping cypress, see *Cupressus funebris*
Weeping willow, see *Salix babylonica*
Weigela florida: Ch. 3.2, Ch. 3.5, Ch. 3.1 note 74, Box 3.12 (Introductions), Ch. 9.1, Ch. 9.2, Fig. 3.23
Weigela hortensis : Ch. 8.3
Weigela rosea , see *Weigela florida*
Wheat, see *Triticum*
White-bellied woodpecker, see *Dryocopus javensis*
White-breasted woodswallow, see *Artamus leucorhynchus*
White loosestrife, see *Lysimachia candida*
White-throated kingfisher, see *Halcyon smyrnensis fusca*
Wild boar, see *Sus scrofa*
Winter daphne, see *Daphne odora*
Winter hazel, see *Corylopsis spicata*
Winter honeysuckle, see *Lonicera fragrantissima*
Winter jasmine, see *Jasminum nudiflorum*
Wistaria , see *Wisteria*
Wisteria alba , see *Wisteria sinensis* ‘Alba’
Wisteria floribunda : Ch. 8.3
Wisteria sinensis : Box 2.3 (Reeves), Ch. 3.1, Ch. 3.2, Ch. 5.8, Ch. 8.3 note 150, Ch. 9.2, Fig. 9.45

Wisteria sinensis forma *alba* , see *Wisteria sinensis* ‘ Alba’
***Wisteria sinensis* ‘ Alba’** : Ch. 3.2, Ch. 3.1 note 74, Box 3.12 (Introductions), Fig. 3.28
Wisteria sinensis var. *albiflora* , see *Wisteria sinensis* ‘ Alba’
Woad, see *Isatis tinctoria*
Woodcock, see *Scolopax rusticola*
Wood pigeon, see *Columba palumbus*
Woodwardia japonica : Box 8.15 (Introductions)
Woodwardia orientalis: Box 8.15 (Introductions)
Wormwood, see *Artemisia*

Yam, see *Dioscorea*
Yangmae, see *Myrica rubra*
Yángméi, see *Myrica rubra*
Yeddo Hawthorn, see *Rhaphiolepis umbellata*
Yedo Vine, see *Parthenocissus tricuspidata*
Yew, see *Taxus*
Yoshino cherry, see *Cerasus* x *yedoensis*
Youngia denticulata , see *Crepidiastrum denticulatum*
Yucca : Box 6.2 (Glendinning)
Yulania liliiflora : Ch. 5.3
Yulania stellata : Box 8.1 (Veitch)
Yumberry, see *Myrica rubra*

Zambac, see *Jasminum sambac*
Zea mays : Ch. 5.4, Ch. 5.8, Ch. 5.9, Ch. 5.11, Box 5.8 (Jameson)
Zelkova serrata : Ch. 8.2 note 67
Zingiber officinale : Ch. 3.1, Ch. 3.4, Ch. 8.3 note 176
Ziziphus jujuba : Ch. 3.4 note 191, Ch. 9.2, Fig. 9.50
Ziziphus vulgaris , see *Ziziphus jujuba*

Illustration Credits

Frontispiece. Cox, 1945: facing p. 46; **Fig. 1.1.** Portrait by Antoine-Jean Gros (1771 – 1835) in the Hermitage (www.arthermitage.org/baron-Antoine-Jean-Gros/Napoleon-Bonaparte-on-the-Bridge-at Arcole); **Fig. 1.2.** Huxley, 1943; **Fig. 1.3.** Portrait by Luis López Piquer (1802 – 1865) from https://commons.wikipedia.org/wiki/File:Retrato_de_Fernando_VII_by_(Luis_López_Piquer).JPG;

Fig. 1.4. https://en.wikipedia.org/wiki/File:James_Madison.jpg; **Fig. 1.5.** Portrait by Sir Henry Raeburn (1756 – 1823) about 1791, Courtesy of Jenny Liddle, National Trust Images (www. nationaltrustimages. org. uk/image/1009107); **Fig. 1.6.** D. K. Ferguson, 18 July 2014; **Fig. 1.7.** D. K. Ferguson, 18 July 2014; **Fig. 1.8.** D. K. Ferguson, 29 August 2007; **Fig. 1.9.** D. K. Ferguson, 29 August 2007; **Fig. 1.10.** D. K. Ferguson, 29 August 2007; **Fig. 1.11.** Kelloe Farm, D. K. Ferguson, 29 August 2007; **Fig. 1.12.** D. K. Ferguson, 20 July 2014; **Fig. 1.13.** D. K. Ferguson, 15 July 2014; **Fig. 1.14.** D. K. Ferguson, 15 July 2014; **Fig. 1.15.** D. K. Ferguson, 15 July 2014; **Fig. 1.16.** Anon., 1908; **Fig. 1.17.** D. K. Ferguson, 22 July 2014.

Fig. 2.1. Portrait by Eden Upton Eddis (1812 – 1901) in 1862, Courtesy of the Royal Horticultural Society; **Fig. 2.2.** https://de. wikipedia. org/wiki/Datei: Lin _ Zexu _ 1. jpg; **Fig. 2. 3.** Drawing based on a painting by George Chinnery (1774 – 1852) in Fitzwilliam Museum, Cambridge; **Fig. 2. 4.** A portrait in Knowsley Hall, Prescot. Licence granted courtesy of The Rt Hon. The Earl of Derby, 2016; **Fig. 2.5.** D. K. Ferguson, 14 July 2014; **Fig. 2. 6.** Courtesy of the Royal Horticultural Society Archives; **Fig. 2.7.** Portrait by Arthur Ackland Hunt (1841 – 1914), image by kind permission of Nicholas Wood, The Worshipful Society of Apothecaries, London; **Fig. 2.8.** D. K. Ferguson, 7 July 2014; **Fig. 2.9.** Courtesy of the Royal Horticultural Society Archives.

Fig. 3.1. D. K. Ferguson, 3 July 2013; **Fig. 3.2.** D. K. Ferguson, 27 May 2012; **Fig. 3.3.** Original by LǏ J.-F., July 2016; **Fig. 3.4.** D. K. Ferguson, 8 September 2014; **Fig. 3.5.** R. Gold, 28 September 2014; **Fig. 3.6.** HÚ Y.-Q., 10 November 2008; **Fig. 3.7.** Williamson, 1848: Frontispiece; **Fig. 3.8.** D. K. Ferguson, 27 July 2014; **Fig. 3.9.** D. K. Ferguson, 8 September 2014; **Fig. 3.10.** Courtesy of Stephen Johnson, Surrey Infantry Museum, Woking, Surrey (www.queensroyalsurreys.org.uk/colonels/026.html); **Fig. 3.11.** D. K. Ferguson, 27 May 2008; **Fig. 3.12.** LǏ Y.-M., 12 November 2007; **Fig. 3.13.** D. K. Ferguson, November 2007; **Fig. 3.14.** D. K. Ferguson, 19 May 2012; **Fig. 3. 15.** D. K. Ferguson, 7 July 2014; **Fig. 3.16.** D. K. Ferguson, 7 July 2014; **Fig. 3.17.** LǏ Y.-M., 12 November 2007; **Fig. 3.18.** D. K. Ferguson, 12 November 2007; **Fig. 3.19.** Rebecca Ewing Peterson (www.findagrave.com/cgi-bin/fg.cgi? page=gr&Grid=122825443); **Fig. 3.20.** LǏ J.-F., 10 November 2007; **Fig. 3.21.** R. Fortune, 1847c; **Fig. 3. 22.** D. K. Ferguson, 7 July 2014; **Fig. 3.23.** D. K. Ferguson, 4 June 2015; **Fig. 3.24.** LǏ J.-F., 10 November 2007; **Fig. 3.25.** D. K. Ferguson, 10 November 2007; **Fig. 3.26.** D. K. Ferguson, 14 July 2014; **Fig. 3.27.** D. K. Ferguson, 20 May 2008; **Fig. 3.28.** D. K. Ferguson, 15 April 2014; **Fig. 3.29.** Curtis's Botanical Magazine 1852: Tab. 4679; **Fig. 3.30.** D. K. Ferguson, 27 May 2012; **Fig. 3.31.** LǏ Y.-M., 11 November 2007; **Fig. 3.32.** D. K. Ferguson, 11 November 2007; **Fig. 3.33.** LǏ J.-F., 11 November 2007; **Fig. 3.34.** D. K. Ferguson, 11 November 2007; **Fig. 3.35.** LǏ J.-F., 11 November 2007; **Fig. 3.36.** Photograph by Maull & Polyblank ca. 1855, Courtesy of the National Portrait Gallery London (www.npg.org.uk/collections/search/portraitLarge/mw01670/Hugh-Cuming); **Fig. 3.37.** D. K. Ferguson, 6 November 2008; **Fig. 3.38.** D. K. Ferguson, 5 November 2008; **Fig. 3.39.** D. K. Ferguson, 15 September 2016; **Fig. 3.40.** Liverpool World Museum, D. K. Ferguson, 14 July 2014; **Fig. 3.41.** The Wellcome Library, London, Courtesy of Betty Yao (Yao, 2015: 97); **Fig. 3.42.** D. K. Ferguson, 16 November 2008; **Fig. 3.43.** HÚ Y.-Q., 14 November 2008; **Fig. 3.44.** Portrait by Richard Woodman in 1831 from Lay, 1952: facing p. 6; **Fig. 3.45.** Photograph by Lock & Whitfield ca. 1877, Courtesy of the National Portrait Gallery London (www.

npg.org.uk/collections/search/portrait/mw135204/Sir-John-Rutherford-Alcock); **Fig. 3.46.** D. K. Ferguson, 26 November 2013; **Fig. 3.47.** Photograph by John Thomson in 1871, The Wellcome Library, London, Courtesy of Betty Yao (Yao, 2015: 99); **Fig. 3.48.** D. K. Ferguson, 15 September 2009; **Fig. 3.49.** http://en.wikipedia.org/wiki/William_Lockhart_(surgeon); **Fig. 3.50.** D. K. Ferguson, September 2013; **Fig. 3.51.** D. K. Ferguson, 9 August 2013; **Fig. 3.52.** D. K. Ferguson, 29 March 2010; **Fig. 3.53.** D. K. Ferguson, 1 October 2014; **Fig. 3.54.** D. K. Ferguson, 13 April 2013; **Fig. 3.55.** D. K. Ferguson, 17 April 2014; **Fig. 3.56.** D. K. Ferguson, 4 April 2014; **Fig. 3.57.** LǏ J.-F., 11 November 2007; **Fig. 3.58.** HÚ Y.-Q., 27 May 2008; **Fig. 3.59.** D. K. Ferguson, March 2016.

Fig. 4.1. Liverpool Central Library Archives, D. K. Ferguson, 15 July 2014; **Fig. 4.2.** D. K. Ferguson, 7 July 2014; **Fig. 4.3.** D. K. Ferguson, 7 July 2014; **Fig. 4.4.** Lithograph by George Henry Ford (1808－1876), https://en.wikipedia.org/wiki/File:JFRoyle.jpg; **Fig. 4.5.** Liverpool Central Library Archives, D. K. Ferguson, 15 July 2014; **Fig. 4.6.** D. K. Ferguson, 5 September 2007; **Fig. 4.7.** D. K. Ferguson, 5 September 2007.

Fig. 5.1. Grieve et al., 1850; **Fig. 5.2.** P&O Heritage Collection (www.poheritage.com); **Fig. 5.3.** D. K. Ferguson, 23 February 2010; **Fig. 5.4.** D. K. Ferguson, 15 February 2014; **Fig. 5.5.** https://commons.wikipedia.org/wiki/File:PyramidDatePalms.jpg; **Fig. 5.6.** Grieve et al., 1850; **Fig. 5.7.** Grieve et al., 1850; **Fig. 5.8.** Oil painting by Stephen Dadd Skillet (1817－1866), P&O Heritage Collection (www.poheritage.com); **Fig. 5.9.** Original by LǏ J.-F., July 2016; **Fig. 5.10.** R. Gold, 3 May 2008; **Fig. 5.11.** R. Gold, 4 May 2008; **Fig. 5.12.** R. Gold, 4 May 2008; **Fig. 5.13.** LǏ J.-F., 13 November 2007; **Fig. 5.14.** R. Gold, 13 May 2008; **Fig. 5.15.** D. K. Ferguson, 13 November 2015; **Fig. 5.16.** D. K. Ferguson, 8 October 2009; **Fig. 5.17.** D. K. Ferguson, 2 October 2009; **Fig. 5.18.** D. K. Ferguson, 1 October 2009; **Fig. 5.19.** D. K. Ferguson, 1 September 2007; **Fig. 5.20.** D. K. Ferguson, 30 September 2009; **Fig. 5.21.** http://news.xinhuanet.com/photo/2008－09/18/content_10071460.htm; **Fig. 5.22.** Original by LǏ J.-F., July 2016; **Fig. 5.23.** HÚ Y.-Q., 30 May 2008; **Fig. 5.24.** Illustration by William Alexander (1767－1816) in Staunton, 1797; **Fig. 5.25.** HÚ Y.-Q., 27 May 2008; **Fig. 5.26.** HÚ Y.-Q., 13 November 2008; **Fig. 5.27.** HÚ Y.-Q., 13 November 2008; **Fig. 5.28.** HÚ Y.-Q., 13 November 2008; **Fig. 5.29.** HÚ Y.-Q., 13 November 2008; **Fig. 5.30.** HÚ Y.-Q., 13 November 2008; **Fig. 5.31.** HÚ Y.-Q., 13 November 2008; **Fig. 5.32.** HÚ Y.-Q., 13 November 2008; **Fig. 5.33.** HÚ Y.-Q., 13 November 2008; **Fig. 5.34.** R. Fortune, 1852: 139; **Fig. 5.35.** D. K. Ferguson, 6 April 2014; **Fig. 5.36.** D. K. Ferguson, 10 November 2007; **Fig. 5.37.** D. K. Ferguson, 10 November 2007; **Fig. 5.38.** HÚ Y.-Q., 29 May 2008; **Fig. 5.39.** HÚ Y.-Q., 29 May 2008; **Fig. 5.40.** D. K. Ferguson, 5 May 2008; **Fig. 5.41.** D. K. Ferguson, 11 May 2008; **Fig. 5.42.** R. Gold, 5 May 2008; **Fig. 5.43.** R. Gold, 6 May 2008; **Fig. 5.44.** D. K. Ferguson, 6 May 2008; **Fig. 5.45.** R. Gold, 6 May 2008; **Fig. 5.46.** R. Fortune, 1852: 202; **Fig. 5.47.** R. Gold, 9 May 2008; **Fig. 5.48.** R. Gold, 9 May 2008; **Fig. 5.49.** D. K. Ferguson, 9 May 2008; **Fig. 5.50.** http://en.wikipedia.org/wiki/Hugh_Falconer#/media/File:Hugh_Falconer.jpeg; **Fig. 5.51.** R. Gold, 9 May 2008; **Fig. 5.52.** R. Gold, 10 May 2008; **Fig. 5.53.** R. Gold, 10 May 2008; **Fig. 5.54.** R. Gold, 10 May 2008; **Fig. 5.55.** D. K. Ferguson, 11 May 2008; **Fig. 5.56.** R. Gold, 11 May 2008; **Fig. 5.57.** R. Fortune, 1852: 348; **Fig. 5.58.** D. K. Ferguson, 11 November 2007; **Fig. 5.59.** D. K. Ferguson, 11 November 2007; **Fig. 5.60.** Detail of engraving by William Alfred Delamotte (1775－1863), National Maritime Museum, Greenwich, London; **Fig. 5.61.** Original by JIǍ H., based on an outline map provided by A. Trivedi, July 2016; **Fig. 5.62.** D. K. Ferguson, 27 April 2009; **Fig. 5.63.** D. K. Ferguson, 27 April 2009; **Fig. 5.64.** R. Gold, 12 May 2008; **Fig. 5.65.** R. Gold, 12 May 2008; **Fig. 5.66.** R. Gold, 12 May 2008; **Fig. 5.67.** https://en.wikipedia.org/wiki/File:Makers_of_British_botany,_Plate_15_(William_Griffith).png; **Fig. 5.68.** D. K. Ferguson, 27 April 2009; **Fig. 5.69.** D. K. Ferguson, 26 April 2009; **Fig. 5.70.** http://tcktcktck.org/2013/09/save-sundarbans-support-bangladeshi-campaigners-long-march-coal/; **Fig. 5.71.** Francis, 1848: facing p. 17, Courtesy of Alexander Roob, Melton Prior Institute, Düsseldorf; **Fig. 5.72.** Royle, 1840; **Fig. 5.73.** D. K. Ferguson, 6 April 2009; **Fig. 5.74.** D. K. Ferguson, 10 April 2009; **Fig. 5.75.** D. K. Ferguson, 11 April 2009; **Fig. 5.76.** D. K. Ferguson, 11 April 2009; **Fig. 5.77.** D. K.

Ferguson, 11 April 2009; **Fig. 5.78.** D. K. Ferguson, 11 April 2009; **Fig. 5.79.** D. K. Ferguson, 13 April 2009; **Fig. 5.80.** D. K. Ferguson, 13 April 2009; **Fig. 5.81.** D. K. Ferguson, 16 April 2009; **Fig. 5.82.** R. Gold, 12 May 2008; **Fig. 5.83.** http://en.wikipedia.org/wiki/Sir_Henry_Ramsay; **Fig. 5.84.** D. K. Ferguson, 15 April 2009; **Fig. 5.85.** D. K. Ferguson, 19 April 2009; **Fig. 5.86.** D. K. Ferguson, 19 April 2009; **Fig. 5.87.** D. K. Ferguson, 19 April 2009; **Fig. 5.88.** D. K. Ferguson, 18 April 2009; **Fig. 5.89.** D. K. Ferguson, 5 April 2009; **Fig. 5.90.** D. K. Ferguson, 27 April 2009; **Fig. 5.91.** https://upload.wikimedia.org/wikipedia/commons/d/d3/Joseph_Paxton.png; **Fig. 5.92.** The Gardeners' Chronicle, 1875 (2): 229; **Fig. 5.93.** D. K. Ferguson, 6 April 2014; **Fig. 5.94.** D. K. Ferguson, 8 September 2014; **Fig. 5.95.** Flore des Serres, 8: Tab. 811, 1853; **Fig. 5.96.** L'Illustration horticole, 1: Planche 36, 1854; **Fig. 5.97.** Flore des Serres, 9: Tab. 887, 1853; **Fig. 5.98.** Flore des Serres, 9: Tab. 885, 1853; **Fig. 5.99.** L'Illustration horticole, 1: Planche 8, 1854; **Fig. 5.100.** Flore des Serres, 9: Tab. 886, 1853; **Fig. 5.101.** Curtis's Botanical Magazine, 1860: Tab. 5164; **Fig. 5.102.** Original by LǏ J.-F., based on Stanford's Library Map of London and its Suburbs, 1862; **Fig. 5.103.** D. K. Ferguson, 4 September 2007; **Fig. 5.104.** Courtesy of David McClay, The John Murray Archive, National Library of Scotland, Edinburgh; **Fig. 5.105.** Courtesy of David McClay, The John Murray Archive, National Library of Scotland, Edinburgh; **Fig. 5.106.** London Stereoscopic Co., 1865; **Fig. 5.107.** D. K. Ferguson, 6 July 2014; **Fig. 5.108.** D. K. Ferguson, 7 July 2014; **Fig. 5.109.** D. K. Ferguson, 8 July 2014; **Fig. 5.110.** D. K. Ferguson, 8 July 2014; **Fig. 5.111.** D. K. Ferguson, 8 July 2014; **Fig. 5.112.** D. K. Ferguson, 9 July 2014; **Fig. 5.113.** Courtesy of David McClay, The John Murray Archive, National Library of Scotland, Edinburgh; **Fig. 5.114.** R. Fortune, 1852; **Fig. 5.115.** The Illustrated London News, 4 December 1852; **Fig. 5.116.** https://commons.wikipedia.org/wiki/File:Sir_Samuel_George_Bonham.jpg; **Fig. 5.117.** Lindley, 1866; **Fig. 5.118.** Bust in the Presidential Palace, Nánjīng, D. K. Ferguson, 12 November 2016; **Fig. 5.119.** LǏ J.-F., 10 November 2007; **Fig. 5.120.** LǏ Y.-M., 10 November 2007; **Fig. 5.121.** LǏ Y.-M., 10 November 2007; **Fig. 5.122.** HÚ Y.-Q., 29 May 2008; **Fig. 5.123.** Montalto de Jesus, 1909: facing p. 60; **Fig. 5.124.** D. K. Ferguson, 27 May 2016; **Fig. 5.125.** D. K. Ferguson, 9 September 2010; **Fig. 5.126.** D. K. Ferguson, 9 September 2010; **Fig. 5.127.** D. K. Ferguson, 9 September 2010; **Fig. 5.128.** D. K. Ferguson, 9 September 2010; **Fig. 5.129.** D. K. Ferguson, 9 September 2010; **Fig. 5.130.** D. K. Ferguson, 9 September 2010; **Fig. 5.131.** LǏ J.-F., 9 November 2007; **Fig. 5.132.** LǏ Y.-M., 9 November 2007; **Fig. 5.133.** LǏ Y.-M., 9 November 2007; **Fig. 5.134.** D. K. Ferguson, 9 September 2010; **Fig. 5.135.** D. K. Ferguson, 7 September 2010; **Fig. 5.136.** D. K. Ferguson, 7 September 2010; **Fig. 5.137.** D. K. Ferguson, 7 September 2010; **Fig. 5.138.** D. K. Ferguson, 7 September 2010; **Fig. 5.139.** D. K. Ferguson, 8 September 2010; **Fig. 5.140.** D. K. Ferguson, 7 September 2010; **Fig. 5.141.** HÚ Y.-Q., 15 November 2008; **Fig. 5.142.** D. K. Ferguson, 2 May 2012; **Fig. 5.143.** D. K. Ferguson, 23 September 2010; **Fig. 5.144.** D. K. Ferguson, 6 September 2010; **Fig. 5.145.** D. K. Ferguson, 6 September 2010; **Fig. 5.146.** D. K. Ferguson, 8 September 2010; **Fig. 5.147.** D. K. Ferguson, 8 September 2010; **Fig. 5.148.** Courtesy of the National Portrait Gallery London (www.npg.org.uk/collections/search/portrait/mw184886/John-Shaw-Burdon); **Fig. 5.149.** D. K. Ferguson, 28 May 2008; **Fig. 5.150.** www.cultural-china.com/chinaWH; **Fig. 5.151.** D. K. Ferguson, 3 May 2008; **Fig. 5.152.** D. K. Ferguson, 3 September 2014; **Fig. 5.153.** D. K. Ferguson, 18 November 2008; **Fig. 5.154.** HÚ Y.-Q., 20 November 2008; **Fig. 5.155.** HÚ Y.-Q., 20 November 2008; **Fig. 5.156.** D. K. Ferguson, 20 November 2008; **Fig. 5.157.** D. K. Ferguson, 20 November 2008; **Fig. 5.158.** D. K. Ferguson, 20 November 2008; **Fig. 5.159.** D. K. Ferguson, 19 November 2008; **Fig. 5.160.** D. K. Ferguson, 19 November 2008; **Fig. 5.161.** D. K. Ferguson, 5 October 2009; **Fig. 5.162.** Gouache by Richard Henry Neville-Cuming in 1887, P&O Heritage Collection (www.poheritage.com); **Fig. 5.163.** D. K. Ferguson, 7 September 2010; **Fig. 5.164.** D. K. Ferguson, 8 September 2010; **Fig. 5.165.** https://en.wikipedia.org/wiki/File:Dalhousie.jpg; **Fig. 5.166.** Painting by Eppo Doeve (www.artnet.com).

Fig. 6.1. Photograph by Felice Beato in 1860, Courtesy of the National Portrait Gallery London (www.npg.org.uk/collections/search/portraitLarge/mw245098/James-Bruce-8th-Earl-of-Elgin); **Fig. 6.2.** https://upload.wikimedia.org/

wikipedia/commons/d/d7/Ye_Mingchen; **Fig. 6.3A.** D. K. Ferguson, 13 April 2013; **Fig. 6.3B.** D. K. Ferguson, 24 September 2015; **Fig. 6.3C.** D. K. Ferguson, 29 May 2013; **Fig. 6.3D.** D. K. Ferguson, 8 September 2014; **Fig. 6. 3E.** D. K. Ferguson, 7 July 2014; **Fig. 6.3F.** D. K. Ferguson, 4 April 2014; **Fig. 6.4.** Courtesy of David McClay, The John Murray Archive, National Library of Scotland, Edinburgh; **Fig. 6.5.** Illustration at Martyrs Monument, Meerut, D. K. Ferguson, 5 April 2009; **Fig. 6.6.** Illustration at Martyrs Monument, Meerut, D. K. Ferguson, 5 April 2009; **Fig. 6.7.** D. K. Ferguson, 6 July 2014; **Fig. 6.8.** D. K. Ferguson, 5 September 2007.

Fig. 7.1. Engraving by John Andrew in Ballou's Pictorial Drawing-Room Companion (26 July 1856), Courtesy of Yale University Art Gallery; **Fig. 7.2.** Pond, 1941: facing p. 87; **Fig. 7.3.** D. K. Ferguson, 9 April 2010; **Fig. 7.4.** D. K. Ferguson, 9 April 2010; **Fig. 7.5.** Faville et al., 1979: 1; **Fig. 7.6.** www.mrlincolnswhitehouse.org/content_inside.asp?; **Fig. 7.7A.** D. K. Ferguson, 27 May 2008; **Fig. 7.7B.** D. K. Ferguson, 10 July 2013; **Fig. 7.7C.** LIÚ B., 18 April 2014; **Fig. 7.7D.** D. K. Ferguson, 6 May 2008; **Fig. 7.7E.** D. K. Ferguson, 2 June 2013; **Fig. 7.7F.** D. K. Ferguson, 18 May 2013.

Fig. 8.1. Veitch, 1906: facing p. 28; **Fig. 8.2.** Original by JIǍ H., July 2016; **Fig. 8.3.** D. K. Ferguson, 12 June 2012; **Fig. 8.4.** D. K. Ferguson, 11 June 2012; **Fig. 8.5.** D. K. Ferguson, 11 June 2012; **Fig. 8.6.** D. K. Ferguson, 11 June 2012; **Fig. 8.7.** D. K. Ferguson, 12 June 2012; **Fig. 8.8.** D. K. Ferguson, 12 June 2012; **Fig. 8.9.** D. K. Ferguson, 12 June 2012; **Fig. 8.10.** Engraving based on a portrait painted in 1808 by Per Krafft Jr. (1777 – 1863); **Fig. 8.11.** D. K. Ferguson, 12 June 2012; **Fig. 8.12.** D. K. Ferguson, 12 June 2012; **Fig. 8.13.** Das Bayerland, 7: 353, 1896; **Fig. 8.14.** D. K. Ferguson, 12 June 2012; **Fig. 8.15.** D. K. Ferguson, 12 June 2012; **Fig. 8.16.** Courtesy of the Yokohama Archives of History (Anon., 2008: 36); **Fig. 8.17.** Griffis, 1913: facing p. 200; **Fig. 8.18.** Courtesy of Fred Notehelfer (Notehelfer, 2001: 39); **Fig. 8.19.** D. K. Ferguson, 26 March 2010; **Fig. 8.20.** D. K. Ferguson, 14 September 2009; **Fig. 8.21.** D. K. Ferguson, 14 September 2009; **Fig. 8.22.** D. K. Ferguson, 15 September 2009; **Fig. 8.23.** D. K. Ferguson, 14 September 2009; **Fig. 8.24.** D. K. Ferguson, 14 September 2009; **Fig. 8.25.** D. K. Ferguson, 14 September 2009; **Fig. 8.26.** President and Fellows of Harvard College, Arnold Arboretum Archives; **Fig. 8.27.** Courtesy of Fred Notehelfer (Notehelfer, 2001: 69); **Fig. 8.28.** D. K. Ferguson, 10 September 2009; **Fig. 8.29.** D. K. Ferguson, 10 September 2009; **Fig. 8.30.** D. K. Ferguson, 9 September 2009; **Fig. 8.31.** D. K. Ferguson, 9 September 2009; **Fig. 8.32.** D. K. Ferguson, 9 September 2009; **Fig. 8.33.** D. K. Ferguson, 8 September 2009; **Fig. 8.34.** D. K. Ferguson, 13 April 2013; **Fig. 8.35.** http://commons.wikipedia.org/wiki/File:Prudence_Seraphin-Barthelemy_Girard, MEP.JPG; **Fig. 8.36.** D. K. Ferguson, 8 September 2009; **Fig. 8. 37.** D. K. Ferguson, 8 September 2009; **Fig. 8.38.** D. K. Ferguson, 15 September 2009; **Fig. 8.39.** D. K. Ferguson, 10 September 2009; **Fig. 8.40.** D. K. Ferguson, 10 September 2009; **Fig. 8.41.** D. K. Ferguson, 15 September 2009; **Fig. 8.42.** D. K. Ferguson, 14 September 2009; **Fig. 8.43.** D. K. Ferguson, 14 September 2009; **Fig. 8.44.** Courtesy of Kevin Plimley, Taiken Japan (https://taiken. co/single/celebrating-the-joy-of-kites-hidaka-citys-handmade-kite-flying-meet); **Fig. 8.45.** Courtesy of Valerie McAtear, The Royal Entomological Society; **Fig. 8.46.** D. K. Ferguson, 15 April 2014; **Fig. 8.47.** Photograph taken by Mathew Brady (1822 – 1896) in 1863 (https://travelsinshizuoka.wordpress.com); **Fig. 8.48.** Kobe City Museum. Based on a sketch by HIBATA Ōsuke (1813 – 1870) made in Yokohama on 8 March 1854; **Fig. 8.49.** Photograph by Felice Beato in 1861, Courtesy of Fred Notehelfer (Notehelfer, 2001: 159); **Fig. 8.50.** D. K. Ferguson, 10 September 2009; **Fig. 8.51.** Hesselink, 1994: 339; **Fig. 8.52.** D. K. Ferguson, 9 September 2009; **Fig. 8.53.** D. K. Ferguson, 9 September 2009; **Fig. 8.54.** HÚ Y.-Q., 27 May 2008; **Fig. 8.55.** HÚ Y.-Q., 27 May 2008; **Fig. 8.56.** D. K. Ferguson, 25 April 2014; **Fig. 8.57.** D. K. Ferguson, 9 September 2009; **Fig. 8.58.** D. K. Ferguson, 12 September 2009; **Fig. 8.59.** D. K. Ferguson, 12 September 2009; **Fig. 8.60.** D. K. Ferguson, 12 September 2009; **Fig. 8.61.** D. K. Ferguson, 12 September 2009; **Fig. 8.62.** D. K. Ferguson, 12 September 2009; **Fig. 8.63.** D. K. Ferguson, 12 September 2009; **Fig. 8.64.** D. K. Ferguson, 12 September 2009; **Fig. 8.65.** D. K. Ferguson, 10 September 2009; **Fig. 8.66.** Davies, 2008: facing p.

52; **Fig. 8.67.** D. K. Ferguson, 15 September 2009; **Fig. 8.68.** D. K. Ferguson, 15 September 2009.

Fig. 9.1. The Illustrated London News (https://commons.wikipedia.org/wiki/File:Signing_of_the_Treaty_of_Tientsin-2.jpg); **Fig. 9.2.** D. K. Ferguson, 22 August 2015; **Fig. 9.3.** D. K. Ferguson, 22 August 2015; **Fig. 9.4.** D. K. Ferguson, 22 August 2015; **Fig. 9.5.** D. K. Ferguson, 23 August 2015; **Fig. 9.6.** D. K. Ferguson, 22 August 2015; **Fig. 9.7.** www.staveley-genealogy.com/cwd_staveley.htm; **Fig. 9.8.** Gordon, 1898: Frontispiece; **Fig. 9.9.** D. K. Ferguson, 30 October 2008; **Fig. 9.10.** D. K. Ferguson, 31 October 2008; **Fig. 9.11.** D. K. Ferguson, 31 October 2008; **Fig. 9.12.** D. K. Ferguson, 31 October 2008; **Fig. 9.13.** Painting by Paul Frenzeny (1840 – 1902); **Fig. 9.14.** D. K. Ferguson, 2 October 2009; **Fig. 9.15.** D. K. Ferguson, 31 October 2008; **Fig. 9.16.** D. K. Ferguson, 16 October 2008; **Fig. 9.17.** D. K. Ferguson, 16 October 2008; **Fig. 9.18.** Redrawn from Lockhart (1866), with modifications by LǏ J.-F.; **Fig. 9.19.** Engravings by Émile Théodore Thérond in Poussielgue, 1864: 117, 128; **Fig. 9.20.** https://prologuepiecesofhistory.files.wordpress/2013/08/05 – 1352; **Fig. 9.21.** LǏ J.-F., 2 November 2007; **Fig. 9.22.** D. K. Ferguson, 2 November 2007; **Fig. 9.23.** LǏ Y.-M., 2 November 2007; **Fig. 9.24.** D. K. Ferguson, 2 November 2007; **Fig. 9.25.** D. K. Ferguson, 2 November 2007; **Fig. 9.26.** D. K. Ferguson, 2 November 2007; **Fig. 9.27.** D. K. Ferguson, 31 October 2007; **Fig. 9.28.** D. K. Ferguson, 31 October 2007; **Fig. 9.29.** D. K. Ferguson, 2 November 2007; **Fig. 9.30.** LǏ Y.-M., 31 October 2007; **Fig. 9.31.** D. K. Ferguson, 31 October 2007; **Fig. 9.32.** LǏ J.-F., 31 October 2007; **Fig. 9.33.** D. K. Ferguson, 28 July 2016; **Fig. 9.34.** D. K. Ferguson, 13 April 2013; **Fig. 9.35.** Painting by Richard Lauchert (1823 – 1869) in 1868 (https://en.wikipedia.org/wiki/Hugh_Wyndham_(diplomat)#/media/File:George_Hugh_Wyndham); **Fig. 9.36.** Photograph by YÙ Xūn-líng (ca. 1880 – 1943) in 1904; **Fig. 9.37.** The Illustrated London News, 38; **Fig. 9.38.** D. K. Ferguson, 2 November 2007; **Fig. 9.39.** LǏ J.-F., 2 November 2007; **Fig. 9.40.** The Illustrated London News, 37: 615, 1860; **Fig. 9.41.** LǏ J.-F., 2 November 2007; **Fig. 9.42.** https://commons.wikipedia.org/wiki/File:Sir_James_Hope_Grant_by_Sir_Francis_Grant.jpg; **Fig. 9.43.** LǏ J.-F., 31 October 2007; **Fig. 9.44.** D. K. Ferguson, 28 October 2007; **Fig. 9.45.** D. K. Ferguson, 28 October 2007; **Fig. 9.46.** D. K. Ferguson, 28 October 2007; **Fig. 9.47.** D. K. Ferguson, 28 October 2007; **Fig. 9.48.** D. K. Ferguson, 28 October 2007; **Fig. 9.49.** D. K. Ferguson, 28 October 2007; **Fig. 9.50.** D. K. Ferguson, 29 May 2013; **Fig. 9.51.** D. K. Ferguson, 16 October 2008; **Fig. 9.52.** The Illustrated London News, 37: 582, 1860; **Fig. 9.53.** Revue de Horticulture Belge et Etrangère, 15: Plate 10, 1889; **Fig. 9.54.** Coloured lithograph by Thomas Dutton (1819 – 1891), P&O Heritage Collection (www.poheritage.com); **Fig. 9.55.** D. K. Ferguson, 14 July 2016.

Fig. 10.1. D. K. Ferguson, 9 July 2014; **Fig. 10.2.** Veitch, 1906: Frontispiece; **Fig. 10.3.** D. K. Ferguson, 6 July 2014; **Fig. 10.4.** D. K. Ferguson, 8 July 2014; **Fig. 10.5.** D. K. Ferguson, 8 July 2014; **Fig. 10.6.** D. K. Ferguson, 7 July 2014; **Fig. 10.7.** R. Fortune, 1857: 254; **Fig. 10.8.** Courtesy of Sue Medway, Chelsea Physic Garden; **Fig. 10.9.** D. K. Ferguson, 7 July 2014; **Fig. 10.10.** D. K. Ferguson, 8 July 2014; **Fig. 10.11.** https://en.wikipedia.org/wiki/File:1862_international_exhibition_01.jpg; **Fig. 10.12.** Painting by Sir George Reid (1841 – 1913) in Murray, 1919: Frontispiece; **Fig. 10.13.** D. K. Ferguson, 8 July 2014; **Fig. 10.14.** D. K. Ferguson, 8 July 2014; **Fig. 10.15.** Photograph by Felice Beato in 1862 (https://upload.wikimedia.org/wikipedia/commons/2/20/CharlesRichardson.jpg); **Fig. 10.16.** Le Monde Illustré (https://en.wikipedia.org/wiki/File:KagoshimaBirdView.jpg); **Fig. 10.17.** D. K. Ferguson, 5 September 2007; **Fig. 10.18.** Courtesy of the National Portrait Gallery London (www.npg.org.uk/collections/search/portrait/mw43413/Lady-Dorothy-Fanny-Nevill-ne-Walpole); **Fig. 10.19.** Courtesy of the National Portrait Gallery London (www.npg.org.uk/collections/search/portraitLarge/mw07437/William-Buller-Fullerton-Elphinstone-15th-Lord-Elphinstone); **Fig. 10.20.** D. K. Ferguson, 17 July 2014; **Fig. 10.21.** D. K. Ferguson, 17 July 2014; **Fig. 10.22.** D. K. Ferguson, 17 July 2014; **Fig. 10.23.** https://en.wikipedia.org/wiki/File:JosephBazalgettePortrait.jpg; **Fig. 10.24.** Flore des Serres, 16: Planche 1609, 1865; **Fig. 10.25.** D. K. Ferguson, 7 July 2014; **Fig. 10.26.** https://en.wikipedia.org/wiki/File:Greaves_Old_Battersea_Bridge_1874.jpg; **Fig. 10.27.**

Friends of Battersea Park, 1993: 35; **Fig. 10.28.** D. K. Ferguson, 7 July 2014; **Fig. 10.29.** D. K. Ferguson, 7 July 2014; **Fig. 10.30.** D. K. Ferguson, 7 July 2014; **Fig. 10.31.** D. K. Ferguson, 6 July 2014; **Fig. 10.32.** L'Illustration, 29 September 1894; **Fig. 10.33.** www.japanvisitor.com/japanese-culture/history/fushimi-toba; **Fig. 10.34.** D. K. Ferguson, 8 September 2009; **Fig. 10.35.** https://en.wikipedia.org/wiki/Battle_of_Ueno#/media/File: Battle_of_Ueno_at_Ueno_temple.JPG; **Fig. 10.36.** D. K. Ferguson, 8 September 2009; **Fig. 10.37.** D. K. Ferguson, 12 September 2009; **Fig. 10.38.** Veselovsky (1899), reproduced in Kim, 2004: 74; **Fig. 10.39.** D. K. Ferguson, 31 October 2008; **Fig. 10.40.** D. K. Ferguson, 31 October 2008; **Fig. 10.41.** Photograph by John Thomson in 1871, The Wellcome Library, London, Courtesy of Betty Yao (Yao, 2015: 77); **Fig. 10.42.** Photograph by John Thomson in 1871, The Wellcome Library, London, Courtesy of Betty Yao (Yao, 2015: 78); **Fig. 10.43.** https://en.wikipedia.org/wiki/File: Albert_Fytche.PNG; **Fig. 10.44.** https://de.wikipedia.org/w/index.php? title = Datei: Sir_Thomas_Francis_Wade_Sinologe.jpg; **Fig. 10.45.** Margary, 1876: Frontispiece; **Fig. 10.46.** Photograph by LÍANG Shí-tài in 1878. The Getty Research Institute, Los Angeles, reproduced in Cody & Terpak, 2011: Plate 6; **Fig. 10.47.** https://en.wikipedia.org/wiki/File:Portrait_of_William_Gill_by_T.B.Wirgman_(18481925)_with_Gill's_signature_below.jpg; **Fig. 10.48.** LǏ J.-F., 12 November 2007; **Fig. 10.49.** D. K. Ferguson, 6 September 2007; **Fig. 10.50.** D. K. Ferguson, 7 July 2014; **Fig. 10.51.** D. K. Ferguson, 6 September 2007; **Fig. 10.52.** D. K. Ferguson, 7 July 2014; **Fig. 10.53.** Liverpool Central Library Archives, D. K. Ferguson, 15 July 2014.

Fig. 11.1. D. K. Ferguson, 30 August 2007; **Fig. 11.2.** D. K. Ferguson, 18 July 2014; **Fig. 11.3.** Courtesy of Bill Wilson, John Gray Centre, Haddington (www.eastlothianmuseums.org/exhibitions/tranent/big/Ai1756.htm); **Fig. 11.4.** D. K. Ferguson, 25 August 2007; **Fig. 11.5.** D. K. Ferguson, 30 August 2007; **Fig. 11.6.** D. K. Ferguson, 30 August 2007; **Fig. 11.7.** Ewing, 1925: facing p. 588; **Fig. 11.8.** D. K. Ferguson, 16 July 2014; **Fig. 11.9.** D. K. Ferguson, 17 July 2014; **Fig. 11.10.** D. K. Ferguson, 20 July 2014; **Fig. 11.11.** D. K. Ferguson, 18 July 2014; **Fig. 11.12.** D. K. Ferguson, 18 July 2014; **Fig. 11.13.** D. K. Ferguson, 17 July 2014; **Fig. 11.14.** Tabraham, 2004: 14; **Fig. 11.15.** D. K. Ferguson, 10 July 2014; **Fig. 11.16.** D. K. Ferguson, 5 September 2007; **Fig. 11.17.** D. K. Ferguson, 10 July 2014; **Fig. 11.18.** D. K. Ferguson, 10 July 2014.